Journey to Math Literacy

RICHARD N. AUFMANN
PALOMAR COLLEGE

JOANNE S. LOCKWOOD
NASHUA COMMUNITY COLLEGE

CENGAGE

Australia • Brazil • Mexico • Singapore • United Kingdom • United States

CENGAGE

Journey to Math Literacy
Richard N. Aufmann, Joanne S. Lockwood

Product Manager: Danielle Derbenti

Sr. Content Developer: Liza Neustaetter

Sr. Content Developer: Guanglei Zhang

Associate Content Developer: Samantha Gomez

Marketing Manager: Mark Linton

Sr. Content Project Manager: Alison Eigel Zade

Art Director: Vernon Boes

Manufacturing Planner: Becky Cross

Production Service and Compositor: Graphic World Inc.

Copy Editor: Jean Bermingham

Illustrator: Graphic World Inc.

Text Designer: Diane Beasley

Cover Designer: Irene Morris

Cover Image: Layered Map of Madrid: Shutterstock/TShooter; World map circuit: Shutterstock/Suriyaphoto; Globe: Shutterstock/rtguest; Background: Shutterstock/Kachan Eduard; Map Pointers: Shutterstock/iro4ka

Library of Congress Control Number: 2015950598

ISBN-13: 978-0-357-03268-8
ISBN-10: 0-357-03268-3

Cengage
20 Channel Street
Boston, MA 02210
USA

Printed at CLDPC, USA, 05-18

Brief Contents

Contents

MODULE 2

Problem Solving with Fractions and Decimals 1

MODULE 3

Applications of Equations and Inequalities 1

MODULE 7

The Mathematics of Finance 1

MODULE

A AIM for Success: How to Succeed in This Course

Get Ready

We are committed to your success in learning mathematics and have developed many tools and resources to support you along the way.

DO YOU WANT TO EXCEL IN THIS COURSE?

Read on to learn about the skills you'll need and how best to use this text to get the results you want.

We have written this text in an *interactive* style. More about this later but, in short, this means that you are supposed to *interact* with the text. Do not just read the text! Work along with it. Ready? Let's begin!

WHY ARE YOU TAKING THIS COURSE?

Did you interact with the text, or did you just read the last question? Get some paper and a pencil or pen and answer the question. Really—you will have more success in math and other courses you take if you **actively participate.** Now, **interact.** On your sheet of paper, write down one reason you are taking this course.

Of course, we have no idea what you just wrote, but experience has shown us that many of you wrote something along the lines of "I have to take it to graduate" or "It is a prerequisite to another course I have to take" or "It is required for my major." Those reasons are perfectly fine. Every teacher has had to take courses that were not directly related to his or her major.

WHY DO YOU WANT TO SUCCEED IN THIS COURSE?

Think about why you want to succeed in this course. List the reasons (not in your head . . . on the sheet of paper!).

One reason you may have listed is that math skills are important in order to be successful in your chosen career. That is certainly an important reason. Here are some other reasons.

- Math is a skill that applies across careers, which is certainly a benefit in our world of changing job requirements. A good foundation in math may enable you to more easily make a career change.
- Math can help you learn critical thinking skills, an attribute all employers want.
- Math can help you see relationships between ideas and identify patterns.

Take Note ➤
Motivation alone won't lead to success. For example, suppose a person who cannot swim is rowed out to the middle of a lake and thrown overboard. That person has a lot of motivation to swim, but most likely will drown without some help. You'll need motivation *and* learning in order to succeed.

Motivate Yourself

You'll find many real-life problems in this text, relating to sports, money, cars, music, and more. We hope that these topics will help you understand how mathematics is used in everyday life. To learn all of the necessary skills and to understand how you can apply them to your life outside of this course, **motivate** yourself to learn.

One of the reasons we asked you why you are taking this course was to provide motivation for you to succeed. When there is a reason to do something, that task is easier to accomplish. We understand that you may not want to be taking this course but, to achieve your career goal, this is a necessary step. Let your career goal be your motivation for success.

MAKE THE COMMITMENT TO SUCCEED!

With practice, you will improve your math skills. Skeptical? Think about when you first learned to drive a car, ride a skateboard, dance, paint, surf, or any other talent that you now have. You may have felt self-conscious or concerned that you might fail. But with time and practice, you learned the skill.

Write down a situation in which you accomplished your goal by spending time practicing and perfecting your skills (such as learning to play the piano or to play basketball).

You do not get "good" at something by doing it once a week. **Practice** is the backbone of any successful endeavor—including math!

Develop a "Can Do" Attitude Toward Math

You can do math! When you first learned the skills you just listed, you may not have done them well. With practice, you got better. With practice, you will get better at math. Stay focused, motivated, and committed to success.

We cannot emphasize enough how important it is to overcome the "I Can't Do Math" syndrome. If you listen to interviews of very successful athletes after a particularly bad performance, you will note that they focus on the positive aspects of what they did, not the negative. Sports psychologists encourage athletes always to be positive—to have a "can do" attitude. Develop this attitude toward math and you will succeed.

Change your conversation about mathematics. Do not say "I can't do math," "I hate math," or "Math is too hard." These comments just give you an excuse to fail. You don't want to fail, and we don't want you to fail. Write it down now: **I can do math!**

Strategies for Success

PREPARE TO SUCCEED

There are a number of things that may be worrisome to you as you begin a new semester. List some of those things now.

Here are some of the concerns that have been expressed by our students.

- **Tuition**
  Will I be able to afford school?
- **Job**
  I must work. Will my employer give me a schedule that will allow me to go to school?
- **Anxiety**
  Will I succeed?
- **Child care**
  What will I do with my kids while I'm in class or when I need to study?
- **Time**
  Will I be able to find the time to attend class and study?
- **Degree goals**
  How long will it take me to finish school and earn my degree?

These are all important and valid concerns. Whatever your concerns, acknowledge them. Choose an education path that allows you to accommodate your concerns. Make sure they don't prevent you from succeeding.

SELECT A COURSE

Many schools offer math assessment tests. These tests evaluate your present math skills. They don't evaluate how smart you are, so don't worry about your score on the test. If you are unsure about where you should start in the math curriculum, these tests can show you where to begin. You are better off starting at a level that is appropriate for you than starting with a more advanced class and then dropping it because you can't keep up. Dropping a class is a waste of time and money.

If you have difficulty with math, avoid short courses that compress the class into a few weeks. If you have struggled with math in the past, this environment does not give you the time to process math concepts. Similarly, avoid classes that meet once a week. The time delay between classes makes it difficult to make connections between concepts.

Some career goals require a number of math courses. If that is true of your major, try to take a math course every semester until you complete the requirements. Think about it this way. If you take, say, French I, and then wait two semesters before taking French II, you may forget a lot of material. Math is much the same. You must keep the concepts fresh in your mind.

Time Management

One of the most important requirements in completing any task is to acknowledge the amount of time it will take to finish the job successfully. Before a construction company starts to build a skyscraper, the company spends months looking at how much time each of the phases of construction will take. This is done so that resources can be allocated when appropriate. For instance, it would not make sense to schedule the electricians to run wiring until the walls are up.

MANAGE YOUR TIME!

We know how busy you are outside of school. Do you have a full-time or part-time job? Do you have children? Do you visit your family often? Do you play school sports or participate in the school orchestra or theater company? It can be stressful to balance all of the important activities and responsibilities in your life. Creating a time management plan will help you schedule enough time to do everything you need to do. Let's get started.

First, you need a calendar. You can use a daily planner, a calendar for a smartphone, or an online calendar, such as the ones offered by Google, MSN, or Yahoo. It is best to have a calendar on which you can fill in daily activities and be able to see a weekly or monthly view as well.

Start filling in your calendar now, even if it means stopping right here and finding a calendar. Some of the things you might include are:

- The hours each class meets
- Time for driving to and from work or school
- Leisure time, an important aspect of a healthy lifestyle
- Time for study. Plan at least one hour of study for each hour in class. This is a *minimum!*
- Time to eat
- Your work schedule
- Time for extracurricular activities such as sports, music lessons, or volunteer work
- Time for family and friends
- Time for sleep
- Time for exercise

Take Note ➤
Be realistic about how much time you have. One gauge is that working 10 hours per week is approximately equivalent to taking one three-unit course. If your college considers 15 units a full load and you are working 10 hours per week, you should consider taking 12 units. The more you work, the fewer units you should take.

We really hope you did this. If not, please reconsider. One of the best pathways to success is understanding how much time it takes to succeed. When you finish your calendar, if it does not allow you enough time to stay physically and emotionally healthy, rethink some of your school or work activities. We don't want you to lose your job because you have to study math. On the other hand, we don't want you to fail in math because of your job.

If math is particularly difficult for you, consider taking fewer course units during the semesters you take math. This applies equally to any other subject that you may find difficult. There is no rule that you must finish college in four years. It is a myth—discard it now.

Now extend your calendar for the entire semester. Many of the entries will repeat, such as the time a class meets. In your extended calendar, include significant events that may disrupt your normal routine. These might include holidays, family outings, birthdays, anniversaries, or special events such as a concert or a football game. In addition to these events, be sure to include the dates of tests, the date of the final exam, and dates that projects or papers are due. These are all important semester events. Having them on your calendar will remind you that you need to make time for them.

CLASS TIME

To be successful, **attend class.** You should consider your commitment to attend class as serious as your commitment to your job or to keeping an appointment with a close friend. It is difficult to overstate the importance of attending class. If you miss work, you don't get paid. If you miss class, you are not getting the full benefit of your tuition dollar. You are losing money.

If, by some unavoidable situation, you cannot attend class, find out as soon as possible what was covered in class. You might:

- Ask a friend for notes and the assignment.
- Contact your instructor and get the assignment. Missing class is no excuse for not being prepared for the next class.
- Determine whether there are online resources that you can use to help you with the topics and concepts that were discussed in the class you missed.

Going to class is important. Once you are there, **participate in class.** Stay involved and active. When your instructor asks a question, try to at least mentally answer the question. If you have a question, ask. Your instructor expects questions and wants you to understand the concept being discussed.

HOMEWORK TIME

In addition to attending class, you must **do homework.** Homework is the best way to reinforce the ideas presented in class. You should plan on at least one to two hours of homework and study for each hour you are in class. We've had many students tell us that one to two hours seems like a lot of time. That may be true, but if you want to attain your goals, you must be willing to devote the time to being successful in this math course.

You should schedule study time just as if it were class time. To do this, write down where and when you study best. For instance, do you study best at home, in the library, at the math center, under a tree, or somewhere else? Some psychologists who research successful study strategies suggest that just by varying where you study, you can increase the effectiveness of a study session. While you are considering where you prefer to study, also think about the time of day during which your study period will be most productive. Write down your thoughts.

Look at what you have written, and be sure that you can consistently be in your favorite study environment at the time you have selected. Studying and homework are extremely important. Just as you should not miss class, **do not miss study time.**

Before we leave this important topic, we have a few suggestions. If at all possible, create a study hour right after class. The material will be fresh in your mind, and the immediate review, along with your homework, will help reinforce the concepts you are learning.

If you can't study right after class, make sure that you set aside some time *on the day of the class* to review notes and begin the homework. The longer you wait, the more difficult it will be to recall some of the important points covered during class. Study math in small chunks—one hour a day (perhaps not enough for most of us), every day, is better than seven hours in one sitting. If you are studying for an extended period of time, break up your study session by studying one subject for a while and then moving on to another subject. Try to alternate between similar or related courses. For instance, study math for a while, then science, and then back to math. Or study history for a while, then political science, and then back to history.

Meet some of the people in your class and try to **put together a study group.** The group could meet two or three times a week. During those meetings, you could quiz each other, prepare for a test, try to explain a concept to someone else in the group, or get help on a topic that is difficult for you.

After reading these suggestions, you may want to rethink where and when you study best. If so, do that now. Remember, however, that it is your individual style that is important. Choose what works for *you,* and stick to it.

Habits of Successful Students

There are a number of habits that successful students use. Think about what these might be, and write them down.

What you have written is very important. The habits you have listed are probably the things you know you must do to succeed. Here is a list of some responses from successful students we have known.

- **Set priorities.** You will encounter many distractions during the semester. Do not allow them to prevent you from reaching your goal.
- **Take responsibility.** Your instructor, this text, tutors, math centers, and other resources are there to help you succeed. Ultimately, however, you must choose to learn. You must *choose* success.
- **Hang out with successful students.** Success breeds success. When you work and study with successful students, you are in an environment that will help you succeed. Seek out people who are committed to their goals.
- **Study regularly.** We have mentioned this before, but it is too important not to be repeated.
- **Self test.** Once every few days, select homework exercises from previous assignments and use them to test your understanding. Try to do these exercises without getting help from examples in the text. These self tests will help you gain confidence that you can do these types of problems on a test given in class.
- **Try different strategies.** If you read the text and are still having difficulty understanding a concept, consider going a step further. Contact the instructor or find a tutor. Many campuses have some free tutorial services. Go to the math or learning center. Consult another text. Be active and get the help you need.
- **Make flash cards.** This is one of the strategies that some math students do not think to try. Flash cards are a very important part of learning math. For instance, your instructor may use words or phrases such as *linear, quadratic, exponent, base, rational,* and many others. If you don't know the meanings of these words, you will not know what is being discussed.

  In this text, you will see important concepts and rules set off in boxes. Here is one about multiplication. These rules are good candidates for flash cards.

Flash Card

Rule for Multiplying Exponential Expressions

If m and n are integers, then $x^m \cdot x^n = x^{m+n}$.

Examples:

$x^4 \cdot x^7 = x^{4+7} = x^{11}$

$y \cdot y^5 = y^{1+5} = y^6$

$a^2 \cdot a^6 \cdot a = a^{2+6+1} = a^9$

Rule for Multiplying Exponential Expressions

If m and n are integers, then $x^m \cdot x^n = x^{m+n}$.

EXAMPLES

In each example below, we are multiplying two exponential expressions with the same base. Simplify the expression by adding the exponents.

1. $x^4 \cdot x^7 = x^{4+7} = x^{11}$
2. $y \cdot y^5 = y^{1+5} = y^6$
3. $a^2 \cdot a^6 \cdot a = a^{2+6+1} = a^9$

- **Plod along.** Your education is not a race. The primary goal is to finish. Taking too many classes and then dropping some does not get you to the end any faster. Take only as many classes as you can successfully manage.

Use the Interactive Method

We want you to be actively involved in learning mathematics, and we have given you many suggestions for getting "hands-on" with this text.

Basic examples introduce a concept (in this case, solving a percent equation) and include a step-by-step solution of the type of exercise you will find in the homework problems.

60% of what number is 300?

$$\text{Percent} \cdot \text{base} = \text{amount}$$ • Use the basic percent equation.

$$0.60 \cdot n = 300$$ • Percent = 60% = 0.60, base = n, amount = 300

$$\frac{0.60n}{0.60} = \frac{300}{0.60}$$ • Solve for n by dividing each side of the equation by 0.60.

$$n = 500$$

60% of 500 is 300.

Focus on/Check your understanding pairs help you get the practice you need to succeed.

Focus on solving the basic percent equation

a. Find 9.4% of 240.

SOLUTION

a. Use the basic percent equation.

Percent = 9.4% = 0.094, base = 240, amount = n

$$\text{Percent} \cdot \text{base} = \text{amount}$$
$$0.094 \cdot 240 = n$$
$$22.56 = n$$

9.4% of 240 is 22.56.

Check your understanding 1

a. Find $66\frac{2}{3}\%$ of 45.

SOLUTION See pages S-2–S-3.

You'll see that each **Focus on** is fully worked out. Study the **Focus on** example by carefully working through each step. Then, try to complete the **Check your understanding.** Use the solution to the **Focus on** as a model for solving the **Check your understanding.** If you get stuck, solutions to the **Check your understandings** are provided at the end of each module. There is a page number directly following the **Check your understanding** exercises that shows you where you can find the completely-worked-out solution. Use the solution to get a hint for the step on which you are stuck. Then, try again! When you've arrived at your answer, check your work against the solution at the end of each module. The solution to **Check your understanding 1** above is given below.

Check your understanding 1

a. STRATEGY

To find the amount, use the basic percent equation. Percent = $66\frac{2}{3}\% = \frac{2}{3}$, base = 45, amount = n

SOLUTION

$$\text{Percent} \cdot \text{base} = \text{amount}$$
$$\frac{2}{3}(45) = n$$
$$30 = n$$

30 is $66\frac{2}{3}\%$ of 45.

Use a Strategy to Solve Word Problems

One of the reasons you are studying math is to learn to solve word problems. Solving word problems involves combining all of the critical thinking skills you have learned to solve practical problems.

Try not to be intimidated by word problems. Basically, what you need is a strategy that will help you come up with the equation you will need to solve the problem. When you are looking at a word problem, try the following:

- **Read the problem.** This may seem pretty obvious, but we mean really **read** it. Don't just scan it. Read the problem slowly and carefully.
- **Write down what is known and what is unknown.** Now that you have read the problem, go back and write down everything that is known. Next, write down what it is you are trying to find. *Write* this—don't just think it! Be as specific as you can. For instance, if you are asked to find a distance, don't just write "I need to find the distance." Be specific and write "I need to find the distance between Earth and the moon."
- **Think of a method to find the unknown.** For instance, is there a formula that relates the known and unknown quantities? This is certainly the most difficult step. Eventually, you must write an equation to be solved.
- **Solve the equation.** Be careful as you solve the equation. There is no sense in getting to this point and then making a careless mistake. The unknown in most word problems will include a unit such as feet, dollars, or miles per hour. When you write your answer, include the unit. An answer such as 20 doesn't mean much. Is it 20 feet, 20 dollars, 20 miles per hour, or something else?
- **Check your solution.** Now that you have an answer, go back to the problem and ask yourself whether it makes sense. This is an important step. For instance, if, according to your answer, the cost of a car is $2.51, you know that something went wrong.

In this text, the solution to every word problem is broken down into two steps, **Strategy** and **Solution.** The Strategy consists of the first three steps discussed above. The Solution is the last two steps.

Focus on solving a percent concentration application

To make a certain color of blue, 4 oz of cyan must be contained in 1 gal of paint. What is the percent concentration of cyan in the paint?

STRATEGY

The amount of cyan is given in ounces and the amount of paint is given in gallons. We must convert ounces to gallons or gallons to ounces. For this problem, we will convert gallons to ounces: 1 gal $=$ 128 oz. Solve $Q = Ar$ for r, with $Q = 4$ and $A = 128$.

SOLUTION

$Q = Ar$ • Use the percent mixture equation.

$4 = 128r$ • $Q = 4, A = 128$

$$\frac{4}{128} = \frac{128r}{128}$$

$0.03125 = r$

The percent concentration of cyan is 3.125%.

Check your understanding 5

The concentration of sugar in a certain breakfast cereal is 25%. If there are 2 oz of sugar contained in a bowl of cereal, how many ounces of cereal are in the bowl?

SOLUTION See page S-4.

When you have finished studying a section, **do the exercises your instructor has selected.** Math is not a spectator sport. You must practice every day. Do the homework and do not get behind.

Ace the Test

There are a number of features in this text that will help you prepare for a test. These features will help you even more if you do just one simple thing: When you are doing your homework, go back to each previous homework assignment for the current module and rework two exercises. That's right—just *two* exercises. You will be surprised at how much better prepared you will be for a test by doing this.

Ready, Set, Succeed!

It takes hard work and commitment to succeed, but we know you can do it! Doing well in mathematics is just one step you'll take on your path to success. Good luck. We wish you success.

MODULE

1 Problem Solving with Integers

SECTION 1.1 Introduction to Integers

Objective 1.1A Use inequality symbols with integers

Mathematicians place objects with similar properties in groups called *sets.* A **set** is a collection of objects. The objects in a set are called the **elements of the set.**

The **roster method** of writing a set encloses a list of the elements in braces. Thus the set of sections within an orchestra is written {brass, percussion, string, woodwind}. When the elements of a set are listed, each element is listed only once. For instance, if the list of numbers 1, 2, 3, 2, 3 were placed in a set, the set would be $\{1, 2, 3\}$.

The symbol $\in$ means "is an element of." $2 \in B$ is read "2 is an element of set B."

Given $C = \{3, 5, 9\}$, then $3 \in C$, $5 \in C$, and $9 \in C$. $7 \notin C$ is read "7 is not an element of set C."

The numbers that we use to count objects, such as the students in a classroom or the horses on a ranch, are the *natural numbers.*

Natural numbers $= \{1, 2, 3, 4, 5, 6, 7, 8, 9, 10, \ldots\}$

The three dots mean that the list of natural numbers continues on and on, and that there is no largest natural number.

A **prime number** is a natural number that can be divided evenly only by 1 and itself. For instance, 13 is a prime number. The only numbers that divide evenly into 13 are 1 and 13. The first 10 prime numbers are 2, 3, 5, 7, 11, 13, 17, 19, 23, and 29.

A number that can be divided evenly by numbers other than 1 and itself is called a **composite number.** For instance, 12 is a composite number. It can be divided evenly by 1, 2, 3, 4, 6, and 12. The first few composite numbers are 4, 6, 8, 9, and 10. The number 1 is neither prime nor composite.

The natural numbers alone do not provide all the numbers that are useful in applications. For instance, a meteorologist also needs the number zero and numbers below zero.

Integers $= \{\ldots, -5, -4, -3, -2, -1, 0, 1, 2, 3, 4, 5, \ldots\}$

Each integer can be shown on a number line. The integers to the left of zero on the number line are called **negative integers.** The integers to the right of zero are called **positive integers,** or natural numbers. Zero is neither a positive nor a negative integer.

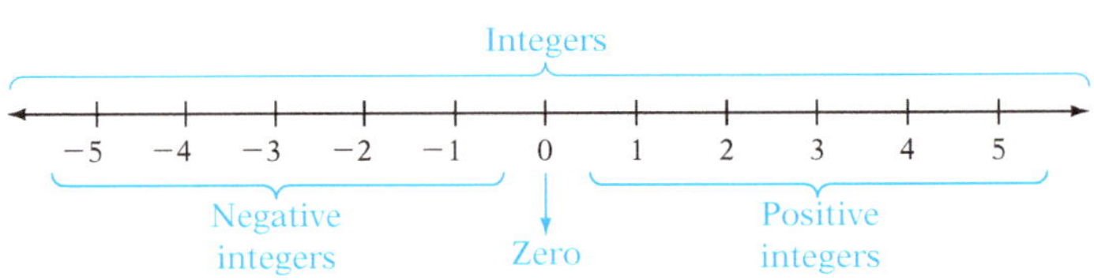

The **graph** of an integer is shown by placing a heavy dot on the number line directly above the number. The graphs of −3 and 4 are shown on the number line below.

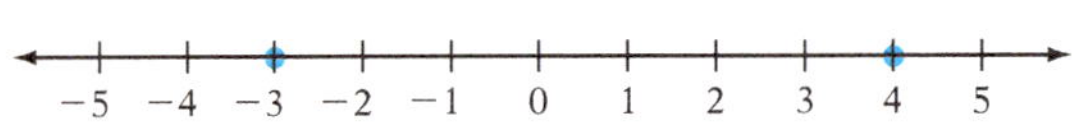

Consider the following sentences.

> The quarterback threw the football and the receiver caught *it.*
>
> A student purchased a computer and used *it* to write history papers.

In the first sentence, *it* is used to mean the football; in the second sentence, *it* means the computer. In language, the word *it* can stand for many different objects. Similarly, in mathematics, a letter of the alphabet can be used to stand for a number. Such a letter is called a **variable.** Variables are used in the following definition of inequality symbols.

Inequality Symbols

If a and b are two numbers and a is to the left of b on the number line, then a **is less than** b. This is written $a < b$.

If a and b are two numbers and a is to the right of b on the number line, then a **is greater than** b. This is written $a > b$.

EXAMPLES

1. $-2 < 1$ — -2 is to the left of 1 on the number line. -2 is less than 1.

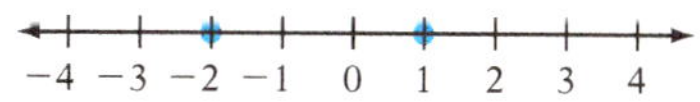

2. $3 > 0$ — 3 is to the right of 0 on the number line. 3 is greater than 0.

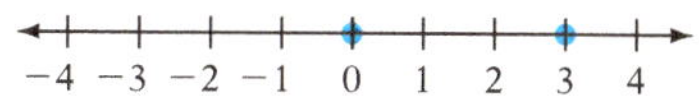

Apply the Concept

Which is the colder temperature, $-8°C$ or $-3°C$?

SOLUTION

The lesser number corresponds to the colder temperature.

$$-8 < -3$$

The colder temperature is $-8°C$.

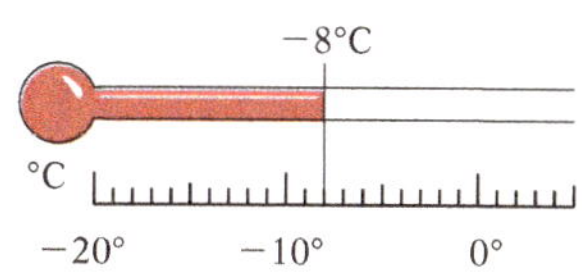

There are also inequality symbols for **is less than or equal to (≤)** and **is greater than or equal to (≥).**

$7 \le 15$	7 is less than or equal to 15. This is true because $7 < 15$.	$6 \le 6$	6 is less than or equal to 6. This is true because $6 = 6$.

Focus on writing a set using the roster method

Use the roster method to write the set of negative integers greater than or equal to -4.

SOLUTION

$A = \{-4, -3, -2, -1\}$ • A set is designated by a capital letter.

Check your understanding 1

Use the roster method to write the set of positive integers less than 7.

SOLUTION See page S-1.

Focus on inequalities

Given $A = \{-6, -2, 0\}$, which elements of set A are less than or equal to -2?

SOLUTION

Find the order relation between each element of set A and -2.

$-6 < -2$

$-2 = -2$

$0 > -2$

The elements -6 and -2 are less than or equal to -2.

Check your understanding 2

Given $B = \{-5, -1, 5\}$, which elements of set B are greater than -1?

SOLUTION See page S-1.

Objective 1.1A Practice

1. Place the correct symbol, $<$ or $>$, between the numbers below.
   $-14 \quad 16$
2. Place the correct symbol, $<$ or $>$, between the numbers below.
   $0 \quad -31$
3. Use the roster method to write the set of positive integers less than 4.
4. Given $C = \{-33, -24, -10, 0\}$, which elements of set C are less than -10?

Solutions on p. S-7.

Objective 1.1B Simplify expressions with absolute value

Two numbers that are the same distance from zero on the number line but are on opposite sides of zero are **opposite numbers,** or **opposites.** The opposite of a number is also called its **additive inverse.**

The opposite of 5 is -5.

The opposite of -5 is 5.

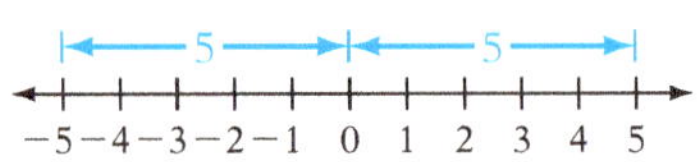

The negative sign can be read "the opposite of."

$-(2) = -2$ The opposite of 2 is -2.

$-(-2) = 2$ The opposite of -2 is 2.

Focus on finding an additive inverse

Given $A = \{-12, 0, 4\}$, find the additive inverse of each element of set A.

SOLUTION

$-(-12) = 12$

$-0 = 0$ • Zero is neither positive nor negative.

$-(4) = -4$

Check your understanding 3

Given $B = \{-11, 0, 8\}$, find the additive inverse of each element of set B.

SOLUTION See page S-1.

The **absolute value of a number** is its distance from zero on the number line. Therefore, the absolute value of a number is a positive number or zero. The symbol for absolute value is two vertical bars, $| \, |$.

The distance from 0 to 3 is 3. Therefore, the absolute value of 3 is 3.

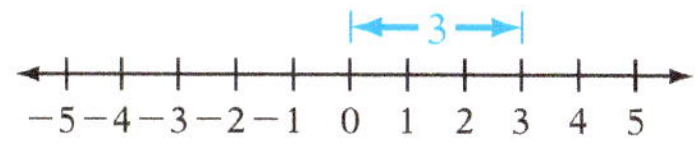

$$|3| = 3$$

The distance from 0 to −3 is 3. Therefore, the absolute value of −3 is 3.

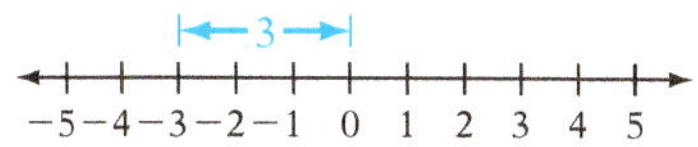

$$|-3| = 3$$

Absolute Value

The absolute value of a positive number is the number itself.

The absolute value of a negative number is the opposite of the number.

The absolute value of zero is zero.

EXAMPLES

1. $|6| = 6$ **2.** $|-8| = 8$ **3.** $|0| = 0$

Using variables, the definition of absolute value is

$$|x| = \begin{cases} x, & x > 0 \\ 0, & x = 0 \\ -x, & x < 0 \end{cases}$$

Focus on evaluating an absolute value expression

Evaluate $|-4|$ and $-|-10|$.

SOLUTION

$|-4| = 4$

$-|-10| = -10$

Check your understanding 4

Evaluate $|-5|$ and $-|-23|$.

SOLUTION See page S-1.

Objective 1.1B Practice

1. Find the additive inverse of 4.
2. Find the additive inverse of −9.
3. Find the additive inverse of −36.
4. Evaluate $-(-40)$.
5. Evaluate $|-74|$.
6. Evaluate $-|81|$.

Solutions on p. S-7.

Objective 1.1C Create stem-and-leaf diagrams

One way to analyze data is to draw a graph or display of the data. To display every number in a data set and still organize the data in a visually helpful way, we can use a **stem-and-leaf plot.** Each data value is split into two parts, a **stem** and a **leaf.** The leaf is usually the rightmost digit of a number; the stem is the remaining digits. The stems and leaves for the numbers 26 and 349 are shown below.

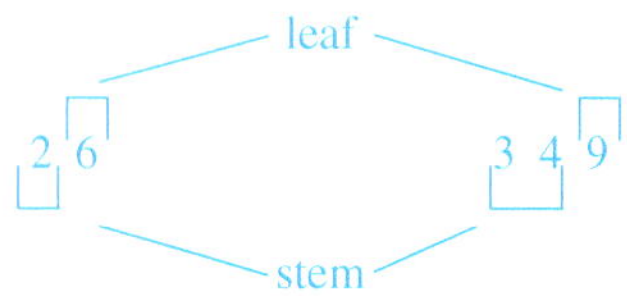

Data are placed into stem-and-leaf plots as a means of quickly showing the frequency of a particular event. The following stem-and-leaf plot shows the round trip commuting distances for 29 employees of a digital art production company.

Stem	Leaves
0	7 9
1	5 3 8
2	2 7 2 8 5
3	7 7 3 4 1 0 8
4	8 7 3 0 6 5
5	1 4 5 3 9
6	6

A legend gives the meaning of a number in the stem-and-leaf plot.

5 | 4 is 54.

Looking at this display and the stem labeled 4, we can determine that because there are six numbers to the right of the vertical bar, there are six employees with a round-trip commute between 40 and 49 miles. Compare this visual plot to a list of the raw data: 28, 25, 37, 37, 33, 34, 31, 7, 9, 15, 51, 54, 55, 53, 59, 66, 13, 18, 22, 30, 38, 48, 47, 43, 27, 22, 40, 46, 45. From the list, it is difficult to see any pattern to the data. However, the stem-and-leaf plot gives a good general idea of how far employees must commute to work.

Focus on drawing a stem-and-leaf plot

The data below show the number of touchdown passes thrown by each team in a recent National Football League season. Draw a stem-and-leaf diagram for the data.

27, 39, 43, 30, 33, 38, 39, 29, 53, 33, 52, 34, 51, 24, 36, 29, 47, 49, 43, 27, 26, 43, 43, 37, 40, 31, 28, 26, 33

SOLUTION

The smallest value in this set is 24, and the largest is 53. Therefore, stems from 2 (the first digit of 24) through 5 (the first digit of 53) are used. Draw a vertical line and list the stems to the left of the line. The beginning of the stem-and-leaf plot is shown below at the left.

Touchdowns	
2	
3	
4	
5	

Touchdowns	
2	46677899
3	01333467899
4	0333379
5	123

To the right of the vertical line, list the leaves. To represent 27, the first data value, place a 7 to the right of 2. For 39, place a 9 to the right of 3. Continue until you have used all the data values. When placing leaf values, list them from smallest to largest. The result is shown at the right above.

Check your understanding 5

The table below gives the approximate annual rainfall totals (in inches) for selected cities in the United States. Make a stem-and-leaf plot for these data. For each city, use the first two digits of the number of inches of rainfall as the stem.

Little Rock, AR	104	Boston, MA	126
Norfolk, VA	115	Pittsburgh, PA	154
Des Moines, IA	107	Nashville, TN	119
Miami, FL	129	Houston, TX	104
Detroit, MI	135	Portland, ME	128
Mobile, AL	122	Philadelphia, PA	117
Spokane, WA	113	Milwaukee, WI	125
Hartford, CT	127	New York City, NY	142
Cleveland, OH	156	Washington, D.C.	120
Chicago, IL	126	Kansas City, MO	104
Concord, NH	125	Atlanta, GA	115
Duluth, MN	134	Portland, OR	152

SOLUTION See page S-1.

Objective 1.1C Practice

1. The winning scores for 32 National Football League games are listed below. Draw a stem-and-leaf plot for the data.

36, 17, 26, 33, 19, 34, 30, 34, 23, 23, 37, 20, 28, 18, 35, 31, 26, 29, 41, 26, 25, 30, 26, 24, 24, 19, 21, 30, 24, 24, 20, 27

Solutions on p. S-7.

SECTION 1.2 Addition and Subtraction of Integers

Objective 1.2A Add integers

A number can be represented anywhere along the number line by an arrow. A positive number is represented by an arrow pointing to the right, and a negative number is represented by an arrow pointing to the left. The size of the number is represented by the length of the arrow.

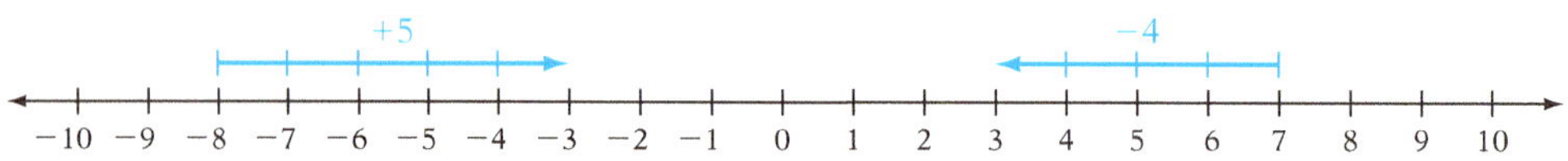

Addition is the process of finding the total of two numbers. The numbers being added are called **addends.** The total is called the **sum.** Addition of integers can be shown on the number line. To add integers, start at zero and draw, above the number line, an arrow representing the first number. At the tip of the first arrow, draw a second arrow representing the second number. The sum is below the tip of the second arrow.

$4 + 2 = 6$

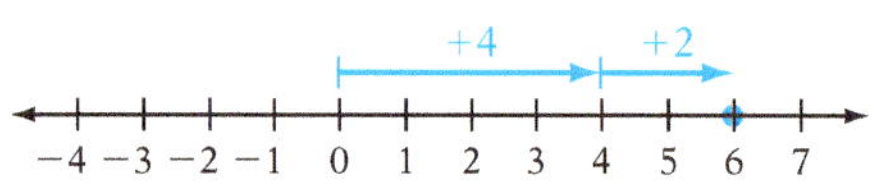

$-4 + (-2) = -6$

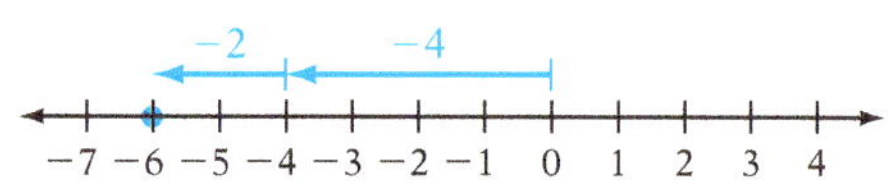

$-4 + 2 = -2$

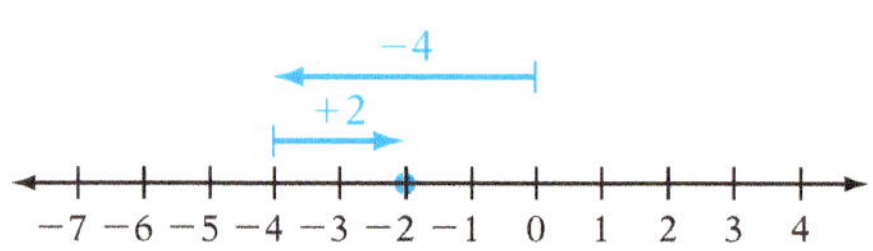

$4 + (-2) = 2$

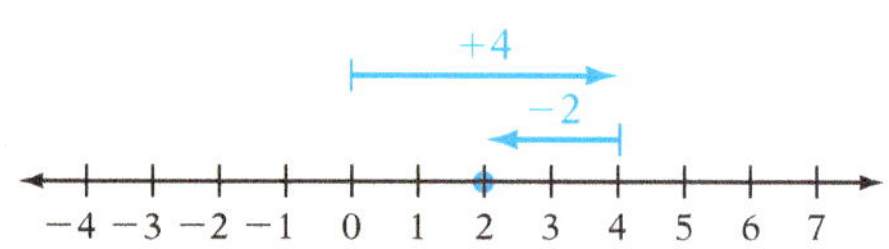

The pattern for addition shown on the number lines above is summarized in the following rules for adding integers.

Addition of Integers

To add two numbers with the same sign, add the absolute values of the numbers. Then attach the sign of the addends.

To add two numbers with different signs, find the absolute value of each number. Subtract the smaller of the absolute values from the larger. Then attach the sign of the number with the larger absolute value.

EXAMPLES

1. Add: $-12 + (-26)$ • The signs are the same.

$-12 + (-26) = -38$ • Add the absolute values of the numbers $(12 + 26)$. Attach the sign of the addends.

2. Add: $-19 + 8$ • The signs are different.

$|-19| = 19; |8| = 8$ • Find the absolute value of each number.

$19 - 8 = 11$ • Subtract the smaller absolute value from the larger.

$-19 + 8 = -11$ • Attach the sign of the number with the larger absolute value.

Focus on adding integers

Add: $-52 + (-39)$

SOLUTION

$-52 + (-39) = -91$

Check your understanding 1

Add: $100 + (-43)$

SOLUTION See page S-1.

Tips for Success ➤
One of the key instructional features of this text is the Focus On/Check Your Understanding pairs. Each Focus On is completely worked. You are to solve the Check Your Understanding problems. When you are ready, check your solution against the one given in the Solutions section at the end of the module.

Focus on adding integers

Add: $37 + (-52) + (-14)$

SOLUTION

$$37 + (-52) + (-14) = -15 + (-14)$$
$$= -29$$

Check your understanding 2

Add: $-51 + 42 + 17 + (-102)$

SOLUTION See page S-1.

Apply the Concept

Suppose you wake up in St. Paul, Minnesota, and the temperature is -12°F. By noon, the temperature has risen 7°F. What is the temperature at noon?

SOLUTION

To find the temperature at noon, add 7° to -12°.

$$-12 + 7 = -5$$

The temperature at noon is -5°F.

Find the sum of -23, 47, -18, and -10.

Recall that a sum is the answer to an addition problem.

$-23 + 47 + (-18) + (-10)$ • To add more than two numbers, add the first two numbers. Then add the sum to the third number. Continue until all the numbers are added.

$= 24 + (-18) + (-10)$

$= 6 + (-10) = -4$

All of the phrases below indicate addition.

added to	-6 added to 9	$9 + (-6) = 3$
more than	3 more than -8	$-8 + 3 = -5$
the sum of	the sum of -2 and -8	$-2 + (-8) = -10$
increased by	-7 increased by 5	$-7 + 5 = -2$
the total of	the total of 4 and -9	$4 + (-9) = -5$
plus	6 plus -10	$6 + (-10) = -4$

Focus on translating an expression

Find 11 more than -23.

SOLUTION

$-23 + 11 = -12$

Check your understanding 3

Find -8 increased by 7.

SOLUTION See page S-1.

Objective 1.2A Practice

1. Add: $-6 + (-9)$
2. Add: $-17 + 17$
3. Add: $-27 + (-42) + (-18)$
4. Add: $-6 + (-8) + 14 + (-4)$
5. What is 4 more than -8?
6. What is -8 added to -21?

Solutions on p. S-7.

Objective 1.2B Subtract integers

Look at the expressions below. Note that each expression equals the same number.

$8 - 3 = 5$ 8 minus 3 is 5.

$8 + (-3) = 5$ 8 plus the opposite of 3 is 5.

This example suggests the following.

Subtraction of Integers

To subtract one number from another, add the opposite of the second number to the first number.

EXAMPLES

1. Subtract: $-21 - (-40)$

$-21 - (-40) = -21 + 40$ • Rewrite the subtraction as addition of the opposite.

$= 19$ • Add.

2. Subtract: $15 - 51$

$15 - 51 = 15 + (-51)$ • Rewrite the subtraction as addition of the opposite.

$= -36$ • Add.

Chemical Element	Boiling Point	Melting Point
Mercury	357	−39
Radon	−62	−71
Xenon	−108	−112

Apply the Concept

The table at the left shows the boiling point and the melting point, in degrees Celsius, of three chemical elements. Find the difference between the boiling point and the melting point of mercury.

SOLUTION

To find the difference, subtract the melting point of mercury from the boiling point of mercury.

$$357 - (-39) = 357 + 39$$
$$= 396$$

The difference is 396°C.

Focus on subtracting integers

Subtract: $-12 - (-21) - 15$

SOLUTION

$$-12 - (-21) - 15 = -12 + 21 + (-15)$$
$$= 9 + (-15) = -6$$

• Rewrite each subtraction as addition of the opposite. Then add.

Check your understanding 4

Subtract: $-9 - (-12) - 17 - 4$

SOLUTION See page S-2.

A *difference* is the answer to a subtraction problem.

Find the difference between -8 and 7.

$$-8 - 7 = -8 + (-7)$$
$$= -15$$

• Rewrite the subtraction as addition of the opposite. Then add.

All of the following phrases indicate subtraction.

Take Note ➤
Note the order in which numbers are subtracted when the phrase *less than* is used. If you have $10 and your friend has $6 *less than* you do, then your friend has $6 less than $10, or $10 − $6 = $4.

minus	−5 minus 11	$-5 - 11 = -16$
less	−3 less 5	$-3 - 5 = -8$
less than	−8 less than −2	$-2 - (-8) = 6$
the difference between	the difference between −5 and 4	$-5 - 4 = -9$
decreased by	−4 decreased by 9	$-4 - 9 = -13$
subtract . . . from	subtract 8 from −3	$-3 - 8 = -11$

Focus on translating an expression

Find 9 less than −4.

SOLUTION

$$-4 - 9 = -4 + (-9) = -13$$

Check your understanding 5

Subtract −12 from −11.

SOLUTION See page S-2.

Objective 1.2B Practice

1. Subtract: $16 - 8$
2. Subtract: $6 - (-12)$
3. Subtract: $-19 - (-19) - 18$
4. What is 9 less than −12?
5. Find −21 decreased by 19.

Solutions on p. S-7.

Objective 1.2C Solve application problems

Focus on solving an application

The average temperature on Mercury's sunlit side is 950°F. The average temperature on Mercury's dark side is −346°F. Find the difference between these two average temperatures.

STRATEGY

To find the difference, subtract the average temperature on the dark side (−346) from the average temperature on the sunlit side (950).

SOLUTION

$$\begin{aligned} 950 - (-346) &= 950 + 346 \\ &= 1296 \end{aligned}$$

The difference between the average temperatures is 1296°F.

Check your understanding 6

The average daytime temperature on Mars is −17°F. The average nighttime temperature on Mars is −130°F. Find the difference between these two average temperatures.

SOLUTION See page S-2.

Objective 1.2C Practice

1. The elevation, or height, of places on Earth is measured in relation to sea level, or the average level of the ocean's surface. The table below shows height above sea level as a positive number and depth below sea level as a negative number. Use the table below to find the difference in elevation between Mt. Kilimanjaro and the Qattara Depression.

Continent	Highest Elevation (in Meters)		Lowest Elevation (in Meters)	
Africa	Mt. Kilimanjaro	5895	Qattara Depression	−133
Asia	Mt. Everest	8850	Dead Sea	−400
Europe	Mt. Elbrus	5634	Caspian Sea	−28
America	Mt. Aconcagua	6960	Death Valley	−86

2. The elevation, or height, of places on Earth is measured in relation to sea level, or the average level of the ocean's surface. The table below shows height above sea level as a positive number and depth below sea level as a negative number. Using the table below, for which continent shown is the difference between the highest and lowest elevations smallest?

Continent	Highest Elevation (in Meters)		Lowest Elevation (in Meters)	
Africa	Mt. Kilimanjaro	5895	Qattara Depression	−133
Asia	Mt. Everest	8850	Dead Sea	−400
Europe	Mt. Elbrus	5634	Caspian Sea	−28
America	Mt. Aconcagua	6960	Death Valley	−86

3. The graph below shows the depths of Earth's three deepest ocean trenches and the heights of its three tallest mountains. What is the difference between the depth of the Philippine Trench and the depth of the Mariana Trench?

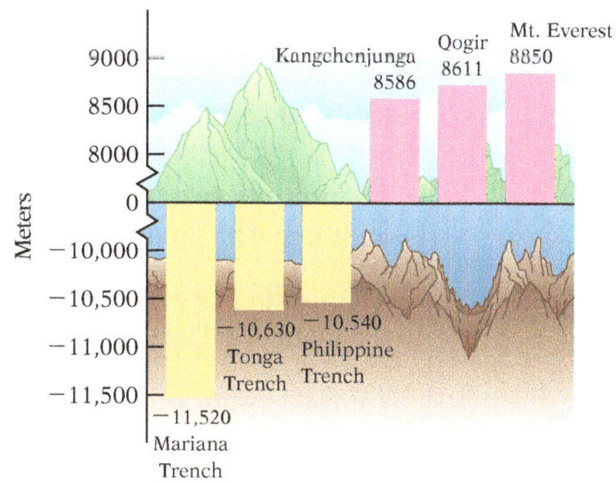

Solutions on pp. S-7–S-8.

SECTION 1.3

Multiplication and Division of Integers

Objective 1.3A Multiply integers

Several different symbols are used to indicate multiplication. The numbers being multiplied are called **factors;** for instance, 3 and 2 are factors in each of the examples at the right. The result is called the **product.** Note that when parentheses are used and there is no operation symbol, the operation is multiplication.

$$3 \times 2 = 6$$
$$3 \cdot 2 = 6$$
$$(3)(2) = 6$$
$$3(2) = 6$$
$$(3)2 = 6$$

Multiplication is repeated addition of the same number. The product 3 × 5 is shown on the number line below.

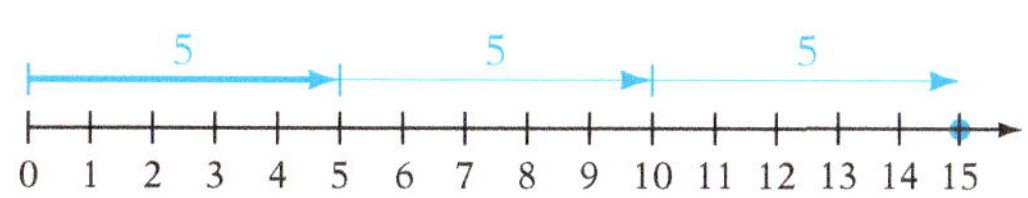

$$3 \times 5 = \overbrace{5 + 5 + 5}^{\text{5 is added 3 times.}} = 15$$

Now consider the product of a positive and a negative number.

$$3(-5) = \underbrace{(-5) + (-5) + (-5)}_{-5 \text{ is added 3 times.}} = -15$$

This example suggests that the product of a positive number and a negative number is negative. Here are a few more examples.

$$4(-7) = -28 \qquad -6 \cdot 7 = -42 \qquad (-8)7 = -56$$

To find the product of two negative numbers, look at the pattern at the right. As −5 multiplies a sequence of decreasing integers, the products increase by 5.

These numbers decrease by 1. These numbers increase by 5.

$$-5 \times 3 = -15$$
$$-5 \times 2 = -10$$
$$-5 \times 1 = -5$$
$$-5 \times 0 = 0$$

The pattern can be continued by requiring that the product of two negative numbers be positive.

$$-5 \times (-1) = 5$$
$$-5 \times (-2) = 10$$
$$-5 \times (-3) = 15$$

Multiplication of Integers

To multiply two numbers with the same sign, multiply the absolute values of the numbers. The product is positive.

To multiply two numbers with different signs, multiply the absolute values of the numbers. The product is negative.

EXAMPLES

1. $-4(-8) = 32$ • The signs are the same. The product is positive.

2. $5(-12) = -60$ • The signs are different. The product is negative.

Focus on multiplying more than two integers

Multiply: $-2(5)(-7)(-4)$

SOLUTION

$$\begin{aligned} -2(5)(-7)(-4) &= -10(-7)(-4) \\ &= 70(-4) \\ &= -280 \end{aligned}$$

• To multiply more than two numbers, multiply the first two numbers. Then multiply the product by the third number. Continue until all the numbers are multiplied.

Check your understanding 1

Multiply: $(-2)3(-8)7$

SOLUTION See page S-2.

Consider the products shown at the right. Note that when there is an even number of negative factors, the product is positive. When there is an odd number of negative factors, the product is negative.

$$(-3)(-5) = 15$$
$$(-2)(-5)(-6) = -60$$
$$(-4)(-3)(-5)(-7) = 420$$
$$(-3)(-3)(-5)(-4)(-5) = -900$$
$$(-6)(-3)(-4)(-2)(-10)(-5) = 7200$$

This idea can be summarized by the following useful rule: The product of an even number of negative factors is positive; the product of an odd number of negative factors is negative.

All of the phrases below indicate multiplication.

times	-7 times -9	$-7(-9) = 63$
the product of	the product of 12 and -8	$12(-8) = -96$
multiplied by	-15 multiplied by 11	$-15(11) = -165$
twice	twice -14	$2(-14) = -28$

Focus on translating an expression

Find the product of 15 and -9.

SOLUTION

$15(-9) = -135$

Check your understanding 2

What is -9 times 34?

SOLUTION See page S-2.

Objective 1.3A Practice

1. Multiply: $(-13)(-9)$
2. Multiply: $17(-13)$
3. Multiply: $-7(-6)(-5)$
4. Multiply: $-12(-4)7(-2)$
5. Find the product of -2, -3, -4, and -5.

Solutions on p. S-8.

Objective 1.3B Divide integers

Take Note ➤
Think of the fraction bar as meaning "divided by." Thus $\frac{8}{2}$ means 8 divided by 2. The number 2 is the **divisor.** The number 8 is the **dividend.** The result of the division, 4, is called the **quotient.**

For every division problem there is a related multiplication problem.

$$\frac{8}{2} = 4 \quad \text{because} \quad 4 \cdot 2 = 8.$$

Division — Related multiplication

This fact and the rules for multiplying integers can be used to illustrate the rules for dividing integers.

Note in the following examples that the quotient of two numbers with the same sign is positive.

$$\frac{12}{3} = 4 \text{ because } 4 \cdot 3 = 12. \qquad \frac{-12}{-3} = 4 \text{ because } 4(-3) = -12.$$

The next two examples illustrate that the quotient of two numbers with different signs is negative.

$$\frac{12}{-3} = -4 \text{ because } (-4)(-3) = 12. \qquad \frac{-12}{3} = -4 \text{ because } (-4)3 = -12.$$

Division of Integers

To divide two numbers with the same sign, divide the absolute values of the numbers. The quotient is positive.

To divide two numbers with different signs, divide the absolute values of the numbers. The quotient is negative.

EXAMPLES

1. $(-18) \div (-2) = 9$ • The signs are the same. The quotient is positive.

2. $-36 \div 9 = -4$ • The signs are different. The quotient is negative.

The fact that $\frac{-12}{3} = -4$, $\frac{12}{-3} = -4$, and $-\frac{12}{3} = -4$ suggests the following rule.

If a and b are integers, and $b \neq 0$, then $\frac{-a}{b} = \frac{a}{-b} = -\frac{a}{b}$.

Focus on dividing integers

Divide: $(-120) \div (-8)$

SOLUTION

$(-120) \div (-8) = 15$ • The signs are the same. The quotient is positive.

Check your understanding 3

Divide: $(-135) \div (-9)$

SOLUTION See page S-2.

Focus on dividing integers

Divide: $\frac{95}{-5}$

SOLUTION

$\frac{95}{-5} = -19$ • The signs are different. The quotient is negative.

Check your understanding 4

Divide: $\frac{-72}{4}$

SOLUTION See page S-2.

Focus on simplifying a division expression

Simplify: $-\frac{-81}{3}$

SOLUTION

$-\frac{-81}{3} = -(-27) = 27$

Check your understanding 5

Simplify: $-\dfrac{36}{-12}$

SOLUTION See page S-2.

All of the phrases below indicate division.

divided by	15 divided by -3	$15 \div (-3) = -5$
the quotient of	the quotient of -56 and -8	$(-56) \div (-8) = 7$
the ratio of	the ratio of 45 and -5	$45 \div (-5) = -9$
divide . . . by . . .	divide -100 by -20	$-100 \div (-20) = 5$

Focus on translating an expression

Find the quotient of 98 and -14.

SOLUTION

$98 \div (-14) = -7$

Check your understanding 6

What is the ratio of -72 and -8?

SOLUTION See page S-2.

The properties of division are stated below. In these statements, the symbol $\neq$ is read "is not equal to."

Properties of Zero and One in Division

If $a \neq 0$, $\dfrac{0}{a} = 0$.	Zero divided by any number other than zero is zero.
If $a \neq 0$, $\dfrac{a}{a} = 1$.	Any number other than zero divided by itself is 1.
$\dfrac{a}{1} = a$	A number divided by 1 is the number.
$\dfrac{a}{0}$ is undefined.	Division by zero is not defined.

EXAMPLES

1. $\dfrac{0}{-5} = 0$ **2.** $\dfrac{-4}{-4} = 1$ **3.** $\dfrac{-7}{1} = -7$ **4.** $\dfrac{-12}{0}$ is undefined.

Objective 1.3B Practice

1. Divide: $18 \div (-3)$
2. Divide: $57 \div (-3)$
3. Divide: $\dfrac{85}{-5}$
4. Divide: $0 \div (-14)$
5. What is 15 divided by -15?

Solutions on p. S-8.

Objective 1.3C Solve application problems

Consider a computer store that is offering special prices on its most popular laptop models. A customer can choose from two sizes of hard drive, D1 and D2; three different screen sizes, S1, S2, and S3; and two memory configurations, M1 and M2. This information can be organized in a **tree diagram.**

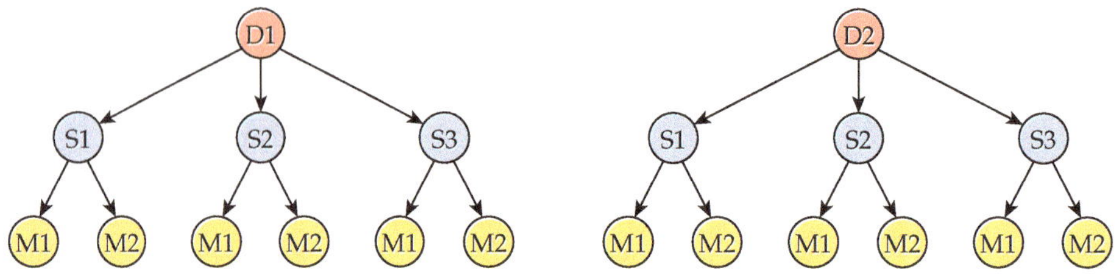

Looking at this diagram, a customer could, for example, choose a laptop that has drive 2 (D2), screen size 3 (S3), and memory size M1. By counting the circles along the bottom row, we can determine that there are 12 possible laptop configurations from which to choose. The number of configurations can also be determined by using the *counting principle.*

Counting Principle

To find the number of possible outcomes when making a series of choices, multiply the numbers of options available for each choice.

We can apply the counting principle to find the number of possible laptop configurations.

Number of hard drives	×	number of screen sizes	×	number of memory options	=	number of laptop configurations
2	×	3	×	2	=	12

Focus on using the counting principle

A coin is tossed four times. How many different sequences of heads (H) and tails (T) are possible? For instance, one possible sequence is HTTH.

SOLUTION

For each toss, there are two options, H or T. Because the coin is tossed four times, the number of possible sequences is $2 \cdot 2 \cdot 2 \cdot 2 = 16$.

Check your understanding 7

In horse racing, choosing a *trifecta* means choosing the exact order of the first three horses across the finish line. If there are eight horses in a race, how many trifectas are possible, assuming no ties?

SOLUTION See page S-2.

Focus on using the counting principle

A password is formed by first choosing three letters from A, B, C, D, E and then choosing two numbers from 1, 2, 3.

a. How many passwords can be formed if a letter or number can be used more than once?
b. How many passwords can be formed if a letter or number can be used only once?

SOLUTION

a. There are five letter options and three number options. Because the password must contain three letters and two numbers, use the counting principle to find the number of possible passwords.

$5 \cdot 5 \cdot 5 \cdot 3 \cdot 3 = 1125$

If a letter or number can be used more than once, there are 1125 possible passwords.

b. When a letter or number can be used only once, then there are five options for the first letter, four options for the second letter (one of the five has already been taken), and three options for the third letter. There are three options for the first number and two options for the second number. Use the counting principle to find the number of possible passwords.

$5 \cdot 4 \cdot 3 \cdot 3 \cdot 2 = 360$

If a letter or number cannot be used more than once, there are 360 possible passwords.

Check your understanding 8

An identification tag for a package is created by choosing two letters from W, X, Y, Z and four numbers from 4, 5, 6, 7, 8, 9.

a. How many different identification tags can be formed if a letter or number can be used more than once?

b. How many different identification tags can be formed if a letter or number can be used only once?

SOLUTION See page S-3.

In part (a) of the preceding Focus On problem, counting is done *with replacement*. In this example, this means that the same letter or number can be used more than once in a password. The option that was chosen is *replaced* (returned to the basket of choices) and can be used again. In part (b), no number or letter can be used more than once in a given password. This type of counting is done *without replacement*. Once a letter or number is used, it is not replaced as an option.

Here are some additional applications involving operations on integers.

Focus on solving an application

The daily high temperatures (in degrees Celsius) for six days in Anchorage, Alaska, were $-14°$, $3°$, $0°$, $-8°$, $2°$, and $-1°$. Find the average daily high temperature.

STRATEGY

To find the average daily high temperature:

- Add the six temperature readings.
- Divide the sum by 6.

SOLUTION

$-14 + 3 + 0 + (-8) + 2 + (-1) = -18$

$-18 \div 6 = -3$

The average daily high temperature was $-3°C$.

Check your understanding 9

The daily low temperatures (in degrees Celsius) during one week were recorded as $-6°$, $-7°$, $0°$, $-5°$, $-8°$, $-1°$, and $-1°$. Find the average daily low temperature.

SOLUTION See page S-3.

Focus on solving an application

To discourage random guessing on a multiple-choice exam, a professor assigns 7 points for a correct answer, −3 points for an incorrect answer, and 0 points for leaving the question blank. What is the score for a student who had 17 correct answers, had 8 incorrect answers, and left 2 questions blank?

STRATEGY

- Multiply each correct answer by 7, multiply each incorrect answer by −3, and multiply each question left blank by 0.
- Add the results.

SOLUTION

There were 17 correct answers: $7(17) = 119$
There were 8 incorrect answers: $-3(8) = -24$
There were 2 questions left blank: $0(2) = 0$

$119 + (-24) + 0 = 95$

The score for the student was 95.

Check your understanding 10

To discourage random guessing on a multiple-choice exam, a professor assigns 5 points for a correct answer, −2 points for an incorrect answer, and 0 points for leaving the question blank. What was the score for a student who had 23 correct answers, had 7 incorrect answers, and left 5 questions blank?

SOLUTION See pages S-3.

Objective 1.3C Practice

In Exercises 1 to 4, use the counting principle to determine the number of ways in which the task can be completed.

1. Two digits are selected without replacement from the digits 1, 2, 3, 4.
2. Two digits are selected with replacement from the digits 1, 2, 3, 4.
3. A gold, a silver, and a bronze medal are awarded in an Olympic event. In how many possible ways can the medals be awarded for a 200-meter sprint in which there are nine runners?
4. Find the number of six-character passwords that can be formed using one letter from a to h followed by five numbers from 1 to 9. Assume a letter or number cannot be used more than once in any six-character password.
5. The daily high temperatures for a six-day period in Barrow, Alaska, were −23°F, −29°F, −21°F, −28°F, −28°F, and −27°F. Calculate the average daily high temperature.
6. To discourage random guessing on a multiple-choice exam, a professor assigns 7 points for a correct answer, −3 points for an incorrect answer, and −1 point for leaving the question blank. What was the score for a student who had 17 correct answers, had 8 incorrect answers, and left 2 questions blank?

Solutions on pp. S-8–S-9.

SECTION 1.4

Exponents and the Order of Operations Agreement

Objective 1.4A Simplify expressions containing exponents

Repeated multiplication of the same factor can be written using an exponent.

$$2 \cdot 2 \cdot 2 \cdot 2 \cdot 2 = 2^5 \leftarrow \text{Exponent (2 is the Base)} \qquad a \cdot a \cdot a \cdot a = a^4 \leftarrow \text{Exponent } (a \text{ is the Base})$$

The **exponent** indicates how many times the factor, which is called the **base,** occurs in the multiplication. The multiplication $2 \cdot 2 \cdot 2 \cdot 2 \cdot 2$ is in **factored form.** The expression 2^5 is in **exponential form.**

2^1 is read "2 to the first power" or just "2." Usually the exponent 1 is not written.

2^2 is read "2 to the second power" or "2 squared."

2^3 is read "2 to the third power" or "2 cubed."

2^4 is read "2 to the fourth power."

a^4 is read "*a* to the fourth power."

The first three natural-number powers can be interpreted geometrically as length, area, and volume, respectively.

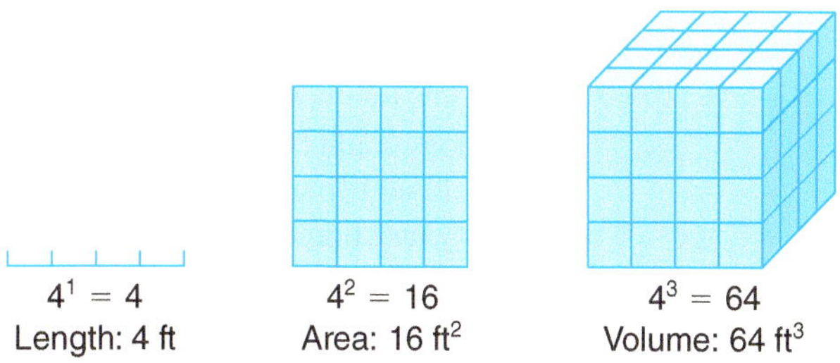

To evaluate an exponential expression, write each factor as many times as indicated by the exponent. Then multiply.

Take Note ➤
Note the difference between $(-2)^4$ and -2^4. $(-2)^4$ is the fourth power of -2:

$(-2)^4 = 16$

-2^4 is the opposite of the fourth power of 2:

$-2^4 = -16$

Evaluate $(-2)^4$.

$(-2)^4 = (-2)(-2)(-2)(-2)$ • Write -2 as a factor 4 times.

$= 16$ • Multiply.

Evaluate -2^4.

$-2^4 = -(2 \cdot 2 \cdot 2 \cdot 2)$ • Write 2 as a factor 4 times.

$= -16$ • Multiply.

Focus on evaluating an exponential expression

Evaluate.

a. $(-3)^2 \cdot 2^3$ **b.** $(-1)^6$ **c.** $-2 \cdot (-3)^2 \cdot (-1)^9$

SOLUTION

a. $(-3)^2 \cdot 2^3 = (-3)(-3) \cdot (2)(2)(2)$
$= 9 \cdot 8 = 72$

b. The product of an even number of negative factors is positive. Therefore, $(-1)^6 = 1$.

c. $-2 \cdot (-3)^2 \cdot (-1)^9 = -2 \cdot 9 \cdot (-1)$ • $(-3)^2 = 9; (-1)^9 = -1$
$= 18$

Check your understanding 1

Evaluate.

a. $(3^3)(-2)^3$ **b.** $(-1)^7$ **c.** $-2^2 \cdot (-1)^{12} \cdot (-3)^2$

SOLUTION See page S-3.

Objective 1.4A Practice

Evaluate.

1. $(-3)^2$
2. -4^4
3. $(-1)^9 \cdot 3^3$
4. $(-3) \cdot 2^2$
5. $(-7) \cdot 4^2 \cdot 3^2$
6. $(-2)^3 \cdot (-3)^2 \cdot (-1)^7$

Solutions on p. S-9.

Objective 1.4B Use the Order of Operations Agreement to simplify expressions

Let's evaluate $2 + 3 \cdot 5$.

There are two arithmetic operations, addition and multiplication, in this expression. The operations could be performed in different orders. We could multiply first and then add, or we could add first and then multiply. To prevent there being more than one answer when simplifying a numerical expression, an Order of Operations Agreement has been established.

The Order of Operations Agreement

Step 1. Perform operations inside grouping symbols. Grouping symbols include parentheses ( ), brackets [ ], braces { }, the absolute value symbol | |, and the fraction bar.

Step 2. Simplify exponential expressions.

Step 3. Do multiplication and division as they occur from left to right.

Step 4. Do addition and subtraction as they occur from left to right.

EXAMPLE

Simplify: $(-4)^2 - 2(3 - 8)$

$(-4)^2 - 2(3 - 8)$
$= (-4)^2 - 2(-5)$ • Perform operations inside parentheses.
$= 16 - 2(-5)$ • Simplify exponential expressions.
$= 16 + 10$ • Do multiplication and division from left to right.
$= 26$ • Do addition and subtraction from left to right.

One or more of the steps may not be needed to evaluate an expression. In that case, proceed to the next step in the Order of Operations Agreement.

Focus on using the Order of Operations Agreement

Evaluate $6 \div [4 - (6 - 8)] - 2^3$.

SOLUTION

$$\begin{aligned} 6 \div [4 - (6 - 8)] - 2^3 &= 6 \div [4 - (-2)] - 2^3 && \text{• Perform operations inside grouping symbols.} \\ &= 6 \div 6 - 2^3 \\ &= 6 \div 6 - 8 && \text{• Simplify exponential expressions.} \\ &= 1 - 8 && \text{• Do multiplication and division from left to right.} \\ &= -7 && \text{• Do addition and subtraction from left to right.} \end{aligned}$$

Check your understanding 2

Evaluate $7 - 2[2 \cdot 3 - 7 \cdot 2]^2$.

SOLUTION See page S-4.

Focus on using the Order of Operations Agreement

Evaluate $4 - 3[4 - 2(6 - 3)] \div 2$.

SOLUTION

$$\begin{aligned} 4 - 3[4 - 2(6 - 3)] \div 2 &= 4 - 3[4 - 2 \cdot 3] \div 2 && \text{• Perform operations inside grouping symbols.} \\ &= 4 - 3[4 - 6] \div 2 \\ &= 4 - 3[-2] \div 2 \\ &= 4 + 6 \div 2 && \text{• Do multiplication and division from left to right.} \\ &= 4 + 3 \\ &= 7 && \text{• Do addition and subtraction from left to right.} \end{aligned}$$

Check your understanding 3

Evaluate $18 - 5[8 - 2(2 - 5)] \div 10$.

SOLUTION See page S-4.

Focus on using the Order of Operations Agreement

Evaluate $27 \div (5 - 2)^2 + (-3)^2 \cdot 4$.

SOLUTION

$$\begin{aligned} 27 \div (5 - 2)^2 + (-3)^2 \cdot 4 &= 27 \div 3^2 + (-3)^2 \cdot 4 && \text{• Perform operations inside grouping symbols.} \\ &= 27 \div 9 + 9 \cdot 4 && \text{• Simplify exponential expressions.} \\ &= 3 + 9 \cdot 4 && \text{• Do multiplication and division from left to right.} \\ &= 3 + 36 \\ &= 39 && \text{• Do addition and subtraction from left to right.} \end{aligned}$$

Check your understanding 4

Evaluate $36 \div (8 - 5)^2 - (-3)^2 \cdot 2$.

SOLUTION See page S-4.

Objective 1.4B Practice

Evaluate using the Order of Operations Agreement.

1. $4 - 8 \div 2$
2. $24 - 18 \div 3 + 2$
3. $27 - 18 \div (-3^2)$
4. $16 + 15 \div (-5) - 2$
5. $6 + \dfrac{16 - 4}{2^2 + 2} - 2$
6. $18 \div 2 - 4^2 - (-3)^2$

Solutions on pp. S-9–S-10.

SECTION 1.5 Inductive and Deductive Reasoning

Objective 1.5A Use inductive reasoning

The type of reasoning in which a conclusion is formed by examining specific examples is called *inductive reasoning*. The conclusion reached through inductive reasoning is often called a **conjecture,** since it may or may not be correct.

Inductive Reasoning

Inductive reasoning is the process of reaching a general conclusion by examining specific examples.

Focus on using inductive reasoning

Use inductive reasoning to predict the next number in each of the following lists.

a. 3, 6, 9, 12, 15, ? **b.** 1, 3, 6, 10, 15, ?

SOLUTION

a. Each successive number is 3 larger than the preceding number. Thus we predict that the next number in the list is 3 larger than 15, which is 18.

b. The first two numbers differ by 2. The second and the third numbers differ by 3. It appears that the difference between any two numbers is always 1 more than the preceding difference. Since 10 and 15 differ by 5, we predict that the next number in the list will be 6 larger than 15, which is 21.

Check your understanding 1

Use inductive reasoning to predict the next number in each of the following lists.

a. 5, 10, 15, 20, 25, ? **b.** 2, 5, 10, 17, 26, ?

SOLUTION See page S-4.

Focus on using inductive reasoning

Consider the following procedure: Pick a number. Multiply the number by 8, add 6 to the product, divide the sum by 2, and subtract 3.

Complete the procedure for several different numbers. Use inductive reasoning to make a conjecture about the relationship between the size of the resulting number and the size of the original number.

SOLUTION

Suppose we pick 5 as our original number. Then the procedure would produce the following results:

Original number:	5
Multiply by 8:	$8 \times 5 = 40$
Add 6:	$40 + 6 = 46$
Divide by 2:	$46 \div 2 = 23$
Subtract 3:	$23 - 3 = 20$

We started with 5 and followed the procedure to produce 20. Starting with 6 as our original number produces a final result of 24. Starting with 10 produces a final result of 40. Starting with 100 produces a final result of 400. In each of these cases, the resulting number is four times the original number. We *conjecture* that the given procedure produces a number that is four times the original number.

Check your understanding 2

Consider the following procedure: Pick a number. Multiply the number by 9, add 15 to the product, divide the sum by 3, and subtract 5.

Complete the above procedure for several different numbers. Use inductive reasoning to make a conjecture about the relationship between the size of the resulting number and the size of the original number.

SOLUTION See page S-4.

Galileo Galilei (1564–1642) used inductive reasoning to discover that the time required for a pendulum to complete one swing, called the *period* of the pendulum, depends on the length of the pendulum. The following table shows some results obtained for pendulums of various lengths. For convenience, a length of 10 inches has been designated as 1 unit.

Length of Pendulum, in Units (1 Unit = 10 Inches)	Period of Pendulum, in Heartbeats
1	1
4	2
9	3
16	4
25	5
36	6

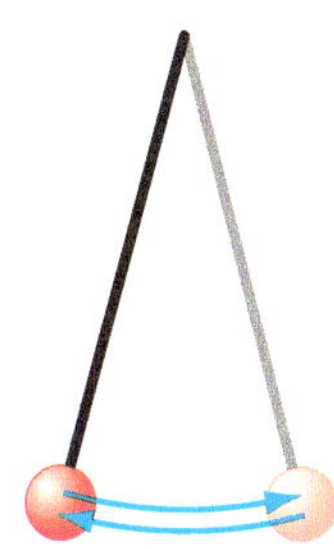

The period of a pendulum is the time it takes for the pendulum to swing from left to right and back to its original position.

Focus on using inductive reasoning

Use the data in the preceding table and inductive reasoning to answer each of the following questions.

a. If a pendulum has a length of 49 units, what is its period?
b. If the length of a pendulum is quadrupled, what happens to its period?

SOLUTION

a. In the table, each pendulum has a period that is the square root of its length. Thus we conjecture that a pendulum with a length of 49 units will have a period of 7 heartbeats.

b. In the table, a pendulum with a length of 4 units has a period that is twice that of a pendulum with a length of 1 unit. A pendulum with a length of 16 units has a period that is twice that of a pendulum with a length of 4 units. It appears that quadrupling the length of a pendulum doubles its period.

Check your understanding 3

Velocity of Tsunami, in Feet Per Second	Height of Tsunami, in Feet
6	4
9	9
12	16
15	25
18	36
21	49
24	64

A tsunami is a sea wave produced by an underwater earthquake. The height of a tsunami as it approaches land depends on the velocity of the tsunami. Use the table at the left and inductive reasoning to answer each of the following questions.

a. What happens to the height of a tsunami when its velocity is doubled?

b. What should be the height of a tsunami if its velocity is 30 ft/s?

SOLUTION See page S-4.

In mathematics, a statement is considered a true statement if and only if it is true in all cases. If you can find even one case for which a statement is not true, called a **counterexample,** then the statement is false. Consider the statement "All prime numbers are odd." A counterexample is the number 2, which is an even prime number. It is the case that proves the statement false.

Focus on finding a counterexample

Find a counterexample to the statement "All numbers ending in 4 can be divided evenly by 4."

SOLUTION

The statement is true for some numbers that end in 4. For instance, 84 ends in 4, and $84 \div 4 = 21$. However, 54 ends in 4, and 4 does not divide into 54 evenly. 54 is a counterexample to the statement. There are other counterexamples, such as 14 and 34. To prove a statement false, you need only one counterexample.

Check your understanding 4

Find a counterexample to the statement "The sum of the same two numbers is always greater than one of the numbers." (*Suggestion:* Think about negative numbers.)

SOLUTION See page S-5.

Objective 1.5A Practice

1. Use inductive reasoning to predict the next number in the list:
3, 5, 9, 15, 23, 33, ?

2. Use inductive reasoning to predict the next number in the list:
2, 7, −3, 2, −8, −3, −13, −8, −18, ?

3. Consider the following procedure: Pick a counting number. Multiply the number by 8, subtract 4 from the product, divide the difference by 2, and add 2 to the quotient.
Complete the above procedure for several different numbers. Use inductive reasoning to make a conjecture about the relationship between the resulting number and the original number.

4. Refer to the table given in Check your understanding 3 of this section. What should be the height of a tsunami with a velocity of 36 ft/s?

5. Find a counterexample to the statement "Any number cubed is greater than the original number."

Solutions on p. S-10.

Objective 1.5B Use deductive reasoning

Another type of reasoning is called *deductive reasoning.* Deductive reasoning is the process of reaching a conclusion by applying general principles and procedures. The conclusion represents a specific case of the general assumption.

Deductive Reasoning

Deductive reasoning is the process of reaching a conclusion by applying general assumptions, procedures, or principles.

Focus on using deductive reasoning

Use deductive reasoning to create a true statement from the following information:
At General Hospital, all nurses are paid a base rate of $26/hour for the first year of employment. Harry is a nurse in his first year of employment at General Hospital.

SOLUTION

From the given information, we can deduce that Harry is paid $26/hour.

Check your understanding 5

Use deductive reasoning to write a true statement, given the following information:
All students attending Silver University are required to take a placement exam. Only students who score above 90 are eligible for honors housing. Maria scored 85 on the placement exam.

SOLUTION See page S-5.

Focus on using deductive reasoning

Use deductive reasoning to answer the following question:
If $\partial\partial\partial\partial = \Sigma\Sigma\Sigma$ and $\Omega\Omega = \Sigma\Sigma\Sigma$, how many ∂'s = $\Omega\Omega$?

SOLUTION

Because four "∂" symbols are equal to $\Sigma\Sigma\Sigma$ and two "Ω" symbols are also equal to $\Sigma\Sigma\Sigma$, we can use deductive reasoning to state that four ∂ symbols are equal to two Ω symbols.

Check your understanding 6

Use deductive reasoning to answer the following question:
If $\partial\partial\partial\partial = \Sigma\Sigma\Sigma$ and $\Omega\Omega = \Sigma\Sigma\Sigma$, how many ∂'s = Ω?

SOLUTION See page S-5.

With the help of a table or chart, we can use deductive reasoning to solve a logic puzzle.

Focus on using deductive reasoning to solve a logic puzzle

Each of three neighbors, Raj, Viola, and Andy, has a different occupation: banker, chef, or teacher. From the following clues, determine the occupation of each neighbor.

1. Andy gets home from work after Raj, who is not the banker.
2. Raj gets home from work before the chef.
3. The chef gets home from work before Andy.

SOLUTION

We can create a table to help us solve the problem. Make a row for each neighbor and a column for each occupation.

	Banker	Chef	Teacher
Raj			
Viola			
Andy			

From the first clue, we know that Raj is not the banker. Put "X1" in the Banker column of Raj's row, to stand for "ruled out by Clue 1." From the second clue, we know that Raj is not the chef. Put an "X2" in the Chef column of Raj's row, to stand for "ruled out by Clue 2." From the table, we can see that Raj must be the teacher. Put a check mark in the Teacher column of Raj's row.

	Banker	Chef	Teacher
Raj	X1	X2	√
Viola			
Andy			

Now, since Raj is the teacher, neither Viola nor Andy can be the teacher. Put "X2" in the Teacher column of both Viola's and Andy's rows.

	Banker	Chef	Teacher
Raj	X1	X2	√
Viola			X2
Andy			X2

From the last clue, we know that Andy cannot be the chef. Put an X3 in the Chef column of Andy's row, to stand for "ruled out by Clue 3." Since Andy is neither the teacher nor the chef, he must be the banker. Put a check mark in the Banker column of Andy's row.

	Banker	Chef	Teacher
Raj	X1	X2	√
Viola			X2
Andy	√	X3	X2

We have determined that Andy is the banker, so Viola cannot be the banker. Put an "X3" in the Banker column of Viola's row. We can see from the table that Viola's occupation must be chef. Put a check mark in the Chef column of Viola's row.

	Banker	Chef	Teacher
Raj	X1	X2	√
Viola	X3	√	X2
Andy	√	X3	X2

From the completed table, Raj is the teacher, Viola is the chef, and Andy is the banker.

Check your understanding 7

Each of three siblings, Raylan, Boyd, and Meg, takes care of a different pet: Siamese cat, Black Lab, or gecko. From the following clues, determine which pet each sibling is responsible for.

1. Boyd does not take care of the gecko or the Black Lab.
2. Meg is the oldest sibling.
3. The middle sibling takes care of the gecko.

SOLUTION See page S-5.

Objective 1.5B Practice

1. Use deductive reasoning to create a true statement from the following information: *Under the Dome* is a novel written by Stephen King. Dumaka has read every novel written by Stephen King.
2. Use deductive reasoning to create a true statement from the following information: Brandon earns \$14 per hour working as an apprentice electrician. Last week, Brandon worked 35 hours at his electrical job.
3. Use deductive reasoning to complete the statement: If OO = XXX and XXX = QQQQQQ, then one O = _?_ Q's.
4. Use deductive reasoning to answer the following question: If three X's are equal to six O's, and two O's are equal to four Q's, how many Q's are equal to one X?
5. Is the following argument a correct example of deductive reasoning? Anna loves to read. The students who attend the Dutile elementary school all love to read. Therefore, Anna attends the Dutile elementary school.
6. Is the following argument a correct example of deductive reasoning? Kanye is on the swim team at his college. All members of the swim team like to play chess. Therefore, Kanye likes to play chess.
7. Use deductive reasoning to solve the following logic problem: Each of three students, Hannah, Robert, and Keisha, takes a different form of transportation to school: bus, car, or subway. Use the following clues to determine which student takes each form of transportation.
   1. Robert gets to school after Keisha, who does not take the bus.
   2. Keisha gets to school before the person who takes the subway.
   3. The person who takes the subway gets to school before Robert.

Solutions on pp. S-10–S-11.

Objective 1.5C Choose between inductive and deductive reasoning

We can analyze arguments to determine whether they use inductive or deductive reasoning.

Focus on determining types of reasoning

Determine whether each of the following arguments is an example of inductive reasoning or deductive reasoning.

a. During the past 10 years, a tree has produced plums every other year. Last year the tree did not produce plums, so this year the tree will produce plums.

b. All home improvements cost more than the estimate. The contractor estimated that my home improvement will cost \$35,000. Thus my home improvement will cost more than \$35,000.

SOLUTION

a. This argument reaches a conclusion based on specific examples, so it is an example of inductive reasoning.

b. Because the conclusion is a specific case of a general assumption, this argument is an example of deductive reasoning.

Check your understanding 8

Determine whether each of the following arguments is an example of inductive reasoning or deductive reasoning.

a. All Janet Evanovich novels are worth reading. The novel *Twelve Sharp* is a Janet Evanovich novel. Thus *Twelve Sharp* is worth reading.

b. I know I will win a jackpot on this slot machine in the next 10 tries because it has not paid out any money during the last 45 tries.

SOLUTION See page S-6.

Objective 1.5C Practice

In Exercises 1 to 5, determine whether the argument is based on inductive or deductive reasoning.

1. All pentagons have exactly five sides. Figure A is a pentagon. Therefore, Figure A has five sides.
2. Every English setter likes to hunt. Duke is an English setter, so Duke likes to hunt.
3. The Atlanta Braves won six games in a row. Therefore, the Atlanta Braves will win their next game.
4. The New England Patriots have never won fewer than nine games in a regular season. Therefore, the New England Patriots will win nine games or more next season.
5. A number is a *neat* number if the sum of the cubes of its digits equals the number. Therefore, 153 is a neat number.

Solutions on p. S-11.

SECTION 1.6 Basic Concepts of Sets

Objective 1.6A Write sets and use Venn diagrams

Recall from Section 1.1 that a collection of objects is called a set, and the objects in the set are called elements, or members, of the set. The roster method of writing a set encloses a list of the elements of the set in braces. The order in which the elements are listed is not important.

Focus on writing a set using the roster method

Use the roster method to represent the set of the days in a week.

SOLUTION {Sunday, Monday, Tuesday, Wednesday, Thursday, Friday, Saturday}

Check your understanding 1

Use the roster method to represent the set of months that start with the letter A.

SOLUTION See page S-6.

The set of all elements that are being considered for a particular problem is called the **universal set.** We will use the letter U to denote the universal set. For instance, if we wanted to investigate the positive even numbers, then the universal set would be $U = \{2n$, where n is any natural number$\}$. Some of the elements of this universal set are 2, 4, 6, and 8. These elements correspond to $n = 1, 2, 3$, and 4. If we wanted to study the perfect squares of the natural numbers, then the universal set would be $U = \{n^2$, where n is any natural number$\}$. In this case, some of the elements of U are 1, 4, 9, 16, 25, and 36, corresponding to $n = 1$, 2, 3, 4, 5, and 6.

The Complement of a Set

The **complement** of a set A, denoted by A', is the set of all elements of the universal set U that are not elements of A.

Focus on finding the complement of a set

Let $U = \{1, 2, 3, 4, 5, 6, 7, 8, 9, 10\}$, $S = \{2, 4, 6, 7\}$, and $T = \{1, 3, 5, 7, 9\}$. Find the following sets.

a. S' **b.** T'

SOLUTION

a. The elements of the universal set are 1, 2, 3, 4, 5, 6, 7, 8, 9, and 10. From these elements, we wish to exclude the elements of S, which are 2, 4, 6, and 7. Therefore, $S' = \{1, 3, 5, 8, 9, 10\}$.

b. $T = \{1, 3, 5, 7, 9\}$. Excluding the elements of T from U gives us $T' = \{2, 4, 6, 8, 10\}$.

Check your understanding 2

Let $U = \{0, 2, 3, 4, 6, 7, 17\}$, $M = \{0, 4, 6, 17\}$, and $P = \{2, 4, 6\}$. Find the following sets.

a. M' **b.** P'

SOLUTION See page S-6.

The **empty set,** denoted by $\varnothing$, is the set containing no elements. There are two fundamental results concerning the universal set and the empty set. Because the universal set contains all elements under consideration, the complement of the universal set is the empty set. Conversely, the complement of the empty set is the universal set, because the empty set has no elements and the universal set contains all the elements under consideration.

The Complement of the Universal Set and the Complement of the Empty Set

$U' = \varnothing$ and $\varnothing' = U$

Consider the set of letters of the alphabet and the set of vowels {a, e, i, o, u}. Every element of the set of vowels is an element of the set of letters of the alphabet. The set of vowels is said to be a *subset* of the set of letters of the alphabet.

A Subset of a Set

Set A is a **subset** of set B, denoted by $A \subseteq B$, if and only if every element of A is also an element of B.

Focus on applying the definition of a subset

Determine whether each statement is true or false.

a. $\{5, 10, 15, 20\} \subseteq \{0, 5, 10, 15, 20, 25, 30\}$
b. $\{2, 4, 6\} \subseteq \{2, 4, 6\}$

SOLUTION

a. True; every element of the first set is an element of the second set.
b. True. In this case, the two sets are equal. Every set is a subset of itself.

Check your understanding 3

Determine whether each statement is true or false.

a. $\{1, 3, 5\} \subseteq \{1, 5, 9\}$
b. $\{2, 3, 5\} \subseteq$ {prime numbers}

SOLUTION See page S-6.

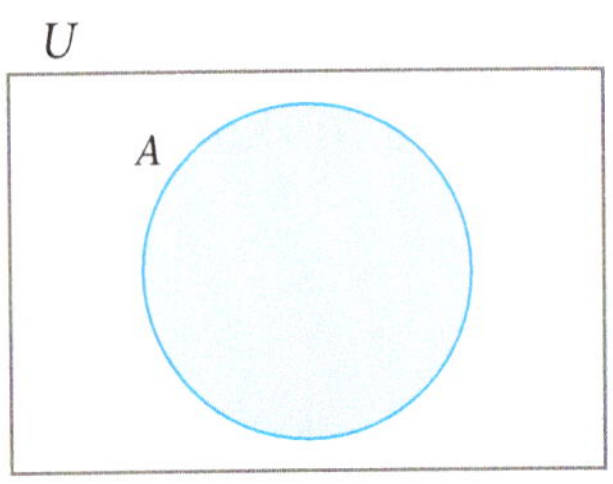

A Venn diagram

The English logician John Venn (1834–1923) developed diagrams, which we now refer to as *Venn diagrams,* that can be used to illustrate sets and relationships between sets. In a **Venn diagram,** the universal set is represented by a rectangular region, and subsets of the universal set are generally represented by oval or circular regions drawn inside the rectangle. The Venn diagram at the left shows a universal set and one of its subsets, labeled as set A. The size of the circle is not important. The region outside the circle but inside the rectangle represents the set A'.

In the Venn diagram at the right, the universal set consists of the set of natural numbers from 1 to 10. The subset containing the prime numbers less than 10 is shown as a circle inside the rectangle representing the universal set.

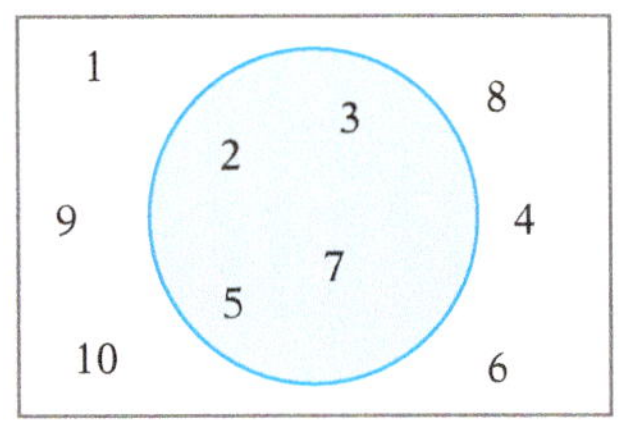

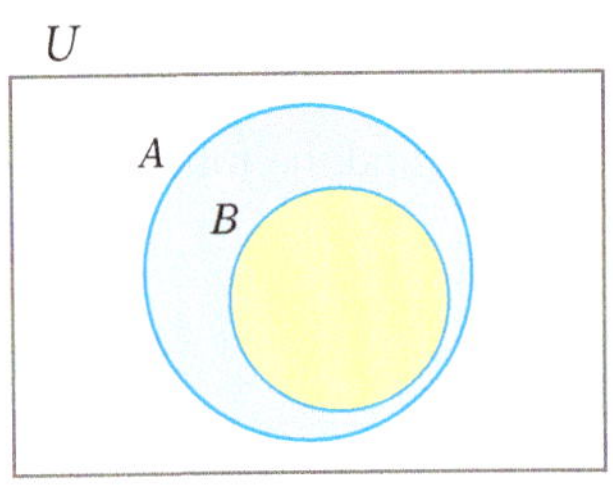

B is a subset of *A*.

Venn diagrams can be used to represent relationships between subsets. For instance, if set B is a subset of set A, then we illustrate this relationship in a Venn diagram by drawing a circle labeled B inside a circle labeled A. See the Venn diagram at the left.

Objective 1.6A Practice

In Exercises 1 and 2, find the complement of the set, given that $U = \{0, 1, 2, 3, 4, 5, 6, 7, 8\}$.

1. $\{2, 4, 6, 7\}$
2. $\varnothing$

In Exercises 3 to 5, let $U = \{p, q, r, s, t\}$, $D = \{p, r, s, t\}$, $E = \{q, s\}$, and $F = \{p, t\}$. Determine whether each statement is true or false.

3. $F \subseteq D$
4. $E \subseteq F'$
5. $D' \subseteq E$

Solutions on p. S-11.

Objective 1.6B Find the intersection and union of sets

In this objective, we define the set operations of *intersection* and *union*.

Intersection of Sets

The **intersection** of sets A and B, denoted by $A \cap B$, is the set of elements common to both A and B.

$$A \cap B = \{x | x \in A \quad \text{and} \quad x \in B\}$$

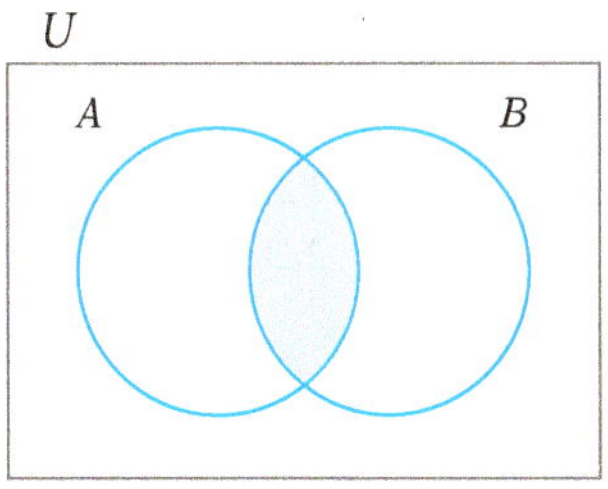

$A \cap B$

In the Venn diagram at the left, the region shown in blue represents the intersection of sets A and B.

Focus on finding the intersection of two sets

Let $A = \{1, 4, 5, 7\}$, $B = \{2, 3, 4, 5, 6\}$, and $C = \{3, 6, 9\}$. Find:

a. $A \cap B$ **b.** $A \cap C$

SOLUTION

a. The elements common to A and B are 4 and 5.

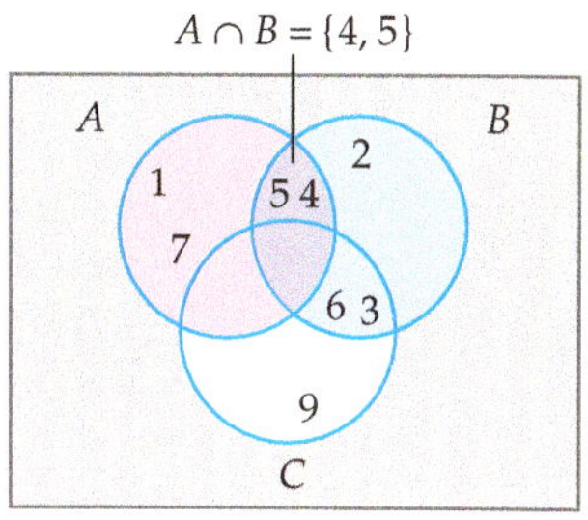

$$A \cap B = \{1, 4, 5, 7\} \cap \{2, 3, 4, 5, 6\}$$
$$= \{4, 5\}$$

b. Sets A and C have no common elements. Thus $A \cap C = \emptyset$.

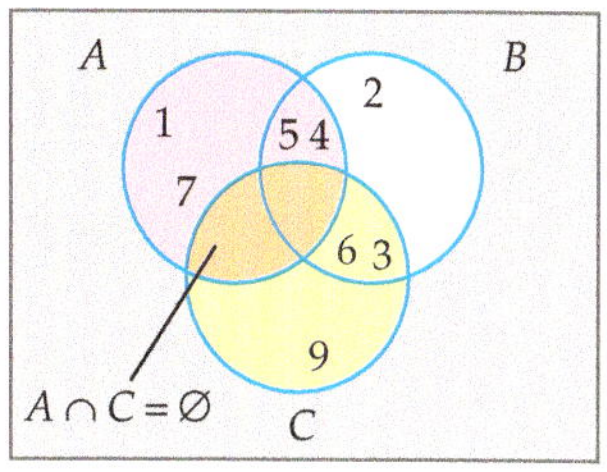

Check your understanding 4

Let $D = \{0, 3, 8, 9\}$, $E = \{3, 4, 8, 9, 11\}$, and $F = \{0, 2, 6, 8\}$. Find:

a. $D \cap E$ b. $D \cap F$

SOLUTION See page S-6.

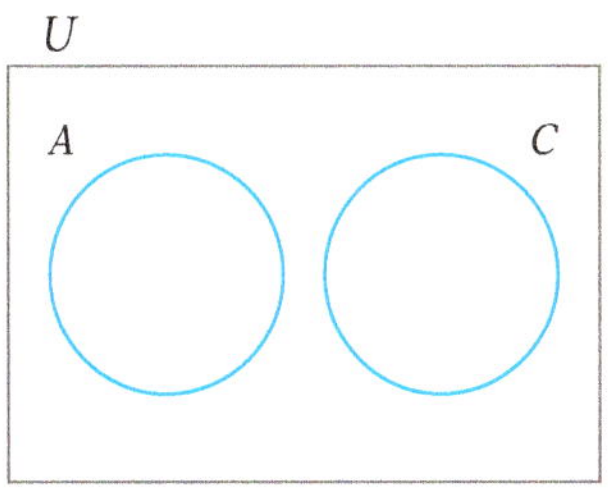

$A \cap C = \emptyset$

Two sets are **disjoint** if their intersection is the empty set. Sets A and C in the preceding Focus On example are disjoint. The Venn diagram at the left illustrates two disjoint sets.

In everyday usage, the word "union" refers to the act of uniting or joining together. The union of two sets has a similar meaning.

Union of Sets

The **union** of sets A and B, denoted by $A \cup B$, is the set that contains all the elements that belong to A or to B, or to both.

$$A \cup B = \{x | x \in A \quad \text{or} \quad x \in B\}$$

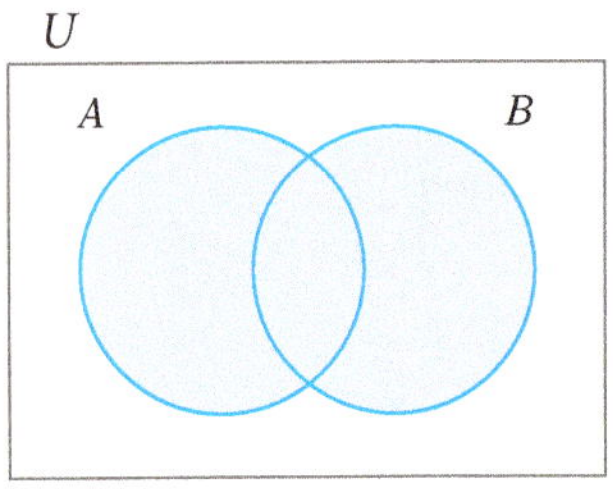

$A \cup B$

In the Venn diagram at the left, the region shown in blue represents the union of sets A and B.

Focus on finding the union of two sets

Let $A = \{1, 4, 5, 7\}$, $B = \{2, 3, 4, 5, 6\}$, and $C = \{3, 6, 9\}$. Find:

a. $A \cup B$ **b.** $A \cup C$

SOLUTION

a. List all the elements of set A, which are 1, 4, 5, and 7. Then add to your list the elements of set B that have not already been listed—in this case, 2, 3, and 6. Enclose all elements with a pair of braces. Thus

$$A \cup B = \{1, 4, 5, 7\} \cup \{2, 3, 4, 5, 6\}$$
$$= \{1, 2, 3, 4, 5, 6, 7\}$$

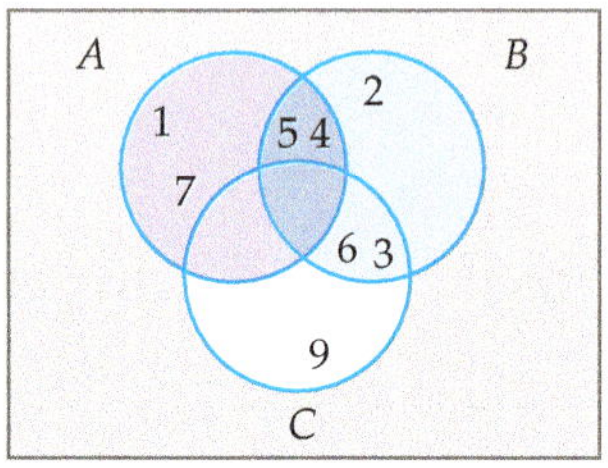

$A \cup B = \{1, 2, 3, 4, 5, 6, 7\}$

b. $A \cup C = \{1, 4, 5, 7\} \cup \{3, 6, 9\}$
$= \{1, 3, 4, 5, 6, 7, 9\}$

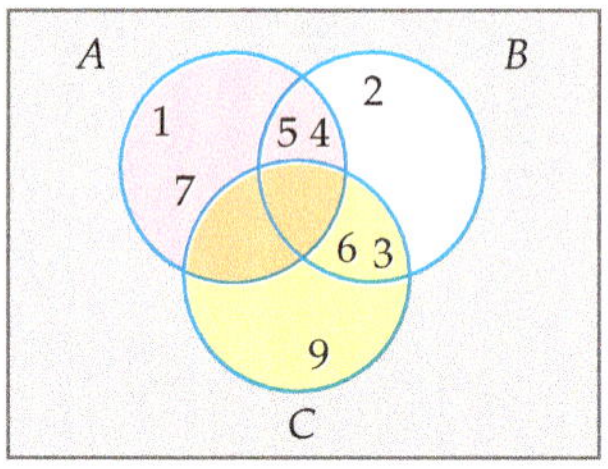

$A \cup C = \{1, 3, 4, 5, 6, 7, 9\}$

Check your understanding 5

Let $D = \{0, 4, 8, 9\}$, $E = \{1, 4, 5, 7\}$, and $F = \{2, 6, 8\}$. Find

a. $D \cup E$ **b.** $D \cup F$

SOLUTION See page S-6.

Focus on finding the intersection or union of sets

Let the universal set $U = \{1, 2, 3, 4, 5, 6, 7, 8, 9, 10, 11, 12, 13, 14, 15\}$, $A = \{2, 4, 6, 8, 10, 12, 14\}$, $B = \{2, 3, 5, 7, 11, 13\}$, and $C = \{1, 3, 5, 7, 9, 11, 12, 14, 15\}$. Find each of the following.

a. $A \cup (B \cap C)$ **b.** $(A \cap C) \cap B'$

SOLUTION

a. First find $B \cap C$, which is the set of elements that are common to B and C. $B \cap C = \{3, 5, 7, 11\}$.
Then $A \cup (B \cap C) = A \cup \{3, 5, 7, 11\} = \{2, 3, 4, 5, 6, 7, 8, 10, 11, 12, 14\}$.

b. First find $A \cap C$, which is the set of elements that are common to A and C. $A \cap C = \{12, 14\}$. Now find B', the elements in U that are not in B. $B' = \{1, 4, 6, 8, 9, 10, 12, 14, 15\}$. Then $(A \cap C) \cap B' = \{12, 14\} \cap \{1, 4, 6, 8, 9, 10, 12, 14, 15\} = \{12, 14\}$.

Check your understanding 6

Let the universal set
$U = \{1, 2, 3, 4, 5, 6, 7, 8, 9, 10, 11, 12, 13, 14, 15, 16, 17, 18, 19, 20\}$,
$A = \{5, 9, 15, 16, 19\}$, $B = \{1, 2, 3, 7, 8, 12, 13, 19\}$, and $C = \{4, 6, 9, 11, 15, 16\}$.
Find each of the following.

a. $(A \cup C) \cap B$ **b.** $B' \cup (A \cap C)$

SOLUTION See page S-6.

Objective 1.6B Practice

In Exercises 1 to 5, let $U = \{1, 2, 3, 4, 5, 6, 7, 8\}$, $A = \{2, 4, 6\}$, $B = \{1, 2, 5, 8\}$, and $C = \{1, 3, 7\}$. Find each of the following.

1. $A \cup B$
2. $A \cap B'$
3. $A \cap (B \cap C)$
4. $A' \cup (B \cap C)$
5. $(A \cap B') \cap C'$

Solutions on p. S-11.

SOLUTIONS TO MODULE 1

Solutions to Check Your Understanding

Section 1.1

Check your understanding 1

$A = \{1, 2, 3, 4, 5, 6\}$

Check your understanding 2

$-5 < -1$

$-1 = -1$

$5 > -1$

The element 5 is greater than -1.

Check your understanding 3

$-(-11) = 11$

$-0 = 0$

$-(8) = -8$

Check your understanding 4

$|-5| = 5$

$-|-23| = -23$

Check your understanding 5

The smallest value in the data set is 104, and the largest is 156. Therefore, the stems range from 10 (the first two digits of 104) to 15 (the first two digits of 156). Draw a vertical line. List the stems to the left of the line and the leaves to the right. When placing leaf values, list them from smallest to largest.

Annual Rainfall Totals	
10	4, 4, 4, 7
11	3, 5, 5, 7, 9
12	0, 2, 5, 5, 6, 6, 7, 8, 9
13	4, 5
14	2
15	2, 4, 6

Section 1.2

Check your understanding 1

$100 + (-43) = 57$

Check your understanding 2

$$\begin{aligned} &-51 + 42 + 17 + (-102) \\ &\quad = -9 + 17 + (-102) \\ &\quad = 8 + (-102) \\ &\quad = -94 \end{aligned}$$

Check your understanding 3

$-8 + 7 = -1$

Check your understanding 4

$$\begin{aligned}
&-9-(-12)-17-4\\
&\quad=-9+12+(-17)+(-4)\\
&\quad=3+(-17)+(-4)\\
&\quad=-14+(-4)\\
&\quad=-18
\end{aligned}$$

Check your understanding 5

$-11-(-12)=-11+12=1$

Check your understanding 6

STRATEGY

To find the difference between the two average temperatures, subtract the smaller number (-130) from the larger number (-17).

SOLUTION

$-17-(-130)=-17+130=113$

The difference is 113°F.

Section 1.3

Check your understanding 1

$$\begin{aligned}
(-2)3(-8)7&=-6(-8)7\\
&=48(7)\\
&=336
\end{aligned}$$

Check your understanding 2

$-9(34)=-306$

Check your understanding 3

$(-135)\div(-9)=15$

Check your understanding 4

$$\frac{-72}{4}=-18$$

Check your understanding 5

$$\begin{aligned}
-\frac{36}{-12}&=-(-3)\\
&=3
\end{aligned}$$

Check your understanding 6

$$\frac{-72}{-8}=9$$

Check your understanding 7

Any one of the eight horses can be first, so there are 8 choices for first place. Because a horse cannot finish both first and second, there are seven horses that can finish second; thus there are 7 choices for second place. Similarly, there are six horses that can finish third, so there are 6 choices for third place. By the counting principle, there are $8\cdot7\cdot6=336$ possible trifectas.

Check your understanding 8

a. There are four letter options and six number options. Each identification tag consists of two letters and four numbers. Use the counting principle to find the number of possible identification tags.

$4 \cdot 4 \cdot 6 \cdot 6 \cdot 6 \cdot 6 = 20{,}736$

If a letter or number can be used more than once, there are 20,736 possible identification tags.

b. When a letter or number can be used only once, then there are four options for the first letter and three options for the second letter (the letter chosen first is no longer available). In a similar manner, there are six options for the first number, five options for the second number, four options for the third number, and three options for the fourth number. Use the counting principle to find the number of possible tags.

$4 \cdot 3 \cdot 6 \cdot 5 \cdot 4 \cdot 3 = 4320$

If a letter or number cannot be used more than once, there are 4320 possible tags.

Check your understanding 9

STRATEGY

To find the average daily low temperature:

- Add the seven temperature readings.
- Divide the sum by 7.

SOLUTION

$-6 + (-7) + 0 + (-5) + (-8) + (-1) + (-1) = -28$

$-28 \div 7 = -4$

The average daily low temperature was $-4°C$.

Check your understanding 10

STRATEGY

- Multiply each correct answer (23) by 5, multiply each incorrect answer (7) by -2, and multiply each question left blank (5) by 0.
- Add the results.

SOLUTION

There were 23 correct answers: $5(23) = 115$

There were 7 incorrect answers: $-2(7) = -14$

There were 5 questions left blank: $0(5) = 0$

$115 + (-14) + 0 = 101$

The score for the student was 101.

Section 1.4

Check your understanding 1

a. $(3^3)(-2)^3 = (3)(3)(3) \cdot (-2)(-2)(-2)$
$= 27(-8) = -216$

b. The product of an odd number of negative factors is negative. Therefore, $(-1)^7 = -1$.

c. $-2^2 \cdot (-1)^{12} \cdot (-3)^2$
$= -(2 \cdot 2) \cdot 1 \cdot (-3) \cdot (-3)$
$= -4 \cdot 1 \cdot 9 = -36$

Check your understanding 2

$7 - 2[2 \cdot 3 - 7 \cdot 2]^2$

$= 7 - 2[6 - 14]^2$	• Perform operations inside grouping symbols.
$= 7 - 2[-8]^2$	
$= 7 - 2[64]$	• Simplify exponential expressions.
$= 7 - 128$	• Multiply or divide from left to right.
$= -121$	• Subtract.

Check your understanding 3

$18 - 5[8 - 2(2 - 5)] \div 10$

$= 18 - 5[8 - 2(-3)] \div 10$	• Perform operations inside grouping symbols.
$= 18 - 5[8 + 6] \div 10$	
$= 18 - 5[14] \div 10$	
$= 18 - 70 \div 10$	• Multiply or divide from left to right.
$= 18 - 7$	
$= 11$	• Subtract.

Check your understanding 4

$36 \div (8 - 5)^2 - (-3)^2 \cdot 2$

$= 36 \div (3)^2 - (-3)^2 \cdot 2$	• Perform operations inside grouping symbols.
$= 36 \div 9 - 9 \cdot 2$	• Multiply or divide from left to right.
$= 4 - 9 \cdot 2$	
$= 4 - 18$	
$= -14$	• Subtract.

Section 1.5

Check your understanding 1

a. Each successive number is 5 larger than the preceding number. Thus we predict that the next number in the list is 5 larger than 25, which is 30.

b. The first two numbers differ by 3. The second and third numbers differ by 5. It appears that the difference between any two numbers is always 2 more than the preceding difference. Thus we predict that the next number will be 11 more than 26, which is 37.

Check your understanding 2

If the original number is 2, then $\frac{2 \times 9 + 15}{3} - 5 = 6$, which is three times the original number.

If the original number is 7, then $\frac{7 \times 9 + 15}{3} - 5 = 21$, which is three times the original number.

If the original number is -12, then $\frac{-12 \times 9 + 15}{3} - 5 = -36$, which is three times the original number.

It appears, by inductive reasoning, that the procedure produces a number that is three times the original number.

Check your understanding 3

a. It appears that when the velocity of a tsunami is doubled, its height is quadrupled.

b. A tsunami with a velocity of 30 ft/s will have a height that is four times that of a tsunami with a speed of 15 ft/s. Thus we predict a height of $4 \times 25 = 100$ ft for a tsunami with a velocity of 30 ft/s.

Check your understanding 4

The statement is true for all positive numbers; for instance, $3 + 3 = 6$. However, if we choose -3 for our original number, we get $-3 + (-3) = -6$. Since -6 is not greater than -3, we have found a counterexample. In fact, any negative number will provide a counterexample.

Check your understanding 5

From the given information, we can deduce that Maria is not eligible for honors housing at Silver University.

Check your understanding 6

Because four ∂ symbols are equal to $\Sigma\Sigma\Sigma$ and two Ω symbols are also equal to $\Sigma\Sigma\Sigma$, we can write $\Sigma\Sigma\Sigma = \partial\partial\partial\partial = \Omega\Omega$. From the statement $\partial\partial\partial\partial = \Omega\Omega$, we can deduce that each Ω symbol is equal to two ∂ symbols.

Check your understanding 7

We can solve the problem by creating a table. Make a row for each sibling and a column for each pet.

	Siamese cat	Black Lab	Gecko
Raylan			
Boyd			
Meg			

From Clue 1, we can put "X1" in the Black Lab and gecko columns of Boyd's row. Because the only pet left for Boyd is the Siamese cat, put a check in that column of Boyd's row. If Boyd takes care of the cat, neither Raylan nor Meg is responsible for the cat. Put X1 in the Siamese cat column of Raylan's and Meg's rows.

	Siamese cat	Black Lab	Gecko
Raylan	X1		
Boyd	√	X1	X1
Meg	X1		

From Clues 2 and 3, we can deduce that Meg does not take care of the gecko. Put X3 in the gecko column of Meg's row. Now, since neither Boyd nor Meg takes care of the gecko, we know that Raylan must take care of it. Put a check mark in the gecko column of Raylan's row.

	Siamese cat	Black Lab	Gecko
Raylan	X1		√
Boyd	√	X1	X1
Meg	X1		X3

If Raylan takes care of the gecko, he does not take care of the Black Lab. Put X3 in the Black Lab column of Raylan's row. Finally, since Meg does not take care of the cat or the gecko, we know she must take care of the Black Lab. Put a check mark in the Black Lab column of Meg's row.

	Siamese cat	**Black Lab**	**Gecko**
Raylan	X1	X3	√
Boyd	√	X1	X1
Meg	X1	√	X3

From the table, we can see that Boyd takes care of the Siamese cat, Meg takes care of the Black Lab, and Raylan takes care of the gecko.

Check your understanding 8

a. The conclusion is a specific case of a general assumption, so the argument is an example of deductive reasoning.

b. The argument reaches a conclusion based on specific examples, so the argument is an example of inductive reasoning.

Section 1.6

Check your understanding 1

The only months that start with the letter A are April and August. Using the roster method, the set is written {April, August}.

Check your understanding 2

a. $M = \{0, 4, 6, 17\}$. The set of elements in $U = \{0, 2, 3, 4, 6, 7, 17\}$ but not in M is $M' = \{2, 3, 7\}$.

b. $P = \{2, 4, 6\}$. The set of elements in $U = \{0, 2, 3, 4, 6, 7, 17\}$ but not in P is $P' = \{0, 3, 7, 17\}$.

Check your understanding 3

a. False. The number 3 is an element of the first set but not an element of the second set. Therefore, the first set is not a subset of the second set.

b. True. The numbers 2, 3, and 5 are all prime numbers.

Check your understanding 4

a. $D \cap E = \{0, 3, 8, 9\} \cap \{3, 4, 8, 9, 11\}$
$= \{3, 8, 9\}$

b. $D \cap F = \{0, 3, 8, 9\} \cap \{0, 2, 6, 8\}$
$= \{0, 8\}$

Check your understanding 5

a. $D \cup E = \{0, 4, 8, 9\} \cup \{1, 4, 5, 7\}$
$= \{0, 1, 4, 5, 7, 8, 9\}$

b. $D \cup F = \{0, 4, 8, 9\} \cup \{2, 6, 8\}$
$= \{0, 2, 4, 6, 8, 9\}$

Check your understanding 6

a. First find $A \cup C$: $A \cup C = \{4, 5, 6, 9, 11, 15, 16, 19\}$.
Then find the intersection of this set with set B:
$\{4, 5, 6, 9, 11, 15, 16, 19\} \cap \{1, 2, 3, 7, 8, 12, 13, 19\} = \{19\}$.

b. First find B', which consists of the elements of U that are not in B:
$B' = \{4, 5, 6, 9, 10, 11, 14, 15, 16, 17, 18, 20\}$.

Next find $A \cap C$: $A \cap C = \{9, 15, 16\}$.

Then find the union of these two sets:
$\{4, 5, 6, 9, 10, 11, 14, 15, 16, 17, 18, 20\} \cup \{9, 15, 16\}$
$= \{4, 5, 6, 9, 10, 11, 14, 15, 16, 17, 18, 20\}$

Solutions to Objective Practice Exercises

Objective 1.1A

1. $-14 < 16$
2. $0 > -31$
3. $A = \{1, 2, 3\}$
4. $-33 < -10$
   $-24 < -10$
   $-10 = -10$
   $0 > -10$
   The elements -33 and -24 are less than -10.

Objective 1.1B

1. -4
2. 9
3. 36
4. 40
5. 74
6. -81

Objective 1.1C

1.

Winning Football Scores
1 | 7899
2 | 00133444456666789
3 | 0001344567
4 | 1

Objective 1.2A

1. $-6 + (-9) = -15$
2. $-17 + 17 = 0$
3. $-27 + (-42) + (-18) = -69 + (-18) = -87$
4. $-6 + (-8) + 14 + (-4) = -14 + 14 + (-4)$
   $= 0 + (-4) = -4$
5. $-8 + 4 = -4$
6. $-21 + (-8) = -29$

Objective 1.2B

1. $16 - 8 = 16 + (-8) = 8$
2. $6 - (-12) = 6 + 12 = 18$
3. $-19 - (-19) - 18 = -19 + 19 + (-18)$
   $= 0 + (-18) = -18$
4. $-12 - 9 = -12 + (-9) = -21$
5. $-21 - 19 = -21 + (-19) = -40$

Objective 1.2C

1. STRATEGY
   To find the difference, subtract the elevation of the Qattara Depression (-133 m) from the elevation of Mt. Kilimanjaro (5895 m).

SOLUTION

$5895 - (-133) = 5895 + 133 = 6028$

The difference in elevations is 6028 m.

2. STRATEGY

To find which continent has the smallest difference in elevations, find the difference for each continent and compare the differences.

SOLUTION

Africa: $5895 - (-133) = 5895 + 133 = 6028$ m

Asia: $8850 - (-400) = 8850 + 400 = 9250$ m

Europe: $5634 - (-28) = 5634 + 28 = 5662$ m

America: $6960 - (-86) = 6960 + 86 = 7046$ m

The continent with the smallest difference between the highest and lowest elevations is Europe.

3. STRATEGY

To find the difference, subtract the depth of the Mariana Trench (−11,520 m) from the depth of the Philippine Trench (−10,540 m).

SOLUTION

$-10{,}540 - (-11{,}520) = -10{,}540 + 11{,}520 = 980$

The difference is 980 m.

Objective 1.3A

1. $(-13)(-9) = 117$
2. $17(-13) = -221$
3. $-7(-6)(-5) = 42(-5) = -210$
4. $-12(-4)7(-2) = (48)7(-2) = 336(-2) = -672$
5. $-2(-3)(-4)(-5) = 6(-4)(-5) = -24(-5) = 120$

Objective 1.3B

1. $18 \div (-3) = -6$
2. $57 \div (-3) = -19$
3. $\dfrac{85}{-5} = -17$
4. $0 \div (-14) = 0$
5. $15 \div (-15) = -1$

Objective 1.3C

1. There are four options for the first digit. Once a digit is chosen, it is no longer available, so there are three options for the second digit. By the counting principle, there are $4 \cdot 3 = 12$ possible two-digit numbers if we are choosing without replacement.
2. There are four options for the first digit. Because the selected digit is replaced, there are also four options for the second digit. By the counting principle, there are $4 \cdot 4 = 16$ possible two-digit numbers using replacement.
3. Any one of the nine runners could come in first. Then there are eight runners who could be second and seven runners who could be third. By the counting principle, there are $9 \cdot 8 \cdot 7 = 504$ ways that the three Olympic medals could be awarded.

4. There are eight letters available for the first character of the password. There are nine numbers available for the second character. Because a number cannot be used more than once, there are eight options for the third character, seven options for the fourth character, six options for the fifth character, and five options for the last character. By the counting principle, there are $8 \cdot 9 \cdot 8 \cdot 7 \cdot 6 \cdot 5 = 120{,}960$ possible passwords.

5. STRATEGY

To find the average daily high temperature:

- Add the six temperature readings.
- Divide the sum by 6.

SOLUTION

$-23 + (-29) + (-21) + (-28) + (-28) + (-27) = -156$

$-156 \div 6 = -26$

The average daily high temperature was $-26°$F.

6. STRATEGY

To find the score:

- Multiply the number of correct answers by 7.
- Multiply the number of incorrect answers by -3.
- Multiply the number of blanks by -1.
- Add the results.

SOLUTION

$17(7) = 119$

$8(-3) = -24$

$2(-1) = -2$

$119 + (-24) + (-2) = 95 + (-2) = 93$

The student's score was 93.

Objective 1.4A

1. $(-3)^2 = (-3)(-3) = 9$
2. $-4^4 = -(4 \cdot 4 \cdot 4 \cdot 4) = -256$
3. $(-1)^9 \cdot 3^3 = -1 \cdot 3 \cdot 3 \cdot 3 = -1 \cdot 27 = -27$
4. $(-3) \cdot 2^2 = (-3) \cdot 2 \cdot 2 = (-3) \cdot 4 = -12$
5. $(-7) \cdot 4^2 \cdot 3^2 = (-7) \cdot 4 \cdot 4 \cdot 3 \cdot 3 = (-7) \cdot 16 \cdot 9$
   $= -112 \cdot 9 = -1008$
6. $(-2)^3(-3)^2(-1)^7 = (-2)(-2)(-2)(-3)(-3)(-1)$
   $= -8(9)(-1) = -72(-1) = 72$

Objective 1.4B

1. $4 - 8 \div 2 = 4 - 4 = 0$
2. $24 - 18 \div 3 + 2 = 24 - 6 + 2 = 18 + 2 = 20$
3. $27 - 18 \div (-3^2) = 27 - 18 \div (-9) = 27 + 2 = 29$
4. $16 + 15 \div (-5) - 2 = 16 + (-3) - 2 = 13 - 2 = 11$

5. $$6 + \frac{16 - 4}{2^2 + 2} - 2 = 6 + \frac{12}{4 + 2} - 2$$
$$= 6 + \frac{12}{6} - 2$$
$$= 6 + 2 - 2$$
$$= 8 - 2$$
$$= 6$$

6. $$18 \div 2 - 4^2 - (-3)^2 = 18 \div 2 - 16 - 9$$
$$= 9 - 16 - 9$$
$$= -7 - 9$$
$$= -16$$

Objective 1.5A

1. The difference between the first two numbers is 2; the difference between the second and third numbers is 4; the difference between the third and fourth numbers is 6; and so on. It appears that the difference between any two numbers is always two more than the preceding difference. Since the difference between 23 and 33 is 10, we predict that the next number in the list will be 12 larger than 33, which is 45.
2. The difference between the first two numbers is 5; the difference between the next two numbers is -10. The pattern continues: Add 5 to get the next digit, and then subtract 10 to get the digit after that. Since $-8 + (-10) = -18$, we add 5 to -18 to get -13 as the next digit in the list.
3. Perform the procedure on a few randomly chosen numbers:

   If we pick 2, we get $2 \times 8 = 16$; $16 - 4 = 12$; $12 \div 2 = 6$; $6 + 2 = 8$.
   If we pick 3, we get $3 \times 8 = 24$; $24 - 4 = 20$; $20 \div 2 = 10$; $10 + 2 = 12$.
   If we pick 4, we get $4 \times 8 = 32$; $32 - 4 = 28$; $28 \div 2 = 14$; $14 + 2 = 16$.

   From these examples, we conjecture that the resulting number is 4 times the original number.
4. A tsunami with a velocity of 36 ft/s will have a height that is 4 times that of a tsunami with a velocity of 18 ft/s. From the table in Check your understanding 3, a tsunami with a velocity of 18 ft/s has a height of 36 ft. Thus $4 \times 36 = 144$ ft.
5. A counterexample is any number that proves the statement false. One example is -3: $(-3)^3 = (-3)(-3)(-3) = -27$. Any negative number except -1 will provide a counterexample.

Objective 1.5B

1. From the given information, we can deduce that Dumaka has read *Under the Dome.*
2. From the given information, we can deduce that Brandon earned $35 \times 14 = \$490$ working at his electrical job last week.
3. Since OO and QQQQQQ are both equal to XXX, we can write XXX = OO = QQQQQQ. From OO = QQQQQQ, we can deduce that one O is equal to three Q's.
4. From the first given statement, we can write XXX = OOOOOO. From this equation, we can deduce that X = OO. The second given statement tells us that OO = QQQQ. Since X = OO and OO = QQQQ, we can write X = QQQQ. Thus one X equals 4 Q's.
5. The argument is *not* a correct example of deductive reasoning. Anna could love to read and could also attend a different elementary school.
6. Yes, this argument is a valid example of deductive reasoning.

7. From the given statements, we can create the following table:

	Bus	**Car**	**Subway**
Hannah	X3	X3	√
Robert	√		X3
Keisha	X1	√	X2

Thus Hannah takes the subway, Robert takes the bus, and Keisha takes the car.

Objective 1.5C

1. Because the conclusion is a specific case of a general assumption, the argument is an example of deductive reasoning.
2. Because the conclusion is a specific case of a general assumption, the argument is an example of deductive reasoning.
3. Because the conclusion is based on specific examples, the argument is an example of inductive reasoning.
4. Because the conclusion is based on specific examples, the argument is an example of inductive reasoning.
5. Because $1^3 + 5^3 + 3^3 = 1 + 125 + 27 = 153$, the conclusion represents a specific case of a general assumption, and therefore the argument is an example of deductive reasoning.

Objective 1.6A

1. $\{0, 1, 3, 5, 8\}$
2. $U = \{0, 1, 2, 3, 4, 5, 6, 7, 8\}$
3. Since both members of set F are also members of set D, the statement is true.
4. $F' = \{q, r, s\}$. Since both members of set E are also members of set F', the statement is true.
5. $D' = \{q\}$. Since every member of set D' is also a member of set E, the statement is true.

Objective 1.6B

1. $A \cup B = \{1, 2, 4, 5, 6, 8\}$
2. Since $B' = \{3, 4, 6, 7\}$, $A \cap B' = \{2, 4, 6\} \cap \{3, 4, 6, 7\} = \{4, 6\}$.
3. Since $B \cap C = \{1\}$, $A \cap (B \cap C) = \{2, 4, 6\} \cap \{1\} = \varnothing$.
4. Since $B \cap C = \{1\}$ and $A' = \{1, 3, 5, 7, 8\}$, $A' \cup (B \cap C) = \{1, 3, 5, 7, 8\} \cup \{1\} = \{1, 3, 5, 7, 8\}$.
5. Since $B' = \{3, 4, 6, 7\}$, $A \cap B' = \{2, 4, 6\} \cap \{3, 4, 6, 7\} = \{4, 6\}$. Also, $C' = \{2, 4, 5, 6, 8\}$.

   Therefore, $(A \cap B') \cap C' = \{4, 6\} \cap \{2, 4, 5, 6, 8\} = \{4, 6\}$.

MODULE

2 Problem Solving with Fractions and Decimals

SECTION

2.1 The Least Common Multiple and Greatest Common Factor

Objective 2.1A Factor numbers and find the prime factorization of numbers

Natural number factors of a number divide that number evenly (there is no remainder).

1, 2, 3, and 6 are natural number factors of 6 because they divide 6 evenly.

Note that both the divisor and the quotient are factors of the dividend.

$$1\overline{)6}^{\,6} \quad 2\overline{)6}^{\,3} \quad 3\overline{)6}^{\,2} \quad 6\overline{)6}^{\,1}$$

To find the factors of a number, try dividing the number by 1, 2, 3, 4, 5, ... Those numbers that divide the number evenly are its factors. Continue this process until the factors start to repeat.

The following rules are helpful in finding the factors of a number.

2 is a factor of a number if the digit in the ones place of the number is 0, 2, 4, 6, or 8.	436 ends in 6. Therefore, 2 is a factor of 436. $(436 \div 2 = 218)$
3 is a factor of a number if the sum of the digits of the number is divisible by 3.	The sum of the digits of 489 is $4 + 8 + 9 = 21$. 21 is divisible by 3. Therefore, 3 is a factor of 489. $(489 \div 3 = 163)$
4 is a factor of a number if the last two digits of the number are divisible by 4.	556 ends in 56. 56 is divisible by 4 $(56 \div 4 = 14)$. Therefore, 4 is a factor of 556. $(556 \div 4 = 139)$
5 is a factor of a number if the ones digit of the number is 0 or 5.	520 ends in 0. Therefore, 5 is a factor of 520. $(520 \div 5 = 104)$

Focus on finding the factors of a number

Find all the factors of 42.

SOLUTION

$42 \div 1 = 42$	1 and 42 are factors of 42.
$42 \div 2 = 21$	2 and 21 are factors of 42.
$42 \div 3 = 14$	3 and 14 are factors of 42.
$42 \div 4$	4 will not divide 42 evenly.
$42 \div 5$	5 will not divide 42 evenly.
$42 \div 6 = 7$	6 and 7 are factors of 42.
$42 \div 7 = 6$	7 and 6 are factors of 42.

The factors are repeating. All the factors of 42 have been found.

The factors of 42 are 1, 2, 3, 6, 7, 14, 21, and 42.

Check your understanding 1

Find all the factors of 30.

SOLUTION See page S-1.

A **prime number** is a natural number greater than 1 that has exactly two natural number factors, 1 and the number itself. 7 is prime because its only factors are 1 and 7. If a number is not prime, it is a **composite number.** Because 6 has factors of 2 and 3, 6 is a composite number. The prime numbers less than 50 are

2, 3, 5, 7, 11, 13, 17, 19, 23, 29, 31, 37, 41, 43, 47

The **prime factorization** of a number is the expression of the number as a product of its prime factors. To find the prime factors of 140, begin with the smallest prime number as a trial divisor and continue with prime numbers as trial divisors until the final quotient is prime.

Focus on finding the prime factorization of a number

Find the prime factorization of 140.

SOLUTION

$\begin{array}{r} 70 \\ 2\overline{)140} \end{array}$	$\begin{array}{r} 35 \\ 2\overline{)70} \\ 2\overline{)140} \end{array}$	$\begin{array}{r} 7 \\ 5\overline{)35} \\ 2\overline{)70} \\ 2\overline{)140} \end{array}$
140 is divisible by 2. Divide 140 by 2.	70 is divisible by 2. Divide 70 by 2.	35 is not divisible by 2 or 3 but is divisible by 5. Divide 35 by 5. The quotient is 7, which is prime.

The prime factorization of 140 is $2 \cdot 2 \cdot 5 \cdot 7$, or $2^2 \cdot 5 \cdot 7$.

Check your understanding 2

Find the prime factorization of 88.

SOLUTION See page S-1.

Finding the prime factorization of larger numbers can be more difficult. Try each prime number as a trial divisor. Stop when the square of the trial divisor is greater than the number being factored.

Focus on finding the prime factorization of a number

Find the prime factorization of 201.

SOLUTION

201 is not divisible by 2 but is divisible by 3. Divide 201 by 3.

$\begin{array}{r} 67 \\ 3\overline{)201} \end{array}$ 67 cannot be divided evenly by 2, 3, 5, 7, or 11. Prime numbers greater than 11 need not be tried because $11^2 = 121$ and $121 > 67$.

The prime factorization of 201 is $3 \cdot 67$.

Check your understanding 3

Find the prime factorization of 295.

SOLUTION See page S-1.

Objective 2.1A Practice

1. Find all the factors of 12.
2. Find all the factors of 56.
3. Find all the factors of 48.
4. Find the prime factorization of 16.
5. Find the prime factorization of 40.
6. Find the prime factorization of 37.

Solutions on pp. S-7–S-8.

Objective 2.1B Find the least common multiple (LCM)

The **multiples of a number** are the products of that number and the numbers 1, 2, 3, 4, 5, ...

$3 \times 1 = 3$
$3 \times 2 = 6$
$3 \times 3 = 9$
$3 \times 4 = 12$ The multiples of 3 are 3, 6, 9, 12, 15, ...
$3 \times 5 = 15$
.
.
.

A number that is a multiple of two or more numbers is a **common multiple** of those numbers.

The multiples of 4 are 4, 8, 12, 16, 20, 24, 28, 32, 36, ...
The multiples of 6 are 6, 12, 18, 24, 30, 36, 42, ...
Some common multiples of 4 and 6 are 12, 24, and 36.

The **least common multiple (LCM)** is the smallest common multiple of two or more numbers.

The least common multiple of 4 and 6 is 12.

Listing the multiples of each number is one way to find the LCM. Another way to find the LCM uses the prime factorization of each number.

To find the LCM of 450 and 600, find the prime factorization of each number and write the factorization of each number in a table. Circle the greatest product in each column. The LCM is the product of the circled numbers.

	2	3	5
450 =	2	(3 · 3)	(5 · 5)
600 =	(2 · 2 · 2)	3	5 · 5

• In the column headed by 5, the products are equal. Circle just one product.

The LCM is the product of the circled numbers.
The LCM $= 2 \cdot 2 \cdot 2 \cdot 3 \cdot 3 \cdot 5 \cdot 5 = 1800$.

Focus on finding the LCM

Find the LCM of 24, 36, and 50.

SOLUTION

	2	3	5
24 =	(2 · 2 · 2)	3	
36 =	2 · 2	(3 · 3)	
50 =	2		(5 · 5)

The LCM $= 2 \cdot 2 \cdot 2 \cdot 3 \cdot 3 \cdot 5 \cdot 5 = 1800$.

Check your understanding 4

Find the LCM of 12, 42, and 45.

SOLUTION See page S-1.

Objective 2.1B Practice

1. Find the LCM of 8 and 12.
2. Find the LCM of 8 and 14.
3. Find the LCM of 9 and 36.
4. Find the LCM of 102 and 184.
5. Find the LCM of 4, 8, and 12.
6. Find the LCM of 3, 5, and 10.

Solutions on pp. S-8–S-9.

Objective 2.1C Find the greatest common factor (GCF)

Recall that a number that divides another number evenly is a factor of that number. The number 64 can be evenly divided by 1, 2, 4, 8, 16, 32, and 64, so the numbers 1, 2, 4, 8, 16, 32, and 64 are factors of 64.

A number that is a factor of two or more numbers is a **common factor** of those numbers.

The factors of 30 are 1, 2, 3, 5, 6, 10, 15, and 30.

The factors of 105 are 1, 3, 5, 7, 15, 21, 35, and 105.

The common factors of 30 and 105 are 1, 3, 5, and 15.

The **greatest common factor (GCF)** is the largest common factor of two or more numbers.

The greatest common factor of 30 and 105 is 15.

Listing the factors of each number is one way of finding the GCF. Another way to find the GCF is to use the prime factorization of each number.

To find the GCF of 126 and 180, find the prime factorization of each number and write the factorization of each number in a table. Circle the least product in each column that does not have a blank. The GCF is the product of the circled numbers.

	2	3	5	7
126 =	(2)	(3 · 3)		7
180 =	2 · 2	3 · 3	5	

• In the column headed by 3, the products are equal. Circle just one product. Columns 5 and 7 have a blank, so 5 and 7 are not common factors of 126 and 180. Do not circle any number in these columns.

The GCF is the product of the circled numbers.
The GCF = 2 · 3 · 3 = 18.

Focus on finding the GCF

Find the GCF of 90, 168, and 420.

SOLUTION

	2	3	5	7
90 =	(2)	3 · 3	5	
168 =	2 · 2 · 2	(3)		7
420 =	2 · 2	3	5	7

The GCF = 2 · 3 = 6.

Check your understanding 5

Find the GCF of 36, 60, and 72.

SOLUTION See page S-1.

Focus on finding the GCF

Find the GCF of 7, 12, and 20.

SOLUTION

	2	3	5	7
7 =				7
12 =	$2 \cdot 2$	3		
20 =	$2 \cdot 2$		5	

Because no numbers are circled, the GCF = 1.

Check your understanding 6

Find the GCF of 11, 24, and 30.

SOLUTION See page S-1.

Objective 2.1C Practice

1. Find the GCF of 6 and 9.
2. Find the GCF of 25 and 100.
3. Find the GCF of 32 and 51.
4. Find the GCF of 3, 5, and 11.
5. Find the GCF of 24, 40, and 72.
6. Find the GCF of 32, 56, and 72.

Solutions on p. S-9.

SECTION 2.2 Writing Equivalent Fractions

Objective 2.2A Write an improper fraction as a mixed number or a whole number, and a mixed number as an improper fraction

Recall that an **improper fraction** is a fraction in which the numerator is greater than or equal to the denominator. An improper fraction is always greater than or equal to 1. A **mixed number** consists of a whole-number part and a fractional part.

Note from the diagram that the mixed number $2\frac{3}{5}$ and the improper fraction $\frac{13}{5}$ both represent the shaded portion of the circles, so $2\frac{3}{5} = \frac{13}{5}$.

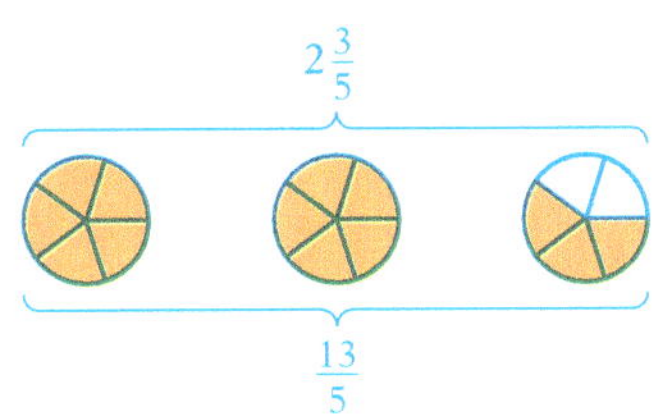

An improper fraction can be written as a mixed number or a whole number.

Focus on writing an improper fraction as a mixed number

Write $\frac{13}{5}$ as a mixed number.

SOLUTION

Divide the numerator by the denominator.

$$5\overline{)13} = 2 \text{ remainder } 3 \quad (13 - 10 = 3)$$

To write the fractional part of the mixed number, write the remainder over the divisor.

$$5\overline{)13} = 2\tfrac{3}{5} \quad (13 - 10 = 3)$$

Write the answer.

$$\frac{13}{5} = 2\frac{3}{5}$$

Check your understanding 1

Write $\frac{22}{5}$ as a mixed number.

SOLUTION See page S-2.

Focus on writing an improper fraction as a whole number

Write $\frac{18}{6}$ as a whole number.

SOLUTION

$$\frac{18}{6} = 18 \div 6 = 3$$

Check your understanding 2

Write $\frac{28}{7}$ as a whole number.

SOLUTION See page S-2.

To write a mixed number as an improper fraction, multiply the denominator of the fractional part by the whole-number part. The sum of this product and the numerator of the fractional part is the numerator of the improper fraction. The denominator remains the same.

Focus on writing a mixed number as an improper fraction

Write $7\frac{3}{8}$ as an improper fraction.

SOLUTION

$$7\frac{3}{8} = \frac{(8 \times 7) + 3}{8} = \frac{56 + 3}{8} = \frac{59}{8} \qquad 7\frac{3}{8} = \frac{59}{8}$$

Check your understanding 3

Write $14\frac{5}{8}$ as an improper fraction.

SOLUTION See page S-2.

Objective 2.2A Practice

1. Write $\frac{11}{4}$ as a mixed or whole number.
2. Write $\frac{72}{8}$ as a mixed or whole number.
3. Write $6\frac{1}{2}$ as an improper fraction.
4. Write $5\frac{3}{11}$ as an improper fraction.

Solutions on p. S-10.

Objective 2.2B Write a fraction in simplest form

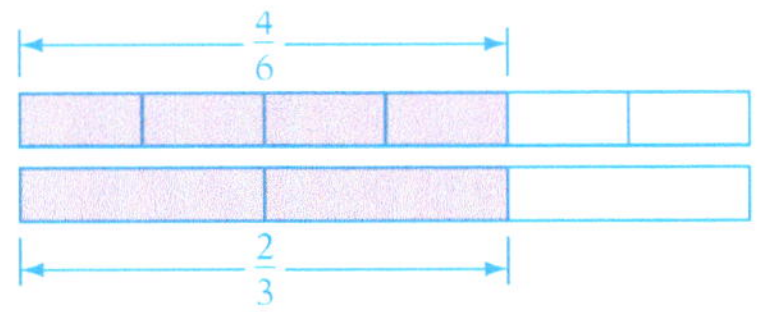

Writing the **simplest form of a fraction** means writing it so that the numerator and denominator have no common factors other than 1.

The fractions $\frac{4}{6}$ and $\frac{2}{3}$ are equivalent fractions.

$\frac{4}{6}$ is written in simplest form as $\frac{2}{3}$.

The Multiplication Property of One can be used to write fractions in simplest form. Write the numerator and denominator of the given fraction as a product of factors. Write factors common to both the numerator and denominator as an improper fraction equivalent to 1.

$$\frac{4}{6} = \frac{2 \cdot 2}{2 \cdot 3} = \frac{2}{2} \cdot \frac{2}{3} = \frac{2}{2} \cdot \frac{2}{3} = 1 \cdot \frac{2}{3} = \frac{2}{3}$$

The process of eliminating common factors is displayed with slashes through the common factors as shown at the right.

$$\frac{4}{6} = \frac{\overset{1}{\cancel{2}} \cdot 2}{\underset{1}{\cancel{2}} \cdot 3} = \frac{2}{3}$$

To write a fraction in simplest form, eliminate the common factors.

$$\frac{18}{30} = \frac{\overset{1}{\cancel{2}} \cdot \overset{1}{\cancel{3}} \cdot 3}{\underset{1}{\cancel{2}} \cdot \underset{1}{\cancel{3}} \cdot 5} = \frac{3}{5}$$

An improper fraction can be changed to a mixed number.

$$\frac{22}{6} = \frac{\overset{1}{\cancel{2}} \cdot 11}{\underset{1}{\cancel{2}} \cdot 3} = \frac{11}{3} = 3\frac{2}{3}$$

Focus on writing a fraction in simplest form

a. Write $\frac{15}{40}$ in simplest form.

b. Write $\frac{6}{42}$ in simplest form.

c. Write $\frac{8}{9}$ in simplest form.

d. Write $\frac{30}{12}$ as a fraction in simplest form.

SOLUTION

a. $\frac{15}{40} = \frac{3 \cdot \overset{1}{\cancel{5}}}{2 \cdot 2 \cdot 2 \cdot \underset{1}{\cancel{5}}} = \frac{3}{8}$

b. $\frac{6}{42} = \frac{\overset{1}{\cancel{2}} \cdot \overset{1}{\cancel{3}}}{\underset{1}{\cancel{2}} \cdot \underset{1}{\cancel{3}} \cdot 7} = \frac{1}{7}$

c. $\frac{8}{9} = \frac{2 \cdot 2 \cdot 2}{3 \cdot 3} = \frac{8}{9}$

$\frac{8}{9}$ is already in simplest form because there are no common factors in the numerator and denominator.

d. $\frac{30}{12} = \frac{\overset{1}{\cancel{2}} \cdot \overset{1}{\cancel{3}} \cdot 5}{\underset{1}{\cancel{2}} \cdot 2 \cdot \underset{1}{\cancel{3}}} = \frac{5}{2}$

Check your understanding 4

a. Write $\frac{16}{24}$ in simplest form.

b. Write $\frac{8}{56}$ in simplest form.

c. Write $\frac{15}{32}$ in simplest form.

d. Write $\frac{48}{36}$ as a fraction in simplest form.

SOLUTION See page S-2.

Objective 2.2B Practice

1. Write $\frac{2}{12}$ in simplest form.
2. Write $\frac{16}{84}$ in simplest form.
3. Write $\frac{12}{35}$ in simplest form.
4. Write $\frac{144}{36}$ in simplest form.

Solutions on p. S-10.

Objective 2.2C Find equivalent fractions by raising to higher terms

Equal fractions with different denominators are called **equivalent fractions.**

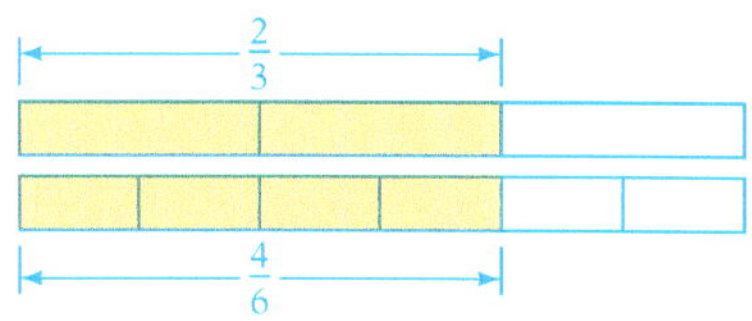

$\frac{4}{6}$ is equivalent to $\frac{2}{3}$.

Remember that the Multiplication Property of One states that the product of a number and 1 is the number. This is true for fractions as well as whole numbers. This property can be used to write equivalent fractions.

$$\frac{2}{3} \times 1 = \frac{2}{3} \times \frac{1}{1} = \frac{2 \cdot 1}{3 \cdot 1} = \frac{2}{3}$$

$$\frac{2}{3} \times 1 = \frac{2}{3} \times \frac{2}{2} = \frac{2 \cdot 2}{3 \cdot 2} = \frac{4}{6} \qquad \frac{4}{6} \text{ is equivalent to } \frac{2}{3}.$$

$$\frac{2}{3} \times 1 = \frac{2}{3} \times \frac{4}{4} = \frac{2 \cdot 4}{3 \cdot 4} = \frac{8}{12} \qquad \frac{8}{12} \text{ is equivalent to } \frac{2}{3}.$$

$\frac{2}{3}$ was rewritten as the equivalent fractions $\frac{4}{6}$ and $\frac{8}{12}$.

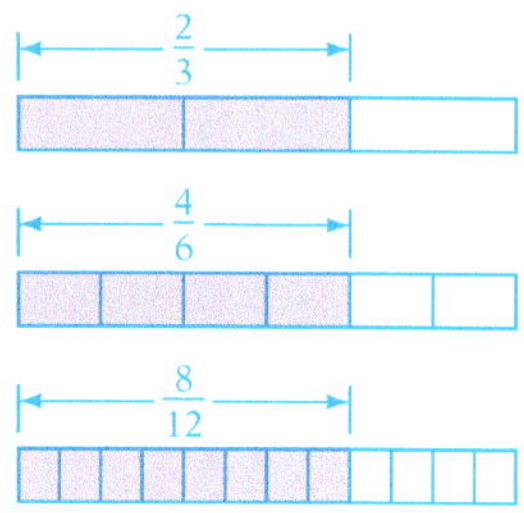

Write a fraction that is equivalent to $\frac{5}{8}$ and has a denominator of 32.

$32 \div 8 = 4$ • Divide the larger denominator by the smaller.

$\frac{5}{8} = \frac{5 \cdot 4}{8 \cdot 4} = \frac{20}{32}$ • Multiply the numerator and denominator of the given fraction by the quotient (4).

$\frac{20}{32}$ is equivalent to $\frac{5}{8}$.

Focus on writing equivalent fractions

Write $\frac{2}{3}$ as an equivalent fraction with a denominator of 42.

SOLUTION

$42 \div 3 = 14$ • Divide the larger denominator by the smaller.

$\frac{2}{3} = \frac{2 \cdot 14}{3 \cdot 14} = \frac{28}{42}$ • Multiply the numerator and denominator of the given fraction by the quotient (14).

$\frac{28}{42}$ is equivalent to $\frac{2}{3}$.

Check your understanding 5

Write $\frac{3}{5}$ as an equivalent fraction with a denominator of 45.

SOLUTION See page S-2.

Focus on writing equivalent fractions

Write 4 as a fraction with a denominator of 12.

SOLUTION

Write 4 as $\frac{4}{1}$.

$12 \div 1 = 12$ • Divide the larger denominator by the smaller.

$\frac{4}{1} = \frac{4 \cdot 12}{1 \cdot 12} = \frac{48}{12}$ • Multiply the numerator and denominator of the given fraction by the quotient (12).

4 is equivalent to $\frac{48}{12}$.

Check your understanding 6

Write 6 as a fraction with a denominator of 18.

SOLUTION See page S-2.

Objective 2.2C Practice

1. Write $\frac{3}{4}$ as an equivalent fraction with a denominator of 16.

2. Write $\frac{5}{7}$ as an equivalent fraction with a denominator of 49.

3. Write $\frac{5}{9}$ as an equivalent fraction with a denominator of 18.

4. Write $\frac{15}{16}$ as an equivalent fraction with a denominator of 64.

Solutions on p. S-10.

Objective 2.2D Identify the order relation between two fractions

Recall that whole numbers can be graphed as points on the number line. Fractions can also be graphed as points on the number line.

The graph of $\frac{3}{4}$ on the number line

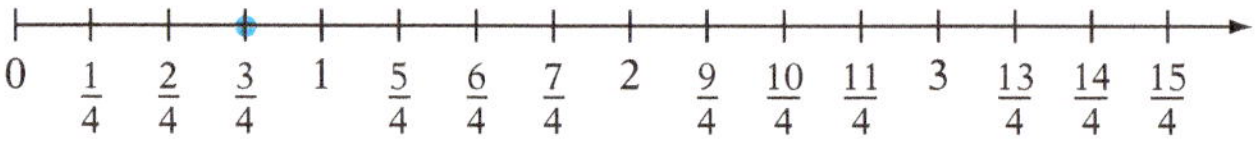

The number line can be used to determine the order relation between two fractions. A fraction that appears to the left of a given fraction is less than the given fraction. A fraction that appears to the right of a given fraction is greater than the given fraction.

$\frac{1}{8} < \frac{3}{8}$ $\quad \frac{6}{8} > \frac{3}{8}$

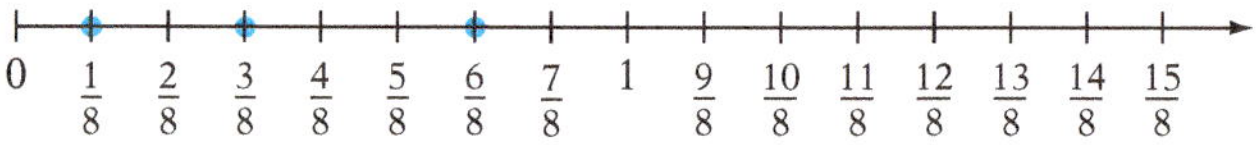

To find the order relation between two fractions with the same denominator, compare the numerators. The fraction that has the smaller numerator is the smaller fraction. When the denominators are different, begin by writing equivalent fractions with a common denominator; then compare the numerators. The **least common denominator (LCD)** of two fractions is the LCM of their denominators.

Focus on finding the order relation between two fractions

Find the order relation between $\frac{11}{18}$ and $\frac{5}{8}$.

SOLUTION

The LCD of the fractions is 72.

$\frac{11}{18} = \frac{44}{72}$ ← Smaller numerator

$\frac{5}{8} = \frac{45}{72}$ ← Larger numerator

$\frac{11}{18} < \frac{5}{8}$ or $\frac{5}{8} > \frac{11}{18}$

Check your understanding 7

Place the correct symbol, < or >, between the two numbers.

$\frac{9}{14} \quad \frac{13}{21}$

SOLUTION See page S-2.

Objective 2.2D Practice

1. Place the correct symbol, < or >, between the two numbers.
$\frac{11}{40} \quad \frac{19}{40}$

2. Place the correct symbol, < or >, between the two numbers.
$\frac{2}{5} \quad \frac{3}{8}$

3. Place the correct symbol, < or >, between the two numbers.
$\frac{13}{18} \quad \frac{7}{12}$

4. Place the correct symbol, < or >, between the two numbers.
$\frac{5}{12} \quad \frac{7}{15}$

Solutions on p. S-10.

SECTION 2.3 Operations on Positive and Negative Fractions

Objective 2.3A Multiply and divide positive and negative fractions

Recall that to multiply two fractions, we multiply the numerators and multiply the denominators.

Take Note ➤
Note that fractions do not need to have the same denominator in order to be multiplied.

Multiplication of Fractions

The product of two fractions is the product of the numerators over the product of the denominators.

$$\frac{a}{b} \cdot \frac{c}{d} = \frac{ac}{bd}, \quad \text{where} \quad b \neq 0 \quad \text{and} \quad d \neq 0$$

EXAMPLES

1. $\frac{2}{5} \cdot \frac{1}{3} = \frac{2 \cdot 1}{5 \cdot 3} = \frac{2}{15}$

2. $\frac{3}{4} \cdot \frac{5}{8} = \frac{3 \cdot 5}{4 \cdot 8} = \frac{15}{32}$

The sign rules for multiplying positive and negative fractions are the same rules used to multiply integers.

The product of two numbers with the same sign is positive.
The product of two numbers with different signs is negative.

Multiply: $-\frac{3}{4} \cdot \frac{8}{15}$

The signs are different. The product is negative.

$$-\frac{3}{4} \cdot \frac{8}{15} = -\frac{3 \cdot 8}{4 \cdot 15}$$

• Multiply the numerators. Multiply the denominators.

$$= -\frac{3 \cdot 2 \cdot 2 \cdot 2}{2 \cdot 2 \cdot 3 \cdot 5}$$

• Write the product in simplest form.

$$= -\frac{2}{5}$$

The **reciprocal** of a fraction is the fraction with the numerator and denominator interchanged. Division is defined as multiplication by the reciprocal. Therefore, "divided by 2" is the same as "times 1/2." This relationship is used when dividing fractions.

Division of Fractions

To divide two fractions, multiply by the reciprocal of the divisor.

$$\frac{a}{b} \div \frac{c}{d} = \frac{a}{b} \cdot \frac{d}{c}, \quad \text{where } b \neq 0,\ c \neq 0, \text{ and } d \neq 0$$

EXAMPLE

Divide: $\frac{2}{5} \div \frac{3}{4}$

$$\frac{2}{5} \div \frac{3}{4} = \frac{2}{5} \cdot \frac{4}{3}$$

• Rewrite the division as multiplication by the reciprocal.

$$= \frac{2 \cdot 4}{5 \cdot 3} = \frac{8}{15}$$

• Multiply the fractions.

The sign rules for dividing positive and negative fractions are the same rules used to divide integers.

The quotient of two numbers with the same sign is positive.
The quotient of two numbers with different signs is negative.

Divide: $-\frac{7}{10} \div \left(-\frac{14}{15}\right)$

The signs are the same. The quotient is positive.

$$-\frac{7}{10} \div \left(-\frac{14}{15}\right) = \frac{7}{10} \div \frac{14}{15}$$

$$= \frac{7}{10} \cdot \frac{15}{14}$$ • Rewrite the division as multiplication by the reciprocal.

$$= \frac{7 \cdot 15}{10 \cdot 14}$$ • Multiply the fractions.

$$= \frac{7 \cdot 3 \cdot 5}{2 \cdot 5 \cdot 2 \cdot 7}$$ • Write the product in simplest form.

$$= \frac{3}{4}$$

Focus on multiplying positive and negative fractions

Multiply: $-\frac{3}{4}\left(\frac{1}{2}\right)\left(-\frac{8}{9}\right)$

SOLUTION

$$-\frac{3}{4}\left(\frac{1}{2}\right)\left(-\frac{8}{9}\right) = \frac{3 \cdot 1 \cdot 8}{4 \cdot 2 \cdot 9}$$ • The product is positive. Multiply the numerators. Multiply the denominators.

$$= \frac{3 \cdot 1 \cdot 2 \cdot 2 \cdot 2}{2 \cdot 2 \cdot 2 \cdot 3 \cdot 3}$$ • Write the product in simplest form.

$$= \frac{1}{3}$$

Check your understanding 1

Multiply: $-\frac{1}{3}\left(-\frac{5}{12}\right)\left(\frac{8}{15}\right)$

SOLUTION See page S-3.

Focus on evaluating a variable expression

Evaluate the variable expression xy for $x = 1\frac{4}{5}$ and $y = -\frac{5}{6}$.

SOLUTION

xy

$$1\frac{4}{5}\left(-\frac{5}{6}\right) = -\left(\frac{9}{5} \cdot \frac{5}{6}\right)$$ • Replace x by $1\frac{4}{5}$ and y by $-\frac{5}{6}$. Write $1\frac{4}{5}$ as an improper fraction.

$$= -\frac{9 \cdot 5}{5 \cdot 6}$$ • Multiply numerators and multiply denominators.

$$= -\frac{3 \cdot 3 \cdot 5}{5 \cdot 2 \cdot 3}$$ • Write the product in simplest form.

$$= -\frac{3}{2} = -1\frac{1}{2}$$ • The answer can be written as an improper fraction or a mixed number.

Check your understanding 2

Evaluate the variable expression xy for $x = -5\frac{1}{8}$ and $y = \frac{2}{3}$.

SOLUTION See page S-3.

Focus on dividing positive and negative mixed numbers

Divide: $3\frac{4}{15} \div \left(-2\frac{1}{10}\right)$

SOLUTION

$$3\frac{4}{15} \div \left(-2\frac{1}{10}\right) = -\left(\frac{49}{15} \div \frac{21}{10}\right)$$

• The signs are different. The quotient is negative. Write the mixed numbers as improper fractions.

$$= -\left(\frac{49}{15} \cdot \frac{10}{21}\right)$$

$$= -\frac{49 \cdot 10}{15 \cdot 21}$$

$$= -\frac{7 \cdot 7 \cdot 2 \cdot 5}{3 \cdot 5 \cdot 3 \cdot 7}$$

• Write the product in simplest form.

$$= -\frac{14}{9} = -1\frac{5}{9}$$

Check your understanding 3

Divide: $-4\frac{3}{8} \div 3\frac{1}{2}$

SOLUTION See page S-3.

Focus on dividing positive and negative fractions

What is the quotient of 6 and $-\frac{3}{5}$?

SOLUTION

$$6 \div \left(-\frac{3}{5}\right) = -\left(\frac{6}{1} \div \frac{3}{5}\right)$$

• The signs are different. The quotient is negative. Write 6 as $\frac{6}{1}$.

$$= -\left(\frac{6}{1} \cdot \frac{5}{3}\right)$$

• Multiply by the reciprocal.

$$= -\frac{6 \cdot 5}{1 \cdot 3}$$

$$= -\frac{2 \cdot 3 \cdot 5}{1 \cdot 3}$$

• Write the product in simplest form.

$$= -\frac{10}{1} = -10$$

Check your understanding 4

Find the quotient of 4 and $-\frac{6}{7}$.

SOLUTION See page S-3.

Objective 2.3A Practice

1. Multiply: $\frac{5}{6} \cdot \left(-\frac{2}{5}\right)$

2. Multiply: $\left(-\frac{3}{4}\right)\left(-\frac{2}{9}\right)$

3. Multiply: $-\frac{3}{4} \cdot \frac{1}{2}$

4. Multiply: $-1\frac{2}{3} \cdot \left(-\frac{3}{5}\right)$

5. Evaluate xy for $x = -\frac{5}{16}$ and $y = \frac{7}{15}$.

6. Divide: $-\frac{5}{7} \div \left(-\frac{5}{6}\right)$

7. Divide: $\frac{3}{8} \div \left(-\frac{5}{12}\right)$

8. Divide: $5\frac{3}{5} \div \left(-\frac{7}{10}\right)$

9. Find the quotient of $\frac{3}{5}$ and $\frac{12}{25}$.

10. Evaluate $x \div y$ for $x = -\frac{5}{8}$ and $y = -\frac{15}{2}$.

Solutions on pp. S-10–S-12.

Objective 2.3B Add and subtract positive and negative fractions

To add a fraction with a negative sign, rewrite the fraction with the negative sign in the numerator. Then add the numerators and place the sum over the common denominator.

Take Note ➤ $\frac{-1}{12} = \frac{1}{-12} = -\frac{1}{12}$ Although the sum could have been left as $\frac{-1}{12}$, all answers in this text are written with the negative sign in front of the fraction.

Add: $-\frac{5}{6} + \frac{3}{4}$

$-\frac{5}{6} + \frac{3}{4} = \frac{-5}{6} + \frac{3}{4}$ • Rewrite the first fraction with the negative sign in the numerator.

$= \frac{-10}{12} + \frac{9}{12}$ • Rewrite each fraction as an equivalent fraction using the LCD (12) of the fractions.

$= \frac{-10 + 9}{12}$ • Add the fractions.

$= \frac{-1}{12} = -\frac{1}{12}$ • Simplify the numerator and write the negative sign in front of the fraction.

Add: $-\frac{2}{3} + \left(-\frac{4}{5}\right)$

$-\frac{2}{3} + \left(-\frac{4}{5}\right) = \frac{-2}{3} + \frac{-4}{5}$ • Rewrite each negative fraction with the negative sign in the numerator.

$= \frac{-10}{15} + \frac{-12}{15}$ • Rewrite each fraction as an equivalent fraction using the LCD (15) of the fractions.

$= \frac{-10 + (-12)}{15}$ • Add the fractions.

$= \frac{-22}{15} = -1\frac{7}{15}$

To subtract fractions with negative signs, first rewrite the fractions with the negative signs in the numerators.

Simplify: $-\frac{2}{9} - \frac{5}{12}$

$-\frac{2}{9} - \frac{5}{12} = \frac{-2}{9} - \frac{5}{12}$ • Rewrite the negative fraction with the negative sign in the numerator.

$= \frac{-8}{36} - \frac{15}{36}$ • Write the fractions as equivalent fractions with a common denominator.

$= \frac{-8 - 15}{36} = \frac{-23}{36}$ • Subtract the numerators and place the difference over the common denominator.

$= -\frac{23}{36}$ • Write the negative sign in front of the fraction.

Subtract: $\frac{2}{3} - \left(-\frac{4}{5}\right)$

$\frac{2}{3} - \left(-\frac{4}{5}\right) = \frac{2}{3} + \frac{4}{5}$

• Rewrite subtraction as addition of the opposite.

$= \frac{10}{15} + \frac{12}{15}$

• Write the fractions as equivalent fractions with a common denominator.

$= \frac{10 + 12}{15}$

• Add the fractions.

$= \frac{22}{15} = 1\frac{7}{15}$

Focus on adding signed fractions

Add: $-\frac{3}{8} + \frac{3}{4} + \left(-\frac{5}{6}\right)$

SOLUTION

$-\frac{3}{8} + \frac{3}{4} + \left(-\frac{5}{6}\right) = \frac{-3}{8} + \frac{3}{4} + \frac{-5}{6}$

• Rewrite each negative fraction with the negative sign in the numerator.

$= \frac{-9}{24} + \frac{18}{24} + \frac{-20}{24}$

• Rewrite each fraction as an equivalent fraction using the LCD (24) of the fractions.

$= \frac{-9 + 18 + (-20)}{24}$

• Add the fractions.

$= \frac{-11}{24} = -\frac{11}{24}$

Check your understanding 5

Add: $-\frac{5}{12} + \frac{5}{8} + \left(-\frac{1}{6}\right)$

SOLUTION See page S-4.

Focus on subtracting signed fractions

Subtract: $-\frac{5}{6} - \left(-\frac{3}{8}\right)$

SOLUTION

$-\frac{5}{6} - \left(-\frac{3}{8}\right) = -\frac{5}{6} + \frac{3}{8}$

• Rewrite subtraction as addition of the opposite.

$= \frac{-20}{24} + \frac{9}{24}$

• Rewrite the fractions using the LCD, 24.

$= \frac{-20 + 9}{24} = \frac{-11}{24}$

$= -\frac{11}{24}$

Check your understanding 6

Subtract: $-\frac{5}{6} - \frac{7}{9}$

SOLUTION See page S-4.

Objective 2.3B Practice

1. Add: $-\frac{3}{4} + \frac{2}{3}$

2. Add: $-\frac{7}{12} + \frac{5}{8}$

3. Add: $\frac{2}{5} + \left(-\frac{11}{15}\right)$

4. Add: $\frac{3}{8} + \left(-\frac{1}{2}\right) + \frac{7}{12}$

5. Evaluate $x + y$ when $x = \frac{3}{10}$ and $y = -\frac{7}{15}$.

6. Subtract: $-\frac{1}{2} - \frac{3}{8}$

7. Subtract: $-\frac{5}{12} - \left(-\frac{2}{3}\right)$

8. Subtract: $-\frac{5}{9} - \left(-\frac{11}{12}\right)$

9. Subtract: $-\frac{5}{8} - \left(-\frac{7}{12}\right)$

10. Evaluate $x - y$ when $x = -\frac{5}{12}$ and $y = -\frac{5}{9}$.

Solutions on pp. S-12–S-14.

SECTION 2.4 Introduction to Decimals

Objective 2.4A Round a decimal to a given place value

Take Note ➤
In decimal notation, the part of the number that appears to the left of the decimal point is the **whole-number part.** The part of the number that appears to the right of the decimal point is the **decimal part.** The **decimal point** separates the whole-number part from the decimal part.

The price tag on a sweater reads \$61.88. The number 61.88 is in **decimal notation.** A number written in decimal notation is often called simply a **decimal.**

A number written in decimal notation has three parts.

61	.	88
Whole-number part	**Decimal point**	**Decimal part**

The decimal part of the number represents a number less than 1. For example, \$.88 is less than \$1. The decimal point (.) separates the whole-number part from the decimal part.

The position of a digit in a decimal determines the digit's place value. The place-value chart is extended to the right to show the place values of digits to the right of a decimal point.

In the decimal 458.302719, the position of the digit 7 determines that its place value is ten-thousandths.

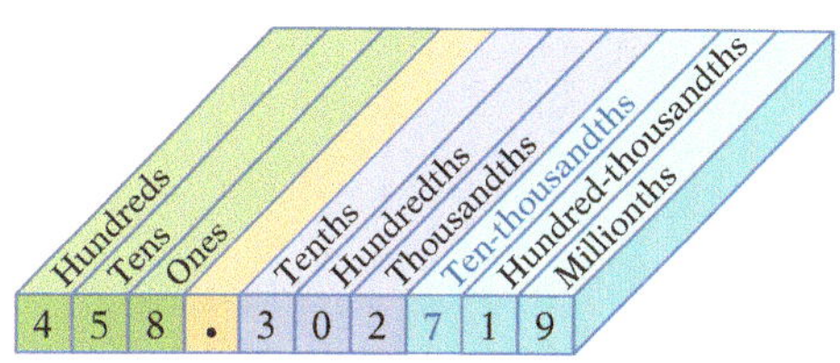

To write a decimal in words, write the decimal part of the number as though it were a whole number, and then name the place value of the last digit.

0.9684 Nine thousand six hundred eighty-four ten-thousandths

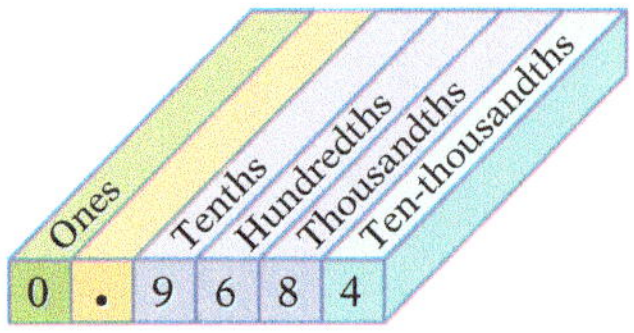

The decimal point in a decimal is read as "and."

372.516 Three hundred seventy-two and five hundred sixteen thousandths

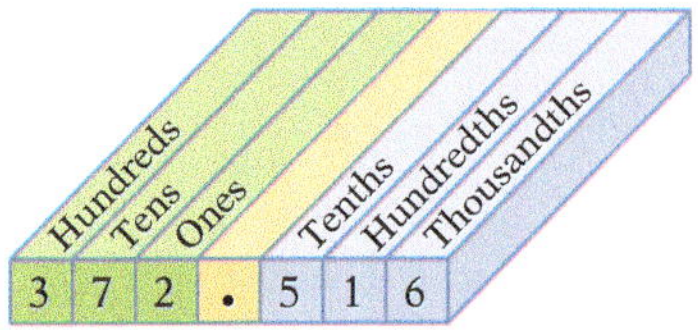

Focus on naming a place value

Name the place value of the digit 8 in the number 45.687.

SOLUTION

The digit 8 is in the hundredths place.

Check your understanding 1

Name the place value of the digit 4 in the number 907.1342.

SOLUTION See page S-4.

Tips for Success ➤
Have you considered joining a study group? Getting together regularly with other students in the class to go over material and quiz each other can be very beneficial. See *AIM for Success.*

In general, rounding decimals is similar to rounding whole numbers except that the digits to the right of the given place value are dropped instead of being replaced by zeros.

If the digit to the right of the given place value is less than 5, that digit and all digits to the right are dropped.

Round 6.9237 to the nearest hundredth.

Given place value (hundredths)

6.9237

$3 < 5$ Drop the digits 3 and 7.

6.9237 rounded to the nearest hundredth is 6.92.

If the digit to the right of the given place value is greater than or equal to 5, increase the digit in the given place value by 1, and drop all digits to its right.

Round 12.385 to the nearest tenth.

Given place value (tenths)

12.385

$8 > 5$ Increase 3 by 1 and drop all digits to the right of 3.

12.385 rounded to the nearest tenth is 12.4.

Focus on rounding to a given place value

a. Round 0.9375 to the nearest thousandth.
b. Round 2.5963 to the nearest hundredth.
c. Round 72.416 to the nearest whole number.

SOLUTION

a. Given place value

0.9375

$5 = 5$

0.9375 rounded to the nearest thousandth is 0.938.

b. Given place value

2.5963

6 > 5

2.5963 rounded to the nearest hundredth is 2.60.

c. Given place value

72.416

4 < 5

72.416 rounded to the nearest whole number is 72.

Check your understanding 2

a. Round 3.675849 to the nearest ten-thousandth.
b. Round 48.907 to the nearest tenth.
c. Round 31.8652 to the nearest whole number.

SOLUTION See page S-4.

Objective 2.4A Practice

1. Name the place value of the digit 5 in the number 432.09157.
2. Round 6.249 to the nearest tenth.
3. Round 18.40937 to the nearest hundredth.
4. Round 936.2905 to the nearest thousandth.
5. Round 47.3192 to the nearest whole number.
6. Round 2.975268 to the nearest hundred-thousandth.

Solutions on p. S-14.

Objective 2.4B Compare decimals

There is a relationship between numbers written in decimal notation and fractions.

7 tenths	59 hundredths	19 thousandths	27 ten-thousandths
$0.7 = \frac{7}{10}$	$0.59 = \frac{59}{100}$	$0.019 = \frac{19}{1000}$	$0.0027 = \frac{27}{10{,}000}$

This relationship can be used to compare two decimals.

Focus on comparing two decimals

Place the correct symbol, < or >, between the numbers.
0.02 0.0135

SOLUTION

$0.02 \qquad 0.0135$

$\frac{2}{100} \qquad \frac{135}{10{,}000}$ • Write the numbers as fractions.

$\frac{200}{10{,}000} \qquad \frac{135}{10{,}000}$ • Write the fractions with a common denominator.

$\frac{200}{10{,}000} > \frac{135}{10{,}000}$ • Compare the fractions.

$0.02 > 0.0135$

Check your understanding 3

Place the correct symbol, < or >, between the numbers.
0.15 0.107

SOLUTION See page S-4.

Objective 2.4B Practice

1. Place the correct symbol, < or >, between the numbers.
   0.278 0.203
2. Place the correct symbol, < or >, between the numbers.
   0.045 0.038
3. Place the correct symbol, < or >, between the numbers.
   0.037 0.13
4. Place the correct symbol, < or >, between the numbers.
   0.031 0.00987
5. Place the correct symbol, < or >, between the numbers.
   0.02883 0.0305

Solutions on p. S-14.

SECTION 2.5 Operations on Decimals

Objective 2.5A Add and subtract decimals

To add decimals, write the numbers so that the decimal points are on a vertical line. Add as you would with whole numbers. Then write the decimal point in the sum directly below the decimal points in the addends.

Add: 0.326 + 4.8 + 57.23

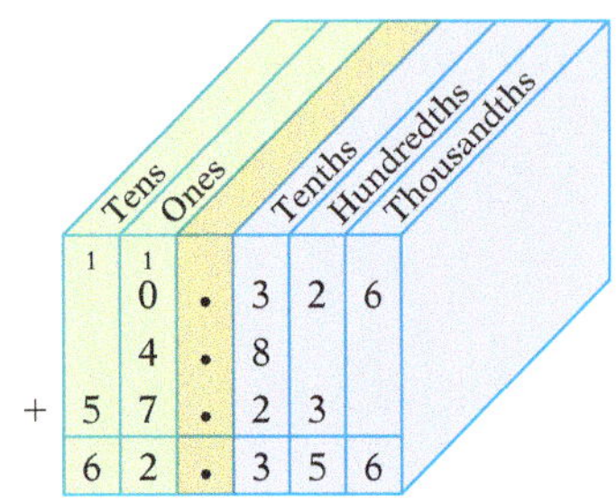

Note that placing the decimal points on a vertical line ensures that digits of the same place value are added.

The sum is 62.356.

To subtract decimals, write the numbers so that the decimal points are on a vertical line. Subtract as you would with whole numbers. Then write the decimal point in the difference directly below the decimal point in the subtrahend.

Subtract and check: 31.642 − 8.759

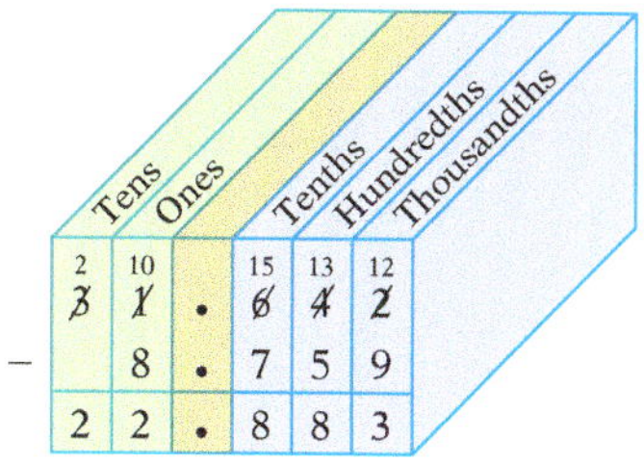

Note that placing the decimal points on a vertical line ensures that digits of the same place value are subtracted.

The difference is 22.883.

Check: Subtrahend 8.759
+ Difference + 22.883
= Minuend 31.642

Subtract and check: 5.4 − 1.6832

Insert zeros in the minuend so that it has the same number of decimal places as the subtrahend.

$$\begin{array}{r} 5.4000 \\ -\ 1.6832 \\ \hline \end{array}$$

Subtract and then check.

$$\begin{array}{r} \scriptstyle 4\ 13\ 9\ 9\ 10 \\ \not{5}.\not{4}\not{0}\not{0}\not{0} \\ -\ 1.6832 \\ \hline 3.7168 \end{array}$$

Check: 1.6832
+ 3.7168
5.4000

The sign rules for adding and subtracting decimals are the same rules used to add and subtract integers.

Take Note ➤
Recall that the absolute value of a number is its distance from zero on the number line. The absolute value of a number is a positive number or zero.

$|54.29| = 54.29$

$|-36.087| = 36.087$

Simplify: −36.087 + 54.29

The signs of the addends are different. Subtract the smaller absolute value from the larger absolute value.

54.29 − 36.087 = 18.203

Attach the sign of the number with the larger absolute value.

$|54.29| > |-36.087|$

The sum is positive. −36.087 + 54.29 = 18.203

Focus on adding and subtracting decimals

a. Add: 35.8 + 182.406 + 71.0934
b. What is −251.49 more than −638.7?
c. Subtract and check: 73 − 8.16

SOLUTION

a.
$$\begin{array}{r} \scriptstyle 1\ 1\ \ \ \ \ \ \ \ \ \\ 35.8 \\ 182.406 \\ +\ 71.0934 \\ \hline 289.2994 \end{array}$$

b. $-638.7 + (-251.49) = -890.19$

c.

$$\begin{array}{r} \overset{6\,12}{\cancel{7}\cancel{3}}.\overset{9\,10}{\cancel{0}\cancel{0}} \\ -\ \ 8.16 \\ \hline 64.84 \end{array} \qquad \textit{Check:} \quad \begin{array}{r} 8.16 \\ +\ 64.84 \\ \hline 73.00 \end{array}$$

Check your understanding 1

a. Add: $8.64 + 52.7 + 0.39105$
b. What is 4.002 minus 9.378?
c. Subtract and check: $25 - 4.91$

SOLUTION See page S-5.

Objective 2.5A Practice

1. Subtract: $-42.1 - 8.6$
2. Add: $-9.37 + 3.465$
3. Find 382.9 more than -430.6.
4. What is 4.793 less than -6.82?

Solutions on pp. S-14–S-15.

Objective 2.5B Multiply and divide decimals

To multiply decimals, multiply the numbers as you would whole numbers. Then write the decimal point in the product so that the number of decimal places in the product is the sum of the numbers of decimal places in the factors.

Take Note ➤
Estimating the product of 32.41 and 7.6 shows that the decimal point has been placed correctly.

$$\begin{array}{rcr} 32.41 & \longrightarrow & 30 \\ 7.6 & \longrightarrow & \times\ 8 \\ \hline & & 240 \end{array}$$

240 is an estimate of $(32.41)(7.6)$. It is close to the actual product, 246.316.

Multiply: $(32.41)(7.6)$

$$\begin{array}{rl} 32.41 & \quad 2 \text{ decimal places} \\ \times\ \ 7.6 & \quad 1 \text{ decimal place} \\ \hline 19446 & \\ 22687\ \ & \\ \hline 246.316 & \quad 3 \text{ decimal places} \end{array}$$

Multiply: $0.061(0.08)$

$$\begin{array}{rl} 0.061 & \quad 3 \text{ decimal places} \\ \times\ \ 0.08 & \quad 2 \text{ decimal places} \\ \hline 0.00488 & \quad 5 \text{ decimal places} \end{array}$$

• Insert two zeros between the 4 and the decimal point so that there are 5 decimal places in the product.

The sign rules for multiplying decimals are the same rules used to multiply integers.

The product of two numbers with the same sign is positive.

The product of two numbers with different signs is negative.

Focus on multiplying decimals

Multiply: $0.00073(0.052)$

SOLUTION

$$\begin{array}{rl} 0.00073 & \longleftarrow 5 \text{ decimal places} \\ \times\ \ \ \ 0.052 & \longleftarrow 3 \text{ decimal places} \\ \hline 146 & \\ 365\ \ & \\ \hline 0.00003796 & \longleftarrow 8 \text{ decimal places} \end{array}$$

Check your understanding 2

Multiply: $0.000081(0.025)$

SOLUTION See page S-5.

Focus on multiplying signed decimals

Multiply: $-3.42(6.1)$

SOLUTION

$-3.42(6.1) = -20.862$

Check your understanding 3

Multiply: $(-0.7)(-5.8)$

SOLUTION See page S-5.

To divide decimals, move the decimal point in the divisor to the right so that the divisor is a whole number. Move the decimal point in the dividend the same number of places to the right. Place the decimal point in the quotient directly above the decimal point in the dividend. Then divide as you would with whole numbers.

Divide: $29.585 \div 4.85$

$4.85.\overline{)29.58.5}$

Move the decimal point 2 places to the right in the divisor. Move the decimal point 2 places to the right in the dividend. Place the decimal point in the quotient. Then divide as shown at the right.

$$\begin{array}{r} 6.1 \\ 485.\overline{)\,2958.5} \\ -2910 \\ \hline 48\ 5 \\ -48\ 5 \\ \hline 0 \end{array}$$

Moving the decimal point the same number of places in the divisor and the dividend does not change the quotient because the process is the same as multiplying the numerator and denominator of a fraction by the same number. For the example above,

$$4.85\overline{)29.585} = \frac{29.585}{4.85} = \frac{29.585 \cdot 100}{4.85 \cdot 100} = \frac{2958.5}{485} = 485\overline{)2958.5}$$

In division of decimals, rather than writing the quotient with a remainder, we usually round the quotient to a specified place value.

Divide and round to the nearest tenth: $0.86 \div 0.7$

$$\begin{array}{r} 1.22 \approx 1.2 \\ 0.7.\overline{)0.8.60} \\ -7 \\ \hline 1\ 6 \\ -1\ 4 \\ \hline 20 \\ -14 \\ \hline 6 \end{array}$$

• Recall that the symbol ≈ is read "approximately equal to."

To round the quotient to the nearest tenth, the division must be carried to the hundredths place. Therefore, zeros must be inserted in the dividend so that the quotient has a digit in the hundredths place.

The sign rules for dividing decimals are the same rules used to divide integers.

The quotient of two numbers with the same sign is positive.

The quotient of two numbers with different signs is negative.

Divide: $-1.16 \div 2.9$

$-1.16 \div 2.9 = -0.4$

• The signs are different.
The quotient is negative.
Divide the absolute values of the numbers.

Focus on dividing decimals

Divide: $431.97 \div 7.26$

SOLUTION

```
           5 9.5
7.26.)4 3 1.9 7.0
     -3 6 3 0
        6 8 9 7
       -6 5 3 4
          3 6 3 0
         -3 6 3 0
                0
```

• Move the decimal point 2 places to the right.

Check your understanding 4

Divide: $314.746 \div 6.53$

SOLUTION See page S-5.

Focus on dividing and rounding to a place value

Divide and round to the nearest hundredth: $448.2 \div 53$

SOLUTION

```
      8.4 5 6 ≈ 8.46
53)4 4 8.2 0 0
  -4 2 4
     2 4 2
    -2 1 2
       3 0 0
      -2 6 5
         3 5 0
        -3 1 8
           3 2
```

Check your understanding 5

Divide and round to the nearest thousandth: $519.37 \div 86$

SOLUTION See page S-5.

Focus on dividing signed decimals

Divide and round to the nearest tenth: $-6.94 \div (-1.5)$

SOLUTION

The quotient is positive.

$-6.94 \div (-1.5) \approx 4.6$

Check your understanding 6

Divide and round to the nearest tenth: $-25.7 \div 0.31$

SOLUTION See page S-5.

Objective 2.5B Practice

1. Multiply: $(3.4)(0.5)$
2. Multiply: $(-6.3)(-2.4)$
3. Multiply: $1.31(-0.006)$
4. Find the product of 2.7, -16, and 3.04.
5. Divide: $7.02 \div 3.6$
6. Divide: $(-3.312) \div (-0.8)$
7. Divide: $84.66 \div (-1.7)$

Solutions on p. S-15.

SECTION 2.6 Comparing and Converting Fractions and Decimals

Objective 2.6A Convert fractions to decimals

To convert a fraction to a decimal, divide the numerator by the denominator.

Convert $\frac{5}{8}$ to a decimal.

Divide the numerator by the denominator.

$$\begin{array}{r} 0.625 \\ 8\overline{)\,5.000} \\ -\underline{4\,8} \\ 20 \\ -\underline{16} \\ 40 \\ -\underline{40} \\ 0 \end{array}$$

In this case, the remainder is 0. This fraction can be written as a **terminating decimal,** which is a decimal whose remainder is eventually 0.

$$\frac{5}{8} = 0.625$$

Convert $\frac{7}{11}$ to a decimal.

Divide the numerator by the denominator.

$$\begin{array}{r} 0.6363 \\ 11\overline{)\,7.0000} \\ -\,6.6 \\ \hline 40 \\ -\,33 \\ \hline 70 \\ -\,66 \\ \hline 40 \\ -\,33 \\ \hline 7 \end{array}$$

In this case, the remainder is never 0. This fraction can be written as a **repeating decimal,** which is a decimal whose remainder is never 0. When a remainder begins to repeat (as 4 does in this example), the digits in the quotient will begin to repeat.

It is common practice to write a bar over the repeating digits of a decimal.

$$\frac{7}{11} = 0.6363... = 0.\overline{63}$$

Take Note ➤
The decimal representation of every fraction either terminates or repeats. However, the repeating cycle may be quite long. For instance, there are 22 digits in the repeating cycle of the decimal representation of $\frac{9}{23}$.

$$\frac{9}{23} = 0.\overline{3913043478260869565217}$$

Focus on converting a fraction to a decimal

a. Convert $\frac{5}{12}$ to a decimal.

b. Convert $\frac{5}{16}$ to a decimal.

c. Convert $\frac{32}{15}$ to a decimal.

SOLUTION

a.

$$\begin{array}{r} 0.4166 \\ 12\overline{)\,5.0000} \\ -\,4\,8 \\ \hline 20 \\ -\,12 \\ \hline 80 \\ -\,72 \\ \hline 80 \end{array}$$

• Divide the numerator by the denominator. Continue to divide until the remainder is 0 or the remainder repeats.

$$\frac{5}{12} = 0.41\overline{6}$$

b.

$$\begin{array}{r} 0.3125 \\ 16\overline{)\,5.0000} \\ -\,4\,8 \\ \hline 20 \\ -\,16 \\ \hline 40 \\ -\,32 \\ \hline 80 \\ -\,80 \\ \hline 0 \end{array}$$

• Divide the numerator by the denominator. Continue to divide until the remainder is 0 or the remainder repeats.

$$\frac{5}{16} = 0.3125$$

c.

$$\begin{array}{r} 2.133 \\ 15\overline{)\,32.000} \\ -\,30 \\ \hline 2\,0 \\ -\,1\,5 \\ \hline 50 \\ -\,45 \\ \hline 50 \end{array}$$

• Divide the numerator by the denominator. Continue to divide until the remainder is 0 or the remainder repeats.

$$\frac{32}{15} = 2.1\overline{3}$$

Check your understanding 1

a. Convert $\frac{11}{18}$ to a decimal.

b. Convert $\frac{7}{25}$ to a decimal.

c. Convert $\frac{13}{6}$ to a decimal.

SOLUTION See page S-6.

Objective 2.6A Practice

1. Convert $\frac{5}{9}$ to a terminating or repeating decimal. Place a bar over any repeating digits.
2. Convert $\frac{8}{11}$ to a terminating or repeating decimal. Place a bar over any repeating digits.
3. Convert $\frac{71}{111}$ to a terminating or repeating decimal. Place a bar over any repeating digits.
4. Convert $\frac{17}{22}$ to a terminating or repeating decimal. Place a bar over any repeating digits.
5. Convert $\frac{3}{40}$ to a terminating or repeating decimal. Place a bar over any repeating digits.
6. Convert $\frac{29}{4}$ to a terminating or repeating decimal. Place a bar over any repeating digits.

Solutions on pp. S-15–S-16.

Objective 2.6B Convert decimals to fractions

To convert a decimal to a fraction, remove the decimal point and place the decimal part over a denominator equal to the place value of the last digit in the decimal. Then write the fraction in simplest form.

Convert 0.275 to a fraction.

thousandths

$$0.275 = \frac{275}{1000} = \frac{\overset{1}{\cancel{5}} \cdot \overset{1}{\cancel{5}} \cdot 11}{2 \cdot 2 \cdot 2 \cdot \underset{1}{\cancel{5}} \cdot \underset{1}{\cancel{5}} \cdot 5} = \frac{11}{40}$$

Convert 0.33 to a fraction.

hundredths

$$0.33 = \frac{33}{100} = \frac{3 \cdot 11}{2 \cdot 2 \cdot 5 \cdot 5} = \frac{33}{100}$$

Focus on converting a decimal to a mixed number

Convert 7.45 to a mixed number.

SOLUTION

$$7.45 = 7\frac{45}{100} = 7\frac{9}{20}$$

Check your understanding 2

Convert 0.0005 to a fraction.

SOLUTION See page S-6.

Objective 2.6B Practice

1. Convert 0.4 to a fraction.

2. Convert 0.48 to a fraction.

3. Convert 0.485 to a fraction.

4. Convert 3.75 to a mixed number.

5. Convert 0.052 to a fraction.

6. Convert 0.00015 to a fraction.

Solutions on p. S-17.

Objective 2.6C Compare fractions and decimals

One way to determine the order relation between a decimal and a fraction is to write the decimal as a fraction and then compare the fractions.

Place the correct symbol, < or >, between the two numbers. $0.8 \quad \frac{8}{9}$

$0.8 \quad \frac{8}{9}$

$\frac{8}{10} \quad \frac{8}{9}$ • Write the decimal as a fraction.

$\frac{72}{90} \quad \frac{80}{90}$ • Write the fractions with a common denominator.

$\frac{72}{90} < \frac{80}{90}$ • Compare the fractions.

$0.8 < \frac{8}{9}$

Focus on comparing a decimal and a fraction

Place the correct symbol, < or >, between the two numbers.

$0.7 \quad \frac{3}{4}$

SOLUTION

$0.7 \quad \frac{3}{4}$

$\frac{7}{10} \quad \frac{3}{4}$ • Write the decimal as a fraction.

$\frac{14}{20} \quad \frac{15}{20}$ • Write the fractions with a common denominator.

$\frac{14}{20} < \frac{15}{20}$ • Compare the fractions.

$0.7 < \frac{3}{4}$

Check your understanding 3

Place the correct symbol, < or >, between the two numbers.

$\frac{5}{6}$ 0.83

SOLUTION See page S-6.

Objective 2.6C Practice

1. Place the correct symbol, < or >, between the numbers.
0.8 $\frac{3}{4}$

2. Place the correct symbol, < or >, between the numbers.
$\frac{7}{8}$ 0.9

3. Place the correct symbol, < or >, between the numbers.
0.13 $\frac{5}{40}$

4. Place the correct symbol, < or >, between the numbers.
$\frac{12}{55}$ 0.22

5. Place the correct symbol, < or >, between the numbers.
0.55 $\frac{5}{9}$

6. Place the correct symbol, < or >, between the numbers.
$\frac{22}{7}$ 3.14

Solutions on pp. S-17–S-18.

SECTION 2.7 Radical Expressions and Real Numbers

Objective 2.7A Find the square root of a perfect square

Recall that the square of a number is equal to the number multiplied times itself.

$3^2 = 3 \cdot 3 = 9$

The square of an integer is called a **perfect square.**

9 is a perfect square because 9 is the square of 3: $3^2 = 9$.

The numbers 1, 4, 9, 16, 25, 36, 49, 64, 81, and 100 are perfect squares.

$1^2 = 1$
$2^2 = 4$
$3^2 = 9$
$4^2 = 16$
$5^2 = 25$
$6^2 = 36$
$7^2 = 49$
$8^2 = 64$
$9^2 = 81$
$10^2 = 100$

Larger perfect squares can be found by squaring 11, squaring 12, squaring 13, and so on.

Note that squaring the negative integers results in the same list of numbers.

$(-1)^2 = 1$
$(-2)^2 = 4$
$(-3)^2 = 9$
$(-4)^2 = 16$, and so on.

Perfect squares are used in simplifying square roots. The symbol for square root is $\sqrt{\ }$.

Square Root

A **square root** of a positive number x is a number whose square is x.

If $a^2 = x$, then $\sqrt{x} = a$.

Example

The expression $\sqrt{9}$, read "the square root of 9," is equal to the number that, when squared, is equal to 9.

Since $3^2 = 9$, $\sqrt{9} = 3$.

Every positive number has two square roots, one a positive number and one a negative number. The symbol $\sqrt{\ }$ is used to indicate the positive square root of a number. When the negative square root of a number is to be found, a negative sign is placed in front of the square root symbol. For example,

$$\sqrt{9} = 3 \quad \text{and} \quad -\sqrt{9} = -3$$

The square root symbol, $\sqrt{\ }$, is also called a **radical.** The number under the radical is called the **radicand.** In the radical expression $\sqrt{9}$, 9 is the radicand.

Simplify: $\sqrt{49}$

$\sqrt{49} = 7$ • $\sqrt{49}$ is equal to the number that, when squared, equals 49. $7^2 = 49$.

Take Note ➤ The radical is a grouping symbol. Therefore, when simplifying numerical expressions, simplify the radicand as part of Step 1 of the Order of Operations Agreement.

Simplify: $6 + 4\sqrt{9}$

$6 + 4\sqrt{9} = 6 + 4 \cdot 3$ • Simplify $\sqrt{9}$.

$= 6 + 12$ • Use the Order of Operations Agreement.

$= 18$

Focus on simplifying perfect square radical expressions

Simplify: **a.** $\sqrt{121}$ **b.** $\sqrt{\frac{4}{25}}$ **c.** $\sqrt{36} - 9\sqrt{4}$

SOLUTION

a. Since $11^2 = 121$, $\sqrt{121} = 11$.

b. Since $\left(\frac{2}{5}\right)^2 = \frac{4}{25}$, $\sqrt{\frac{4}{25}} = \frac{2}{5}$.

c. $\sqrt{36} - 9\sqrt{4} = 6 - 9 \cdot 2$
$= 6 - 18$
$= 6 + (-18)$
$= -12$

Check your understanding 1

Simplify: **a.** $-\sqrt{144}$ **b.** $\sqrt{\dfrac{81}{100}}$ **c.** $4\sqrt{16} - \sqrt{9}$

SOLUTION See page S-7.

Focus on evaluating an expression

Evaluate $6\sqrt{ab}$ for $a = 2$ and $b = 8$.

SOLUTION

$6\sqrt{ab}$

$6\sqrt{2 \cdot 8} = 6\sqrt{16}$
$= 6(4)$
$= 24$

Check your understanding 2

Evaluate $5\sqrt{a + b}$ for $a = 17$ and $b = 19$.

SOLUTION See page S-7.

Objective 2.7A Practice

1. Simplify: $-\sqrt{1}$
2. Simplify: $\sqrt{81}$
3. Simplify: $\sqrt{144} - \sqrt{25}$
4. Simplify: $\sqrt{144} + 3\sqrt{9}$
5. Evaluate $7\sqrt{x + y}$ for $x = 34$ and $y = 15$.
6. What is 7 decreased by the square root of 16?

Solutions on p. S-18.

Objective 2.7B Approximate the square root of a natural number

If the radicand is not a perfect square, the square root can only be approximated. For example, the radicand in the radical expression $\sqrt{2}$ is 2, and 2 is not a perfect square. The square root of 2 can be approximated to any desired place value.

To the nearest tenth:	$\sqrt{2} \approx 1.4$	$(1.4)^2 = 1.96$
To the nearest hundredth:	$\sqrt{2} \approx 1.41$	$(1.41)^2 = 1.9881$
To the nearest thousandth:	$\sqrt{2} \approx 1.414$	$(1.414)^2 = 1.999396$
To the nearest ten-thousandth:	$\sqrt{2} \approx 1.4142$	$(1.4142)^2 = 1.99996164$

The square of each decimal approximation gets closer and closer to 2 as the number of place values in the approximation increases. But no matter how many place values are used to approximate $\sqrt{2}$, the digits never terminate or repeat. In general, the square root of any number that is not a perfect square can only be approximated.

Recall that a rational number has a decimal representation that terminates or repeats. A number such as $\sqrt{2}$ has a non-terminating, non-repeating decimal representation. Such a number is called an **irrational number.**

Irrational Numbers

An **irrational number** is a number whose decimal representation never terminates or repeats.

EXAMPLES OF IRRATIONAL NUMBERS

1. π **2.** $\sqrt{3}$ **3.** 0.23233233323333 . . .

The rational numbers and the irrational numbers taken together are called the **real numbers.**

Real Numbers

The **real numbers** are all the rational numbers together with all the irrational numbers.

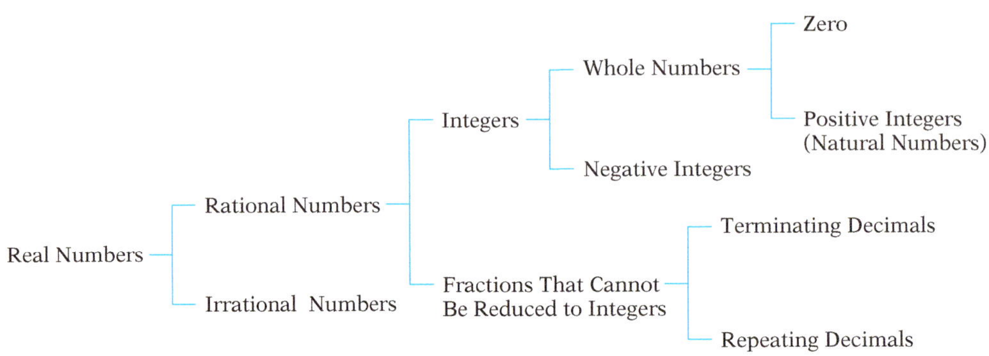

Focus on approximating a square root

Approximate $\sqrt{11}$ to the nearest ten-thousandth.

SOLUTION

11 is not a perfect square.

Use a calculator to approximate $\sqrt{11}$.

$\sqrt{11} \approx 3.3166$

Check your understanding 3

Approximate $3\sqrt{5}$ to the nearest ten-thousandth.

SOLUTION See page S-7.

Focus on finding a natural number inequality for a square root

Between what two whole numbers is the value of $\sqrt{41}$?

SOLUTION

Since the number 41 is between the perfect squares 36 and 49, the value of $\sqrt{41}$ is between $\sqrt{36}$ and $\sqrt{49}$.

Because $\sqrt{36} = 6$ and $\sqrt{49} = 7$, the value of $\sqrt{41}$ is between the whole numbers 6 and 7.

This can be written using inequality symbols as $6 < \sqrt{41} < 7$, which is read "the square root of 41 is greater than 6 and less than 7."

Use a calculator to verify that $\sqrt{41} \approx 6.4$, which is between 6 and 7.

Check your understanding 4

Between what two whole numbers is the value of $\sqrt{57}$?

SOLUTION See page S-7.

Objective 2.7B Practice

1. Approximate $\sqrt{10}$ to the nearest ten-thousandth.
2. Approximate $6\sqrt{15}$ to the nearest ten-thousandth.
3. Approximate $10\sqrt{21}$ to the nearest ten-thousandth.
4. Approximate $-12\sqrt{53}$ to the nearest ten-thousandth.
5. Between what two whole numbers is the value of $\sqrt{29}$?
6. Between what two whole numbers is the value of $\sqrt{130}$?

Solutions on pp. S-18–S-19.

Objective 2.7C Solve application problems

Focus on solving an application

Find the range of a submarine periscope that is 8 ft above the surface of the water. Use the formula $R = 1.4\sqrt{h}$, where R is the range in miles and h is the height in feet of the periscope above the surface of the water. Round to the nearest hundredth.

STRATEGY

To find the range, replace h by 8 in the given formula and solve for R.

SOLUTION

$R = 1.4\sqrt{h}$

$R = 1.4\sqrt{8}$

$R \approx 3.96$ • Use a calculator.

The range of the periscope is approximately 3.96 mi.

Check your understanding 5

The formula $t = \sqrt{\frac{d}{16}}$ gives the time t, in seconds, it takes for an object to fall a distance of d feet. This formula assumes no air resistance. Find the time it would take for an object to fall 96 ft. Round to the nearest tenth of a second.

SOLUTION See pages S-7.

Objective 2.7C Practice

For Exercises 1 and 2, use the formula $t = \sqrt{\frac{d}{16}}$, where t is the time, in seconds, for an object to fall d feet.

1. If an object is dropped from a plane, how long will it take for the object to fall 144 ft?

2. If an object is dropped from a plane, how long will it take for the object to fall 64 ft?

Solutions on p. S-19.

SOLUTIONS TO MODULE 2

Solutions to Check Your Understanding

Section 2.1

Check your understanding 1

$30 \div 1 = 30$
$30 \div 2 = 15$
$30 \div 3 = 10$
$30 \div 4$ • Does not divide evenly
$30 \div 5 = 6$
$30 \div 6 = 5$ • The factors are repeating.

The factors of 30 are 1, 2, 3, 5, 6, 10, 15, and 30.

Check your understanding 2

$$\begin{array}{r} 11 \\ 2\overline{)22} \\ 2\overline{)44} \\ 2\overline{)88} \end{array}$$

$88 = 2 \cdot 2 \cdot 2 \cdot 11 = 2^3 \cdot 11$

Check your understanding 3

$$\begin{array}{r} 59 \\ 5\overline{)295} \end{array}$$

$295 = 5 \cdot 59$

Check your understanding 4

	2	3	5	7
12 =	(2 · 2)	3		
42 =	2	3		(7)
45 =		(3 · 3)	(5)	

The LCM $= 2 \cdot 2 \cdot 3 \cdot 3 \cdot 5 \cdot 7 = 420$.

Check your understanding 5

	2	3	5
36 =	(2 · 2)	3 · 3	
60 =	2 · 2	(3)	5
72 =	2 · 2 · 2	3 · 3	

The GCF $= 2 \cdot 2 \cdot 3 = 12$.

Check your understanding 6

	2	3	5	11
11 =				11
24 =	2 · 2 · 2	3		
30 =	2	3	5	

Because no numbers are circled, the GCF = 1.

Section 2.2

Check your understanding 1

$$5\overline{)22}\quad \text{quotient } 4,\ -20,\ \text{remainder } 2$$

$$\frac{22}{5} = 4\frac{2}{5}$$

Check your understanding 2

$$7\overline{)28}\quad \text{quotient } 4,\ -28,\ \text{remainder } 0$$

$$\frac{28}{7} = 4$$

Check your understanding 3

$$14\frac{5}{8} = \frac{112 + 5}{8} = \frac{117}{8}$$

Check your understanding 4

a. $\dfrac{16}{24} = \dfrac{\cancel{2}\cdot\cancel{2}\cdot\cancel{2}\cdot 2}{\cancel{2}\cdot\cancel{2}\cdot\cancel{2}\cdot 3} = \dfrac{2}{3}$

b. $\dfrac{8}{56} = \dfrac{\cancel{2}\cdot\cancel{2}\cdot\cancel{2}}{\cancel{2}\cdot\cancel{2}\cdot\cancel{2}\cdot 7} = \dfrac{1}{7}$

c. $\dfrac{15}{32} = \dfrac{3\cdot 5}{2\cdot 2\cdot 2\cdot 2\cdot 2} = \dfrac{15}{32}$

d. $\dfrac{48}{36} = \dfrac{\cancel{2}\cdot\cancel{2}\cdot 2\cdot 2\cdot\cancel{3}}{\cancel{2}\cdot\cancel{2}\cdot\cancel{3}\cdot 3} = \dfrac{4}{3}$

Check your understanding 5

$45 \div 5 = 9$ • Divide the larger denominator by the smaller.

$\dfrac{3}{5} = \dfrac{3\cdot 9}{5\cdot 9} = \dfrac{27}{45}$ • Multiply the numerator and denominator of the given fraction by the quotient (9).

$\frac{27}{45}$ is equivalent to $\frac{3}{5}$.

Check your understanding 6

Write 6 as $\frac{6}{1}$.

$18 \div 1 = 18$ • Divide the larger denominator by the smaller.

$6 = \dfrac{6\cdot 18}{1\cdot 18} = \dfrac{108}{18}$ • Multiply the numerator and denominator of the given fraction by the quotient (18).

$\frac{108}{18}$ is equivalent to 6.

Check your understanding 7

$$\frac{9}{14} = \frac{27}{42} \qquad \frac{13}{21} = \frac{26}{42} \qquad \frac{9}{14} > \frac{13}{21}$$

Section 2.3

Check your understanding 1

$$-\frac{1}{3}\left(-\frac{5}{12}\right)\left(\frac{8}{15}\right) = \frac{1}{3}\cdot\frac{5}{12}\cdot\frac{8}{15}$$

• The signs are the same. The product is positive.

$$= \frac{1\cdot 5\cdot 8}{3\cdot 12\cdot 15}$$

$$= \frac{1\cdot 5\cdot 2\cdot 2\cdot 2}{3\cdot 2\cdot 2\cdot 3\cdot 3\cdot 5}$$

$$= \frac{2}{27}$$

Check your understanding 2

xy

$$-5\frac{1}{8}\cdot\frac{2}{3} = -\left(\frac{41}{8}\cdot\frac{2}{3}\right)$$

• The signs are different. The answer is negative. Write $5\frac{1}{8}$ as an improper fraction.

$$= -\left(\frac{41\cdot 2}{8\cdot 3}\right)$$

• Multiply the numerators. Multiply the denominators.

$$= -\left(\frac{41\cdot 2}{2\cdot 2\cdot 2\cdot 3}\right)$$

• Write the product in simplest form.

$$= -\frac{41}{12} = -3\frac{5}{12}$$

• Write the answer as an improper fraction or a mixed number.

Check your understanding 3

$$-4\frac{3}{8}\div 3\frac{1}{2} = -\left(\frac{35}{8}\div\frac{7}{2}\right)$$

• The signs are different. The answer is negative. Write $4\frac{3}{8}$ and $3\frac{1}{2}$ as improper fractions.

$$= -\left(\frac{35}{8}\cdot\frac{2}{7}\right)$$

• Rewrite division as multiplication by the reciprocal.

$$= -\left(\frac{35\cdot 2}{8\cdot 7}\right)$$

• Multiply the numerators. Multiply the denominators.

$$= -\left(\frac{5\cdot 7\cdot 2}{2\cdot 2\cdot 2\cdot 7}\right)$$

• Write the product in simplest form.

$$= -\frac{5}{4} = -1\frac{1}{4}$$

• Write the answer as an improper fraction or a mixed number.

Check your understanding 4

$$4\div\left(-\frac{6}{7}\right) = -\left(\frac{4}{1}\div\frac{6}{7}\right)$$

• The signs are different. The answer is negative. Write $4 = \frac{4}{1}$.

$$= -\left(\frac{4}{1}\cdot\frac{7}{6}\right)$$

• Rewrite division as multiplication by the reciprocal.

$$= -\left(\frac{4\cdot 7}{1\cdot 6}\right)$$

• Multiply the numerators. Multiply the denominators.

$$= -\left(\frac{2\cdot 2\cdot 7}{1\cdot 2\cdot 3}\right)$$

• Write the product in simplest form.

$$= -\frac{14}{3} = -4\frac{2}{3}$$

• Write the answer as an improper fraction or a mixed number.

Check your understanding 5

$$-\frac{5}{12}+\frac{5}{8}+\left(-\frac{1}{6}\right)=\frac{-5}{12}+\frac{5}{8}+\frac{-1}{6}$$

• Rewrite each fraction with the negative sign in the numerator.

$$=\frac{-10}{24}+\frac{15}{24}+\frac{-4}{24}$$

• Write equivalent fractions using the LCD (24) of the fractions. Then add.

$$=\frac{-10+15+(-4)}{24}$$

$$=\frac{1}{24}$$

Check your understanding 6

$$-\frac{5}{6}-\frac{7}{9}=\frac{-5}{6}-\frac{7}{9}$$

• Rewrite $-\frac{5}{6}$ with the negative sign in the numerator.

$$=\frac{-15}{18}-\frac{14}{18}$$

• Write equivalent fractions using the LCD (18) of the fractions.

$$=\frac{-15-14}{18}$$

• Subtract the fractions.

$$=\frac{-29}{18}$$

$$=-\frac{29}{18}=-1\frac{11}{18}$$

• Write the difference as an improper fraction or a mixed number.

Section 2.4

Check your understanding 1

The digit 4 is in the thousandths place.

Check your understanding 2

a. 3.675849 — Given place value: 5; next digit $4 < 5$

3.675849 rounded to the nearest ten-thousandth is 3.6758.

b. 48.907 — Given place value: 9; next digit $0 < 5$

48.907 rounded to the nearest tenth is 48.9.

c. 31.8652 — Given place value: 1; next digit $8 > 5$

31.8652 rounded to the nearest whole number is 32.

Check your understanding 3

$0.15 \qquad 0.107$

$\frac{15}{100} \qquad \frac{107}{1000}$ • Write the numbers as fractions.

$\frac{150}{1000} \qquad \frac{107}{1000}$ • Write the fractions with a common denominator.

$\frac{150}{1000} > \frac{107}{1000}$ • Compare the fractions.

$0.15 > 0.107$

Section 2.5

Check your understanding 1

a.

$$\begin{array}{r} \overset{1\,1\;\;1}{8.64} \\ 52.7 \\ +\ 0.39105 \\ \hline 61.73105 \end{array}$$

b. $4.002 - 9.378 = 4.002 + (-9.378)$
$= -5.376$

c.

$$\begin{array}{r} 2\overset{4}{\cancel{5}}.\overset{9}{\cancel{0}}\overset{10}{\cancel{0}} \\ -\ \ 4.91 \\ \hline 20.09 \end{array}$$

Check:

$$\begin{array}{r} 4.91 \\ +20.09 \\ \hline 25.00 \end{array}$$

Check your understanding 2

$$\begin{array}{rl} 0.000081 & \leftarrow \text{6 decimal places} \\ \times \quad 0.025 & \leftarrow \text{3 decimal places} \\ \hline 405 & \\ 162 & \\ \hline 0.000002025 & \leftarrow \text{9 decimal places} \end{array}$$

Check your understanding 3

$(-0.7)(-5.8) = 4.06$

Check your understanding 4

$$\begin{array}{r} 48.2 \\ 6.53.\overline{)\ 314.74.6} \\ -\ 261\ 2 \\ \hline 53\ 54 \\ -\ 52\ 24 \\ \hline 1\ 30\ 6 \\ -\ 1\ 30\ 6 \\ \hline 0 \end{array}$$

• Move the decimal point 2 places to the right.

Check your understanding 5

$$\begin{array}{r} 6.0391 \approx 6.039 \\ 86\overline{)\ 519.3700} \\ -\ 516 \\ \hline 3\ 3 \\ -\quad 0 \\ \hline 3\ 37 \\ -\ 2\ 58 \\ \hline 790 \\ -\ 774 \\ \hline 160 \\ -\quad 86 \\ \hline 74 \end{array}$$

Check your understanding 6

The quotient is negative.

$-25.7 \div 0.31 \approx -82.9$

Section 2.6

Check your understanding 1

a.

$$\begin{array}{r} 0.611 \\ 18\overline{)\ 11.000} \\ -\,10\,8 \\ \hline 20 \\ -\,18 \\ \hline 20 \\ -\,18 \\ \hline 2 \end{array}$$

• Divide the numerator by the denominator. Continue to divide until the remainder is 0 or the remainder repeats.

$$\frac{11}{18} = 0.6\overline{1}$$

• Use a bar over the repeating digit.

b.

$$\begin{array}{r} 0.28 \\ 25\overline{)\ 7.00} \\ -\,5\,0 \\ \hline 2\,00 \\ -\,2\,00 \\ \hline 0 \end{array}$$

• The remainder is 0.

$$\frac{7}{25} = 0.28$$

c.

$$\begin{array}{r} 2.166 \\ 6\overline{)\ 13.000} \\ -\,12 \\ \hline 1\,0 \\ -\,6 \\ \hline 40 \\ -\,36 \\ \hline 40 \\ -\,36 \\ \hline 4 \end{array}$$

• The remainder repeats.

$$\frac{13}{6} = 2.1\overline{6}$$

Check your understanding 2

$$0.0005 = \frac{5}{10{,}000} = \frac{1}{2000}$$

Check your understanding 3

$$\frac{5}{6} \qquad 0.83$$

$$\frac{5}{6} \qquad \frac{83}{100}$$

• Write the decimal as a fraction.

$$\frac{250}{300} \qquad \frac{249}{300}$$

• Write the fractions with a common denominator.

$$\frac{250}{300} > \frac{249}{300}$$

• Compare the fractions.

$$\frac{5}{6} > 0.83$$

Section 2.7

Check your understanding 1

a. Since $12^2 = 144$, $-\sqrt{144} = -12$.

b. Since $\left(\frac{9}{10}\right)^2 = \frac{81}{100}$, $\sqrt{\frac{81}{100}} = \frac{9}{10}$.

c. $4\sqrt{16} - \sqrt{9} = 4 \cdot 4 - 3$
$= 16 - 3 = 13$

Check your understanding 2

$5\sqrt{a + b}$

$5\sqrt{17 + 19} = 5\sqrt{36}$
$= 5 \cdot 6$
$= 30$

Check your understanding 3

$3\sqrt{5}$ means 3 times $\sqrt{5}$.

Use a calculator to approximate $\sqrt{5}$.

$3\sqrt{5} \approx 6.7082$

Check your understanding 4

57 is between the perfect squares 49 and 64.

$\sqrt{49} = 7$ and $\sqrt{64} = 8$

$7 < \sqrt{57} < 8$

Check your understanding 5

STRATEGY

To find the time, replace d by 96 in the given equation. Use a calculator to approximate the square root of the resulting expression.

SOLUTION

$$t = \sqrt{\frac{d}{16}}$$

$$t = \sqrt{\frac{96}{16}} = \sqrt{6} \approx 2.4$$

It takes approximately 2.4 s for the object to fall 96 ft.

Solutions to Objective Practice Exercises

Objective 2.1A

1. $12 \div 1 = 12$

$12 \div 2 = 6$

$12 \div 3 = 4$

$12 \div 4 = 3$

The factors of 12 are 1, 2, 3, 4, 6, and 12.

2. $56 \div 1 = 56$

$56 \div 2 = 28$

$56 \div 4 = 14$

$56 \div 7 = 8$

$56 \div 8 = 7$

The factors of 56 are 1, 2, 4, 7, 8, 14, 28, and 56.

3. $48 \div 1 = 48$

$48 \div 2 = 24$

$48 \div 3 = 16$

$48 \div 4 = 12$

$48 \div 6 = 8$

$48 \div 8 = 6$

The factors of 48 are 1, 2, 3, 4, 6, 8, 12, 16, 24, and 48.

4.
$$\begin{array}{r} 2 \\ 2\overline{)4} \\ 2\overline{)8} \\ 2\overline{)16} \end{array}$$

$16 = 2 \cdot 2 \cdot 2 \cdot 2 = 2^4$

5.
$$\begin{array}{r} 5 \\ 2\overline{)10} \\ 2\overline{)20} \\ 2\overline{)40} \end{array}$$

$40 = 2 \cdot 2 \cdot 2 \cdot 5 = 2^3 \cdot 5$

6. 37 is a prime number.

Objective 2.1B

1.

	2	3
8 =	$\boxed{2 \cdot 2 \cdot 2}$	
12 =	$2 \cdot 2$	$\boxed{3}$

$\text{LCM} = 2 \cdot 2 \cdot 2 \cdot 3 = 24$

2.

	2	7
8 =	$\boxed{2 \cdot 2 \cdot 2}$	
14 =	2	$\boxed{7}$

$\text{LCM} = 2 \cdot 2 \cdot 2 \cdot 7 = 56$

3.

	2	3
9 =		$3 \cdot 3$
36 =	$\boxed{2 \cdot 2}$	$\boxed{3 \cdot 3}$

$\text{LCM} = 2 \cdot 2 \cdot 3 \cdot 3 = 36$

4.

	2	3	17	23
102 =	2	3	17	
184 =	2 · 2 · 2			23

LCM = 2 · 2 · 2 · 3 · 17 · 23 = 9384

5.

	2	3
4 =	2 · 2	
8 =	2 · 2 · 2	
12 =	2 · 2	3

LCM = 2 · 2 · 2 · 3 = 24

6.

	2	3	5
3 =		3	
5 =			5
10 =	2		5

LCM = 2 · 3 · 5 = 30

Objective 2.1C

1.

	2	3
6 =	2	3
9 =		3 · 3

GCF = 3

2.

	2	5
25 =		5 · 5
100 =	2 · 2	5 · 5

GCF = 5 · 5 = 25

3.

	2	3	17
32 =	2 · 2 · 2 · 2 · 2		
51 =		3	17

GCF = 1

4.

	3	5	11
3 =	3		
5 =		5	
11 =			11

GCF = 1

5.

	2	3	5
24 =	2 · 2 · 2	3	
40 =	2 · 2 · 2		5
72 =	2 · 2 · 2	3 · 3	

GCF = 2 · 2 · 2 = 8

6.

	2	3	7
32 =	2 · 2 · 2 · 2 · 2		
56 =	2 · 2 · 2		7
72 =	2 · 2 · 2	3 · 3	

GCF = 2 · 2 · 2 = 8

Objective 2.2A

1. $\begin{array}{r} 2 \\ 4\overline{)11} \\ \underline{-8} \\ 3 \end{array}$ $\quad \dfrac{11}{4} = 2\dfrac{3}{4}$

2. $\begin{array}{r} 9 \\ 8\overline{)72} \\ \underline{-72} \\ 0 \end{array}$ $\quad \dfrac{72}{8} = 9$

3. $6\dfrac{1}{2} = \dfrac{12 + 1}{2} = \dfrac{13}{2}$

4. $5\dfrac{3}{11} = \dfrac{55 + 3}{11} = \dfrac{58}{11}$

Objective 2.2B

1. $\dfrac{2}{12} = \dfrac{\overset{1}{\cancel{2}}}{\underset{1}{\cancel{2}} \cdot 2 \cdot 3} = \dfrac{1}{6}$

2. $\dfrac{16}{84} = \dfrac{\overset{1}{\cancel{2}} \cdot \overset{1}{\cancel{2}} \cdot 2 \cdot 2}{\underset{1}{\cancel{2}} \cdot \underset{1}{\cancel{2}} \cdot 3 \cdot 7} = \dfrac{4}{21}$

3. $\dfrac{12}{35} = \dfrac{2 \cdot 2 \cdot 3}{5 \cdot 7} = \dfrac{12}{35}$

4. $\dfrac{144}{36} = \dfrac{\overset{1}{\cancel{2}} \cdot \overset{1}{\cancel{2}} \cdot 2 \cdot 2 \cdot \overset{1}{\cancel{3}} \cdot \overset{1}{\cancel{3}}}{\underset{1}{\cancel{2}} \cdot \underset{1}{\cancel{2}} \cdot \underset{1}{\cancel{3}} \cdot \underset{1}{\cancel{3}}} = 4$

Objective 2.2C

1. $16 \div 4 = 4; \dfrac{3 \cdot 4}{4 \cdot 4} = \dfrac{12}{16}$

2. $49 \div 7 = 7; \dfrac{5 \cdot 7}{7 \cdot 7} = \dfrac{35}{49}$

3. $18 \div 9 = 2; \dfrac{5 \cdot 2}{9 \cdot 2} = \dfrac{10}{18}$

4. $64 \div 16 = 4; \dfrac{15 \cdot 4}{16 \cdot 4} = \dfrac{60}{64}$

Objective 2.2D

1. $\dfrac{11}{40} < \dfrac{19}{40}$

2. $\dfrac{2}{5} = \dfrac{16}{40}, \dfrac{3}{8} = \dfrac{15}{40}, \dfrac{2}{5} > \dfrac{3}{8}$

3. $\dfrac{13}{18} = \dfrac{26}{36}, \dfrac{7}{12} = \dfrac{21}{36}, \dfrac{13}{18} > \dfrac{7}{12}$

4. $\dfrac{5}{12} = \dfrac{25}{60}, \dfrac{7}{15} = \dfrac{28}{60}, \dfrac{5}{12} < \dfrac{7}{15}$

Objective 2.3A

1. $$\begin{aligned} \frac{5}{6} \cdot \left(-\frac{2}{5}\right) &= -\left(\frac{5}{6} \cdot \frac{2}{5}\right) \\ &= -\frac{5 \cdot 2}{6 \cdot 5} \\ &= -\frac{5 \cdot 2}{2 \cdot 3 \cdot 5} \\ &= -\frac{1}{3} \end{aligned}$$

2. $\left(-\frac{3}{4}\right)\cdot\left(-\frac{2}{9}\right) = \frac{3}{4}\cdot\frac{2}{9}$

$= \frac{3\cdot 2}{4\cdot 9}$

$= \frac{3\cdot 2}{2\cdot 2\cdot 3\cdot 3}$

$= \frac{1}{6}$

3. $-\frac{3}{4}\cdot\frac{1}{2} = -\left(\frac{3}{4}\cdot\frac{1}{2}\right)$

$= -\frac{3\cdot 1}{4\cdot 2}$

$= -\frac{3\cdot 1}{2\cdot 2\cdot 2}$

$= -\frac{3}{8}$

4. $-1\frac{2}{3}\cdot\left(-\frac{3}{5}\right) = \frac{5}{3}\cdot\frac{3}{5}$

$= \frac{5\cdot 3}{3\cdot 5}$

$= 1$

5. xy

$-\frac{5}{16}\cdot\frac{7}{15} = -\left(\frac{5\cdot 7}{16\cdot 15}\right)$

$= -\frac{5\cdot 7}{2\cdot 2\cdot 2\cdot 2\cdot 3\cdot 5}$

$= -\frac{7}{48}$

6. $\left(-\frac{5}{7}\right)\div\left(-\frac{5}{6}\right) = \frac{5}{7}\div\frac{5}{6}$

$= \frac{5}{7}\cdot\frac{6}{5}$

$= \frac{5\cdot 6}{7\cdot 5}$

$= \frac{5\cdot 2\cdot 3}{7\cdot 5} = \frac{6}{7}$

7. $\frac{3}{8}\div\left(-\frac{5}{12}\right) = -\left(\frac{3}{8}\div\frac{5}{12}\right)$

$= -\left(\frac{3}{8}\cdot\frac{12}{5}\right)$

$= -\frac{3\cdot 12}{8\cdot 5}$

$= -\frac{3\cdot 2\cdot 2\cdot 3}{2\cdot 2\cdot 2\cdot 5}$

$= -\frac{9}{10}$

8. $5\frac{3}{5} \div \left(-\frac{7}{10}\right) = -\left(\frac{28}{5} \div \frac{7}{10}\right)$

$= -\left(\frac{28}{5} \cdot \frac{10}{7}\right)$

$= -\frac{28 \cdot 10}{5 \cdot 7}$

$= -\frac{2 \cdot 2 \cdot 7 \cdot 2 \cdot 5}{5 \cdot 7}$

$= -\frac{8}{1} = -8$

9. $\frac{3}{5} \div \frac{12}{25} = \frac{3}{5} \cdot \frac{25}{12}$

$= \frac{3 \cdot 25}{5 \cdot 12}$

$= \frac{3 \cdot 5 \cdot 5}{5 \cdot 2 \cdot 2 \cdot 3}$

$= \frac{5}{4} = 1\frac{1}{4}$

10. $x \div y$

$-\frac{5}{8} \div \left(-\frac{15}{2}\right) = \frac{5}{8} \div \frac{15}{2}$

$= \frac{5}{8} \cdot \frac{2}{15}$

$= \frac{5 \cdot 2}{8 \cdot 15}$

$= \frac{5 \cdot 2}{2 \cdot 2 \cdot 2 \cdot 3 \cdot 5}$

$= \frac{1}{12}$

Objective 2.3B

1. $-\frac{3}{4} + \frac{2}{3} = \frac{-3}{4} + \frac{2}{3}$

$= \frac{-9}{12} + \frac{8}{12}$

$= \frac{-9 + 8}{12} = \frac{-1}{12} = -\frac{1}{12}$

2. $-\frac{7}{12} + \frac{5}{8} = \frac{-7}{12} + \frac{5}{8}$

$= \frac{-14}{24} + \frac{15}{24}$

$= \frac{-14 + 15}{24} = \frac{1}{24}$

3. $\frac{2}{5} + \left(-\frac{11}{15}\right) = \frac{2}{5} + \frac{-11}{15}$

$= \frac{6}{15} + \frac{-11}{15}$

$= \frac{6 + (-11)}{15} = \frac{-5}{15} = -\frac{1}{3}$

4. $\frac{3}{8} + \left(-\frac{1}{2}\right) + \frac{7}{12} = \frac{3}{8} + \frac{-1}{2} + \frac{7}{12}$

$= \frac{9}{24} + \frac{-12}{24} + \frac{14}{24}$

$= \frac{9 + (-12) + 14}{24}$

$= \frac{11}{24}$

5. $x + y$

$\frac{3}{10} + \left(-\frac{7}{15}\right) = \frac{3}{10} + \frac{-7}{15}$

$= \frac{9}{30} + \frac{-14}{30}$

$= \frac{9 + (-14)}{30} = \frac{-5}{30} = -\frac{1}{6}$

6. $-\frac{1}{2} - \frac{3}{8} = \frac{-1}{2} - \frac{3}{8}$

$= \frac{-4}{8} - \frac{3}{8}$

$= \frac{-4 - 3}{8} = \frac{-7}{8} = -\frac{7}{8}$

7. $-\frac{5}{12} - \left(-\frac{2}{3}\right) = \frac{-5}{12} - \frac{-2}{3}$

$= \frac{-5}{12} - \frac{-8}{12}$

$= \frac{-5 - (-8)}{12}$

$= \frac{-5 + 8}{12} = \frac{3}{12} = \frac{1}{4}$

8. $-\frac{5}{9} - \left(-\frac{11}{12}\right) = \frac{-5}{9} - \frac{-11}{12}$

$= \frac{-20}{36} - \frac{-33}{36}$

$= \frac{-20 - (-33)}{36}$

$= \frac{-20 + 33}{36} = \frac{13}{36}$

9. $-\frac{5}{8} - \left(-\frac{7}{12}\right) = \frac{-5}{8} - \frac{-7}{12}$

$= \frac{-15}{24} - \frac{-14}{24}$

$= \frac{-15 - (-14)}{24}$

$= \frac{-15 + 14}{24} = \frac{-1}{24} = -\frac{1}{24}$

10. $x - y$

$$-\frac{5}{12} - \left(-\frac{5}{9}\right) = \frac{-5}{12} - \frac{-5}{9}$$
$$= \frac{-15}{36} - \frac{-20}{36}$$
$$= \frac{-15 - (-20)}{36}$$
$$= \frac{-15 + 20}{36} = \frac{5}{36}$$

Objective 2.4A

1. The digit 5 is in the ten-thousandths place.

2. 6.249 — Given place value: 2; next digit: $4 < 5$

   6.249 rounded to the nearest tenth is 6.2.

3. 18.40937 — Given place value: 0; next digit: $9 > 5$

   18.40937 rounded to the nearest hundredth is 18.41.

4. 936.2905 — Given place value: 0; next digit: $5 = 5$

   936.2905 rounded to the nearest thousandth is 936.291.

5. 47.3192 — Given place value: 7; next digit: $3 < 5$

   47.3192 rounded to the nearest whole number is 47.

6. 2.975268 — Given place value: 2; next digit: $8 > 5$

   2.975268 rounded to the nearest hundred-thousandth is 2.97527.

Objective 2.4B

1. $0.278 > 0.203$
2. $0.045 > 0.038$
3. $0.037 < 0.13$
4. $0.031 > 0.00987$
5. $0.02883 < 0.0305$

Objective 2.5A

1. $-42.1 - 8.6 = -42.1 + (-8.6)$
   $= -50.7$

2. $-9.37 + 3.465 = -5.905$

3. $382.9 + (-430.6) = -47.7$

4. $-6.82 - 4.793 = -6.82 + (-4.793)$

$= -11.613$

Objective 2.5B

1.
$$\begin{array}{r} 3.4 \\ \times 0.5 \\ \hline 1.70 \end{array}$$

2. $(-6.3)(-2.4) = 15.12$

3. $1.31(-0.006) = -0.00786$

4. $(2.7)(-16)(3.04) = (-43.2)(3.04)$

$= -131.328$

5.
$$\begin{array}{r} 1.95 \\ 3.6.\overline{)\;7.0.20} \\ -3\,6 \\ \hline 3\,4\,2 \\ -3\,2\,4 \\ \hline 1\,80 \\ -\,1\,80 \\ \hline 0 \end{array}$$

6. $(-3.312) \div (-0.8) = 4.14$

7. $84.66 \div (-1.7) = -49.8$

Objective 2.6A

1.
$$\begin{array}{r} 0.55 \\ 9\overline{)5.00} \\ -4\,5 \\ \hline 50 \\ -45 \\ \hline 5 \end{array}$$

$\frac{5}{9} = 0.55... = 0.\overline{5}$

2.
$$\begin{array}{r} 0.7272 \\ 11\overline{)8.0000} \\ -7\,7 \\ \hline 30 \\ -22 \\ \hline 80 \\ -77 \\ \hline 30 \\ -22 \\ \hline 8 \end{array}$$

$\frac{8}{11} = 0.7272... = 0.\overline{72}$

3.

$$
\begin{array}{r}
0.639639 \\
111\overline{)71.000000} \\
\underline{-66\,6} \\
4\,40 \\
\underline{-3\,33} \\
1\,070 \\
\underline{-999} \\
710 \\
\underline{-666} \\
440 \\
\underline{-333} \\
1070 \\
\underline{-999} \\
71
\end{array}
$$

$$\frac{71}{111} = 0.639639... = 0.\overline{639}$$

4.

$$
\begin{array}{r}
0.77272 \\
22\overline{)17.00000} \\
\underline{-15\,4} \\
1\,60 \\
\underline{-1\,54} \\
60 \\
\underline{-44} \\
160 \\
\underline{-154} \\
60 \\
\underline{-44} \\
16
\end{array}
$$

$$\frac{17}{22} = 0.77272... = 0.7\overline{72}$$

5.

$$
\begin{array}{r}
0.075 \\
40\overline{)3.000} \\
\underline{-2\,80} \\
200 \\
\underline{-200} \\
0
\end{array}
$$

$$\frac{3}{40} = 0.075$$

6.

$$
\begin{array}{r}
7.25 \\
4\overline{)29.00} \\
\underline{-28} \\
1\,0 \\
\underline{-8} \\
20 \\
\underline{-20} \\
0
\end{array}
$$

$$\frac{29}{4} = 7.25$$

Objective 2.6B

1. $0.4 = \frac{4}{10} = \frac{2}{5}$

2. $0.48 = \frac{48}{100} = \frac{12}{25}$

3. $0.485 = \frac{485}{1000} = \frac{97}{200}$

4. $3.75 = 3\frac{75}{100} = 3\frac{3}{4}$

5. $0.052 = \frac{52}{1000} = \frac{\overset{1}{\cancel{2}} \cdot \overset{1}{\cancel{2}} \cdot 13}{\underset{1}{\cancel{2}} \cdot \underset{1}{\cancel{2}} \cdot 250} = \frac{13}{250}$

6. $0.00015 = \frac{15}{100{,}000} = \frac{\overset{1}{\cancel{5}} \cdot 3}{\underset{1}{\cancel{5}} \cdot 20{,}000} = \frac{3}{20{,}000}$

Objective 2.6C

1. $0.8 \quad \frac{3}{4}$

$\frac{8}{10} \quad \frac{3}{4}$

$\frac{16}{20} \quad \frac{15}{20}$

$\frac{16}{20} > \frac{15}{20}$

$0.8 > \frac{3}{4}$

2. $\frac{7}{8} \quad 0.9$

$\frac{7}{8} \quad \frac{9}{10}$

$\frac{35}{40} \quad \frac{36}{40}$

$\frac{35}{40} < \frac{36}{40}$

$\frac{7}{8} < 0.9$

3. $0.13 \quad \frac{5}{40}$

$\frac{13}{100} \quad \frac{5}{40}$

$\frac{26}{200} \quad \frac{25}{200}$

$\frac{26}{200} > \frac{25}{200}$

$0.13 > \frac{5}{40}$

4. $\frac{12}{55} \quad 0.22$

$\frac{12}{55} \quad \frac{22}{100}$

$\frac{240}{1100} \quad \frac{242}{1100}$

$\frac{240}{1100} < \frac{242}{1100}$

$\frac{12}{55} < 0.22$

5. $0.55 \quad \frac{5}{9}$

$\frac{55}{100} \quad \frac{5}{9}$

$\frac{495}{900} \quad \frac{500}{900}$

$\frac{495}{900} < \frac{500}{900}$

$0.55 < \frac{5}{9}$

6. $\frac{22}{7} \quad 3.14$

$\frac{22}{7} \quad \frac{314}{100}$

$\frac{2200}{700} \quad \frac{2198}{700}$

$\frac{2200}{700} > \frac{2198}{700}$

$\frac{22}{7} > 3.14$

Objective 2.7A

1. Since $1^2 = 1, \ -\sqrt{1} = -1$.
2. Since $9^2 = 81, \sqrt{81} = 9$.
3. $\sqrt{144} - \sqrt{25} = 12 - 5 = 7$
4. $\sqrt{144} + 3\sqrt{9} = 12 + 3 \cdot 3$
   $= 12 + 9 = 21$
5. $7\sqrt{x + y}$
   $7\sqrt{34 + 15} = 7\sqrt{49}$
   $= 7 \cdot 7 = 49$
6. $7 - \sqrt{16} = 7 - 4 = 3$

Objective 2.7B

1. $\sqrt{10} \approx 3.1623$
2. $6\sqrt{15} \approx 23.2379$
3. $10\sqrt{21} \approx 45.8258$

4. $-12\sqrt{53} \approx -87.3613$

5. 29 is between the perfect squares 25 and 36.

$\sqrt{25} = 5$ and $\sqrt{36} = 6$

$5 < \sqrt{29} < 6$

6. 130 is between the perfect squares 121 and 144.

$\sqrt{121} = 11$ and $\sqrt{144} = 12$

$11 < \sqrt{130} < 12$

Objective 2.7C

1. STRATEGY

To find the time for the object to fall, substitute 144 for d in the given equation and solve for t.

SOLUTION

$t = \sqrt{\dfrac{d}{16}}$

$t = \sqrt{\dfrac{144}{16}}$

$t = \sqrt{9} = 3$

It will take 3 s for the object to fall 144 ft.

2. STRATEGY

To find the time for the object to fall, substitute 64 for d in the given equation and solve for t.

SOLUTION

$t = \sqrt{\dfrac{d}{16}}$

$t = \sqrt{\dfrac{64}{16}}$

$t = \sqrt{4} = 2$

It will take 2 s for the object to fall 64 ft.

MODULE

3 Applications of Equations and Inequalities

SECTION 3.1 **Evaluating and Simplifying Variable Expressions**
- Objective **3.1A** Evaluate a variable expression
- Objective **3.1B** Simplify a variable expression using the Properties of Real Numbers

SECTION 3.2 **Introduction to Equations**
- Objective **3.2A** Solve an equation of the form $x + a = b$ or $ax = b$
- Objective **3.2B** Solve proportions
- Objective **3.2C** Solve the basic percent equation
- Objective **3.2D** Solve applications of the basic percent equation

SECTION 3.3 **General First-Degree Equations**
- Objective **3.3A** Solve an equation of the form $ax + b = c$
- Objective **3.3B** Solve an equation of the form $ax + b = cx + d$
- Objective **3.3C** Solve an equation containing parentheses

SECTION 3.4 **Literal, Absolute Value, and Radical Equations**
- Objective **3.4A** Solve a literal equation
- Objective **3.4B** Solve an absolute value equation
- Objective **3.4C** Solve a radical equation

SECTION 3.5 **Solving First-Degree Inequalities and Absolute Value Inequalities**
- Objective **3.5A** Graph an inequality on a number line
- Objective **3.5B** Solve a first-degree inequality
- Objective **3.5C** Solve an absolute value inequality

SECTION

3.1 Evaluating and Simplifying Variable Expressions

Objective 3.1A Evaluate a variable expression

Often we discuss a quantity without knowing its exact value—for example, the price of gold next month, the cost of a new automobile next year, or the tuition cost for next semester. Recall that a letter of the alphabet, called a **variable,** is used to stand for a quantity that is unknown or that can change, or *vary.* An expression that contains one or more variables is called a **variable expression.**

A variable expression is shown at the right. The expression can be rewritten by writing subtraction as the addition of the opposite.

$$3x^2 - 5y + 2xy - x - 7$$
$$3x^2 + (-5y) + 2xy + (-x) + (-7)$$

Note that the expression has five addends. The **terms** of a variable expression are the addends of the expression. The expression has five terms.

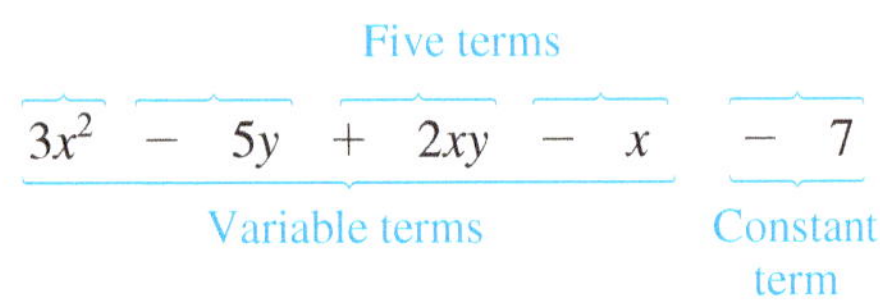

The terms $3x^2$, $-5y$, $2xy$, and $-x$ are **variable terms.**

The term -7 is a **constant term,** or simply a **constant.**

Each variable term is composed of a **numerical coefficient** and a **variable part** (the variable or variables and their exponents).

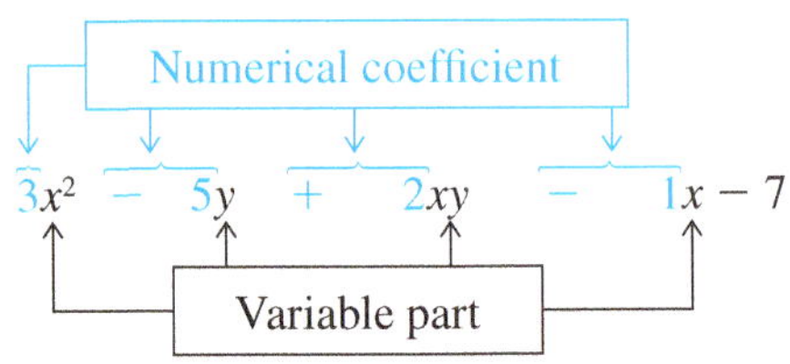

When the numerical coefficient is 1 or -1, the 1 is usually not written ($x = 1x$ and $-x = -1x$).

Variable expressions can be used to model scientific phenomena. In a physics lab, a student may discover that a weight of 1 pound will stretch a spring $\frac{1}{2}$ inch. Two pounds will stretch the spring 1 inch. By experimenting, the student can discover that the distance the spring will stretch is found by multiplying the weight by $\frac{1}{2}$. By letting W represent the weight attached to the spring, the student can represent the distance the spring stretches by the variable expression $\frac{1}{2}W$.

With a weight of W pounds attached, the spring will stretch $\frac{1}{2} \cdot W = \frac{1}{2}W$ inches.

With a weight of 10 pounds attached, the spring will stretch $\frac{1}{2} \cdot 10 = 5$ inches. The number 10 is called the **value of the variable** W.

With a weight of 3 pounds attached, the spring will stretch $\frac{1}{2} \cdot 3 = 1\frac{1}{2}$ inches.

Replacing each variable by its value and then simplifying the resulting numerical expression is called **evaluating a variable expression.**

Focus on naming the terms of an expression

Name the variable terms of the expression $2a^2 - 5a + 7$.

SOLUTION

$2a^2$ and $-5a$

Check your understanding 1

Name the constant term of the expression $6n^2 + 3n - 4$.

SOLUTION See page S-1.

Focus on evaluating a variable expression

a. Evaluate $x^2 - 3xy$ when $x = 3$ and $y = -4$.

b. Evaluate $\frac{a^2 - b^2}{a - b}$ when $a = 3$ and $b = -4$.

c. Evaluate $x^2 - 3(x - y) - z^2$ when $x = 2$, $y = -1$, and $z = 3$.

SOLUTION

a. $x^2 - 3xy$

$$\begin{aligned} 3^2 - 3(3)(-4) &= 9 - 3(3)(-4) \quad \text{• } x = 3, y = -4 \\ &= 9 - 9(-4) \\ &= 9 - (-36) \\ &= 9 + 36 = 45 \end{aligned}$$

b. $\frac{a^2 - b^2}{a - b}$

$$\begin{aligned} \frac{3^2 - (-4)^2}{3 - (-4)} &= \frac{9 - 16}{3 - (-4)} \quad \text{• } a = 3, b = -4 \\ &= \frac{-7}{7} = -1 \end{aligned}$$

c. $x^2 - 3(x - y) - z^2$

$$\begin{aligned} &2^2 - 3[2 - (-1)] - 3^2 \quad \text{• } x = 2, y = -1, z = 3 \\ &= 2^2 - 3(3) - 3^2 \\ &= 4 - 3(3) - 9 \\ &= 4 - 9 - 9 \\ &= -5 - 9 = -14 \end{aligned}$$

Check your understanding 2

a. Evaluate $2xy + y^2$ when $x = -4$ and $y = 2$.

b. Evaluate $\frac{a^2 + b^2}{a + b}$ when $a = 5$ and $b = -3$.

c. Evaluate $x^3 - 2(x + y) + z^2$ when $x = 2$, $y = -4$, and $z = -3$.

SOLUTION See page S-1.

Objective 3.1A Practice

1. Evaluate $3a + 2b$ for $a = 2$ and $b = 3$.
2. Evaluate $bc \div (2a)$ for $a = 2$, $b = 3$, and $c = -4$.
3. Evaluate $(b - 2a)^2 + bc$ for $a = 2$, $b = 3$, and $c = -4$.
4. Evaluate $(d - a)^2 - 3c$ for $a = -2$, $c = -1$, and $d = 3$.
5. Evaluate $-\frac{3}{4}b + \frac{1}{2}(ac + bd)$ for $a = -2$, $b = 4$, $c = -1$, and $d = 3$.

Solutions on p. S-11.

Objective 3.1B Simplify a variable expression using the Properties of Real Numbers

Like terms of a variable expression are terms with the same variable part. (Because $x^2 = x \cdot x$, x^2 and x are not like terms.)

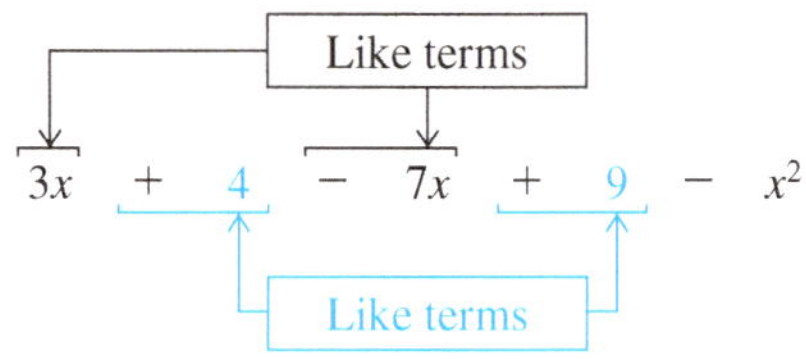

Constant terms are like terms. 4 and 9 are like terms.

The Distributive Property is used to add the numerical coefficients of like variable terms. The variable part remains unchanged. This is called **combining like terms.**

Take Note ➤
Here is an example of the Distributive Property using just numbers.

$2(5 + 9) = 2(5) + 2(9)$
$= 10 + 18 = 28$

This is the same result we would obtain using the Order of Operations Agreement.

$2(5 + 9) = 2(14) = 28$

The usefulness of the Distributive Property will become more apparent as we explore variable expressions.

The Distributive Property

If a, b, and c are real numbers, then $a(b + c) = ab + ac$ or $(b + c)a = ba + ca$.

By the Distributive Property, the term outside the parentheses is multiplied by each term inside the parentheses.

EXAMPLES

1. $2(3 + 4) = 2 \cdot 3 + 2 \cdot 4$
   $2(7) = 6 + 8$
   $14 = 14$

2. $(4 + 5)2 = 4 \cdot 2 + 5 \cdot 2$
   $(9)2 = 8 + 10$
   $18 = 18$

The Distributive Property in the form $(b + c)a = ba + ca$ is used to simplify a variable expression.

To simplify $2x + 3x$, use the Distributive Property to add the numerical coefficients of the like variable terms.

$$2x + 3x = (2 + 3)x$$
$$= 5x$$

Simplify: $5y - 11y$

$5y - 11y = (5 - 11)y$ • Use the Distributive Property.
$= -6y$

Take Note ➤
Simplifying an expression means combining like terms. The constant term 5 and the variable term $7p$ are not like terms and therefore cannot be combined.

Simplify: $5 + 7p$

The terms 5 and $7p$ are not like terms.

The expression $5 + 7p$ is in simplest form.

The following Properties of Addition and Multiplication are used to simplify variable expressions.

The Associative Property of Addition

If a, b, and c are real numbers, then $(a + b) + c = a + (b + c)$.

When three or more terms are added, the terms can be grouped (with parentheses, for example) in any order; the sum is the same.

EXAMPLES

1. $(5 + 7) + 15 = 5 + (7 + 15)$
$12 + 15 = 5 + 22$
$27 = 27$

2. $(3x + 5x) + 9x = 3x + (5x + 9x)$
$8x + 9x = 3x + 14x$
$17x = 17x$

The Commutative Property of Addition

If a and b are real numbers, then $a + b = b + a$.

When two terms are added, the terms can be added in either order; the sum is the same.

EXAMPLES

1. $15 + (-28) = (-28) + 15$
$-13 = -13$

2. $2x + (-4x) = -4x + 2x$
$-2x = -2x$

The Addition Property of Zero

If a is a real number, then $a + 0 = a$ and $0 + a = a$.

The sum of a term and zero is the term.

EXAMPLES

1. $-9 + 0 = -9$ and $0 + (-9) = -9$

2. $0 + 5x = 5x$ and $5x + 0 = 5x$

The Inverse Property of Addition

If a is a real number, then $a + (-a) = 0$ and $(-a) + a = 0$.

The sum of a term and its additive inverse (or opposite) is zero.

EXAMPLES

1. $8 + (-8) = 0$ and $-8 + 8 = 0$ **2.** $-7x + 7x = 0$ and $7x + (-7x) = 0$

Take Note ➤
The Associative Property of Multiplication allows us to multiply a coefficient by a number. Without this property, the expression $2(3x)$ could not be simplified.

The Associative Property of Multiplication

If a, b, and c are real numbers, then $(ab)c = a(bc)$.

When three or more factors are multiplied, the factors can be grouped in any order; the product is the same.

EXAMPLES

1.
$$\begin{aligned} 3(5 \cdot 6) &= (3 \cdot 5)6 \\ 3(30) &= (15)6 \\ 90 &= 90 \end{aligned}$$

2.
$$\begin{aligned} 2(3x) &= (2 \cdot 3)x \\ &= 6x \end{aligned}$$

Take Note ➤
The Commutative Property of Multiplication allows us to rearrange factors. This property, along with the Associative Property of Multiplication, enables us to simplify some variable expressions.

The Commutative Property of Multiplication

If a and b are real numbers, then $ab = ba$.

Two factors can be multiplied in either order; the product is the same.

EXAMPLES

1.
$$\begin{aligned} 5(-7) &= -7(5) \\ -35 &= -35 \end{aligned}$$

2.
$$\begin{aligned} (5x) \cdot 3 &= 3 \cdot (5x) && \text{• Commutative Property of Multiplication} \\ &= (3 \cdot 5)x && \text{• Associative Property of Multiplication} \\ &= 15x \end{aligned}$$

The Multiplication Property of One

If a is a real number, then $a \cdot 1 = a$ and $1 \cdot a = a$.

The product of a term and 1 is the term.

EXAMPLES

1. $9 \cdot 1 = 9$ **2.** $(8x) \cdot 1 = 8x$

The Inverse Property of Multiplication

If a is a real number and a is not equal to zero, then $a \cdot \frac{1}{a} = 1$ and $\frac{1}{a} \cdot a = 1$.

$\frac{1}{a}$ is called the **reciprocal** of a. $\frac{1}{a}$ is also called the **multiplicative inverse** of a.

The product of a number and its reciprocal is 1.

EXAMPLES

1. $7 \cdot \frac{1}{7} = 1$ and $\frac{1}{7} \cdot 7 = 1$ **2.** $x \cdot \frac{1}{x} = 1$ and $\frac{1}{x} \cdot x = 1, \quad x \neq 0$

Take Note ➤
We must state that $x \neq 0$ because division by zero is undefined.

The Distributive Property is also used to remove parentheses from a variable expression.

Simplify: $3(2x + 7)$

$3(2x + 7) = 3(2x) + 3(7)$ • Use the Distributive Property. Multiply each term inside the parentheses by 3.
$= 6x + 21$

Simplify: $-5(4x + 6)$

$-5(4x + 6) = -5(4x) + (-5)(6)$ • Use the Distributive Property.
$= -20x - 30$

Simplify: $-(2x - 4)$

$-(2x - 4) = -1(2x - 4)$ • Use the Distributive Property.
$= -1(2x) - (-1)(4)$
$= -2x + 4$

An extension of the Distributive Property is used when an expression contains more than two terms.

Simplify: $3(4x - 2y - z)$

$3(4x - 2y - z) = 3(4x) - 3(2y) - 3(z)$ • Use the Distributive Property.
$= 12x - 6y - 3z$

Focus on using the Distributive Property to simplify a variable expression

a. Simplify: $-3(-5a + 7b)$
b. Simplify: $3(x^2 - x - 5)$
c. Simplify: $-2(x^2 + 5x - 4)$

SOLUTION

a. Use the Distributive Property.

$-3(-5a + 7b) = 15a - 21b$

b. Use the Distributive Property.

$3(x^2 - x - 5) = 3x^2 - 3x - 15$

c. Use the Distributive Property.

$-2(x^2 + 5x - 4) = -2x^2 - 10x + 8$

Check your understanding 3

a. Simplify: $-8(-2a + 7b)$
b. Simplify: $3(12x^2 - x + 8)$
c. Simplify: $3(-a^2 - 6a + 7)$

SOLUTION See page S-1.

When simplifying more complex variable expressions, use the Distributive Property to remove parentheses and brackets used as grouping symbols.

Simplify: $4(x - y) - 2(-3x + 6y)$

$4(x - y) - 2(-3x + 6y) = 4x - 4y + 6x - 12y$ • Use the Distributive Property.
$= 10x - 16y$ • Combine like terms.

Focus on simplifying a general variable expression

a. Simplify: $2x - 3(2x - 7y)$
b. Simplify: $7(x - 2y) - (-x - 2y)$
c. Simplify: $2x - 3[2x - 3(x + 7)]$

SOLUTION

a. $2x - 3(2x - 7y) = 2x - 6x + 21y$ • Use the Distributive Property.
$= -4x + 21y$ • Combine like terms.

b. $7(x - 2y) - (-x - 2y) = 7x - 14y + x + 2y$ • Use the Distributive Property.
$= 8x - 12y$ • Combine like terms.

c. $2x - 3[2x - 3(x + 7)] = 2x - 3[2x - 3x - 21]$ • Use the Distributive Property.
$= 2x - 3[-x - 21]$ • Combine like terms.
$= 2x + 3x + 63$ • Use the Distributive Property.
$= 5x + 63$ • Combine like terms.

Check your understanding 4

a. Simplify: $3y - 2(y - 7x)$
b. Simplify: $-2(x - 2y) - (-x + 3y)$
c. Simplify: $3y - 2[x - 4(2 - 3y)]$

SOLUTION See page S-1.

Objective 3.1B Practice

1. Simplify: $4(x^2 - 3x + 5)$
2. Simplify: $\frac{3}{4}(2x - 6y + 8)$
3. Simplify: $-(8b^2 - 6b + 9)$
4. Simplify: $4x - 2(3x + 8)$
5. Simplify: $5n - (7 - 2n)$
6. Simplify: $12(y - 2) + 3(7 - 3y)$
7. Simplify: $-2[3x + 2(4 - x)]$

Solutions on p. S-11.

SECTION 3.2 Introduction to Equations

Objective 3.2A Solve an equation of the form $x + a = b$ or $ax = b$

An **equation** expresses the equality of two mathematical expressions. The expressions can be either numerical or variable expressions.

$$\left.\begin{aligned} &9 + 3 = 12\\ &3x - 2 = 10\\ &y^2 + 4 = 2y - 1\\ &z = 2 \end{aligned}\right\} \text{Equations}$$

The equation at the right is true if the variable is replaced by 5.

$x + 8 = 13$

$5 + 8 = 13$ A true equation

The equation is false if the variable is replaced by 7. $7 + 8 = 13$ A false equation

A **solution of an equation** is a number that, when substituted for the variable, results in a true equation. 5 is a solution of the equation $x + 8 = 13$. 7 is not a solution of the equation $x + 8 = 13$.

Focus on determining whether a number is a solution of an equation

Is -2 a solution of $2x + 5 = x^2 - 3$?

SOLUTION

$2x + 5$	$x^2 - 3$
$2(-2) + 5$	$(-2)^2 - 3$
$-4 + 5$	$4 - 3$
1	1

- Replace x by -2.
- Evaluate the numerical expressions.
- If the results are equal, -2 is a solution of the equation. If the results are not equal, -2 is not a solution of the equation.

$1 = 1$

Yes, -2 is a solution of the equation.

Take Note The Order of Operations Agreement applies when evaluating $2(-2) + 5$ and $(-2)^2 - 3$.

Check your understanding 1

Is 5 a solution of $10x - x^2 = 3x - 10$?

SOLUTION See page S-1.

To **solve an equation** means to find a solution of the equation. The simplest equation to solve is an equation of the form *variable* = *constant,* because the constant is the solution.

The solution of the equation $x = 5$ is 5 because $5 = 5$ is a true equation.

Equations that have the same solution are called **equivalent equations.** The equations $x + 2 = 9$, $x + 6 = 13$, and $x - 3 = 4$ are equivalent equations; each equation has 7 as its solution.

Tips for Success ➤
To learn mathematics, you must be an active participant. Listening and watching your professor do mathematics are not enough. Take notes in class, mentally think through every question your instructor asks, and try to answer it even if you are not called on to do so. Ask questions when you have them. See *AIM for Success*.

Addition Property of Equations

The same number can be added to each side of an equation without changing its solution. In symbols, the equation $a = b$ has the same solution as the equation $a + c = b + c$.

EXAMPLE

The equation $x - 3 = 7$ has the same solution as the equation $x - 3 + 3 = 7 + 3$.

In solving an equation, the goal is to rewrite the given equation in the form *variable* = *constant*. The Addition Property of Equations is used to remove a *term* from one side of the equation by adding the opposite of that term to each side of the equation.

Take Note ➤
An equation has some properties that are similar to those of a balance scale. For instance, if a balance scale is in balance and equal weights are added to each side of the scale, then the balance scale remains in balance. If an equation is true, then adding the same number to each side of the equation produces another true equation.

Focus on solving an equation

Solve: $x - 4 = 2$

SOLUTION

$x - 4 = 2$ • The goal is to rewrite the equation in the form *variable* = *constant*.

$x - 4 + 4 = 2 + 4$ • Add 4 to each side of the equation.

$x + 0 = 6$ • Simplify.

$x = 6$ • The equation is in the form *variable* = *constant*.

Check: $x - 4 = 2$

$6 - 4 \mid 2$

$2 = 2$ • A true equation

The solution is 6.

Check your understanding 2

Solve: $26 = y - 14$

SOLUTION See page S-2.

Because subtraction is defined in terms of addition, the Addition Property of Equations also makes it possible to subtract the same number from each side of an equation without changing the solution of the equation.

Focus on solving an equation with fractions

Solve: $y + \frac{3}{4} = \frac{1}{2}$

SOLUTION

$y + \frac{3}{4} = \frac{1}{2}$ • The goal is to rewrite the equation in the form *variable* = *constant*.

$y + \frac{3}{4} - \frac{3}{4} = \frac{1}{2} - \frac{3}{4}$ • Subtract $\frac{3}{4}$ from each side of the equation.

$y + 0 = \frac{2}{4} - \frac{3}{4}$ • Simplify.

$y = -\frac{1}{4}$ • The equation is in the form *variable* = *constant*.

The solution is $-\frac{1}{4}$. You should check this solution.

Check your understanding 3

Solve: $c + \frac{3}{4} = -\frac{1}{4}$

SOLUTION See page S-2.

The equations $2x = 6$, $10x = 30$, and $-8x = -24$ are equivalent equations; each equation has 3 as its solution. These examples suggest that multiplying each side of an equation by the same nonzero number produces an equivalent equation.

Multiplication Property of Equations

Each side of an equation can be multiplied by the same nonzero number without changing the solution of the equation. In symbols, if $c \neq 0$, then the equation $a = b$ has the same solutions as the equation $ac = bc$.

EXAMPLE

The equation $3x = 21$ has the same solution as the equation $\frac{1}{3} \cdot 3x = \frac{1}{3} \cdot 21$.

The Multiplication Property of Equations is used to remove a coefficient by multiplying each side of the equation by the reciprocal of the coefficient.

Solve: $\frac{3}{4}z = 9$

$\frac{3}{4}z = 9$ • The goal is to rewrite the equation in the form *variable = constant*.

$\frac{4}{3} \cdot \frac{3}{4}z = \frac{4}{3} \cdot 9$ • Multiply each side of the equation by $\frac{4}{3}$.

$1 \cdot z = 12$ • Simplify.

$z = 12$ • The equation is in the form *variable = constant*.

The solution is 12. You should check this solution.

Because division is defined in terms of multiplication, each side of an equation can be divided by the same nonzero number without changing the solution of the equation.

Take Note ➤
Remember to check the solution.

Check:

$$\begin{array}{c|c} 6x & 14 \\ \hline 6\left(\frac{7}{3}\right) & 14 \\ 14 & 14 \end{array}$$

Solve: $6x = 14$

$6x = 14$ • The goal is to rewrite the equation in the form *variable = constant*.

$\frac{6x}{6} = \frac{14}{6}$ • Divide each side of the equation by 6.

$x = \frac{7}{3}$ • Simplify. The equation is in the form *variable = constant*.

The solution is $\frac{7}{3}$.

When using the Multiplication Property of Equations, multiply each side of the equation by the reciprocal of the coefficient when the coefficient is a fraction. Divide each side of the equation by the coefficient when the coefficient is an integer or a decimal.

Focus on using the Multiplication Property of Equations to solve an equation

a. Solve: $-9 = \frac{3x}{4}$

b. Solve: $5x - 9x = 12$

SOLUTION

a. $-9 = \frac{3x}{4}$ • $\frac{3x}{4} = \frac{3}{4}x$

$\frac{4}{3}(-9) = \frac{4}{3} \cdot \frac{3}{4}x$ • Multiply each side by $\frac{4}{3}$.

$-12 = x$

The solution is −12.

b. $5x - 9x = 12$

$-4x = 12$ • Combine like terms.

$\frac{-4x}{-4} = \frac{12}{-4}$ • Divide each side by −4.

$x = -3$

The solution is −3.

Check your understanding 4

Solve: $4x - 8x = 16$

SOLUTION See page S-2.

Objective 3.2A Practice

1. Is 4 a solution of $3y - 4 = 2y$?
2. Solve and check: $-9 = 5 + x$
3. Solve and check: $x - \frac{2}{5} = \frac{3}{5}$
4. Solve and check: $\frac{x}{4} = 3$
5. Solve and check: $\frac{2}{5}x = 6$
6. Solve and check: $7d - 4d = 9$

Solutions on p. S-12.

Objective 3.2B Solve proportions

In previous work, we have used quantities with units, such as 12 ft, 3 h, 2¢, and 15 acres. In these examples, the units are feet, hours, cents, and acres.

A **ratio** is the quotient or comparison of two quantities with the *same* unit. We can compare the measure of 3 ft to the measure of 8 ft by writing a quotient.

$$\frac{3\text{ ft}}{8\text{ ft}} = \frac{3}{8} \qquad 3\text{ ft is } \frac{3}{8} \text{ of } 8\text{ ft.}$$

A ratio can be written in three ways:

1. As a fraction $\frac{3}{8}$
2. As two numbers separated by a colon 3 : 8
3. As two numbers separated by the word *to* 3 to 8

The ratio of 15 mi to 45 mi is written as

$$\frac{15 \text{ mi}}{45 \text{ mi}} = \frac{15}{45} = \frac{1}{3} \text{ or } 1:3 \text{ or } 1 \text{ to } 3$$

A ratio is in **simplest form** when the two numbers do not have a common factor. The units are not written in a ratio.

Focus on writing a ratio

a. Write the comparison \$6 to \$8 as a ratio in simplest form using a fraction, a colon, and the word *to*.

b. Write the comparison 18 qt to 6 qt as a ratio in simplest form using a fraction, a colon, and the word *to*.

SOLUTION

a. $\frac{\$6}{\$8} = \frac{6}{8} = \frac{3}{4}$

$\$6 : \$8 = 6 : 8 = 3 : 4$

$\$6 \text{ to } \$8 = 6 \text{ to } 8 = 3 \text{ to } 4$

b. $\frac{18 \text{ qt}}{6 \text{ qt}} = \frac{18}{6} = \frac{3}{1}$

$18 \text{ qt} : 6 \text{ qt} = 18 : 6 = 3 : 1$

$18 \text{ qt to } 6 \text{ qt} = 18 \text{ to } 6$

$= 3 \text{ to } 1$

Check your understanding 5

a. Write the comparison 20 lb to 24 lb as a ratio in simplest form using a fraction, a colon, and the word *to*.

b. Write the comparison 64 mi to 8 mi as a ratio in simplest form using a fraction, a colon, and the word *to*.

SOLUTION See page S-2.

A **rate** is the comparison of two quantities with *different* units.

A catering company prepares 9 gal of coffee for every 50 people at a reception. This rate is written

$$\frac{9 \text{ gal}}{50 \text{ people}}$$

You traveled 200 mi in 6 h. The rate is written

$$\frac{200 \text{ mi}}{6 \text{ h}} = \frac{100 \text{ mi}}{3 \text{ h}}$$

A rate is in **simplest form** when the numbers have no common factors. The units are written as part of the rate.

Take Note ➤
Unit rates make comparisons easier. For example, if you travel 37 mph and I travel 43 mph, we know that I am traveling faster than you are. It is more difficult to compare speeds if we are told that you are traveling $\frac{111 \text{ mi}}{3 \text{ h}}$ and I am traveling $\frac{172 \text{ mi}}{4 \text{ h}}$.

Many rates are written as unit rates. A **unit rate** is a rate in which the number in the denominator is 1. The word *per* generally indicates a unit rate. It means "for each" or "for every." For example,

23 mi per gallon • The unit rate is $\frac{23 \text{ mi}}{1 \text{ gal}}$.

65 mi per hour • The unit rate is $\frac{65 \text{ mi}}{1 \text{ h}}$.

\$4.78 per pound • The unit rate is $\frac{\$4.78}{1 \text{ lb}}$.

Focus on writing a unit rate

a. Write "285 mi in 5 h" as a unit rate.
b. Write "300 ft in 8 s" as a unit rate.

SOLUTION

a. $\frac{285 \text{ mi}}{5 \text{ h}}$ • Write the rate as a fraction.

$285 \div 5 = 57$ • Divide the numerator by the denominator.

The unit rate is 57 mph.

b. $\frac{300 \text{ ft}}{8 \text{ s}}$ $\qquad 8\overline{)300.0} = 37.5$

The unit rate is 37.5 ft/s.

Check your understanding 6

a. Write "$8.96 for 3.5 lb" as a unit rate.
b. Write "260 mi in 8 h" as a unit rate.

SOLUTION See page S-2.

Focus on solving an application

A motorcycle racer completed a 6-mile lap in 0.05 h (3 min). Find the racer's speed in miles per hour.

STRATEGY

To find the speed in miles per hour, divide the length of the lap (6 mi) by the time (0.05 h).

SOLUTION

$$\frac{6 \text{ mi}}{0.05 \text{ h}} = 120 \text{ mi/h}$$

The racer's speed was 120 mi/h.

Check your understanding 7

A landscape technician purchased 40 ft of $\frac{3}{4}$-inch PVC pipe for $22.40. What is the per-foot cost of the pipe?

SOLUTION See page S-3.

A **proportion** states the equality of two ratios or rates.

Definition of Proportion

If $\frac{a}{b}$ and $\frac{c}{d}$ are equal ratios or rates, then $\frac{a}{b} = \frac{c}{d}$ is a proportion.

EXAMPLES OF PROPORTIONS

1. $\frac{3}{6} = \frac{1}{2}$ **2.** $\frac{\$30}{4 \text{ lb}} = \frac{\$90}{12 \text{ lb}}$ **3.** $\frac{250 \text{ mi}}{5 \text{ h}} = \frac{50 \text{ mi}}{1 \text{ h}}$

Each of the four numbers in a proportion is called a **term.** Each term is numbered according to the following diagram.

$$\begin{matrix} \text{first term} \longrightarrow \\ \text{second term} \longrightarrow \end{matrix} \frac{a}{b} = \frac{c}{d} \begin{matrix} \longleftarrow \text{third term} \\ \longleftarrow \text{fourth term} \end{matrix}$$

The first and fourth terms of the proportion are called the **extremes,** and the second and third terms are called the **means.**

Tips for Success ➤
You are learning a new skill in this objective. Be sure to do all you need to do in order to be successful at solving proportions: Read through the introductory material, work through the "Focus on" examples, and do the "Check your understanding" problems. Check your solutions against the solutions given at the end of the module.

If we multiply the proportion by the least common multiple of the denominators, we obtain the following result:

$$\frac{a}{b} = \frac{c}{d}$$

$$bd\left(\frac{a}{b}\right) = bd\left(\frac{c}{d}\right)$$

$$ad = bc$$

ad is the product of the extremes.
bc is the product of the means.

In any true proportion, the product of the means equals the product of the extremes. This is sometimes phrased as "the cross products are equal."

In the true proportion $\frac{3}{4} = \frac{9}{12}$, the cross products are equal.

$$\frac{3}{4} \times \frac{9}{12} \quad \begin{matrix} \longrightarrow 4 \cdot 9 = 36 \longleftarrow \text{Product of the means} \\ \longrightarrow 3 \cdot 12 = 36 \longleftarrow \text{Product of the extremes} \end{matrix}$$

Determine whether the proportion $\frac{47 \text{ mi}}{2 \text{ gal}} = \frac{304 \text{ mi}}{13 \text{ gal}}$ is a true proportion.

The product of the means:

$2 \cdot 304 = 608$

The product of the extremes:

$47 \cdot 13 = 611$

The proportion is not true because $608 \neq 611$.

When three terms of a proportion are given, the fourth term can be found. To solve a proportion for an unknown term, use the fact that the product of the means equals the product of the extremes.

Focus on solving a proportion

a. Solve: $\frac{n}{5} = \frac{9}{16}$

b. Solve $\frac{6}{n} = \frac{45}{124}$. Round to the nearest tenth.

c. Solve $\frac{x+2}{3} = \frac{7}{8}$

SOLUTION

a.

$\frac{n}{5} = \frac{9}{16}$ • Find the number n that will make the proportion true.

$5 \cdot 9 = n \cdot 16$ • The product of the means equals the product of the extremes.

$45 = 16n$ • Solve for n.

$\frac{45}{16} = \frac{16n}{16}$

$2.8125 = n$

b. $\dfrac{6}{n} = \dfrac{45}{124}$

$n \cdot 45 = 6 \cdot 124$ • The product of the means equals the product of the extremes.

$45n = 744$ • Solve for n.

$\dfrac{45n}{45} = \dfrac{744}{45}$

$n \approx 16.5$

c. $\dfrac{x + 2}{3} = \dfrac{7}{8}$

$3 \cdot 7 = (x + 2)8$ • The product of the means equals the product of the extremes.

$21 = 8x + 16$ • Solve for x.

$5 = 8x$

$0.625 = x$

Check your understanding 8

a. Solve: $\dfrac{7}{12} = \dfrac{42}{x}$

b. Solve: $\dfrac{4}{5} = \dfrac{3}{x - 3}$

SOLUTION See page S-3.

Proportions are useful in many types of application problems. In cooking, proportions are used when a larger quantity of food is needed than the original recipe yields. In mixing cement, the amounts of cement, sand, and rock are mixed in the same ratio. A map is drawn on a proportional basis, such as 1 in. representing 50 mi.

In setting up a proportion, keep the same units in the numerators and the same units in the denominators. For example, if *feet* is in the numerator on one side of the proportion, then *feet* must be in the numerator on the other side of the proportion.

Take Note ➤
It is also correct to write the proportion with the costs in the numerators and the numbers of tires in the denominators:

$$\frac{\$266.50}{2 \text{ tires}} = \frac{c}{5 \text{ tires}}$$

The solution will be the same.

A customer sees an ad in a newspaper advertising 2 tires for \$266.50. The customer wants to buy 5 tires and use one as a spare. How much will the 5 tires cost?

$\dfrac{2 \text{ tires}}{\$266.50} = \dfrac{5 \text{ tires}}{c}$ • Let c = the cost of the 5 tires. Write a proportion.

$266.50 \cdot 5 = 2 \cdot c$

$1332.50 = 2c$

$\dfrac{1332.50}{2} = \dfrac{2c}{2}$

$666.25 = c$

The 5 tires will cost \$666.25.

Focus on solving an application using proportions

The dosage of a certain medication is 2 oz for every 50 lb of body weight. How many ounces of this medication are required for a person who weighs 175 lb?

STRATEGY

To find the number of ounces of medication required for a person weighing 175 lb, write and solve a proportion using n to represent the number of ounces of medication required.

SOLUTION

$$\frac{2 \text{ oz}}{50 \text{ lb}} = \frac{n \text{ oz}}{175 \text{ lb}}$$
• Write "oz" in the numerator and "lb" in the denominator.

$$50n = 2 \cdot 175$$
• The product of the means equals the product of the extremes.

$$50n = 350$$
$$\frac{50n}{50} = \frac{350}{50}$$
$$n = 7$$

A 175-pound person requires 7 oz of medication.

Check your understanding 9

An automobile can travel 396 mi on 11 gal of gas. At the same rate, how many gallons of gas would be necessary to travel 832 mi? Round to the nearest tenth.

SOLUTION See page S-3.

Objective 3.2B Practice

1. Write 10 ft in 4 s as a unit rate.
2. Write \$11.05 for 3.4 lb as a unit rate.
3. A hybrid electric vehicle used 15 gal of gasoline to travel 639 mi. How many miles per gallon did the car get?
4. Solve: $\frac{3}{8} = \frac{n}{12}$
5. Solve: $\frac{4}{3} = \frac{n-2}{6}$
6. A life insurance policy costs \$15.22 for every \$1000 of insurance. At this rate, what is the cost of \$75,000 of insurance?
7. If a 56-gram serving of pasta contains 7 g of protein, how many grams of protein are contained in a 454-gram box of the pasta?
8. An account executive bought a new car and drove 22,000 mi in the first 4 months. At the same rate, how many miles will the account executive drive in 3 years?

Solutions on pp. S-12–S-13.

Objective 3.2C Solve the basic percent equation

Percent means "parts of 100." In the figure at the right, there are 100 parts. Because 13 of the 100 parts are shaded, 13% of the figure is shaded. The symbol % is the **percent sign.**

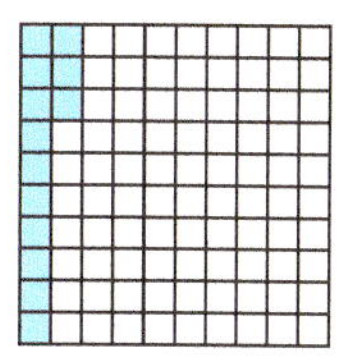

Take Note ➤
Recall that division is defined as multiplication by the reciprocal. Therefore, multiplying by $\frac{1}{100}$ is equivalent to dividing by 100.

In most applied problems involving percents, it is necessary either to rewrite a percent as a decimal or a fraction or to rewrite a fraction or a decimal as a percent.

To write a percent as a decimal, remove the percent sign and multiply by 0.01.

$$13\% = 13 \times 0.01 = 0.13$$

Move the decimal point two places to the left. Then remove the percent sign.

To write a percent as a fraction, remove the percent sign and multiply by $\frac{1}{100}$.

$$13\% = 13 \times \frac{1}{100} = \frac{13}{100}$$

Focus on writing a percent as a decimal and as a fraction

Write each percent as a decimal and as a fraction. **a.** 120% **b.** 4.3% **c.** 0.45%

SOLUTION

a. $120\% = 120 \times 0.01 = 1.2$

$120\% = 120 \times \frac{1}{100} = \frac{120}{100} = 1\frac{1}{5}$

b. $4.3\% = 4.3 \times 0.01 = 0.043$

$4.3\% = 4.3 \times \frac{1}{100}$

$= 4\frac{3}{10} \times \frac{1}{100}$ • $4.3 = 4\frac{3}{10}$

$= \frac{43}{10} \times \frac{1}{100} = \frac{43}{1000}$ • Multiply the fractions.

c. $0.45\% = 0.45 \times 0.01 = 0.0045$

$0.45\% = 0.45 \times \frac{1}{100}$

$= \frac{9}{20} \times \frac{1}{100}$ • $0.45 = \frac{45}{100} = \frac{9}{20}$

$= \frac{9}{2000}$ • Multiply the fractions.

Check your understanding 10

Write each percent as a decimal and as a fraction. **a.** 125% **b.** 8.5% **c.** 0.25%

SOLUTION See pages S-3–S-4.

Focus on writing a percent as a fraction

Write $16\frac{2}{3}\%$ as a fraction.

SOLUTION

$16\frac{2}{3}\% = 16\frac{2}{3} \times \frac{1}{100}$

$= \frac{50}{3} \times \frac{1}{100} = \frac{50}{300} = \frac{1}{6}$

Check your understanding 11

Write $33\frac{1}{3}\%$ as a fraction.

SOLUTION See page S-4.

A decimal or a fraction can be written as a percent by multiplying by 100%.

Write 0.37 as a percent.

$0.37 = 0.37 \times 100\% = 37\%$

Move the decimal point two places to the right. Then write the percent sign.

When changing a fraction to a percent, if the fraction can be written as a terminating decimal, the percent is written in decimal form. If the decimal representation of the fraction is a repeating decimal, the answer is written with a fraction.

Take Note ➤
The decimal form of $\frac{3}{8}$ terminates.

$$\begin{array}{r} 0.375 \\ 8\overline{)\,3.000} \\ \underline{-2\,4} \\ 60 \\ \underline{-56} \\ 40 \\ \underline{-40} \\ 0 \end{array}$$

Write $\frac{3}{8}$ as a percent.

$\frac{3}{8} = \frac{3}{8} \times \frac{100\%}{1}$ • $\frac{3}{8} = 0.375$ is a terminating decimal.

$= \frac{300\%}{8}$

$= 37.5\%$ • The answer is written in decimal form.

Take Note ➤
The decimal form of $\frac{1}{6}$ repeats.

$$\begin{array}{r} 0.16\overline{6} \\ 6\overline{)1.000} \\ \underline{-6} \\ 40 \\ \underline{-36} \\ 40 \\ \underline{-36} \\ 4 \end{array}$$

Write $\frac{1}{6}$ as a percent.

$\frac{1}{6} = \frac{1}{6} \times \frac{100\%}{1}$ • $\frac{1}{6} = 0.1\overline{6}$ is a repeating decimal.

$= \frac{100\%}{6}$

$= 16\frac{2}{3}\%$ • The answer is written with a fraction.

Focus on writing a decimal or a fraction as a percent

a. Write 0.015 and 2.3 as percents.

b. Write $\frac{19}{80}$ as a percent.

c. Write $\frac{2}{3}$ as a percent.

SOLUTION

a. $0.015 = 0.015 \times 100\%$

$= 1.5\%$

$2.3 = 2.3 \times 100\%$

$= 230\%$

b. $\frac{19}{80} \times \frac{100\%}{1} = \frac{1900\%}{80}$

$= 23.75\%$ • Write the answer in decimal form.

c. $\frac{2}{3} = \frac{2}{3} \times \frac{100\%}{1}$

$= \frac{200\%}{3}$

$= 66\frac{2}{3}\%$ • Write the answer with a fraction.

Check your understanding 12

a. Write 0.048 and 3.6 as percents.

b. Write $\frac{5}{16}$ as a percent.

c. Write $\frac{5}{6}$ as a percent.

SOLUTION See page S-4.

Consider the question, "How many squares should be shaded if we want to shade 7.5% of the region shown below?"

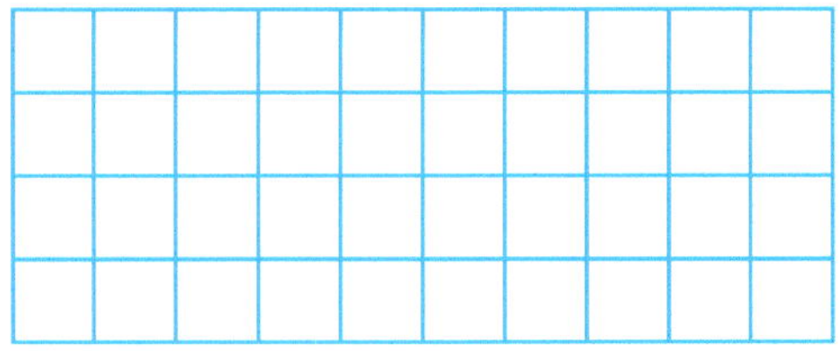

Determining the number of squares to be shaded requires answering the question

"7.5% of 40 is what number?"

This question can be translated into an equation and solved for the unknown number.

of	translates to	$\cdot$	(times)
is	translates to	$=$	(equals)
what	translates to	n	(the unknown number)

Here is the translation of "7.5% of 40 is what number?" Note that the percent is written as a decimal.

7.5%	of	40	is	what number?
↓	↓	↓	↓	↓
0.075	$\cdot$	40	$=$	n

$$0.075 \cdot 40 = n$$ • Solve this equation for n.

$$3 = n$$

If we shade 3 squares, 7.5% of the region will be shaded. See the figure that follows.

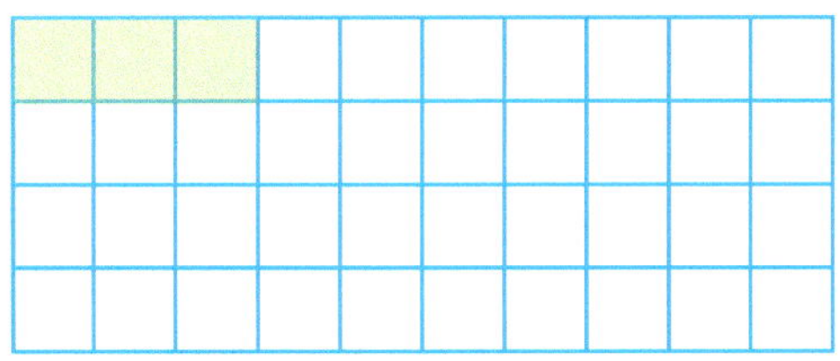

7.5% of the region is shaded.

The Basic Percent Equation

Percent $\cdot$ base $=$ amount

What is 25% of 800?

$$\text{Percent} \cdot \text{base} = \text{amount}$$ • Use the basic percent equation.

$$0.25 \cdot 800 = n$$ • Percent $= 25\% = 0.25$, base $= 800$, amount $= n$

$$200 = n$$

25% of 800 is 200.

Look for the number or phrase that follows the word *of* when determining the base in the basic percent equation.

Apply the Concept

A real estate broker receives a commission of 3% of the selling price of a house. Find the amount the broker receives on the sale of a $275,000 house.

SOLUTION

$\text{Percent} \cdot \text{base} = \text{amount}$ • Use the basic percent equation. The base is the selling price of the house.

$0.03 \cdot 275{,}000 = n$ • Percent = 3% = 0.03, base = 275,000, amount = n

$8250 = n$

The broker receives a commission of $8250.

In most cases, the percent is written as a decimal before the basic percent equation is solved. However, some percents are more easily written as a fraction than as a decimal. For example,

$$33\frac{1}{3}\% = \frac{1}{3} \qquad 66\frac{2}{3}\% = \frac{2}{3} \qquad 16\frac{2}{3}\% = \frac{1}{6} \qquad 83\frac{1}{3}\% = \frac{5}{6}$$

What is $33\frac{1}{3}\%$ of 90?

$\text{Percent} \cdot \text{base} = \text{amount}$ • Use the basic percent equation.

$\frac{1}{3} \cdot 90 = n$ • Percent = $33\frac{1}{3}\% = \frac{1}{3}$, base = 90, amount = n

$30 = n$

$33\frac{1}{3}\%$ of 90 is 30.

The three elements of the basic percent equation are the percent, the base, and the amount. If any two elements of the basic percent equation are given, the third element can be found.

20 is what percent of 32?

$\text{Percent} \cdot \text{base} = \text{amount}$ • Use the basic percent equation.

$n \cdot 32 = 20$ • Percent = n, base = 32, amount = 20

$\frac{32n}{32} = \frac{20}{32}$ • Solve for n by dividing each side of the equation by 32.

$n = 0.625$

$n = 62.5\%$ • Write the decimal as a percent.

20 is 62.5% of 32.

Apply the Concept

You correctly answered 96 of the 120 questions on an exam. What percent of the questions did you answer correctly?

SOLUTION

$\text{Percent} \cdot \text{base} = \text{amount}$ • Use the basic percent equation.

$n \cdot 120 = 96$ • Percent = n, base = 120, amount = 96

$\frac{120n}{120} = \frac{96}{120}$ • Solve for n by dividing each side of the equation by 120.

$n = 0.8$

$n = 80\%$ • Write the decimal as a percent.

You answered 80% of the questions correctly.

Tips for Success ➤ In this objective, we have solved the basic percent equation for each of the three elements: percent, base, and amount. You will need to be able to recognize these three different types of problems. Be sure to practice this skill by doing the exercises.

60% of what number is 300?

$\text{Percent} \cdot \text{base} = \text{amount}$ • Use the basic percent equation.

$0.60 \cdot n = 300$ • Percent = 60% = 0.60, base = n, amount = 300

$\frac{0.60n}{0.60} = \frac{300}{0.60}$ • Solve for n by dividing each side of the equation by 0.60.

$n = 500$

60% of 500 is 300.

Apply the Concept

A used car has a value of $11,250, which is 45% of the car's original value. What was the car's original value?

SOLUTION

$\text{Percent} \cdot \text{base} = \text{amount}$ • Use the basic percent equation. The base is the car's original value.

$0.45 \cdot n = 11{,}250$ • Percent = 45% = 0.45, base = n, amount = 11,250

$\frac{0.45n}{0.45} = \frac{11{,}250}{0.45}$ • Solve for n by dividing each side of the equation by 0.45.

$n = 25{,}000$

The car's original value was $25,000.

Focus on solving the basic percent equation

a. Find 9.4% of 240.
b. What percent of 30 is 12?
c. 60 is 2.5% of what?

SOLUTION

a. Use the basic percent equation.

Percent $= 9.4\% = 0.094$, base $= 240$, amount $= n$

$$\begin{aligned} \text{Percent} \cdot \text{base} &= \text{amount} \\ 0.094 \cdot 240 &= n \\ 22.56 &= n \end{aligned}$$

9.4% of 240 is 22.56.

b. Use the basic percent equation.

Percent $= n$, base $= 30$, amount $= 12$

$$\begin{aligned} \text{Percent} \cdot \text{base} &= \text{amount} \\ n \cdot 30 &= 12 \\ \frac{30n}{30} &= \frac{12}{30} \\ n &= 0.4 = 40\% \end{aligned}$$

12 is 40% of 30.

c. Use the basic percent equation.

Percent $= 2.5\% = 0.025$, base $= n$, amount $= 60$

$$\begin{aligned} \text{Percent} \cdot \text{base} &= \text{amount} \\ 0.025 \cdot n &= 60 \\ \frac{0.025n}{0.025} &= \frac{60}{0.025} \\ n &= 2400 \end{aligned}$$

60 is 2.5% of 2400.

Check your understanding 13

a. Find $66\frac{2}{3}\%$ of 45.

b. 25 is what percent of 40?

c. $16\frac{2}{3}\%$ of what is 15?

SOLUTION See pages S-4–S-5.

Objective 3.2C Practice

1. Write 36% as a decimal and as a fraction.
2. Write 6.2% as a decimal and as a fraction.
3. Write 0.25% as a decimal and as a fraction.
4. Write $12\frac{1}{2}\%$ as a fraction.
5. Write 0.73 as a percent.
6. Write $\frac{17}{20}$ as a percent.
7. Write $\frac{9}{4}$ as a percent.
8. Using the basic percent equation, what percent of 16 is 6?
9. Using the basic percent equation, find 10.7% of 485.
10. Using the basic percent equation, 54 is what percent of 2000?

Solutions on p. S-14.

Objective 3.2D Solve applications of the basic percent equation

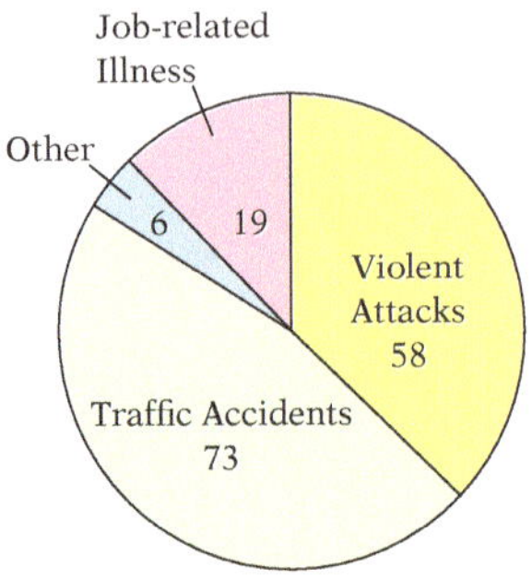

Figure 1 Causes of Death for Police Officers Killed in the Line of Duty
Source: International Union of Police Associations

The circle graph in Figure 1 shows the causes of death for all police officers who died while on duty during a recent year. What percent of the deaths were due to traffic accidents? Round to the nearest tenth of a percent.

STRATEGY

To find the percent:

- Find the total number of officers who died in the line of duty.
- Use the basic percent equation.
  Percent = n, base = total number killed,
  amount = number of deaths due to traffic accidents = 73

SOLUTION

$58 + 73 + 6 + 19 = 156$ • Total number killed

$$\text{Percent} \cdot \text{base} = \text{amount}$$
$$n \cdot 156 = 73$$
$$\frac{156n}{156} = \frac{73}{156}$$
$$n \approx 0.468$$
$$n \approx 46.8\%$$

46.8% of the deaths were due to traffic accidents.

Focus on solving a percent application

During a recent year, 276 billion product coupons were issued by manufacturers. Shoppers redeemed 4.8 billion of these coupons. (*Source:* NCH NuWorld Consumer Behavior Study, America Coupon Council) What percent of the coupons issued were redeemed by customers? Round to the nearest tenth of a percent.

STRATEGY

To find the percent, use the basic percent equation.
Percent = n, base = number of coupons issued = 276 billion, amount = number of coupons redeemed = 4.8 billion

SOLUTION

$$\text{Percent} \cdot \text{base} = \text{amount}$$
$$n \cdot 276 = 4.8$$
$$\frac{276n}{276} = \frac{4.8}{276}$$
$$n \approx 0.017$$
$$n \approx 1.7\%$$

Of the product coupons issued, 1.7% were redeemed by customers.

Check your understanding 14

An instructor receives a monthly salary of \$4330, and \$649.50 is deducted for income tax. Find the percent of the instructor's salary that is deducted for income tax.

SOLUTION See page S-5.

The simple interest that an investment earns is given by the **simple interest equation** $I = Prt$, where I is the simple interest; P is the principal, or amount invested; r is the simple interest rate; and t is the time.

A $1500 investment has an annual simple interest rate of 7%. Find the simple interest earned on the investment after 18 months.

To find the interest, solve $I = Prt$ for I.

The time is given in months but the interest rate is an annual rate. Therefore, we must convert 18 months to years. 18 months $= \frac{18}{12}$ years $= 1.5$ years

$$I = Prt$$
$$I = 1500(0.07)(1.5) \qquad \bullet\ P = 1500,\ r = 0.07,\ t = 1.5$$
$$I = 157.5$$

The investment earned $157.50.

Focus on solving a simple interest application

In April, Marshall Wardell was charged an interest fee of $8.72 on an unpaid credit card balance of $545. Find the annual interest rate on this credit card.

STRATEGY

The interest is $8.72. Therefore, $I = 8.72$. The unpaid balance is $545. This is the principal on which interest is calculated. Therefore, $P = 545$. The time is one month. Because the *annual* interest rate must be found and the time is given as one month, we write one month as $\frac{1}{12}$ year: $t = \frac{1}{12}$. To find the interest rate, solve $I = Prt$ for r.

SOLUTION

$$I = Prt \qquad \bullet\ \text{Use the simple interest equation.}$$
$$8.72 = 545r\left(\frac{1}{12}\right) \qquad \bullet\ I = 8.72,\ P = 545,\ t = \frac{1}{12}$$
$$8.72 = \frac{545}{12}r$$
$$\frac{12}{545}(8.72) = \frac{12}{545}\left(\frac{545}{12}r\right) \qquad \bullet\ \text{Multiply each side by the reciprocal of } \frac{545}{12}.$$
$$0.192 = r$$

The annual interest rate is 19.2%.

Check your understanding 15

Clarissa Adams purchased a municipal bond for $1000. The bond earns an annual simple interest rate of 6.4%. How much must she deposit into an account that earns 8% annual simple interest so that the interest earned from each account after one year is the same?

SOLUTION See page S-5.

Percent increase is used to show how much a quantity has increased over its original value. The statements "health care costs increased 4.8% last year" and "employees received an 8% pay increase" are illustrations of the use of percent increase.

Take Note ➤
In solving percent increase and percent decrease problems, remember that the base in the basic percent equation is always the quantity *before* the increase or decrease. That is, it is the *original value.* For the example at the right, the base is 277,000—the quantity before the increase.

According to the Energy Information Administration, the number of alternative-fuel vehicles increased from approximately 277,000 to approximately 352,000 in four years. Find the percent increase in the number of alternative-fuel vehicles. Round to the nearest percent.

STRATEGY

To find the percent increase:
- Find the increase in the number of alternative-fuel vehicles.
- Use the basic percent equation.
  Percent $= n$, base $= 277{,}000$, amount $=$ the amount of increase

SOLUTION

$352{,}000 - 277{,}000 = 75{,}000$ • The amount of increase

Percent · base = amount

$n \cdot 277{,}000 = 75{,}000$ • The base is the original value. The amount is the amount of increase.

$$\frac{277{,}000n}{277{,}000} = \frac{75{,}000}{277{,}000}$$

$$n \approx 0.27$$

The number of alternative-fuel vehicles increased by approximately 27%.

Focus on solving a percent increase application

A sales associate earned \$11.60 per hour before an 8% increase in pay. What is the new hourly wage? Round to the nearest cent.

STRATEGY

To find the new hourly wage:

- Use the basic percent equation to find the increase in pay.
  Percent = 8% = 0.08, base = 11.60, amount = n
- Add the amount of increase to the original wage.

SOLUTION

Percent · base = amount

$0.08 \cdot 11.60 = n$

$0.93 \approx n$ • Increase in pay

$\$11.60 + \$.93 = \$12.53$ • Add the increase to the original wage.

The new hourly wage is \$12.53.

Check your understanding 16

An automobile manufacturer increased the average fuel efficiency on a car from 17.5 mi/gal to 18.2 mi/gal. Find the percent increase in fuel efficiency.

SOLUTION See pages S-5–S-6.

Percent decrease is used to show how much a quantity has decreased from its original value. The statements "the number of family farms decreased 2% last year" and "the president's approval rating has decreased 9% over the last month" are illustrations of the use of percent decrease.

Take Note ➤ Remember that the base in the basic percent equation is the *original value*—that is, the quantity *before* the decrease. For the example at the right, the base is 60.6.

During a 2-year period, the value of U.S. agricultural exports decreased from approximately \$60.6 billion to approximately \$52.0 billion. Find the percent decrease in the value of U.S. agricultural exports. Round to the nearest tenth of a percent.

STRATEGY

To find the percent decrease:

- Find the decrease in the value of agricultural exports.
- Use the basic percent equation.
  Percent = n, base = 60.6, amount = the amount of decrease

SOLUTION

$60.6 - 52.0 = 8.6$ • The amount of decrease

$$\text{Percent} \cdot \text{base} = \text{amount}$$

$$n \cdot 60.6 = 8.6$$ • The base is the original value. The amount is the amount of decrease.

$$\frac{60.6n}{60.6} = \frac{8.6}{60.6}$$

$$n \approx 0.142$$

The value of agricultural exports decreased by approximately 14.2%.

Focus on solving a percent decrease application

During a recent year, violent crime in a small city decreased from 27 crimes per 1000 people to 24 crimes per 1000 people. Find the percent decrease in violent crime. Round to the nearest tenth of a percent.

STRATEGY

To find the percent decrease in crime:
- Find the decrease in the number of crimes.
- Use the basic percent equation to find the percent decrease in crime.

SOLUTION

$27 - 24 = 3$ • Amount of decrease

$$\text{Percent} \cdot \text{base} = \text{amount}$$

$$n \cdot 27 = 3$$ • The amount is the amount of decrease.

$$n = \frac{3}{27} \approx 0.111$$

Violent crime decreased by approximately 11.1% during the year.

Check your understanding 17

The market value of a luxury car decreased by 24% during the past year. Find the market value of a luxury car that cost \$47,000 last year. Round to the nearest dollar.

SOLUTION See page S-6.

Objective 3.2D Practice

1. According to the annual Summer Vacation Survey conducted by Myvesta, a non-profit consumer education organization, the average summer vacation costs \$2252. If \$1850 of this amount is charged on a credit card, what percent of the vacation cost is charged? Round to the nearest tenth of a percent.
2. If Kachina Caron invested \$1200 in a simple interest account and earned \$72 in 8 months, what is the annual simple interest rate?
3. How much money must Andrea invest for 2 years in an account that earns an annual simple interest rate of 8% if she wants to earn \$300 from the investment?
4. In 1924, the number of events in the Winter Olympics was 14. The 2014 Winter Olympics in Sochi, Russia, included 98 medal events. Find the percent increase in the number of events in the Winter Olympics from 1924 to 2014.
5. The table at the right shows the numbers of millionaire households in the United States for selected years. Find the percent increase in the number of millionaire households from 1975 to 2010.

Year	Number of Households Containing Millionaires
1975	350,000
1997	3,500,000
2010	8,400,000

Source: Affluent Market Institute; CNNMoney.com

6. In the 1990–1991 school year, the average cost for tuition, fees, and room and board at a public college was \$5074. In the 2009–2010 school year, the cost had risen to \$7020. (*Source:* The College Board) Find the percent increase in the average cost of attending a public college from the 1990–1991 school year to the 2009–2010 school year.
7. The graph at the right shows the numbers of passports, in millions, issued to U.S. citizens each year from 2007 to 2010. Find the percent decrease in the number of passports issued from 2007 to 2010.

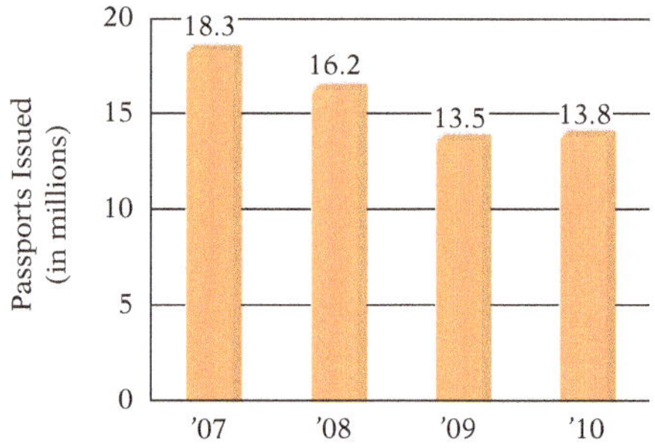

Passports Issued to U.S. Citizens
Source: State Department

8. Due to an increased number of service lines at a grocery store, the average amount of time a customer waits in line has decreased from 3.8 min to 2.5 min. Find the percent decrease.
9. It is estimated that the value of a new car is reduced by 30% after one year of ownership. Find the value after one year of a car that cost \$21,900 new.

Solutions on pp. S-15–S-17.

SECTION 3.3 General First-Degree Equations

Objective 3.3A Solve an equation of the form $ax + b = c$

In solving an equation of the form $ax + b = c$, the goal is to rewrite the equation in the form *variable* = *constant*. This requires the application of both the Addition and Multiplication Properties of Equations.

Solve: $\frac{3}{4}x - 2 = -11$

The goal is to write the equation in the form *variable* = *constant*.

$$\frac{3}{4}x - 2 = -11$$

$$\frac{3}{4}x - 2 + 2 = -11 + 2$$ • Add 2 to each side of the equation.

$$\frac{3}{4}x = -9$$ • Simplify.

$$\frac{4}{3} \cdot \frac{3}{4}x = \frac{4}{3}(-9)$$ • Multiply each side of the equation by $\frac{4}{3}$.

$$x = -12$$ • The equation is in the form *variable* = *constant*.

The solution is −12.

Take Note ➤
Check:

$$\frac{3}{4}x - 2 = -11$$

$\frac{3}{4}(-12) - 2$	-11
$-9 - 2$	-11
$-11 =$	-11

A true equation

Focus on solving an equation

Solve: $5 = 9 - 2x$

SOLUTION

$$5 = 9 - 2x$$

$$5 - 9 = 9 - 9 - 2x \qquad \text{• Subtract 9 from each side.}$$

$$-4 = -2x$$

$$\frac{-4}{-2} = \frac{-2x}{-2} \qquad \text{• Divide each side by } -2.$$

$$2 = x$$

The solution is 2.

Check your understanding 1

Solve: $2 = 11 + 3x$

SOLUTION See page S-6.

It may be easier to solve an equation containing two or more fractions by multiplying each side of the equation by the least common multiple (LCM) of the denominators. The result is an equation that does not contain any fractions. Multiplying each side of an equation that contains fractions by the LCM of the denominators is called **clearing denominators.** It is an alternative method, as we show in the next example, of solving an equation that contains fractions.

Take Note ➤
Observe that after we multiply both sides of the equation by the LCM of the denominators and then simplify, the equation no longer contains fractions.

Clearing denominators is a method of solving equations. The process applies only to equations, never to expressions.

Solve: $\frac{2}{3}x + \frac{1}{2} = \frac{3}{4}$

$$\frac{2}{3}x + \frac{1}{2} = \frac{3}{4}$$

$$12\left(\frac{2}{3}x + \frac{1}{2}\right) = 12\left(\frac{3}{4}\right) \qquad \text{• Multiply each side of the equation by 12, the LCM of 3, 2, and 4.}$$

$$12\left(\frac{2}{3}x\right) + 12\left(\frac{1}{2}\right) = 12\left(\frac{3}{4}\right) \qquad \text{• Use the Distributive Property.}$$

$$8x + 6 = 9 \qquad \text{• Simplify.}$$

$$8x + 6 - 6 = 9 - 6 \qquad \text{• Subtract 6 from each side of the equation.}$$

$$8x = 3$$

$$\frac{8x}{8} = \frac{3}{8} \qquad \text{• Divide each side of the equation by 8.}$$

$$x = \frac{3}{8}$$

The solution is $\frac{3}{8}$.

Focus on solving an equation by clearing denominators

Solve $\frac{4}{5}x - \frac{1}{2} = \frac{3}{4}$ by first clearing denominators.

SOLUTION

The LCM of 5, 2, and 4 is 20.

$$\frac{4}{5}x - \frac{1}{2} = \frac{3}{4}$$

$$20\left(\frac{4}{5}x - \frac{1}{2}\right) = 20\left(\frac{3}{4}\right)$$ • Multiply each side by 20.

$$20\left(\frac{4}{5}x\right) - 20\left(\frac{1}{2}\right) = 20\left(\frac{3}{4}\right)$$ • Use the Distributive Property.

$$16x - 10 = 15$$

$$16x - 10 + 10 = 15 + 10$$ • Add 10 to each side.

$$16x = 25$$

$$\frac{16x}{16} = \frac{25}{16}$$ • Divide each side by 16.

$$x = \frac{25}{16}$$

The solution is $\frac{25}{16}$.

Check your understanding 2

Solve $\frac{2}{3}x + 3 = \frac{7}{2}$ by first clearing denominators.

SOLUTION See page S-6.

Focus on solving an equation related to statistics

Solve: $\frac{x - 1.2}{3.2} = 1.96$

SOLUTION

$$\frac{x - 1.2}{3.2} = 1.96$$

$$3.2\left(\frac{x - 1.2}{3.2}\right) = 3.2(1.96)$$ • Multiply each side of the equation by 3.2.

$$x - 1.2 = 6.272$$ • Simplify.

$$x - 1.2 + 1.2 = 6.272 + 1.2$$ • Add 1.2 to each side of the equation.

$$x = 7.472$$

The solution is 7.472.

Check your understanding 3

Solve: $-2.33 = \frac{x + 1.7}{2.71}$

SOLUTION See page S-7.

Objective 3.3A Practice

Solve.

1. $2x - 5 = -11$

2. $7 - 9a = 4$

3. $\frac{2}{3}x - \frac{5}{6} = -\frac{1}{3}$

4. $\frac{x - 1.5}{1.65} = -1.8$

5. $5y + 9 + 2y = 23$

Solutions on pp. S-17–S-18.

Objective 3.3B Solve an equation of the form $ax + b = cx + d$

Tips for Success ➤
Have you considered joining a study group? Getting together regularly with other students in the class to go over material and quiz each other can be very beneficial. See *AIM for Success*.

In solving an equation of the form $ax + b = cx + d$, the goal is to rewrite the equation in the form *variable* = *constant*. Begin by rewriting the equation so that there is only one variable term in the equation. Then rewrite the equation so that there is only one constant term.

Focus on solving an equation with the variable on both sides

Solve: $2x + 3 = 5x - 9$

SOLUTION

$$2x + 3 = 5x - 9$$

$$2x - 5x + 3 = 5x - 5x - 9$$ • Subtract $5x$ from each side of the equation.

$$-3x + 3 = -9$$ • Simplify. There is only one variable term.

$$-3x + 3 - 3 = -9 - 3$$ • Subtract 3 from each side of the equation.

$$-3x = -12$$ • Simplify. There is only one constant term.

$$\frac{-3x}{-3} = \frac{-12}{-3}$$ • Divide each side of the equation by -3.

$$x = 4$$ • The equation is in the form *variable* = *constant*.

The solution is 4. You should verify this by checking this solution.

Check your understanding 4

Solve: $5x + 4 = 6 + 10x$

SOLUTION See page S-7.

Focus on solving an equation with the variable on both sides

Solve: $3x + 4 - 5x = 2 - 4x$

SOLUTION

$$3x + 4 - 5x = 2 - 4x$$

$$-2x + 4 = 2 - 4x$$ • Combine like terms.

$$-2x + 4x + 4 = 2 - 4x + 4x$$ • Add $4x$ to each side.

$$2x + 4 = 2$$

$$2x + 4 - 4 = 2 - 4$$ • Subtract 4 from each side.

$$2x = -2$$

$$\frac{2x}{2} = \frac{-2}{2}$$ • Divide each side by 2.

$$x = -1$$

The solution is -1.

Check your understanding 5

Solve: $5x - 10 - 3x = 6 - 4x$

SOLUTION See page S-7.

Objective 3.3B Practice

1. Solve and check: $15x - 2 = 4x - 13$
2. Solve and check: $0.2b + 3 = 0.5b + 12$
3. Solve and check: $5 + 7x = 11 + 9x$
4. Solve and check: $7x - 8 = x - 3$
5. If $7x + 3 = 5x - 7$, evaluate $3x - 2$.

Solutions on pp. S-18–S-19.

Objective 3.3C Solve an equation containing parentheses

When an equation contains parentheses, one of the steps in solving the equation is to use the Distributive Property. Recall that the Distributive Property is used to remove parentheses from a variable expression.

Focus on solving an equation containing parentheses

Solve: $4 + 5(2x - 3) = 3(4x - 1)$

SOLUTION

$$4 + 5(2x - 3) = 3(4x - 1)$$

$$4 + 10x - 15 = 12x - 3$$ • Use the Distributive Property. Then simplify.

$$10x - 11 = 12x - 3$$

$$10x - 12x - 11 = 12x - 12x - 3$$ • Subtract $12x$ from each side of the equation.

$$-2x - 11 = -3$$ • Simplify.

$$-2x - 11 + 11 = -3 + 11$$ • Add 11 to each side of the equation.

$$-2x = 8$$ • Simplify.

$$\frac{-2x}{-2} = \frac{8}{-2}$$ • Divide each side of the equation by -2.

$$x = -4$$ • The equation is in the form *variable* = *constant*.

The solution is -4. You should verify this by checking this solution.

Check your understanding 6

Solve: $5x - 4(3 - 2x) = 2(3x - 2) + 6$

SOLUTION See page S-7.

Objective 3.3C Practice

1. Solve and check: $9n - 3(2n - 1) = 15$
2. Solve and check: $5(3 - 2y) + 4y = 3$
3. Solve and check: $0.05(4 - x) + 0.1x = 0.32$
4. Solve and check: $7 - (5 - 8x) = 4x + 3$

Solutions on pp. S-19–S-20.

SECTION 3.4 Literal, Absolute Value, and Radical Equations

Objective 3.4A Solve a literal equation

A **literal equation** is an equation that contains more than one variable. Examples of literal equations are shown at the right.

$$2x + 3y = 6$$
$$4w - 2x + z = 0$$

Formulas are used to express relationships among physical quantities. A **formula** is a literal equation that states a rule about measurements. Examples of formulas are shown at the right.

$\frac{1}{R_1} + \frac{1}{R_2} = \frac{1}{R}$ (Physics)

$s = a + (n - 1)d$ (Mathematics)

$A = P + Prt$ (Business)

The Addition and Multiplication Properties can be used to solve a literal equation for one of the variables. The goal is to rewrite the equation so that the variable being solved for is alone on one side of the equation, and all the other numbers and variables are on the other side.

Focus on solving a literal equation

Solve $3x - 4y = 12$ for y.

SOLUTION

$$3x - 4y = 12$$

$$3x - 3x - 4y = -3x + 12$$ • Subtract $3x$.

$$-4y = -3x + 12$$

$$\frac{-4y}{-4} = \frac{-3x + 12}{-4}$$ • Divide by -4.

$$y = \frac{-3x}{-4} + \frac{12}{-4}$$ • Recall that a fraction bar acts as a grouping symbol. Divide each term in the numerator by -4.

$$y = \frac{3}{4}x - 3$$

Check your understanding 1

Solve $5x - 2y = 10$ for y.

SOLUTION See page S-8.

Focus on solving a literal equation by removing parentheses

Solve $A = P(1 + i)$ for i.

SOLUTION

The goal is to rewrite the equation so that i is on one side of the equation and all other variables are on the other side.

$$A = P(1 + i)$$

$$A = P + Pi \quad \text{• Use the Distributive Property to remove parentheses.}$$

$$A - P = P - P + Pi \quad \text{• Subtract } P \text{ from each side of the equation.}$$

$$A - P = Pi$$

$$\frac{A - P}{P} = \frac{Pi}{P} \quad \text{• Divide each side of the equation by } P.$$

$$\frac{A - P}{P} = i$$

Check your understanding 2

Solve $s = \dfrac{A + L}{2}$ for L.

SOLUTION See page S-8.

In the previous example, the Distributive Property was used to remove parentheses, as can be symbolized by

$$a(x + y) = ax + ay$$

In a similar way, we can use the Distributive Property to rewrite two or more terms containing a common factor with parentheses.

$$ax + ay = a(x + y)$$

In this case, we say that we have *factored* a from the expression. This concept is used in solving some literal equations.

Focus on solving a literal equation by factoring

Solve $S = C - rC$ for C.

SOLUTION

$$S = C - rC \quad \text{• } C \text{ is a common factor.}$$

$$S = C(1 - r) \quad \text{• Use the Distributive Property to factor } C \text{ from the two terms. Recall that } C = C \cdot 1.$$

$$\frac{S}{1 - r} = \frac{C(1 - r)}{1 - r} \quad \text{• Divide by } (1 - r).$$

$$\frac{S}{1 - r} = C$$

Check your understanding 3

Solve $S = rS + C$ for S.

SOLUTION See page S-8.

Focus on solving a formula used in statistics

Solve $z = \dfrac{x - m}{s}$ for m.

SOLUTION

$$z = \frac{x - m}{s}$$

$$s \cdot z = s\left(\frac{x - m}{s}\right)$$ • Multiply each side of the equation by s.

$$sz = x - m$$ • Simplify.

$$sz - x = x - x - m$$ • Subtract x from each side of the equation.

$$sz - x = -m$$

$$-1(sz - x) = -1(-m)$$ • Multiply each side of the equation by -1.

$$-sz + x = m$$

Check your understanding 4

Solve $m = x - \frac{ts}{n}$ for n.

SOLUTION See page S-8.

Objective 3.4A Practice

1. Solve $4x + 3y = 12$ for x.

2. Solve $E = IR$ for R.

3. Solve $F = \frac{9}{5}C + 32$ for C.

4. Solve $s = a(x - vt)$ for t.

5. Solve $a = S - Sr$ for S.

Solutions on pp. S-20–S-21.

Objective 3.4B Solve an absolute value equation

Tips for Success ➤
Before the class meeting in which your professor begins a new section, you should read each objective statement for that section. Next, browse through the objective material. The purpose of browsing through the material is to set the stage for your brain to accept and organize new information when it is presented to you. See *AIM for Success*.

Recall that the *absolute value* of a number is its distance from zero on the number line. Distance is always a positive number or zero. Therefore, the absolute value of a number is always a positive number or zero.

The distance from 0 to 3 or from 0 to -3 is 3 units.

$|3| = 3$ $\quad$ $|-3| = 3$

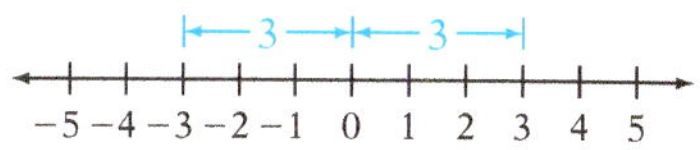

Absolute value can be used to represent the distance between any two points on the number line. The **distance between two points** on the number line is the absolute value of the difference between the coordinates of the two points.

The distance between point a and point b is given by $|b - a|$.

The distance between 4 and -3 on the number line is 7 units. Note that the order in which the coordinates are subtracted does not affect the distance.

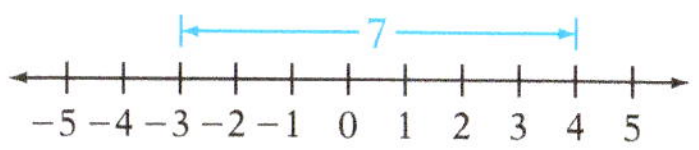

$$\text{Distance} = |-3 - 4| = |-7| = 7$$

$$\text{Distance} = |4 - (-3)| = |7| = 7$$

For any two numbers a and b, $|b - a| = |a - b|$.

An equation containing a variable within an absolute value symbol is called an **absolute value equation.** Here are three examples.

$$|x| = 3 \qquad |x + 2| = 8 \qquad |3x - 4| = 5x - 9$$

Solutions of an Absolute Value Equation

If $a > 0$ and $|x| = a$, then $x = -a$ or $x = a$. If $|x| = 0$, then $x = 0$. If $a < 0$, then $|x| = a$ has no solution.

EXAMPLES

1. If $|x| = 3$, then $x = -3$ or $x = 3$.
2. If $|-x| = 8$, then $x = -8$ or $x = 8$.
3. If $|x + 3| = 4$, then $x + 3 = 4$ or $x + 3 = -4$. The solution of $x + 3 = 4$ is 1. The solution of $x + 3 = -4$ is -7.
4. If $|z| = 0$, then $z = 0$.
5. If $|y| = -2$, then the equation has no solution. The absolute value of any number is greater than or equal to zero.

Take Note ➤
You should always check your answers. Here is the check for examples (1), (2), and (3) at the right.

$$\begin{array}{r|l} |x| & = 3 \\ \hline |-3| & 3 \\ 3 & = 3 \end{array} \qquad \begin{array}{r|l} |x| & = 3 \\ \hline |3| & 3 \\ 3 & = 3 \end{array}$$

$$\begin{array}{r|l} |-x| & = 8 \\ \hline |-(-8)| & 8 \\ |8| & 8 \\ 8 & = 8 \end{array} \qquad \begin{array}{r|l} |-x| & = 8 \\ \hline |-8| & 8 \\ 8 & = 8 \end{array}$$

$$\begin{array}{r|l} |x+3| & = 4 \\ \hline |1+3| & 4 \\ |4| & 4 \\ 4 & = 4 \end{array} \qquad \begin{array}{r|l} |x+3| & = 4 \\ \hline |-7+3| & 4 \\ |-4| & 4 \\ 4 & = 4 \end{array}$$

Solve: $|x + 2| = 8$

$$|x + 2| = 8$$

$x + 2 = 8 \qquad x + 2 = -8$ • Remove the absolute value sign and rewrite as two equations.

$x = 6 \qquad x = -10$ • Solve each equation.

Check:

$$\begin{array}{r|l} |x+2| & = 8 \\ \hline |6+2| & 8 \\ |8| & 8 \\ 8 & = 8 \end{array} \qquad \begin{array}{r|l} |x+2| & = 8 \\ \hline |-10+2| & 8 \\ |-8| & 8 \\ 8 & = 8 \end{array}$$

The solutions are 6 and -10.

Focus on solving an absolute value equation

Solve: $|2 - x| = 12$

SOLUTION

$$|2 - x| = 12$$

$2 - x = 12 \qquad 2 - x = -12$

$-x = 10 \qquad -x = -14$ • Subtract 2.

$x = -10 \qquad x = 14$ • Multiply by -1.

The solutions are -10 and 14.

Check your understanding 5

Solve: $|2x - 3| = 5$

SOLUTION See page S-8.

Focus on solving an absolute value equation

Solve: $|2x| = -4$

SOLUTION

$|2x| = -4$

There is no solution to this equation because the absolute value of a number must be nonnegative.

Check your understanding 6

Solve: $|x - 3| = -2$

SOLUTION See page S-9.

Focus on solving an absolute value equation

Solve: $3 - |2x - 4| = -5$

SOLUTION

$$3 - |2x - 4| = -5$$
$$-|2x - 4| = -8 \quad \text{• Subtract 3.}$$
$$|2x - 4| = 8 \quad \text{• Multiply by } -1.$$

$$2x - 4 = 8 \qquad 2x - 4 = -8$$
$$2x = 12 \qquad 2x = -4$$
$$x = 6 \qquad x = -2$$

The solutions are 6 and -2.

Check your understanding 7

Solve: $5 - |3x + 5| = 3$

SOLUTION See page S-9.

Objective 3.4B Practice

1. Solve: $|a - 2| = 0$

2. Solve: $|2x - 3| + 4 = -4$

3. Solve: $|3t + 2| + 3 = 4$

4. Solve: $5 - |2x + 1| = 5$

5. Solve: $8 - |1 - 3x| = -1$

Solutions on p. S-21.

Objective 3.4C Solve a radical equation

An equation that contains a variable expression in a radicand is a **radical equation.**

$$\left.\begin{array}{l}\sqrt{x} = 4 \\ \sqrt{x+2} = \sqrt{x-7}\end{array}\right\} \text{Radical equations}$$

The following property of equality states that if two numbers are equal, then the squares of the numbers are equal. This property is used to solve radical equations.

Property of Squaring Both Sides of an Equation

If a and b are real numbers and $\boldsymbol{a = b}$, then $\boldsymbol{a^2 = b^2}$.

Take Note ➤
Any time each side of an equation is squared, you must check the proposed solution of the equation.

Solve: $\sqrt{x-2} - 7 = 0$

$\sqrt{x-2} - 7 = 0$ • Rewrite the equation with the radical on one side of the equation and the constant on the other side.

$\sqrt{x-2} = 7$

$(\sqrt{x-2})^2 = 7^2$ • Square both sides of the equation.

$x - 2 = 49$ • Solve the resulting equation.

$x = 51$

Check: $\sqrt{x-2} - 7 = 0$ • Check the solution. When both sides of an equation are squared, the resulting equation may have a solution that is not a solution of the original equation.

$\sqrt{51-2} - 7$	0
$\sqrt{49} - 7$	0
$7 - 7$	0

$0 = 0$

The solution is 51.

Focus on solving an equation with one radical expression

Solve: $\sqrt{3x} + 2 = 5$

SOLUTION

$\sqrt{3x} + 2 = 5$

$\sqrt{3x} = 3$ • Rewrite the equation so that the radical is alone on one side of the equation.

$(\sqrt{3x})^2 = 3^2$ • Square both sides of the equation.

$3x = 9$ • Solve for x.

$x = 3$

Check: $\sqrt{3x} + 2 = 5$ • Both sides of the equation were squared. The solution must be checked.

$\sqrt{3 \cdot 3} + 2$	5
$\sqrt{9} + 2$	5
$3 + 2$	5

$5 = 5$ • This is a true equation. The solution checks.

The solution is 3.

Check your understanding 8

Solve: $\sqrt{4x} + 3 = 7$

SOLUTION See page S-9.

Objective 3.4C Practice

Solve and check.

1. $\sqrt{5x+4} = 3$

2. $\sqrt{5x} + 8 = 23$

Solutions on p. S-22.

SECTION

3.5 Solving First-Degree Inequalities and Absolute Value Inequalities

Objective 3.5A Graph an inequality on a number line

Recall that a *set* is a collection of objects, called the *elements* of the set. A set can be represented using **set-builder notation.** This method of writing a set uses a rule to describe the elements of the set. Using set-builder notation, we represent the set of all positive integers less than 10 as

$\{x|x < 10, x \in \text{positive integers}\}$, which is read "the set of all positive integers x that are less than 10."

Use set-builder notation to write the set of integers less than or equal to 12.

$\{x|x \leq 12, x \in \text{integers}\}$ • This is read "the set of all integers x that are less than or equal to 12."

Use set-builder notation to write the set of real numbers greater than 4.

$\{x|x > 4, x \in \text{real numbers}\}$ • This is read "the set of all real numbers x that are greater than 4."

For the remainder of this section, all variables will represent real numbers. Given this convention, $\{x|x > 4, x \in \text{real numbers}\}$ is written $\{x|x > 4\}$.

Take Note ➤
Set-builder notation is mainly used to represent sets that have an infinite number of elements. The set $\{x|x > 4\}$ has an infinite number of elements and cannot be represented using the roster method, which lists every element of the set.

Some sets of real numbers that are written in set-builder notation can also be written in **interval notation.** For instance, the interval notation $[-3, 2)$ represents the set of real numbers between -3 and 2. The bracket means that -3 is included in the set, and the parenthesis means that 2 is *not* included in the set. Using set-builder notation, the interval $[-3, 2)$ is written

$\{x|-3 \leq x < 2\}$ • This is read "the set of all real numbers x between -3 and 2, including -3 but excluding 2."

To indicate an interval that extends forever in the positive direction, we use the **infinity symbol, ∞;** to indicate an interval that extends forever in the negative direction, we use the **negative infinity symbol, $-\infty$.**

Write $\{x|x \leq -2\}$ in interval notation.

$\{x|x \leq -2\}$ is the set of real numbers less than or equal to -2. This set extends forever in the negative direction. In interval notation, this set is written $(-\infty, -2]$.

When writing a set in interval notation, we always use a parenthesis to the right of ∞ and to the left of $-\infty$. Infinity is not a real number, so it cannot be represented as belonging to the set of real numbers by using a bracket.

Focus on writing a set in interval notation

Write in interval notation.

a. $\{x|x \geq 2\}$ **b.** $\{x|0 \leq x \leq 1\}$

SOLUTION

a. $\{x|x \geq 2\}$ is the set of real numbers greater than or equal to 2. This set extends forever in the positive direction. In interval notation, this set is written $[2, \infty)$.

b. $\{x|0 \leq x \leq 1\}$ is the set of real numbers between 0 and 1, including 0 and 1. In interval notation, this set is written $[0, 1]$.

Check your understanding 1

Write in interval notation.

a. $\{x|x \leq 3\}$ **b.** $\{x|-5 \leq x \leq -3\}$

SOLUTION See page S-9.

Write $[1, 3]$ in set-builder notation.

This is the set of real numbers between 1 and 3, including 1 and 3. In set-builder notation, this set is written $\{x|1 \leq x \leq 3\}$.

Focus on writing a set in set-builder notation

Write in set-builder notation.

a. $(-\infty, 0]$ **b.** $(-3, 3)$

SOLUTION

a. The interval $(-\infty, 0]$ is the set of real numbers less than or equal to 0. In set-builder notation, this set is written $\{x|x \leq 0\}$.

b. The interval $(-3, 3)$ is the set of real numbers between -3 and 3, excluding -3 and 3. In set-builder notation, this set is written $\{x|-3 < x < 3\}$.

Check your understanding 2

Write in set-builder notation.

a. $(-3, \infty)$ **b.** $[0, 4)$

SOLUTION See page S-9.

We can graph sets of real numbers given in set-builder notation or in interval notation.

Graph: $(-\infty, -1)$

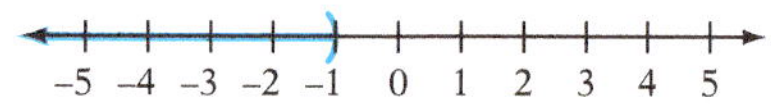

This is the set of real numbers less than -1, excluding -1. The parenthesis at -1 on the number line indicates that -1 is excluded from the set.

Focus on graphing a set of real numbers written in interval notation

Graph: $(-\infty, 5)$

SOLUTION

The graph is the set of real numbers less than 5. Use a parenthesis at 5.

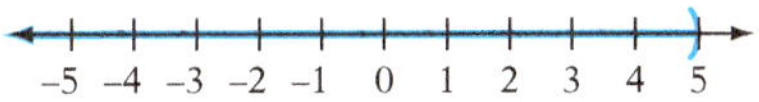

Check your understanding 3

Graph: $[2, 5]$

SOLUTION See page S-9.

Graph: $\{x|x \geq 1\}$

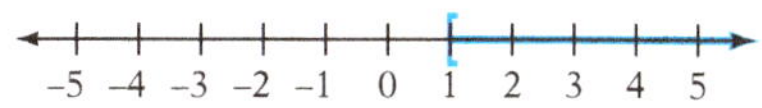

This is the set of real numbers greater than or equal to 1. The bracket at 1 indicates that 1 is included in the set.

Focus on graphing a set of real numbers written in set-builder notation

Graph.

a. $\{x|-2 < x < 1\}$ **b.** $\{x|x < 4\}$

SOLUTION

a. The graph is the set of real numbers between -2 and 1, excluding -2 and 1. Use parentheses at -2 and 1.

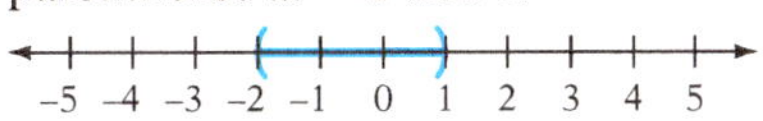

b. The graph is the set of real numbers less than 4. Use a parenthesis at 4.

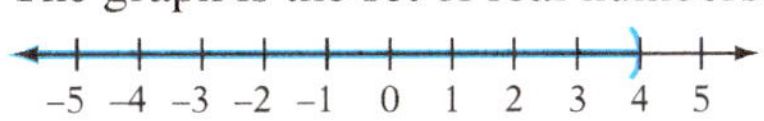

Check your understanding 4

Graph.

a. $\{x|-4 \leq x \leq 4\}$ **b.** $\{x|x > -3\}$

SOLUTION See pp. S-9–S-10.

Objective 3.5A Practice

1. Use set-builder notation to write the set of integers less than -70.

2. Use set-builder notation to write the set of real numbers greater than 8.

3. Write $\{x|x \leq 0\}$ in interval notation.

4. Write $(4, 9]$ in set-builder notation.

5. Graph: $(-\infty, 3]$

6. Graph: $\{x|-4 < x < 1\}$

Solutions on p. S-22.

Objective 3.5B Solve a first-degree inequality

The **solution set of an inequality** is a set of numbers each element of which, when substituted for the variable, results in a true inequality.

The inequality at the right is true if the variable is replaced by (for instance) 3, −1.98, or $\frac{2}{3}$.

$$x - 1 < 4$$
$$3 - 1 < 4$$
$$-1.98 - 1 < 4$$
$$\frac{2}{3} - 1 < 4$$

There are many values of the variable x that will make the inequality $x - 1 < 4$ true. The solution set of the inequality is any number less than 5. The solution set can be written in set-builder notation as $\{x|x < 5\}$.

The graph of the solution set of $x - 1 < 4$ is shown at the right.

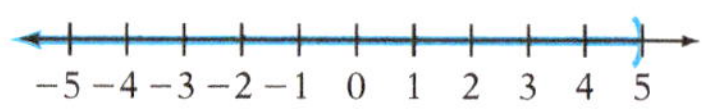

When solving an inequality, we use the **Addition and Multiplication Properties of Inequalities** to rewrite the inequality in the form *variable* < *constant* or in the form *variable* > *constant*.

Addition Property of Inequalities

The same term can be added to each side of an inequality without changing the solution set of the inequality. Symbolically, this is written

If $a < b$, then $a + c < b + c$.
If $a > b$, then $a + c > b + c$.

This property is also true for an inequality that contains ≤ or ≥.

The Addition Property of Inequalities is used to remove a term from one side of an inequality by adding the additive inverse of that term to each side of the inequality. Because subtraction is defined in terms of addition, the same number can be subtracted from each side of an inequality without changing the solution set of the inequality.

Take Note ➤
The solution set of an inequality can be written in set-builder notation or in interval notation.

Solve and graph the solution set: $x + 2 \geq 4$

$$x + 2 \geq 4$$
$$x + 2 - 2 \geq 4 - 2$$ • Subtract 2 from each side of the inequality.
$$x \geq 2$$ • Simplify.

The solution set is $\{x|x \geq 2\}$ or, in interval notation, $[2, \infty)$.

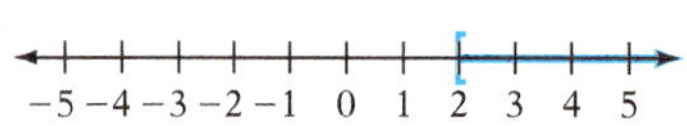

Care must be taken when multiplying each side of an inequality by a nonzero constant. The rule for multiplying each side by a *positive* number is different from the rule for multiplying each side by a *negative* number.

Take Note ➤
$c > 0$ means c is a positive number. Note that the inequality symbol does not change.

$c < 0$ means c is a negative number. Note that the inequality symbol is reversed.

Multiplication Property of Inequalities

Rule 1 Each side of an inequality can be multiplied by the same *positive* constant without changing the solution set of the inequality. Symbolically, this is written

$$\text{If } a < b \text{ and } c > 0, \text{ then } ac < bc.$$
$$\text{If } a > b \text{ and } c > 0, \text{ then } ac > bc.$$

Rule 2 If each side of an inequality is multiplied by the same *negative* constant and the inequality symbol is *reversed*, then the solution set of the inequality is not changed. Symbolically, this is written

$$\text{If } a < b \text{ and } c < 0, \text{ then } ac > bc.$$
$$\text{If } a > b \text{ and } c < 0, \text{ then } ac < bc.$$

This property is also true for an inequality that contains $\le$ or $\ge$.

Here are examples of this property.

Rule 1: Multiply by a *positive* number.		**Rule 2:** Multiply by a *negative* number.	
$2 < 5$	$3 > 2$	$2 < 5$	$3 > 2$
$2(4) < 5(4)$	$3(4) > 2(4)$	$2(-4) > 5(-4)$	$3(-4) < 2(-4)$
$8 < 20$	$12 > 8$	$-8 > -20$	$-12 < -8$

The Multiplication Property of Inequalities is used to remove a coefficient from one side of an inequality by multiplying each side of the inequality by the reciprocal of the coefficient.

Take Note ➤
Each side of the inequality is divided by a negative number; the inequality symbol must be reversed.

Solve $-3x > 9$. Write the solution set in interval notation.

$$-3x > 9$$
$$\frac{-3x}{-3} < \frac{9}{-3}$$

• Divide each side of the inequality by the coefficient -3. Because -3 is a negative number, the inequality symbol must be reversed.

$$x < -3$$

The solution set is $(-\infty, -3)$.

Focus on solving an inequality

Solve $\frac{1}{6} - \frac{3}{4}x > \frac{11}{12}$. Write the solution set in set-builder notation.

SOLUTION

$$\frac{1}{6} - \frac{3}{4}x > \frac{11}{12}$$
$$12\left(\frac{1}{6}\right) - 12\left(\frac{3}{4}x\right) > 12\left(\frac{11}{12}\right)$$

• Clear fractions by multiplying each side of the inequality by 12.

$$2 - 9x > 11$$
$$-9x > 9$$

• Subtract 2 from each side.

$$\frac{-9x}{-9} < \frac{9}{-9}$$

• Divide each side by -9. Because -9 is a negative number, reverse the inequality symbol.

$$x < -1$$

The solution set is $\{x \mid x < -1\}$.

Check your understanding 5

Solve $3x + 2 < -4$. Write the solution set in set-builder notation.

SOLUTION See page S-10.

Focus on solving an inequality

Solve $2x - 9 > 4x + 5$. Write the solution set in set-builder notation.

SOLUTION

$$2x - 9 > 4x + 5$$

$$-2x - 9 > 5 \quad \text{• Subtract } 4x \text{ from each side of the inequality.}$$

$$-2x > 14 \quad \text{• Add 9 to each side of the inequality.}$$

$$\frac{-2x}{-2} < \frac{14}{-2} \quad \text{• Divide each side of the inequality by the coefficient } -2. \text{ Because } -2 \text{ is a negative number, reverse the inequality symbol.}$$

$$x < -7$$

The solution set is $\{x|x < -7\}$.

Check your understanding 6

Solve $2x - 1 < 6x + 7$. Write the solution set in set-builder notation.

SOLUTION See page S-10.

Objective 3.5B Practice

1. Solve and graph the solution set: $-2x > 8$. Write the solution set in set-builder notation.
2. Solve $5x - 2 \le 8$. Write the solution set in set-builder notation.
3. Solve $3x + 1 \le 7x - 15$. Write the solution set in set-builder notation.
4. Solve $\frac{1}{2}x - \frac{3}{4} < \frac{7}{4}x - 2$. Write the solution set in interval notation.

Solutions on pp. S-22–S-23.

Objective 3.5C Solve an absolute value inequality

Recall that absolute value represents the distance between two points. For example, the solutions of the absolute value equation $|x - 1| = 3$ are the numbers whose distance from 1 is 3. Therefore, the solutions are -2 and 4. An **absolute value inequality** is an inequality that contains a variable within an absolute value symbol.

The solutions of the absolute value inequality $|x - 1| < 3$ are the numbers whose distance from 1 is less than 3. Therefore, the solutions are the numbers greater than -2 and less than 4. The solution set is $\{x|-2 < x < 4\}$.

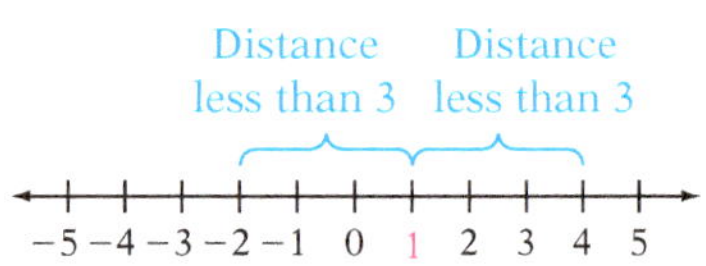

To solve an absolute value inequality of the form $|ax + b| < c$, solve the equivalent compound inequality $-c < ax + b < c$.

Focus on solving an absolute value inequality involving "less than"

Solve: $|3x - 1| < 5$

SOLUTION

$$|3x - 1| < 5$$

$$-5 < 3x - 1 < 5$$ • Solve an equivalent inequality.

$$-5 + 1 < 3x - 1 + 1 < 5 + 1$$

$$-4 < 3x < 6$$

$$\frac{-4}{3} < \frac{3x}{3} < \frac{6}{3}$$

$$-\frac{4}{3} < x < 2$$

The solution set is $\{x \mid -\frac{4}{3} < x < 2\}$.

Take Note ➤ In this objective, we will write all solution sets in set-builder notation.

Check your understanding 7

Solve: $|3x + 2| < 8$

SOLUTION See page S-10.

Focus on solving an absolute value inequality related to statistics

Solve: $\left|\frac{x - 23.7}{5.83}\right| < 2.33$

SOLUTION

$$\left|\frac{x - 23.7}{5.83}\right| < 2.33$$

$$-2.33 < \frac{x - 23.7}{5.83} < 2.33$$

$$5.83(-2.33) < 5.83\left(\frac{x - 23.7}{5.83}\right) < 5.83(2.33)$$ • Multiply each part by 5.83 to clear fractions.

$$-13.5839 < x - 23.7 < 13.5839$$

$$-13.5839 + 23.7 < x - 23.7 + 23.7 < 13.5839 + 23.7$$

$$10.1161 < x < 37.2839$$

The solution set is $\{x \mid 10.1161 < x < 37.2839\}$.

Check your understanding 8

Solve: $\left|\frac{x - 12.36}{4.21}\right| < 1.65$

SOLUTION See page S-10.

The solutions of the absolute value inequality $|x + 1| > 2$ are the numbers whose distance from -1 is greater than 2. Therefore, the solutions are the numbers that are less than -3 or greater than 1. The solution set of $|x + 1| > 2$ is $\{x \mid x < -3\} \cup \{x \mid x > 1\}$.

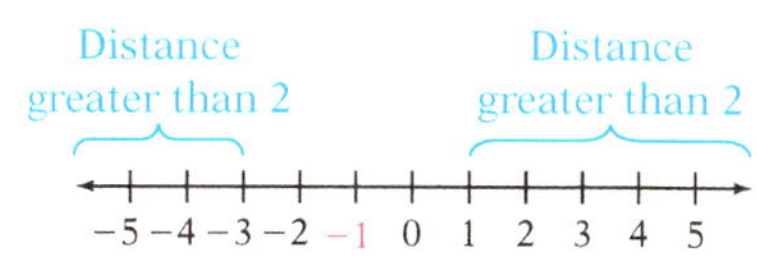

Take Note ➤
Carefully observe the difference between the method used to solve $|ax + b| > c$ shown here and that used to solve $|ax + b| < c$ shown in the preceding example.

To solve an absolute value inequality of the form $|ax + b| > c$, solve the equivalent compound inequality $ax + b < -c$ or $ax + b > c$.

Focus on solving an absolute value inequality involving "greater than"

Solve: $|3 - 2x| > 1$

SOLUTION

$3 - 2x < -1$ or $3 - 2x > 1$

$-2x < -4$ $\qquad -2x > -2$ • Solve each inequality.

$x > 2$ $\qquad x < 1$

$\{x|x > 2\}$ $\qquad \{x|x < 1\}$

The solution set of a compound inequality with the word *or* is the union of the solution sets of the two inequalities.

$\{x|x > 2\} \cup \{x|x < 1\}$

Check your understanding 9

Solve: $|5x + 3| > 8$

SOLUTION See page S-11.

The rules for solving absolute value inequalities are summarized below.

Solutions of Absolute Value Inequalities

To solve an absolute value inequality of the form $|ax + b| < c$, $c > 0$, solve the equivalent compound inequality $-c < ax + b < c$.

To solve an absolute value inequality of the form $|ax + b| > c$, solve the equivalent compound inequality $ax + b < -c$ or $ax + b > c$.

Objective 3.5C Practice

1. Solve: $|2x + 1| < 5$

2. Solve: $|7x - 1| > 13$

3. Solve: $|2x + 7| > -5$

4. Solve: $|4 - 3x| \geq 5$

5. Solve: $|5 - 4x| \leq 13$

6. Solve: $\left|\dfrac{x - 56.4}{20.56}\right| < 1.96$

Solutions on pp. S-23–S-24.

SOLUTIONS TO MODULE 3

Solutions to Check Your Understanding

Section 3.1

Check your understanding 1

-4 is the constant term.

Check your understanding 2

a. $2xy + y^2$

$$\begin{aligned} &2(-4)(2) + (2)^2 && \bullet\ x = -4, y = 2 \\ &= 2(-4)(2) + 4 \\ &= (-8)(2) + 4 \\ &= (-16) + 4 \\ &= -12 \end{aligned}$$

b. $\dfrac{a^2 + b^2}{a + b}$

$$\begin{aligned} \frac{5^2 + (-3)^2}{5 + (-3)} &= \frac{25 + 9}{5 + (-3)} && \bullet\ a = 5, b = -3 \\ &= \frac{34}{2} \\ &= 17 \end{aligned}$$

c. $x^3 - 2(x + y) + z^2$

$$\begin{aligned} &(2)^3 - 2[2 + (-4)] + (-3)^2 && \bullet\ x = 2, y = -4, z = -3 \\ &= 8 - 2(-2) + 9 \\ &= 8 + 4 + 9 \\ &= 12 + 9 \\ &= 21 \end{aligned}$$

Check your understanding 3

a. $-8(-2a + 7b) = 16a - 56b$

b. $3(12x^2 - x + 8) = 36x^2 - 3x + 24$

c. $3(-a^2 - 6a + 7) = -3a^2 - 18a + 21$

Check your understanding 4

a. $3y - 2(y - 7x) = 3y - 2y + 14x$ • Use the Distributive Property.
$= y + 14x$ • Combine like terms.

b. $-2(x - 2y) - (-x + 3y) = -2x + 4y + x - 3y$
$= -x + y$

c. $3y - 2[x - 4(2 - 3y)] = 3y - 2[x - 8 + 12y]$
$= 3y - 2x + 16 - 24y$
$= -2x - 21y + 16$

Section 3.2

Check your understanding 1

$10x - x^2$	$= 3x - 10$	
$10(5) - (5)^2$	$3(5) - 10$	• Replace x by 5.
$50 - 25$	$15 - 10$	• Evaluate.

$25 \neq 5$

No, 5 is not a solution.

Check your understanding 2

$$\begin{aligned} 26 &= y - 14 \\ 26 + 14 &= y - 14 + 14 && \text{• Add 14 to each side.} \\ 40 &= y - 0 && \text{• Simplify.} \\ 40 &= y \end{aligned}$$

The solution is 40.

Check your understanding 3

$$\begin{aligned} c + \frac{3}{4} &= -\frac{1}{4} \\ c + \frac{3}{4} - \frac{3}{4} &= -\frac{1}{4} - \frac{3}{4} && \text{• Subtract } \tfrac{3}{4} \text{ from each side.} \\ c + 0 &= -\frac{4}{4} && \text{• Simplify.} \\ c &= -1 \end{aligned}$$

The solution is -1.

Check your understanding 4

$$\begin{aligned} 4x - 8x &= 16 \\ -4x &= 16 \\ \frac{-4x}{-4} &= \frac{16}{-4} \\ x &= -4 \end{aligned}$$

The solution is -4.

Check your understanding 5

a. $\dfrac{20\text{ lb}}{24\text{ lb}} = \dfrac{20}{24} = \dfrac{5}{6}$

20 lb : 24 lb = 20 : 24 = 5 : 6

20 lb to 24 lb = 20 to 24
= 5 to 6

b. $\dfrac{64\text{ mi}}{8\text{ mi}} = \dfrac{64}{8} = \dfrac{8}{1}$

64 mi : 8 mi = 64 : 8 = 8 : 1

64 mi to 8 mi = 64 to 8 = 8 to 1

Check your understanding 6

a. $\dfrac{\$8.96}{3.5\text{ lb}}$

$8.96 \div 3.5 = 2.56$

The unit rate is $2.56/lb.

b. $\dfrac{260\text{ mi}}{8\text{ h}}$

$$8\overline{)260.0} = 32.5$$

The unit rate is 32.5 mi/h.

Check your understanding 7

STRATEGY

To find the per-foot cost of the pipe, divide the total cost ($22.40) by the number of feet purchased (40 ft).

SOLUTION

$22.40 \div 40 = 0.56$

The cost of the pipe is $.56/ft.

Check your understanding 8

a. $$\frac{7}{12} = \frac{42}{x}$$
$$12 \cdot 42 = 7 \cdot x$$
$$504 = 7x$$
$$72 = x$$

b. $$\frac{4}{5} = \frac{3}{x - 3}$$
$$5 \cdot 3 = 4(x - 3)$$
$$15 = 4x - 12$$
$$27 = 4x$$
$$6.75 = x$$

Check your understanding 9

STRATEGY

To find the number of gallons, write and solve a proportion using n to represent the number of gallons needed to travel 832 mi.

SOLUTION

$$\frac{396 \text{ mi}}{11 \text{ gal}} = \frac{832 \text{ mi}}{n \text{ gal}}$$
$$11 \cdot 832 = 396 \cdot n$$
$$9152 = 396n$$
$$23.1 \approx n$$

To travel 832 mi, approximately 23.1 gal of gas are needed.

Check your understanding 10

a. $125\% = 125 \times 0.01 = 1.25$

$$125\% = 125 \times \frac{1}{100}$$
$$= \frac{125}{100} = \frac{5}{4} = 1\frac{1}{4}$$

b. $8.5\% = 8.5 \times 0.01 = 0.085$

$$8.5\% = 8.5 \times \frac{1}{100}$$
$$= 8\frac{1}{2} \times \frac{1}{100}$$ • $8.5 = 8\frac{5}{10} = 8\frac{1}{2}$
$$= \frac{17}{2} \times \frac{1}{100}$$
$$= \frac{17}{200}$$ • Multiply the fractions.

c. $0.25\% = 0.25 \times 0.01 = 0.0025$

$$0.25\% = 0.25 \times \frac{1}{100}$$

$$= \frac{1}{4} \times \frac{1}{100} \quad \bullet\ 0.25 = \frac{25}{100} = \frac{1}{4}$$

$$= \frac{1}{400} \quad \bullet\ \text{Multiply the fractions.}$$

Check your understanding 11

$$33\frac{1}{3}\% = 33\frac{1}{3} \times \frac{1}{100}$$

$$= \frac{100}{3} \times \frac{1}{100}$$

$$= \frac{100}{300} = \frac{1}{3}$$

Check your understanding 12

a. $0.048 = 0.048 \times 100\% = 4.8\%$

$3.6 = 3.6 \times 100\% = 360\%$

b. $\frac{5}{16} \times \frac{100\%}{1} = \frac{500\%}{16} = 31.25\%$ • Write in decimal form.

c. $\frac{5}{6} = \frac{5}{6} \times \frac{100\%}{1} = \frac{500\%}{6} = 83\frac{1}{3}\%$ • Write the answer with a fraction.

Check your understanding 13

a. STRATEGY

To find the amount, use the basic percent equation. Percent $= 66\frac{2}{3}\% = \frac{2}{3}$, base $= 45$, amount $= n$

SOLUTION

$$\text{Percent} \cdot \text{base} = \text{amount}$$

$$\frac{2}{3}(45) = n$$

$$30 = n$$

30 is $66\frac{2}{3}\%$ of 45.

b. STRATEGY

To find the percent, use the basic percent equation. Percent $= n$, base $= 40$, amount $= 25$

SOLUTION

$$\text{Percent} \cdot \text{base} = \text{amount}$$

$$n \cdot 40 = 25$$

$$\frac{40n}{40} = \frac{25}{40}$$

$$n = 0.625 = 62.5\%$$

25 is 62.5% of 40.

c. STRATEGY

To find the base, use the basic percent equation. Percent $= 16\frac{2}{3}\% = \frac{1}{6}$, base $= n$, amount $= 15$

SOLUTION

$$\begin{aligned} \text{Percent} \cdot \text{base} &= \text{amount} \\ \frac{1}{6} \cdot n &= 15 \\ 6 \cdot \frac{1}{6} n &= 15 \cdot 6 \\ n &= 90 \end{aligned}$$

$16\frac{2}{3}\%$ of 90 is 15.

Check your understanding 14

STRATEGY

To find the percent, use the basic percent equation. Percent $= n$, base $= 4330$, amount $= 649.50$

SOLUTION

$$\begin{aligned} \text{Percent} \cdot \text{base} &= \text{amount} \\ n \cdot 4330 &= 649.50 \\ \frac{4330n}{4330} &= \frac{649.50}{4330} \\ n &= 0.15 \end{aligned}$$

15% of the instructor's salary is deducted for income tax.

Check your understanding 15

STRATEGY

To find how much Clarissa must deposit into the account:

- Find the amount of interest earned on the municipal bond by solving $I = Prt$ for I. Use $P = 1000$, $r = 6.4\% = 0.064$, and $t = 1$.
- Solve $I = Prt$ for P using the amount of interest earned on the municipal bond as I. $r = 8\% = 0.08$, and $t = 1$.

SOLUTION

$$\begin{aligned} I &= Prt \\ &= 1000(0.064)(1) = 64 \end{aligned}$$

The interest earned on the municipal bond was \$64.

$$\begin{aligned} I &= Prt \\ 64 &= P(0.08)(1) && \bullet\ I = 64,\ r = 0.08,\ t = 1 \\ 64 &= 0.08P \\ \frac{64}{0.08} &= \frac{0.08P}{0.08} \\ 800 &= P \end{aligned}$$

Clarissa must invest \$800 in the account earning 8%.

Check your understanding 16

STRATEGY

To find the percent increase in fuel efficiency:

- Find the amount of increase in fuel efficiency.
- Use the basic percent equation. Percent $= n$, base $= 17.5$, amount $=$ amount of increase

SOLUTION

$18.2 - 17.5 = 0.7$

$$\begin{aligned} \text{Percent} \cdot \text{base} &= \text{amount} \\ n \cdot 17.5 &= 0.7 \\ \frac{17.5n}{17.5} &= \frac{0.7}{17.5} \\ n &= 0.04 \end{aligned}$$

The percent increase in fuel efficiency is 4%.

Check your understanding 17

STRATEGY

To find the market value of the car:

- Solve the basic percent equation to find the amount of decrease in value. Percent = 24% = 0.24, base = 47,000, amount = n
- Subtract the amount of decrease from the cost.

SOLUTION

$$\begin{aligned} \text{Percent} \cdot \text{base} &= \text{amount} \\ 0.24 \cdot 47{,}000 &= n \\ 11{,}280 &= n \end{aligned}$$

$47{,}000 - 11{,}280 = 35{,}720$

The market value of the car is \$35,720.

Section 3.3

Check your understanding 1

$$\begin{aligned} 2 &= 11 + 3x \\ 2 - 11 &= 11 - 11 + 3x && \text{• Subtract 11.} \\ -9 &= 3x \\ \frac{-9}{3} &= \frac{3x}{3} && \text{• Divide by 3.} \\ -3 &= x \end{aligned}$$

The solution is -3.

Check your understanding 2

$$\begin{aligned} \frac{2}{3}x + 3 &= \frac{7}{2} \\ 6\left(\frac{2}{3}x + 3\right) &= 6\left(\frac{7}{2}\right) \\ 6\left(\frac{2}{3}x\right) + 6(3) &= 6\left(\frac{7}{2}\right) && \text{• Distributive Property} \\ 4x + 18 &= 21 \\ 4x + 18 - 18 &= 21 - 18 && \text{• Subtract 18.} \\ 4x &= 3 \\ \frac{4x}{4} &= \frac{3}{4} && \text{• Divide by 4.} \\ x &= \frac{3}{4} \end{aligned}$$

The solution is $\frac{3}{4}$.

Check your understanding 3

$$-2.33 = \frac{x + 1.7}{2.71}$$

$$2.71(-2.33) = 2.71\left(\frac{x + 1.7}{2.71}\right)$$ • Multiply each side by 2.71.

$$-6.3143 = x + 1.7$$ • Simplify.

$$-6.3143 - 1.7 = x + 1.7 - 1.7$$ • Subtract 1.7 from each side.

$$-8.0143 = x$$

The solution is -8.0143.

Check your understanding 4

$$5x + 4 = 6 + 10x$$

$$5x - 10x + 4 = 6 + 10x - 10x$$ • Subtract $10x$.

$$-5x + 4 = 6$$

$$-5x + 4 - 4 = 6 - 4$$ • Subtract 4.

$$-5x = 2$$

$$\frac{-5x}{-5} = \frac{2}{-5}$$ • Divide by -5.

$$x = -\frac{2}{5}$$

The solution is $-\frac{2}{5}$.

Check your understanding 5

$$5x - 10 - 3x = 6 - 4x$$

$$2x - 10 = 6 - 4x$$ • Combine like terms.

$$2x + 4x - 10 = 6 - 4x + 4x$$ • Add $4x$.

$$6x - 10 = 6$$

$$6x - 10 + 10 = 6 + 10$$ • Add 10.

$$6x = 16$$

$$\frac{6x}{6} = \frac{16}{6}$$ • Divide by 6.

$$x = \frac{8}{3}$$

The solution is $\frac{8}{3}$.

Check your understanding 6

$$5x - 4(3 - 2x) = 2(3x - 2) + 6$$

$$5x - 12 + 8x = 6x - 4 + 6$$ • Distributive Property

$$13x - 12 = 6x + 2$$

$$13x - 6x - 12 = 6x - 6x + 2$$ • Subtract $6x$.

$$7x - 12 = 2$$

$$7x - 12 + 12 = 2 + 12$$ • Add 12.

$$7x = 14$$

$$\frac{7x}{7} = \frac{14}{7}$$ • Divide by 7.

$$x = 2$$

The solution is 2.

Section 3.4

Check your understanding 1

$$5x - 2y = 10$$
$$5x - 5x - 2y = -5x + 10 \quad \text{• Subtract } 5x.$$
$$-2y = -5x + 10$$
$$\frac{-2y}{-2} = \frac{-5x + 10}{-2} \quad \text{• Divide by } -2.$$
$$y = \frac{5}{2}x - 5$$

Check your understanding 2

$$s = \frac{A + L}{2}$$
$$2 \cdot s = 2\left(\frac{A + L}{2}\right) \quad \text{• Multiply by 2.}$$
$$2s = A + L$$
$$2s - A = A - A + L \quad \text{• Subtract } A.$$
$$2s - A = L$$

Check your understanding 3

$$S = rS + C$$
$$S - rS = rS - rS + C \quad \text{• Subtract } rS.$$
$$S - rS = C$$
$$(1 - r)S = C \quad \text{• Factor.}$$
$$\frac{(1 - r)S}{1 - r} = \frac{C}{1 - r} \quad \text{• Divide by } 1 - r.$$
$$S = \frac{C}{1 - r}$$

Check your understanding 4

$$m = x - \frac{ts}{n}$$
$$n \cdot m = n\left(x - \frac{ts}{n}\right) \quad \text{• Multiply each side by } n.$$
$$nm = nx - ts \quad \text{• Simplify.}$$
$$nm - nx = nx - nx - ts \quad \text{• Subtract } nx \text{ from each side.}$$
$$nm - nx = -ts$$
$$n(m - x) = -ts \quad \text{• Factor.}$$
$$\frac{n(m - x)}{m - x} = \frac{-ts}{m - x} \quad \text{• Divide each side by } m - x.$$
$$n = \frac{-ts}{m - x}$$

Check your understanding 5

$$|2x - 3| = 5$$

$2x - 3 = 5$	$2x - 3 = -5$	
$2x = 8$	$2x = -2$	• Add 3.
$x = 4$	$x = -1$	• Divide by 2.

The solutions are 4 and -1.

Check your understanding 6

$|x - 3| = -2$

There is no solution to this equation because the absolute value of a number must be nonnegative.

Check your understanding 7

$$
\begin{aligned}
5 - |3x + 5| &= 3 \\
-|3x + 5| &= -2 && \text{• Subtract 5.} \\
|3x + 5| &= 2 && \text{• Multiply by } -1.
\end{aligned}
$$

$$
\begin{aligned}
3x + 5 &= 2 & \qquad 3x + 5 &= -2 \\
3x &= -3 & 3x &= -7 \\
x &= -1 & x &= -\frac{7}{3}
\end{aligned}
$$

The solutions are -1 and $-\frac{7}{3}$.

Check your understanding 8

$$
\begin{aligned}
\sqrt{4x} + 3 &= 7 && \text{• Subtract 3 from each side.} \\
\sqrt{4x} &= 4 \\
(\sqrt{4x})^2 &= 4^2 && \text{• Square both sides.} \\
4x &= 16 && \text{• Solve for } x. \\
x &= 4
\end{aligned}
$$

Check:

$$
\begin{array}{r|l}
\sqrt{4x} + 3 & 7 \\
\hline
\sqrt{4 \cdot 4} + 3 & 7 \\
\sqrt{16} + 3 & 7 \\
4 + 3 & 7 \\
7 & = 7 \quad \text{• A true equation}
\end{array}
$$

The solution is 4.

Section 3.5

Check your understanding 1

a. $\{x | x \le 3\}$ is the set of real numbers less than or equal to 3. This set extends forever in the negative direction. In interval notation, this set is written $(-\infty, 3]$.

b. $\{x | -5 \le x \le -3\}$ is the set of real numbers between -5 and -3, including -5 and -3. In interval notation, this set is written $[-5, -3]$.

Check your understanding 2

a. The interval $(-3, \infty)$ is the set of real numbers greater than -3. In set-builder notation, this set is written $\{x | x > -3\}$.

b. The interval $[0, 4)$ is the set of real numbers between 0 and 4, including 0 and excluding 4. In set-builder notation, this set is written $\{x | 0 \le x < 4\}$.

Check your understanding 3

The graph is the set of real numbers between 2 and 5, including 2 and 5. Use brackets at 2 and 5.

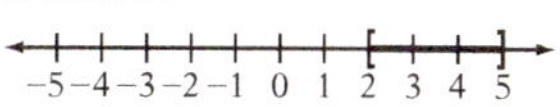

Check your understanding 4

a. The graph is the set of real numbers between -4 and 4, including -4 and 4. Use brackets at -4 and 4.

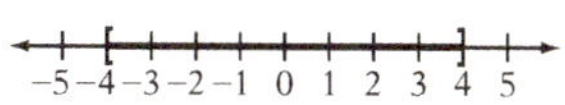

b. The graph is the set of real numbers greater than -3. Use a parenthesis at -3.

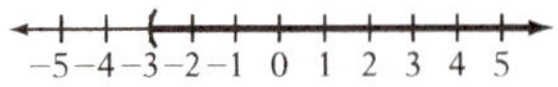

Check your understanding 5

$$3x + 2 < -4$$

$$3x < -6$$ • Subtract 2 from each side.

$$\frac{3x}{3} < \frac{-6}{3}$$ • Divide each side by 3. Because 3 is a positive number, the inequality symbol remains the same.

$$x < -2$$

The solution set is $\{x|x < -2\}$.

Check your understanding 6

$$2x - 1 < 6x + 7$$

$$-4x - 1 < 7$$ • Subtract $6x$ from each side.

$$-4x < 8$$ • Add 1 to each side.

$$\frac{-4x}{-4} > \frac{8}{-4}$$ • Divide each side by -4 and reverse the inequality symbol.

$$x > -2$$

$$\{x|x > -2\}$$

Check your understanding 7

$$|3x + 2| < 8$$

$$-8 < 3x + 2 < 8$$ • Solve the equivalent compound inequality.

$$-8 - 2 < 3x + 2 - 2 < 8 - 2$$

$$-10 < 3x < 6$$

$$\frac{-10}{3} < \frac{3x}{3} < \frac{6}{3}$$

$$-\frac{10}{3} < x < 2$$

$$\left\{x \middle| -\frac{10}{3} < x < 2\right\}$$

Check your understanding 8

$$\left|\frac{x - 12.36}{4.21}\right| < 1.65$$

$$-1.65 < \frac{x - 12.36}{4.21} < 1.65$$

$$4.21(-1.65) < 4.21\left(\frac{x - 12.36}{4.21}\right) < 4.21(1.65)$$ • Multiply each part by 4.21 to clear fractions.

$$-6.9465 < x - 12.36 < 6.9465$$

$$5.4135 < x < 19.3065$$

Check your understanding 9

$|5x + 3| > 8$

$5x + 3 < -8 \quad \text{or} \quad 5x + 3 > 8$ • Solve the equivalent compound inequality.

$5x < -11 \qquad 5x > 5$

$x < -\frac{11}{5} \qquad x > 1$

$\left\{x \middle| x < -\frac{11}{5}\right\} \qquad \{x|x > 1\}$

$\left\{x \middle| x < -\frac{11}{5}\right\} \cup \{x|x > 1\}$

Solutions to Objective Practice Exercises

Objective 3.1A

1. $3a + 2b$

$3(2) + 2(3) = 6 + 6 = 12$

2. $bc \div (2a)$

$3(-4) \div (2 \cdot 2) = -12 \div 4 = -3$

3. $(b - 2a)^2 + bc$

$$\begin{aligned}(3 - 2(2))^2 + 3(-4) &= (3 - 4)^2 + 3(-4)\\ &= (-1)^2 + 3(-4) = 1 + 3(-4)\\ &= 1 + (-12) = -11\end{aligned}$$

4. $(d - a)^2 - 3c$

$$\begin{aligned}[3 - (-2)]^2 - 3(-1) &= 5^2 - 3(-1)\\ &= 25 - (-3) = 28\end{aligned}$$

5. $-\frac{3}{4}b + \frac{1}{2}(ac + bd)$

$$\begin{aligned}-\frac{3}{4}(4) + \frac{1}{2}[(-2)(-1) + 4(3)] &= -\frac{3}{4}(4) + \frac{1}{2}[2 + 12]\\ &= -\frac{3}{4}(4) + \frac{1}{2}(14)\\ &= -3 + 7 = 4\end{aligned}$$

Objective 3.1B

1. $4(x^2 - 3x + 5) = 4x^2 - 12x + 20$

2. $\frac{3}{4}(2x - 6y + 8) = \frac{3}{2}x - \frac{9}{2}y + 6$

3. $-(8b^2 - 6b + 9) = -8b^2 + 6b - 9$

4. $4x - 2(3x + 8) = 4x - 6x - 16 = -2x - 16$

5. $5n - (7 - 2n) = 5n - 7 + 2n = 7n - 7$

6. $$\begin{aligned}12(y - 2) + 3(7 - 3y) &= 12y - 24 + 21 - 9y\\ &= 3y - 3\end{aligned}$$

7. $$\begin{aligned}-2[3x + 2(4 - x)] &= -2[3x + 8 - 2x]\\ &= -2[x + 8] = -2x - 16\end{aligned}$$

Objective 3.2A

1. $$\begin{array}{r|l} \multicolumn{2}{c}{3y - 4 = 2y} \\ \hline 3(4) - 4 & (2)4 \\ 12 - 4 & 8 \end{array}$$
$$8 = 8$$
Yes, 4 is a solution.

2. $$\begin{aligned} -9 &= 5 + x \\ -9 - 5 &= 5 - 5 + x \\ -14 &= x \end{aligned}$$
The solution is -14.

3. $$\begin{aligned} x - \frac{2}{5} &= \frac{3}{5} \\ x - \frac{2}{5} + \frac{2}{5} &= \frac{3}{5} + \frac{2}{5} \\ x &= 1 \end{aligned}$$
The solution is 1.

4. $$\begin{aligned} \frac{x}{4} &= 3 \\ 4\left(\frac{1}{4}x\right) &= 4(3) \\ x &= 12 \end{aligned}$$
The solution is 12.

5. $$\begin{aligned} \frac{2}{5}x &= 6 \\ \frac{5}{2}\left(\frac{2}{5}x\right) &= \frac{5}{2}(6) \\ x &= 15 \end{aligned}$$
The solution is 15.

6. $$\begin{aligned} 7d - 4d &= 9 \\ 3d &= 9 \\ \frac{3d}{3} &= \frac{9}{3} \\ d &= 3 \end{aligned}$$
The solution is 3.

Objective 3.2B

1. $$\frac{10\text{ ft}}{4\text{ s}} = 2.5\text{ ft/s}$$

2. $$\frac{\$11.05}{3.4\text{ lb}} = \$3.25/\text{lb}$$

3. $$\frac{639\text{ mi}}{15\text{ gal}} = 42.6\text{ mi/gal}$$

4. $$\begin{aligned} \frac{3}{8} &= \frac{n}{12} \\ 3 \cdot 12 &= 8 \cdot n \\ 36 &= 8n \\ \frac{36}{8} &= \frac{8n}{8} \\ 4.5 &= n \end{aligned}$$

5. $$\frac{4}{3} = \frac{n-2}{6}$$
$$4 \cdot 6 = (3n - 2)$$
$$24 = 3n - 6$$
$$30 = 3n$$
$$\frac{30}{3} = \frac{3n}{3}$$
$$10 = n$$

6. STRATEGY

To find the cost, write and solve a proportion using c to represent the cost of the life insurance.

SOLUTION

$$\frac{\$15.22}{\$1000} = \frac{c}{\$75{,}000}$$
$$15.22 \cdot 75{,}000 = 1000 \cdot c$$
$$1{,}141{,}500 = 1000c$$
$$1141.50 = c$$

The cost of the life insurance is \$1141.50.

7. STRATEGY

To find the number of grams of protein in 454 g of pasta, write and solve a proportion using n to represent the number of grams of protein.

SOLUTION

$$\frac{7 \text{ g of protein}}{56 \text{ g of pasta}} = \frac{x \text{ g of protein}}{454 \text{ g of pasta}}$$
$$7 \cdot 454 = 56x$$
$$3178 = 56x$$
$$\frac{3178}{56} = \frac{56x}{56}$$
$$56.75 = x$$

There are 56.75 g of protein in 454 g of pasta.

8. STRATEGY

To find the number of miles, write and solve a proportion using n to represent the number of miles.

SOLUTION

3 years is 36 months.

$$\frac{22{,}000 \text{ mi}}{4 \text{ months}} = \frac{n \text{ mi}}{36 \text{ months}}$$
$$22{,}000 \cdot 36 = 4 \cdot n$$
$$792{,}000 = 4n$$
$$\frac{792{,}000}{4} = \frac{4n}{4}$$
$$198{,}000 = n$$

The account executive will drive 198,000 mi.

Objective 3.2C

1. $36\% = 36 \times 0.01 = 0.36$

   $36\% = 36 \times \frac{1}{100} = \frac{36}{100} = \frac{9}{25}$

2. $6.2\% = 6.2 \times 0.01 = 0.062$

   $6.2\% = 6.2 \times \frac{1}{100} = \frac{6.2}{100} = \frac{62}{1000} = \frac{31}{500}$

3. $0.25\% = 0.25 \times 0.01 = 0.0025$

   $0.25\% = 0.25 \times \frac{1}{100} = \frac{0.25}{100} = \frac{25}{10{,}000} = \frac{1}{400}$

4. $12\frac{1}{2}\% = 12\frac{1}{2} \times \frac{1}{100} = \frac{25}{2} \times \frac{1}{100}$

   $= \frac{25}{200} = \frac{1}{8}$

5. $0.73 = 0.73 \times 100\% = 73\%$

6. $\frac{17}{20} = \frac{17}{20} \times 100\% = \frac{1700}{20}\% = 85\%$

7. $\frac{9}{4} = \frac{9}{4} \times 100\% = \frac{900}{4}\% = 225\%$

8. STRATEGY

   To find the percent, solve the basic percent equation.

   Percent $= n$, base $= 16$, amount $= 6$

   SOLUTION

   Percent · base = amount

   $n \cdot 16 = 6$

   $n = \frac{6}{16} = 0.375 = 37.5\%$

   6 is 37.5% of 16.

9. STRATEGY

   To find the amount, solve the basic percent equation.

   Percent $= 10.7\% = 0.107$, base $= 485$, amount $= n$

   SOLUTION

   Percent · base = amount

   $0.107 \cdot 485 = n$

   $51.895 = n$

   10.7% of 485 is 51.895.

10. STRATEGY

    To find the percent, solve the basic percent equation.

    Percent $= n$, base $= 2000$, amount $= 54$

    SOLUTION

    Percent · base = amount

    $n \cdot 2000 = 54$

    $n = \frac{54}{2000} = 0.027 = 2.7\%$

    54 is 2.7% of 2000.

Objective 3.2D

1. STRATEGY

To find the percent, solve the basic percent equation $P \cdot B = A$ using $B = 2252$ and $A = 1850$.

SOLUTION

$$\begin{aligned} P \cdot B &= A \\ P \cdot 2252 &= 1850 \\ P &\approx 0.821 \end{aligned}$$

The percent of the vacation cost that is charged is 82.1%.

2. STRATEGY

To find the simple interest rate, solve the simple interest equation for r, using $I = \$72$, $P = \$1200$, and $t = 8$ months $= \frac{8}{12}$ year.

SOLUTION

$$\begin{aligned} I &= Prt \\ 72 &= (1200)r\left(\frac{8}{12}\right) \\ 72 &= 800r \\ \frac{72}{800} &= \frac{800r}{800} \\ 0.09 &= r \end{aligned}$$

The annual simple interest rate is 9%.

3. STRATEGY

To find the principal, solve the simple interest equation for P, using $I = \$300$, $r = 8\% = 0.08$, and $t = 2$ years.

SOLUTION

$$\begin{aligned} I &= Prt \\ 300 &= P(0.08)(2) \\ 300 &= 0.16P \\ \frac{300}{0.16} &= \frac{0.16P}{0.16} \\ 1875 &= P \end{aligned}$$

Andrea must invest \$1875.

4. STRATEGY

To find the percent increase:

- Find the increase in the number of Olympic events.
- Use the basic percent equation. Percent $= n$, base $= 14$, amount $=$ amount of increase

SOLUTION

$98 - 14 = 84$

$$\begin{aligned} \text{Percent} \cdot \text{base} &= \text{amount} \\ n \cdot 14 &= 84 \\ n &= \frac{84}{14} \\ n &= 6 = 600\% \end{aligned}$$

The percent increase in the number of Olympic events was 600%.

5. STRATEGY

To find the percent increase:

- Find the increase in the number of millionaire households.
- Use the basic percent equation. Percent = n, base = 350,000, amount = amount of increase

SOLUTION

$8{,}400{,}000 - 350{,}000 = 8{,}050{,}000$

$$\text{Percent} \cdot \text{base} = \text{amount}$$
$$n \cdot 350{,}000 = 8{,}050{,}000$$
$$n = \frac{8{,}050{,}000}{350{,}000}$$
$$n = 23 = 2300\%$$

The percent increase in the number of millionaire households was 2300%.

6. STRATEGY

To find the percent increase:

- Find the increase in the cost for public college.
- Use the basic percent equation. Percent = n, base = 5074, amount = amount of increase

SOLUTION

$7020 - 5074 = 1946$

$$\text{Percent} \cdot \text{base} = \text{amount}$$
$$n \cdot 5074 = 1946$$
$$n = \frac{1946}{5074}$$
$$n \approx 0.38352 \approx 38.4\%$$

The percent increase in the average cost of attending a public college was 38.4%.

7. STRATEGY

To find the percent decrease:

- Find the decrease in the number of passports issued.
- Use the basic percent equation. Percent = n, base = 18.3, amount = amount of decrease

SOLUTION

$18.3 - 13.8 = 4.5$

$$\text{Percent} \cdot \text{base} = \text{amount}$$
$$n \cdot 18.3 = 4.5$$
$$n = \frac{4.5}{18.3}$$
$$n \approx 0.2459 \approx 24.6\%$$

The percent decrease in the number of passports was 24.6%.

8. STRATEGY

To find the decrease in the average waiting time:

- Find the decrease in time.
- Use the basic percent equation to find the percent decrease in waiting time.
  Percent = n, base = 3.8, amount = decrease in time

SOLUTION

$3.8 - 2.5 = 1.3$

$$\begin{aligned} \text{Percent} \cdot \text{base} &= \text{amount} \\ n \cdot 3.8 &= 1.3 \\ n &= \frac{1.3}{3.8} \\ n &\approx 0.3421 \approx 34.2\% \end{aligned}$$

The waiting time decreased by approximately 34.2%.

9. STRATEGY

To find the value after one year:

- Find the decrease in value by using the basic percent equation. Percent = 30% = 0.30, base = 21,900, amount = n
- Subtract the decrease in value from 21,900.

SOLUTION

$$\begin{aligned} \text{Percent} \cdot \text{base} &= \text{amount} \\ 0.30 \cdot 21{,}900 &= n \\ 6570 &= n \end{aligned}$$

$21{,}900 - 6570 = 15{,}330$

The value of the car after one year is approximately \$15,330.

Objective 3.3A

1.

$$\begin{aligned} 2x - 5 &= -11 \\ 2x - 5 + 5 &= -11 + 5 \\ 2x &= -6 \\ \frac{2x}{2} &= \frac{-6}{2} \\ x &= -3 \end{aligned}$$

The solution is -3.

2.

$$\begin{aligned} 7 - 9a &= 4 \\ 7 - 7 - 9a &= 4 - 7 \\ -9a &= -3 \\ \frac{-9a}{-9} &= \frac{-3}{-9} \\ a &= \frac{3}{9} = \frac{1}{3} \end{aligned}$$

The solution is $\frac{1}{3}$.

3.

$$\begin{aligned} \frac{2}{3}x - \frac{5}{6} &= -\frac{1}{3} \\ 6\left(\frac{2}{3}x - \frac{5}{6}\right) &= 6\left(-\frac{1}{3}\right) \\ 4x - 5 &= -2 \\ 4x &= 3 \\ x &= \frac{3}{4} \end{aligned}$$

The solution is $\frac{3}{4}$.

4.
$$\begin{aligned}\frac{x-1.5}{1.65} &= -1.8\\ 1.65\left(\frac{x-1.5}{1.65}\right) &= 1.65(-1.8)\\ x-1.5 &= -2.97\\ x-1.5+1.5 &= -2.97+1.5\\ x &= -1.47\end{aligned}$$

The solution is -1.47.

5.
$$\begin{aligned}5y+9+2y &= 23\\ 7y+9 &= 23\\ 7y+9-9 &= 23-9\\ 7y &= 14\\ \frac{7y}{7} &= \frac{14}{7}\\ y &= 2\end{aligned}$$

The solution is 2.

Objective 3.3B

1.
$$\begin{aligned}15x-2 &= 4x-13\\ 15x-4x-2 &= 4x-4x-13\\ 11x-2 &= -13\\ 11x-2+2 &= -13+2\\ 11x &= -11\\ \frac{11x}{11} &= \frac{-11}{11}\\ x &= -1\end{aligned}$$

The solution is -1.

2.
$$\begin{aligned}0.2b+3 &= 0.5b+12\\ 0.2b-0.5b+3 &= 0.5b-0.5b+12\\ -0.3b+3 &= 12\\ -0.3b+3-3 &= 12-3\\ -0.3b &= 9\\ \frac{-0.3b}{-0.3} &= \frac{9}{-0.3}\\ b &= -30\end{aligned}$$

The solution is -30.

3.
$$\begin{aligned}5+7x &= 11+9x\\ 5+7x-9x &= 11+9x-9x\\ 5-2x &= 11\\ 5-5-2x &= 11-5\\ -2x &= 6\\ \frac{-2x}{-2} &= \frac{6}{-2}\\ x &= -3\end{aligned}$$

The solution is -3.

4.

$$\begin{aligned} 7x - 8 &= x - 3 \\ 7x - x - 8 &= x - x - 3 \\ 6x - 8 &= -3 \\ 6x - 8 + 8 &= -3 + 8 \\ 6x &= 5 \\ \frac{6x}{6} &= \frac{5}{6} \\ x &= \frac{5}{6} \end{aligned}$$

The solution is $\frac{5}{6}$.

5.

$$\begin{aligned} 7x + 3 &= 5x - 7 \\ 7x - 5x + 3 &= 5x - 5x - 7 \\ 2x + 3 &= -7 \\ 2x + 3 - 3 &= -7 - 3 \\ 2x &= -10 \\ \frac{2x}{2} &= \frac{-10}{2} \\ x &= -5 \end{aligned}$$

$$\begin{aligned} &3x - 2 \\ &= 3(-5) - 2 \\ &= -15 - 2 \\ &= -17 \end{aligned}$$

The solution is –17.

Objective 3.3C

1.

$$\begin{aligned} 9n - 3(2n - 1) &= 15 \\ 9n - 6n + 3 &= 15 \\ 3n + 3 &= 15 \\ 3n + 3 - 3 &= 15 - 3 \\ 3n &= 12 \\ \frac{3n}{3} &= \frac{12}{3} \\ n &= 4 \end{aligned}$$

The solution is 4.

2.

$$\begin{aligned} 5(3 - 2y) + 4y &= 3 \\ 15 - 10y + 4y &= 3 \\ 15 - 6y &= 3 \\ 15 - 15 - 6y &= 3 - 15 \\ -6y &= -12 \\ \frac{-6y}{-6} &= \frac{-12}{-6} \\ y &= 2 \end{aligned}$$

The solution is 2.

3. $$\begin{aligned}0.05(4 - x) + 0.1x &= 0.32\\ 0.2 - 0.05x + 0.1x &= 0.32\\ 0.2 + 0.05x &= 0.32\\ 0.2 - 0.2 + 0.05x &= 0.32 - 0.2\\ 0.05x &= 0.12\\ \frac{0.05x}{0.05} &= \frac{0.12}{0.05}\\ x &= 2.4\end{aligned}$$

The solution is 2.4.

4. $$\begin{aligned}7 - (5 - 8x) &= 4x + 3\\ 7 - 5 + 8x &= 4x + 3\\ 2 + 8x &= 4x + 3\\ 2 + 8x - 4x &= 4x - 4x + 3\\ 2 + 4x &= 3\\ 2 - 2 + 4x &= 3 - 2\\ 4x &= 1\\ \frac{4x}{4} &= \frac{1}{4}\\ x &= \frac{1}{4}\end{aligned}$$

The solution is $\frac{1}{4}$.

Objective 3.4A

1. $$\begin{aligned}4x + 3y &= 12\\ 4x + 3y - 3y &= -3y + 12\\ 4x &= -3y + 12\\ \frac{4x}{4} &= \frac{-3y + 12}{4}\\ x &= -\frac{3}{4}y + 3\end{aligned}$$

2. $$\begin{aligned}E &= IR\\ \frac{E}{I} &= \frac{IR}{I}\\ \frac{E}{I} &= R\end{aligned}$$

3. $$\begin{aligned}F &= \frac{9}{5}C + 32\\ 5F &= 5\left(\frac{9}{5}C + 32\right)\\ 5F &= 9C + 160\\ 5F - 160 &= 9C + 160 - 160\\ 5F - 160 &= 9C\\ \frac{5F - 160}{9} &= \frac{9C}{9}\\ \frac{5F - 160}{9} &= C\end{aligned}$$

4.
$$s = a(x - vt)$$
$$s = ax - avt$$
$$s - ax = ax - ax - avt$$
$$s - ax = -avt$$
$$\frac{s - ax}{-av} = \frac{-avt}{-av}$$
$$-\frac{s - ax}{av} = t$$

5.
$$a = S - Sr$$
$$a = S(1 - r)$$
$$\frac{a}{1 - r} = \frac{S(1 - r)}{1 - r}$$
$$\frac{a}{1 - r} = S$$

Objective 3.4B

1.
$$|a - 2| = 0$$
$$a - 2 = 0$$
$$a = 2$$

The solution is 2.

2.
$$|2x - 3| + 4 = -4$$
$$|2x - 3| = -8$$

There is no solution to this equation because the absolute value of a number must be nonnegative.

3.
$$|3t + 2| + 3 = 4$$
$$|3t + 2| = 1$$
$$3t + 2 = 1 \quad \text{or} \quad 3t + 2 = -1$$
$$3t = -1 \qquad\qquad 3t = -3$$
$$t = -\frac{1}{3} \qquad\qquad t = -1$$

The solutions are $-\frac{1}{3}$ and -1.

4.
$$5 - |2x + 1| = 5$$
$$-|2x + 1| = 0$$
$$2x + 1 = 0$$
$$2x = -1$$
$$x = -\frac{1}{2}$$

The solution is $-\frac{1}{2}$.

5.
$$8 - |1 - 3x| = -1$$
$$-|1 - 3x| = -9$$
$$|1 - 3x| = 9$$
$$1 - 3x = 9 \quad \text{or} \quad 1 - 3x = -9$$
$$-3x = 8 \qquad\qquad -3x = -10$$
$$x = -\frac{8}{3} \qquad\qquad x = \frac{10}{3}$$

The solutions are $-\frac{8}{3}$ and $\frac{10}{3}$.

Objective 3.4C

1. $$\begin{aligned} \sqrt{5x+4} &= 3 \\ (\sqrt{5x+4})^2 &= 3^2 \\ 5x+4 &= 9 \\ 5x &= 5 \\ x &= 1 \end{aligned}$$

Check:
$$\begin{aligned} \sqrt{5x+4} &= 3 \\ \sqrt{5\cdot 1+4} &\stackrel{?}{=} 3 \\ \sqrt{5+4} &\stackrel{?}{=} 3 \\ \sqrt{9} &\stackrel{?}{=} 3 \\ 3 &= 3 \end{aligned}$$

The solution is 1.

2. $$\begin{aligned} \sqrt{5x}+8 &= 23 \\ \sqrt{5x} &= 15 \\ (\sqrt{5x})^2 &= 15^2 \\ 5x &= 225 \\ x &= 45 \end{aligned}$$

Check:
$$\begin{aligned} \sqrt{5x}+8 &= 23 \\ \sqrt{5\cdot 45}+8 &\stackrel{?}{=} 23 \\ \sqrt{225}+8 &\stackrel{?}{=} 23 \\ 15+8 &\stackrel{?}{=} 23 \\ 23 &= 23 \end{aligned}$$

The solution is 45.

Objective 3.5A

1. $\{x \mid x < -70, x \in \text{integers}\}$

2. $\{x \mid x > 8\}$

3. $(-\infty, 0]$

4. $\{x \mid 4 < x \le 9\}$

5. $(-\infty, 3]$

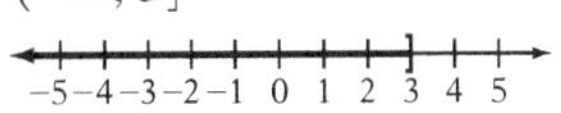

6. $\{x \mid -4 < x < 1\}$

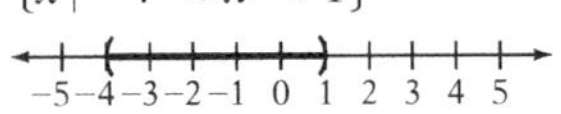

Objective 3.5B

1. $$\begin{aligned} -2x &> 8 \\ \frac{-2x}{-2} &< \frac{8}{-2} \\ x &< -4 \end{aligned}$$

$\{x \mid x < -4\}$

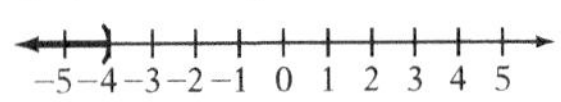

2. $5x - 2 \le 8$

$$5x \le 10$$

$$\frac{5x}{5} \le \frac{10}{5}$$

$$x \le 2$$

$\{x|x \le 2\}$

3. $3x + 1 \le 7x - 15$

$$-4x + 1 \le -15$$

$$-4x \le -16$$

$$\frac{-4x}{-4} \ge \frac{-16}{-4}$$

$$x \ge 4$$

$\{x|x \ge 4\}$

4. $\frac{1}{2}x - \frac{3}{4} < \frac{7}{4}x - 2$

$$4\left(\frac{1}{2}x - \frac{3}{4}\right) < 4\left(\frac{7}{4}x - 2\right)$$

$$2x - 3 < 7x - 8$$

$$-5x - 3 < -8$$

$$-5x < -5$$

$$\frac{-5x}{-5} > \frac{-5}{-5}$$

$$x > 1$$

$(1, \infty)$

Objective 3.5C

1. $|2x + 1| < 5$

$$-5 < 2x + 1 < 5$$

$$-5 - 1 < 2x + 1 - 1 < 5 - 1$$

$$-6 < 2x < 4$$

$$-3 < x < 2$$

$\{x|-3 < x < 2\}$

2. $|7x - 1| > 13$

$7x - 1 > 13$ or $7x - 1 < -13$

$7x > 14$ $\qquad$ $7x < -12$

$x > 2$ $\qquad$ $x < -\frac{12}{7}$

$\{x|x > 2\}$ $\qquad$ $\left\{x\middle|x < -\frac{12}{7}\right\}$

$$\{x|x > 2\} \cup \left\{x\middle|x < -\frac{12}{7}\right\}$$

3. $|2x + 7| > -5$

$$\begin{aligned} 2x + 7 &> -5 & \text{or} \quad 2x + 7 &< 5 \\ 2x &> -12 & 2x &< -2 \\ x &> -6 & x &< -1 \end{aligned}$$

$\{x \mid x > -6\}$ $\qquad$ $\{x \mid x < -1\}$

$\{x \mid x > -6\} \cup \{x \mid x < -1\}$

The solution set is the set of all real numbers.

4. $|4 - 3x| \geq 5$

$$\begin{aligned} 4 - 3x &\geq 5 & \text{or} \quad 4 - 3x &\leq -5 \\ -3x &\geq 1 & -3x &\leq -9 \\ x &\leq -\frac{1}{3} & x &\geq 3 \end{aligned}$$

$\left\{x \,\middle|\, x \leq -\frac{1}{3}\right\}$ $\qquad$ $\{x \mid x \geq 3\}$

$\left\{x \,\middle|\, x \leq -\frac{1}{3}\right\} \cup \{x \mid x \geq 3\}$

5. $|5 - 4x| \leq 13$

$$-13 \leq 5 - 4x \leq 13$$

$$-13 + (-5) \leq 5 + (-5) - 4x \leq 13 + (-5)$$

$$-18 \leq -4x \leq 8$$

$$\frac{18}{4} \geq x \geq -2$$

$\left\{x \,\middle|\, -2 \leq x \leq \frac{9}{2}\right\}$

6. $\left|\dfrac{x - 56.4}{20.56}\right| < 1.96$

$$-1.96 < \frac{x - 56.4}{20.56} < 1.96$$

$$20.56(-1.96) < 20.56\left(\frac{x - 56.4}{20.56}\right) < 20.56(1.96)$$

$$-40.2976 < x - 56.4 < 40.2976$$

$$-40.2976 + 56.4 < x - 56.4 + 56.4 < 40.2976 + 56.4$$

$$16.1024 < x < 96.6976$$

$\{x \mid 16.1024 < x < 96.6976\}$

MODULE

4 Geometry and Right Triangle Trigonometry

SECTION 4.1	**Angles, Lines, and Geometric Figures**
Objective **4.1A**	Define and describe lines and angles
Objective **4.1B**	Define and describe geometric figures
Objective **4.1C**	Solve problems involving angles formed by intersecting lines
SECTION 4.2	**Perimeter of a Plane Geometric Figure**
Objective **4.2A**	Find the perimeter of a plane geometric figure
Objective **4.2B**	Find the perimeter of a composite geometric figure
Objective **4.2C**	Solve application problems
SECTION 4.3	**Area of a Plane Geometric Figure**
Objective **4.3A**	Find the area of a geometric figure
Objective **4.3B**	Find the area of a composite geometric figure
Objective **4.3C**	Solve application problems
SECTION 4.4	**Volume**
Objective **4.4A**	Find the volume of a geometric solid
Objective **4.4B**	Find the volume of a composite geometric solid
Objective **4.4C**	Solve application problems
SECTION 4.5	**The Pythagorean Theorem**
Objective **4.5A**	Apply the Pythagorean Theorem
SECTION 4.6	**Right Triangle Trigonometry**
Objective **4.6A**	Find the values of trigonometric ratios of a right triangle
Objective **4.6B**	Solve applications involving trigonometric ratios

SECTION

4.1 Angles, Lines, and Geometric Figures

Objective 4.1A Define and describe lines and angles

The word *geometry* comes from the Greek words for "earth" (*geo*) and "measure" (*metron*). The original purpose of geometry was to measure land. Today geometry is used in many sciences, such as physics, chemistry, and geology, and in applied fields such as mechanical drawing and astronomy. Geometric forms are used in art and design.

Two basic geometric concepts are plane and space.

A **plane** is a flat surface, such as a table top or a blackboard. Figures that lie totally in a plane are called **plane figures.**

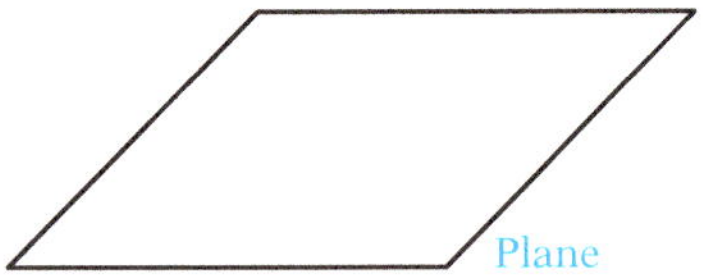

Space extends in all directions. Objects in space, such as trees, ice cubes, and doors, are called **solids.**

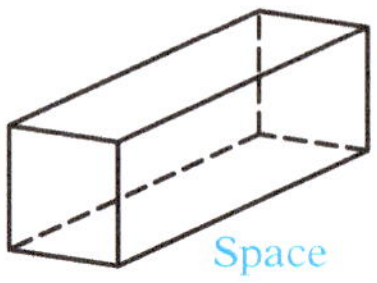

A **line** extends indefinitely in two directions in a plane. A line has no width.

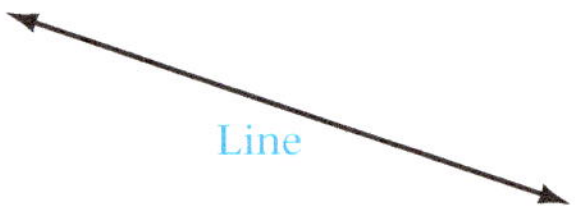

A **line segment** is part of a line and has two endpoints. The line segment, written as $\overline{AB}$, is shown in the figure.

The length of a line segment is the distance between the endpoints of the line segment. The length of a line segment may be expressed as the sum of two or more shorter line segments. For the figure at the right, $AB = 5$, $BC = 3$, and $AC = AB + BC = 5 + 3 = 8$.

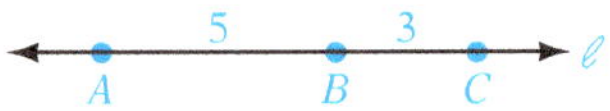

Tips for Success ➤
A great many new vocabulary words are introduced in this module. There are six new terms on this page alone: plane, plane figures, space, solids, line, and line segment. All of these terms are in **bold type.** The bold type indicates that these are concepts you must know to learn the material. Be sure to study each new term as it is presented.

Focus on finding the length of a line segment

Given that $AB = 22$ and $AC = 31$, find BC.

SOLUTION

$$AC = AB + BC$$
$$31 = 22 + BC \quad \text{• Substitute 22 for } AB \text{ and 31 for } AC\text{, and solve for } BC.$$
$$31 - 22 = 22 - 22 + BC$$
$$9 = BC$$

Take Note ➤
When no units, such as feet or meters, are given for the lengths along a line segment, all of the distances are assumed to be in the same unit of length.

Check your understanding 1

Given that $QR = 24$, $ST = 17$, and $QT = 62$, find RS.

SOLUTION See page S-1.

Lines in a plane can be parallel or intersecting. **Parallel lines** never meet; the distance between them is always the same. **Intersecting lines** cross at a point in the plane.

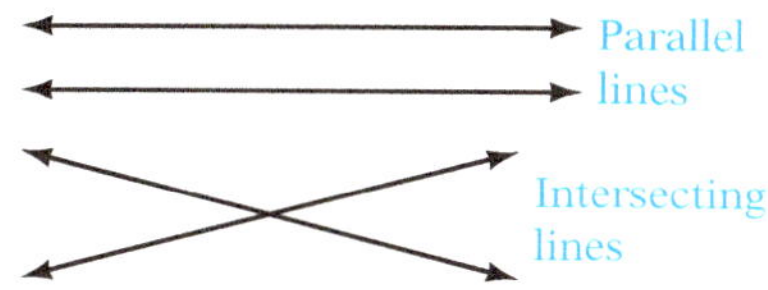

The symbol $\|$ means "is parallel to." In the accompanying figure, $\overline{AB} \| \overline{CD}$ and $p \| q$. Note that line p contains line segment AB and line q contains line segment CD. Parallel lines contain parallel line segments.

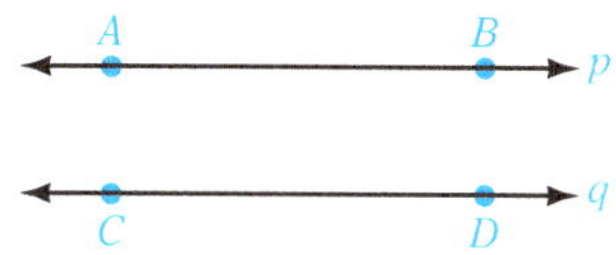

A **ray** starts at a point and extends indefinitely in one direction.

An **angle** is formed when two rays start from the same point. Rays r_1 and r_2 start from point B. The common endpoint is called the **vertex** of the angle.

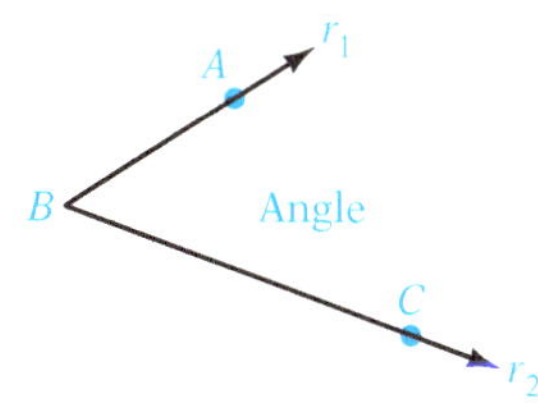

Take Note ➤
For the figure below, we can refer to $\angle ABD$, $\angle ABC$, or $\angle CBD$. Just writing $\angle B$ does not identify a specific angle.

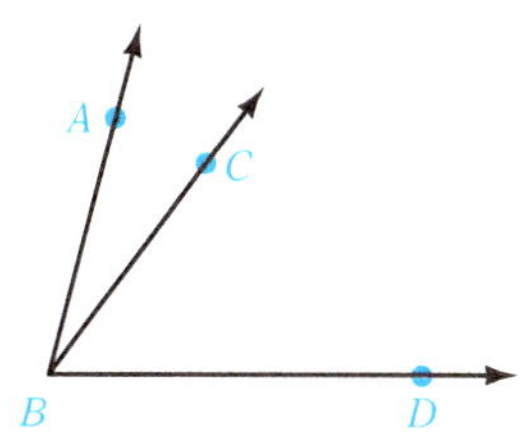

If A and C are points on rays r_1 and r_2 above, respectively, then the angle is called $\angle ABC$, $\angle CBA$, or $\angle B$, where $\angle$ is the symbol for angle. Note that an angle is named by giving three points, with the vertex as the second point listed, or by giving the point at the vertex.

An angle can also be named by writing a variable between the rays, close to the vertex. In the figure, $\angle x = \angle QRS = \angle SRQ$ and $\angle y = \angle SRT = \angle TRS$. Note that in this figure, more than two rays meet at the vertex. In this case, the vertex cannot be used to name the angle.

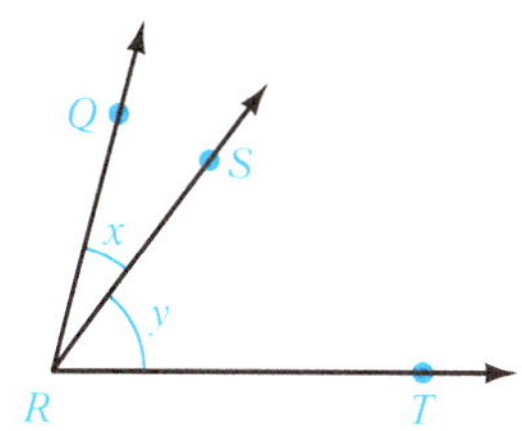

A unit in which angles are measured is the **degree.** The symbol for degree is °. One complete revolution is 360° (360 degrees).

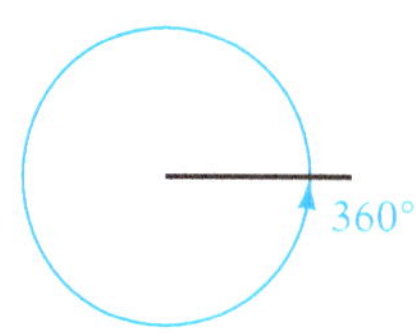

One-quarter of a revolution is 90°. A 90° angle is called a **right angle.** The symbol ⊾ represents a right angle.

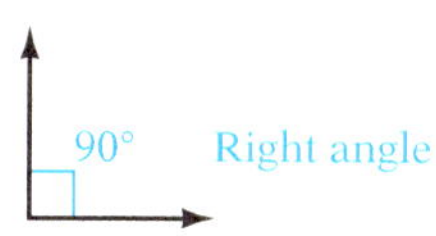

Perpendicular lines are intersecting lines that form right angles.

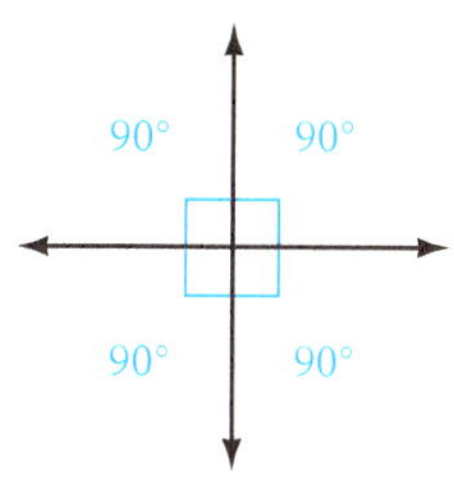

The symbol $\perp$ means "is perpendicular to." In the accompanying figure, $\overline{AB} \perp \overline{CD}$ and $p \perp q$. Note that line p contains line segment AB and line q contains line segment CD. Perpendicular lines contain perpendicular line segments.

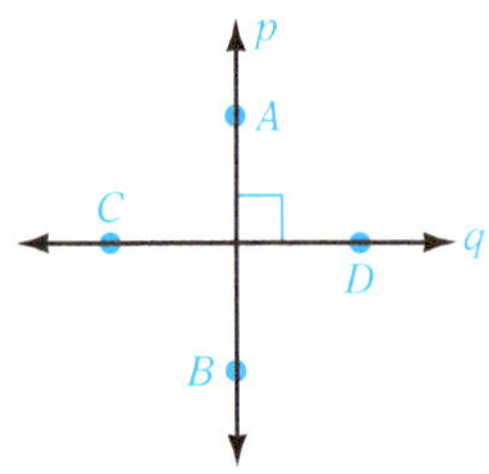

Complementary angles are two angles whose sum is 90°.

$\angle A + \angle B = 70° + 20° = 90°$
$\angle A$ and $\angle B$ are complementary angles.

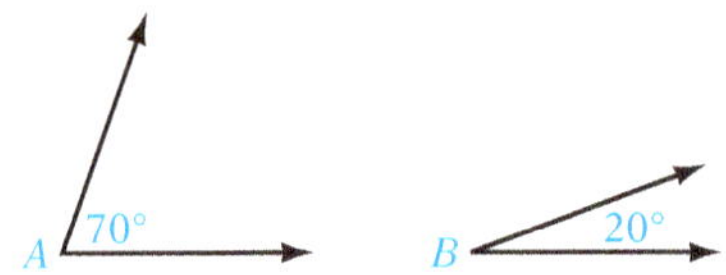

One-half of a revolution is 180°. A 180° angle is called a **straight angle.** $\angle AOB$ in the figure is a straight angle.

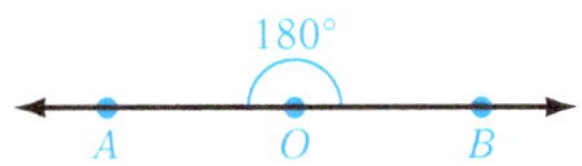

Supplementary angles are two angles whose sum is 180°.

$\angle C + \angle D = 130° + 50° = 180°$
$\angle C$ and $\angle D$ are supplementary angles.

Focus on finding the complement of an angle

Find the complement of a 32° angle.

SOLUTION

Let x represent the complement of 32°.

$$x + 32° = 90°$$ • The sum of complementary angles is 90°.

$$x + 32° - 32° = 90° - 32°$$

$$x = 58°$$

58° is the complement of 32°.

Check your understanding 2

Find the supplement of a 32° angle.

SOLUTION See page S-1.

An **acute angle** is an angle whose measure is between 0° and 90°. $\angle D$ in the figure above the "Focus on" example is an acute angle. An **obtuse angle** is an angle whose measure is between 90° and 180°. $\angle C$ in the same figure is an obtuse angle.

In the accompanying figure, $\angle DAC = 45°$ and $\angle CAB = 55°$.

$$\begin{aligned}\angle DAB &= \angle DAC + \angle CAB \\ &= 45° + 55° \\ &= 100°\end{aligned}$$

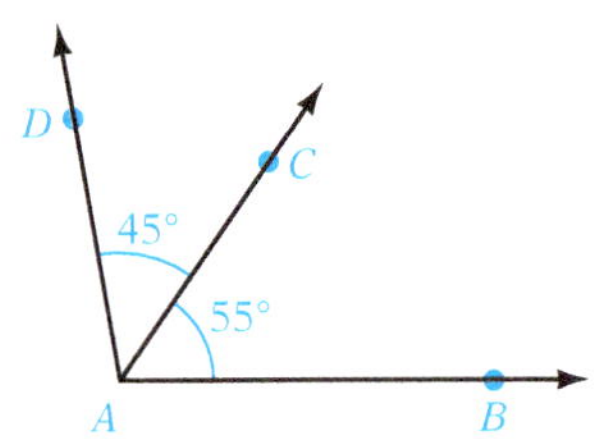

Focus on finding the measure of an angle

Find the measure of $\angle x$.

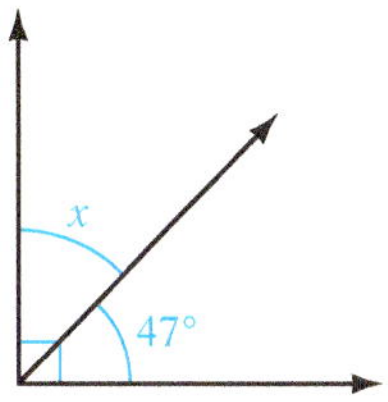

SOLUTION

$$\angle x + 47° = 90°$$
$$\angle x + 47° - 47° = 90° - 47°$$
$$\angle x = 43°$$

Check your understanding 3

Find the measure of $\angle a$.

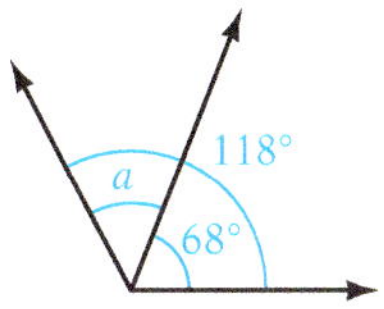

SOLUTION See page S-1.

Objective 4.1A Practice

1. In the figure below, $AB = 21$, $BC = 14$, and $AD = 54$. Find CD.

2. Find the complement of a 62° angle.
3. Find the supplement of 162° angle.
4. Find the measure of angle AOB.

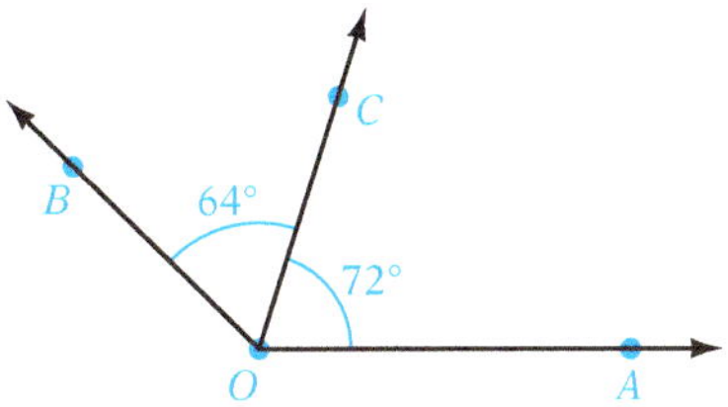

5. Find the measure of angle a.

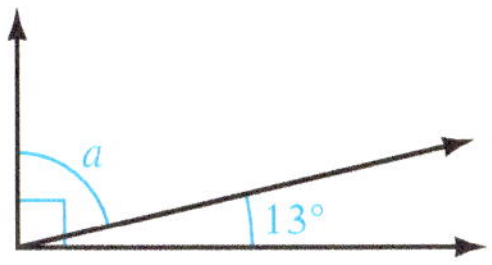

Solutions on pp. S-7–S-8.

Objective 4.1B Define and describe geometric figures

A **triangle** is a closed, three-sided plane figure. Figure ABC is a triangle. $\overline{AB}$ is called the **base.** The line $\overline{CD}$, perpendicular to the base, is called the **height.**

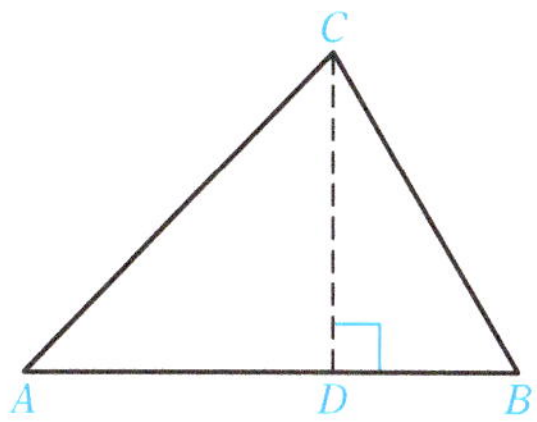

The Angles of a Triangle

The sum of the three angles of a triangle is 180°.

$\angle A + \angle B + \angle C = 180°$

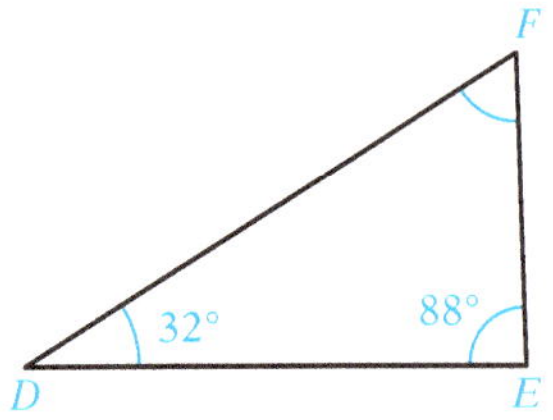

In triangle DEF, $\angle D = 32°$ and $\angle E = 88°$. Find the measure of $\angle F$.

$\angle D + \angle E + \angle F = 180°$ • The sum of the three angles of a triangle is 180°.

$32° + 88° + \angle F = 180°$ • $\angle D = 32°$ and $\angle E = 88°$

$120° + \angle F = 180°$ • Solve for $\angle F$.

$120° - 120° + \angle F = 180° - 120°$

$\angle F = 60°$

Focus on finding the measure of an angle in a triangle

Two angles of a triangle measure 42° and 103°. Find the measure of the third angle.

SOLUTION

The sum of the three angles of a triangle is 180°.

$$\angle A + \angle B + \angle C = 180°$$
$$\angle A + 42° + 103° = 180°$$
$$\angle A + 145° = 180°$$
$$\angle A + 145° - 145° = 180° - 145°$$
$$\angle A = 35°$$

The measure of the third angle is 35°.

Check your understanding 4

Two angles of a triangle measure 62° and 45°. Find the measure of the third angle.

SOLUTION See page S-1.

A **right triangle** contains one right angle. The side opposite the right angle is called the **hypotenuse.** The **legs of a right triangle** are its other two sides. In a right triangle, the two acute angles are complementary.

$\angle A + \angle B = 90°$

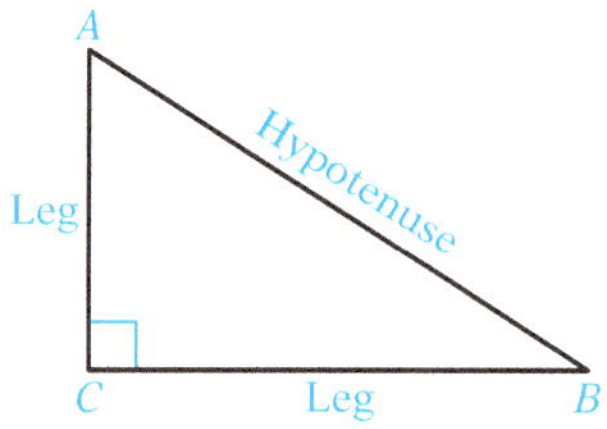

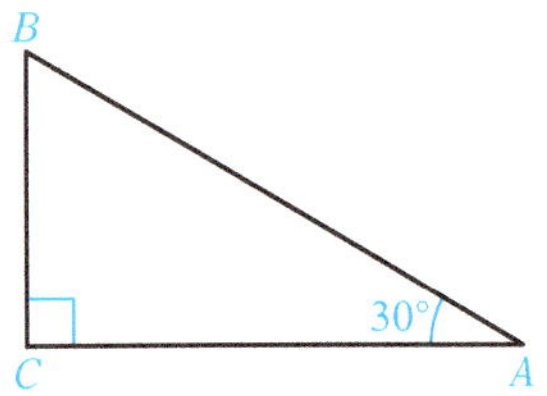

In the right triangle at the left, $\angle A = 30°$. Find the measure of $\angle B$.

$\angle A + \angle B = 90°$ • The two acute angles are complementary.

$30° + \angle B = 90°$ • $\angle A = 30°$

$30° - 30° + \angle B = 90° - 30°$ • Solve for $\angle B$.

$\angle B = 60°$

Focus on finding the measures of the angles in a right triangle

One angle in a right triangle measures 50°. Find the measures of the other two angles.

SOLUTION

In a right triangle, one angle measures 90° and the two acute angles are complementary.

$$\angle A + \angle B = 90°$$
$$\angle A + 50° = 90°$$
$$\angle A + 50° - 50° = 90° - 50°$$
$$\angle A = 40°$$

The other angles measure 90° and 40°.

Check your understanding 5

A right triangle has one angle measuring 7°. Find the measures of the other two angles.

SOLUTION See page S-1

A **quadrilateral** is a closed, four-sided plane figure. Three quadrilaterals with special characteristics are described here.

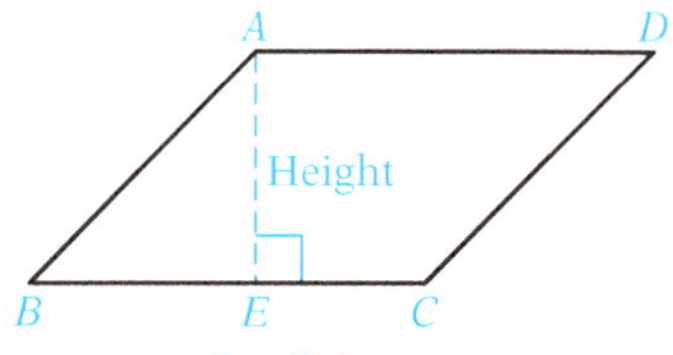

Parallelogram

A **parallelogram** has opposite sides parallel and equal. The perpendicular distance AE between the parallel sides is called the **height.**

A **rectangle** is a parallelogram that has four right angles.

A **square** is a rectangle that has four equal sides.

Rectangle

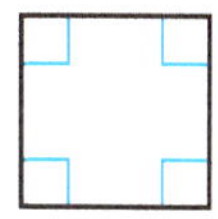
Square

A **circle** is a plane figure in which all points are the same distance from point O, which is called the **center** of the circle.

The **diameter of a circle** (d) is the length of a line segment through the center of the circle with endpoints on the circle. AB is a diameter of the circle shown.

The **radius of a circle** (r) is the length of a line segment from the center to a point on the circle. OC is a radius of the circle shown.

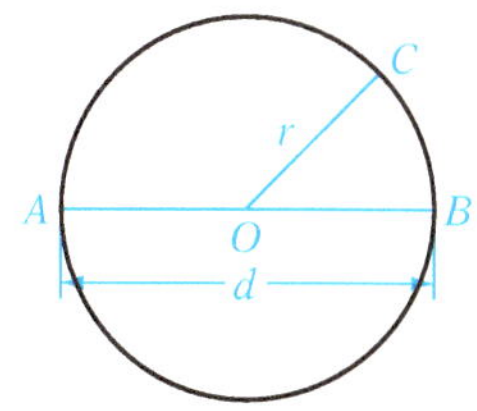

Circle

$d = 2r$ or $r = \frac{1}{2}d$

The line segment AB is a diameter of the circle shown. Find the radius of the circle.

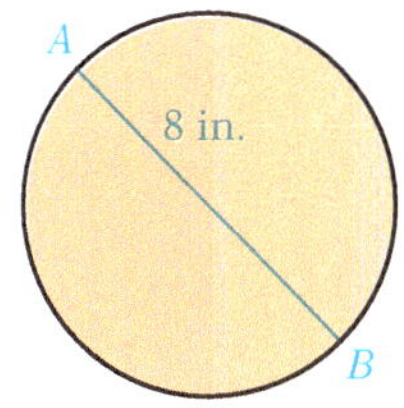

The radius is one-half the diameter. Therefore,

$$r = \frac{1}{2}d$$
$$= \frac{1}{2}(8 \text{ in.}) \qquad \bullet\ d = 8 \text{ in.}$$
$$= 4 \text{ in.}$$

Focus on finding the diameter of a circle

A circle has a radius of 8 cm. Find the diameter.

SOLUTION

$$d = 2r$$
$$= 2 \cdot 8 \text{ cm} = 16 \text{ cm}$$

The diameter is 16 cm.

Check your understanding 6

A circle has a diameter of 6 in. Find the radius.

SOLUTION See page S-1.

A **geometric solid** is a figure in space, or space figure. Four common space figures are the rectangular solid, cube, sphere, and cylinder.

A **rectangular solid** is a solid in which all six faces are rectangles.

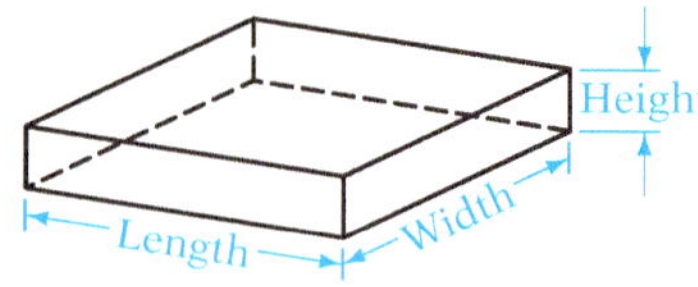

Rectangular solid

A **cube** is a rectangular solid in which all six faces are squares.

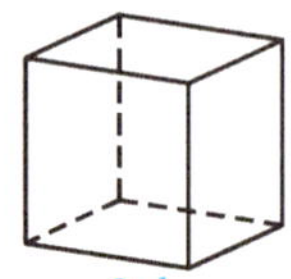
Cube

A **sphere** is a solid in which all points on the surface are the same distance from point O, which is called the **center** of the sphere.

The **diameter of a sphere** is the length of a line segment going through the center with endpoints on the sphere. AB is a diameter of the sphere shown.

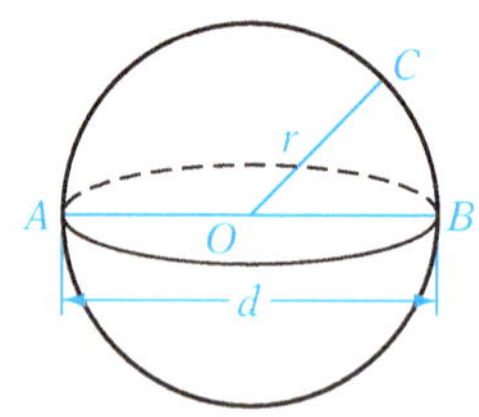

Sphere

The **radius of a sphere** is the length of a line segment from the center to a point on the sphere. OC is a radius of the sphere shown above.

$d = 2r$ or $r = \frac{1}{2}d$

The radius of the sphere shown at the right is 5 cm. Find the diameter of the sphere.

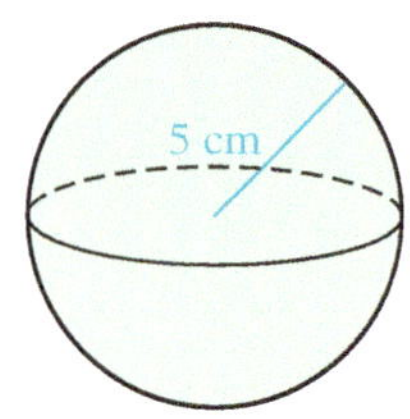

$d = 2r$ • The diameter equals twice the radius.

$= 2(5 \text{ cm})$ • $r = 5$ cm

$= 10 \text{ cm}$

The diameter is 10 cm.

The most common **cylinder** is one in which the bases are circles and are perpendicular to the side.

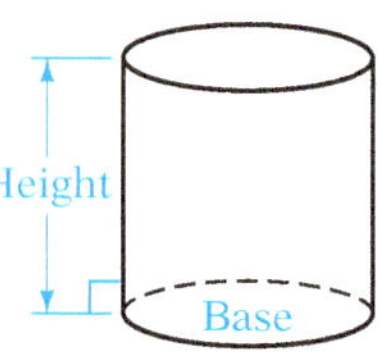

Cylinder

Objective 4.1B Practice

1. A triangle has a 13° angle and a 65° angle. Find the measure of the other angle.
2. A right triangle has a 62° angle. Find the measures of the other two angles.
3. A triangle has a 30° angle and a 45° angle. Find the measure of the other angle.
4. Find the diameter of a circle with a radius of 24 cm.
5. The diameter of a sphere is 1.2 m. Find the radius.

Solutions on pp. S-8–S-9.

Objective 4.1C Solve problems involving angles formed by intersecting lines

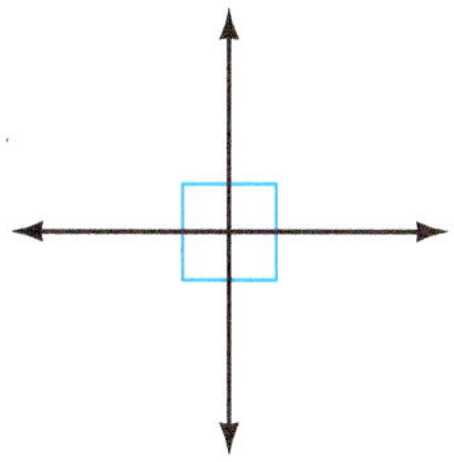

Four angles are formed by the intersection of two lines. If the two lines are perpendicular, then each of the four angles formed is a right angle. If the two lines are not perpendicular, then two of the angles formed are acute angles and two of the angles formed are obtuse angles. The two acute angles are always opposite each other, and the two obtuse angles are always opposite each other.

In the figure, $\angle w$ and $\angle y$ are acute angles. $\angle x$ and $\angle z$ are obtuse angles. Two angles that are on opposite sides of the intersection of two lines are called **vertical angles.** Vertical angles have the same measure. $\angle w$ and $\angle y$ are vertical angles. $\angle x$ and $\angle z$ are vertical angles.

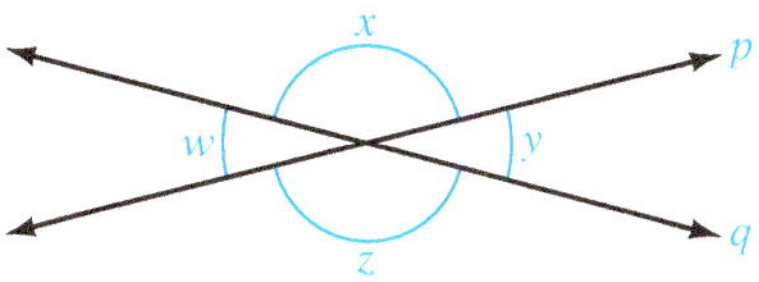

$\angle w = \angle y$

$\angle x = \angle z$

Two angles that share a common side are called **adjacent angles.** In the figure above, $\angle x$ and $\angle y$ are adjacent angles, as are $\angle y$ and $\angle z$, $\angle z$ and $\angle w$, and $\angle w$ and $\angle x$. Adjacent angles of intersecting lines are supplementary angles.

$\angle x + \angle y = 180°$

$\angle y + \angle z = 180°$

$\angle z + \angle w = 180°$

$\angle w + \angle x = 180°$

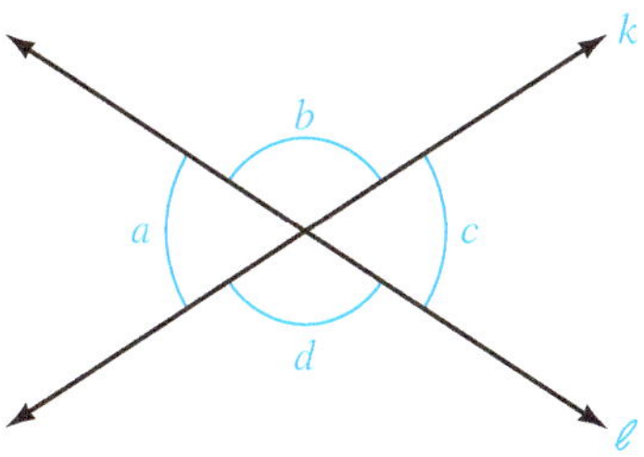

In the figure at the left, $\angle c = 65°$. Find the measures of angles *a*, *b*, and *d*.

$\angle a = 65°$ • $\angle a = \angle c$ because $\angle c$ and $\angle a$ are vertical angles.

$\angle b + \angle c = 180°$ • $\angle c$ is supplementary to $\angle b$ because $\angle c$ and $\angle b$ are adjacent angles.

$\angle b + 65° = 180°$ • $\angle c = 65°$

$\angle b + 65° - 65° = 180° - 65°$

$\angle b = 115°$

$\angle d = 115°$ • $\angle d = \angle b$ because $\angle b$ and $\angle d$ are vertical angles.

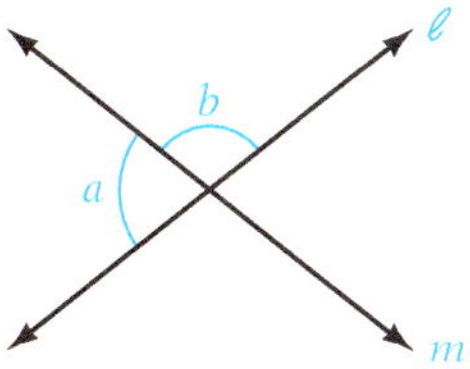

Focus on finding angles formed by intersecting lines

In the figure at the left, $\angle a = 75°$. Find $\angle b$.

SOLUTION

$\angle a + \angle b = 180°$ • $\angle a$ and $\angle b$ are supplementary.

$75° + \angle b = 180°$ • $\angle a = 75°$

$\angle b = 105°$ • Subtract 75° from each side.

Check your understanding 7

In the figure, $\angle a = 125°$. Find $\angle b$.

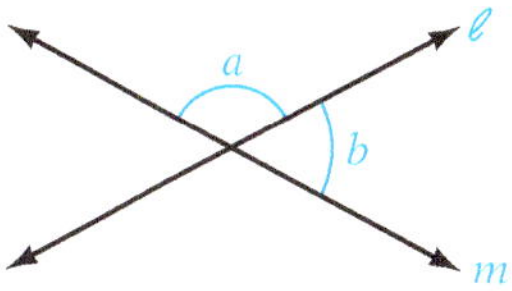

SOLUTION See page S-2.

A line intersecting two other lines at two different points is called a **transversal.**

If the lines cut by a transversal are parallel lines and the transversal is perpendicular to the parallel lines, then all eight angles formed are right angles.

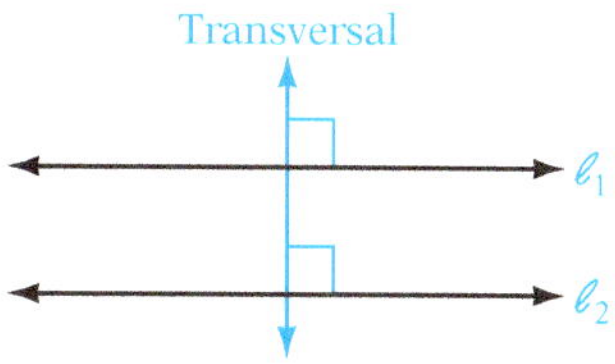

If the lines cut by a transversal are parallel lines and the transversal is not perpendicular to the parallel lines, then all four acute angles have the same measure and all four obtuse angles have the same measure. For the figure at the right, there are two groups of angles with the same measure:

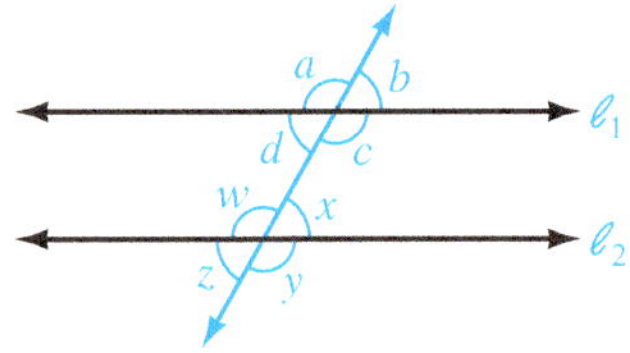

$\angle a = \angle c = \angle w = \angle y$ and $\angle b = \angle d = \angle x = \angle z$

Alternate interior angles are two nonadjacent angles that are on opposite sides of the transversal and between the parallel lines. For the figure above, $\angle c$ and $\angle w$ are alternate interior angles. $\angle d$ and $\angle x$ are alternate interior angles. Alternate interior angles have the same measure.

Alternate exterior angles are two nonadjacent angles that are on opposite sides of the transversal and outside the parallel lines. For the figure above, $\angle a$ and $\angle y$ are alternate exterior angles. $\angle b$ and $\angle z$ are alternate exterior angles. Alternate exterior angles have the same measure.

Corresponding angles are two angles that are on the same side of the transversal and are both acute angles or are both obtuse angles. For the figure above, the following pairs of angles are corresponding angles: $\angle a$ and $\angle w$, $\angle d$ and $\angle z$, $\angle b$ and $\angle x$, $\angle c$ and $\angle y$. Corresponding angles have the same measure.

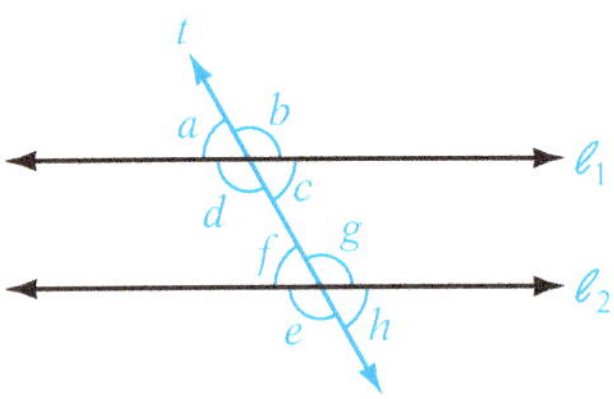

In the figure at the left, $\ell_1 \parallel \ell_2$ and $\angle c = 58°$. Find the measures of $\angle f$, $\angle h$, and $\angle g$.

$\angle f = 58°$ • $\angle f = \angle c$ because $\angle f$ and $\angle c$ are alternate interior angles.

$\angle h = 58°$ • $\angle h = \angle c$ because $\angle c$ and $\angle h$ are corresponding angles.

$\angle g + \angle h = 180°$ • $\angle g$ is supplementary to $\angle h$.

$\angle g + 58° = 180°$ • $\angle h = 58°$

$\angle g = 122°$ • Subtract 58° from each side.

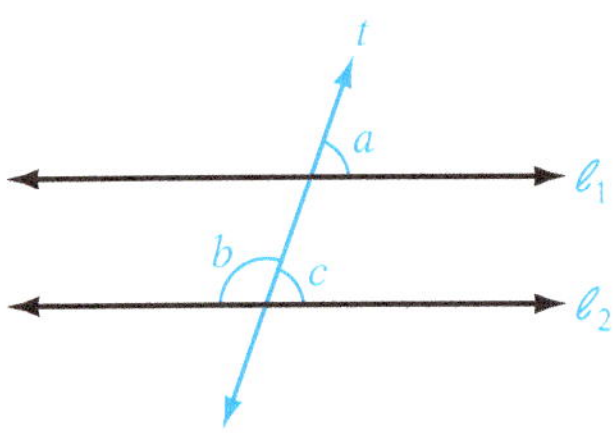

Focus on finding angles formed by a transversal

In the figure, $\ell_1 \parallel \ell_2$ and $\angle a = 70°$. Find $\angle b$.

SOLUTION

$\angle c = \angle a = 70°$ • Corresponding angles are equal.

$\angle b + \angle c = 180°$ • $\angle b$ and $\angle c$ are supplementary.

$\angle b + 70° = 180°$ • $\angle c = 70°$

$\angle b = 110°$ • Subtract 70° from each side.

Check your understanding 8

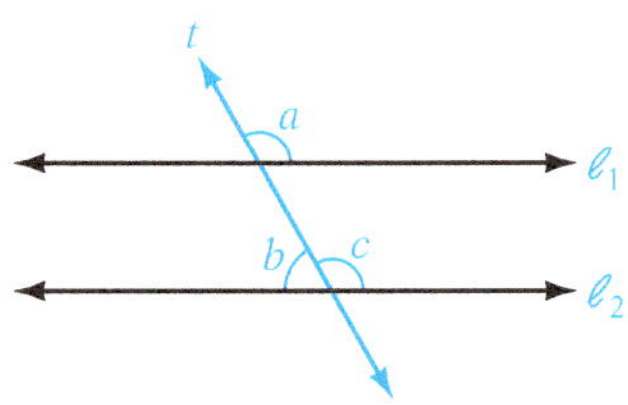

In the figure, $\ell_1 \parallel \ell_2$ and $\angle a = 120°$. Find $\angle b$.

SOLUTION See page S-2.

Objective 4.1C Practice

1. Find the measures of angles a and b.

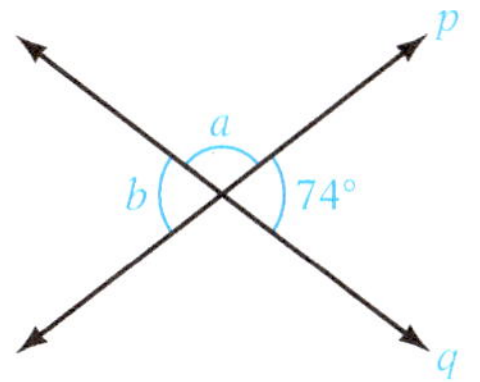

2. Find the measures of angles a and b.

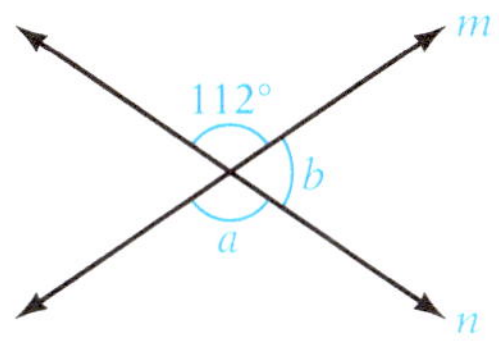

3. Given that $\ell_1 \parallel \ell_2$, find the measures of angles a and b.

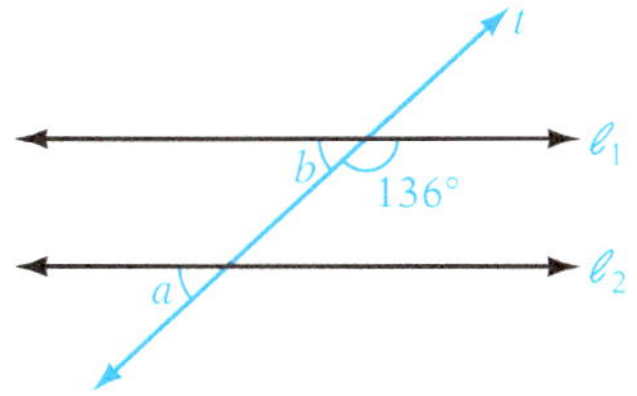

4. Given that $\ell_1 \parallel \ell_2$, find the measures of angles a and b.

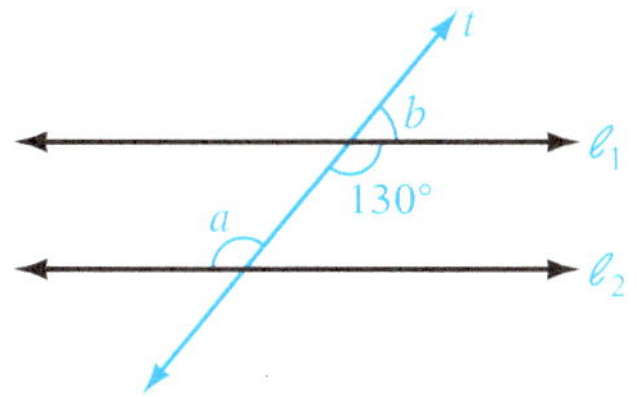

Solutions on p. S-9.

SECTION

4.2 Perimeter of a Plane Geometric Figure

Objective 4.2A Find the perimeter of a plane geometric figure

A **polygon** is a closed figure determined by three or more line segments that lie in a plane. The **sides of a polygon** are the line segments that form the polygon. The following figures are examples of polygons.

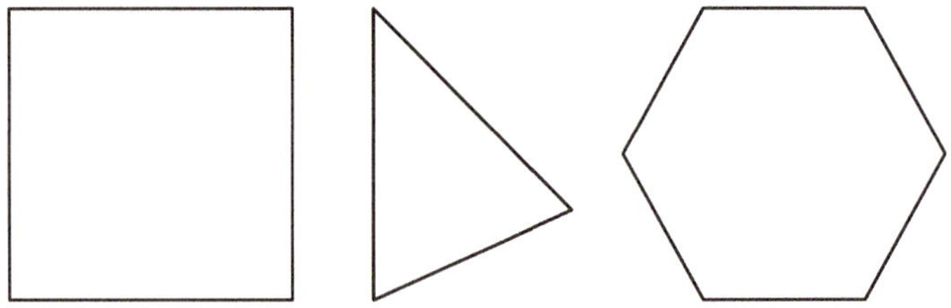

A B C

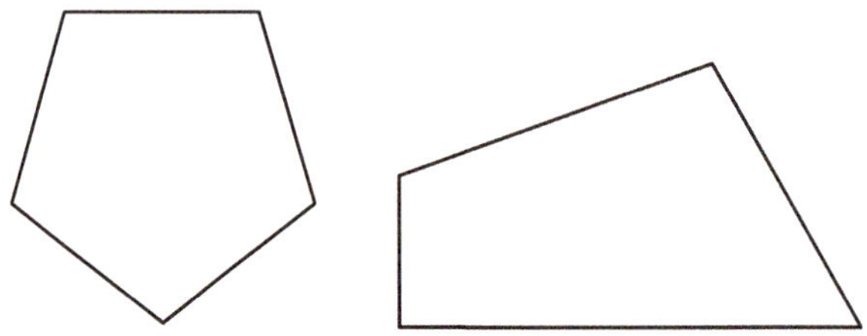

D E

A **regular polygon** is one in which each side has the same length and each angle has the same measure. The polygons in Figures A, C, and D above are regular polygons.

The name of a polygon is based on the number of its sides. The table below lists the names of polygons that have from 3 to 10 sides.

Number of Sides	Name of the Polygon
3	Triangle
4	Quadrilateral
5	Pentagon
6	Hexagon
7	Heptagon
8	Octagon
9	Nonagon
10	Decagon

Triangles and quadrilaterals are two of the most common types of polygons. Triangles are distinguished by the number of equal sides and also by the measures of their angles.

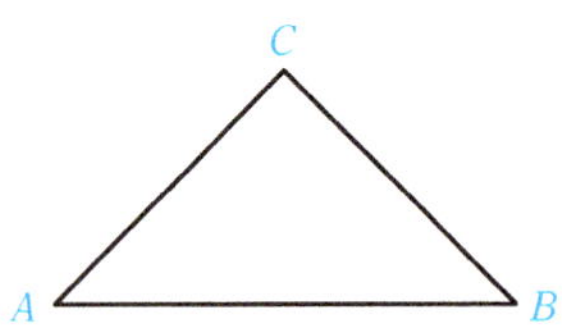

An **isosceles triangle** has two sides of equal length. The angles opposite each of the equal sides are of equal measure.

$AC = BC$
$\angle A = \angle B$

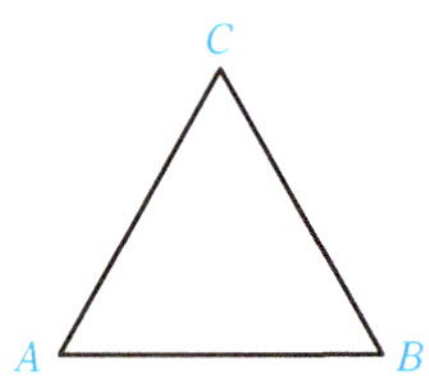

The three sides of an **equilateral triangle** are of equal length. The three angles are of equal measure.

$AB = BC = AC$
$\angle A = \angle B = \angle C$

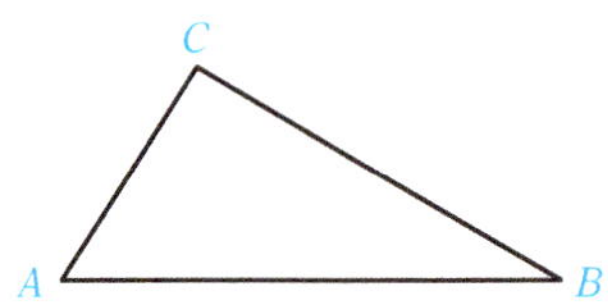

A **scalene triangle** has no two sides of equal length. No two angles are of equal measure.

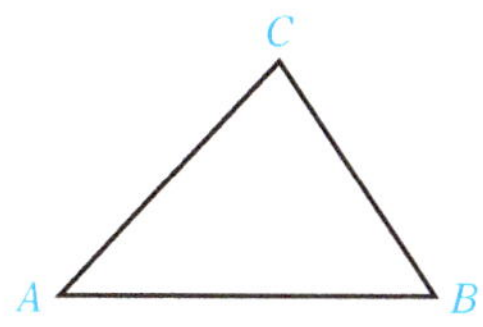

An **acute triangle** has three acute angles.

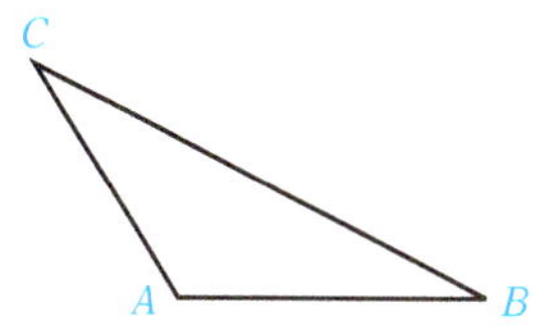

An **obtuse triangle** has one obtuse angle.

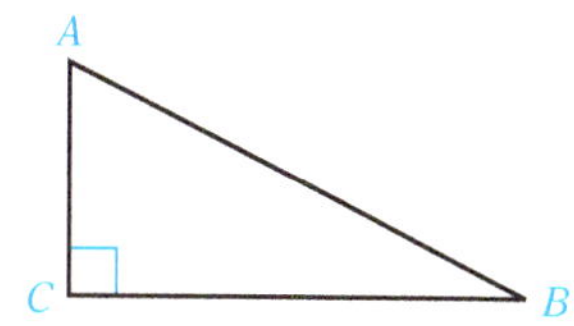

A **right triangle** has a right angle.

Quadrilaterals also are distinguished by their sides and angles, as shown below. Note that a rectangle, a square, and a rhombus are different forms of a parallelogram.

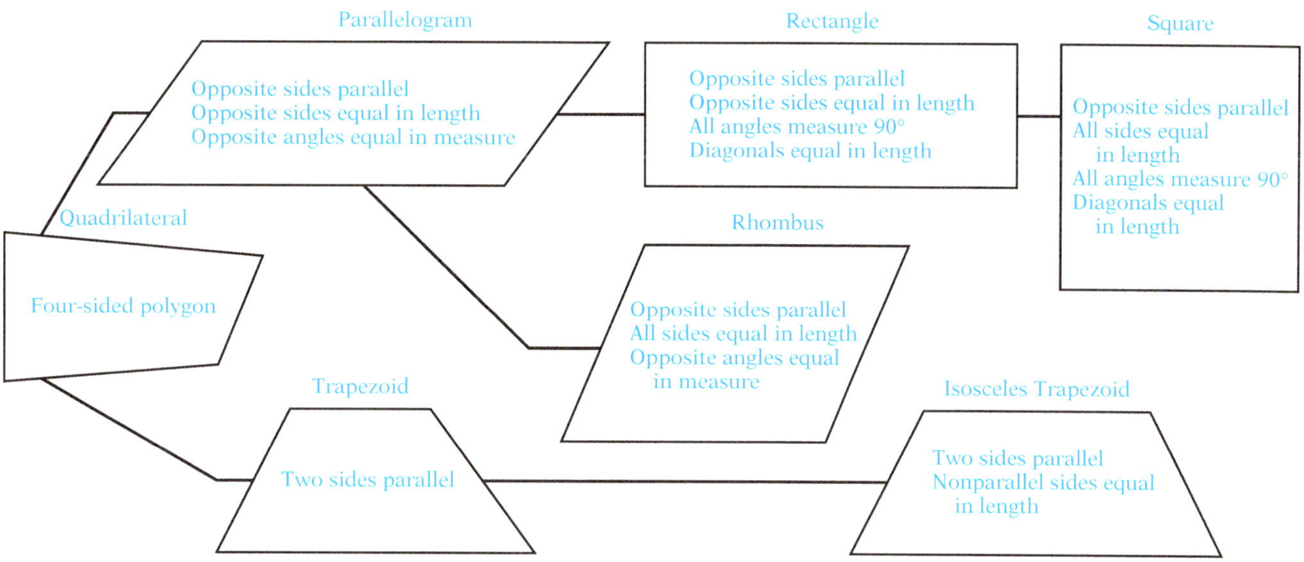

The **perimeter** of a plane geometric figure is a measure of the distance around the figure. The perimeter of a polygon is the sum of the lengths of its sides. Perimeter is used, for example, in buying fencing for a lawn or in determining how much baseboard is needed for a room.

Here are the perimeter formulas for some of the more common geometric figures.

Perimeter of a Triangle

Let a, b, and c be the lengths of the sides of a triangle.

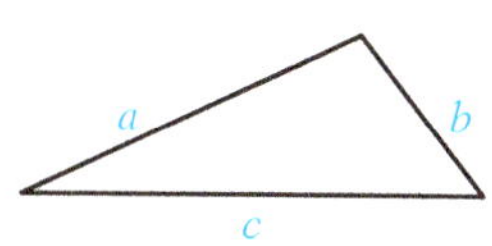

The perimeter of the triangle is $P = a + b + c$.

EXAMPLE

Find the perimeter of the triangle shown at the right.

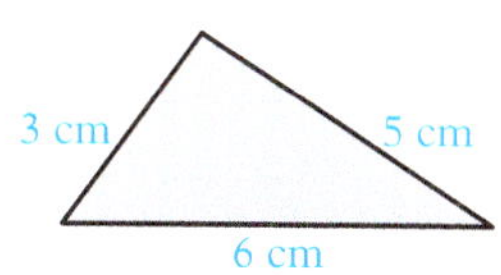

$$P = a + b + c$$
$$= 3\text{ cm} + 5\text{ cm} + 6\text{ cm}$$
$$= 14\text{ cm}$$

The perimeter of the triangle is 14 cm.

Focus on finding the perimeter of a triangle

Find the perimeter of a triangle with sides of lengths 5 in., 7 in., and 8 in.

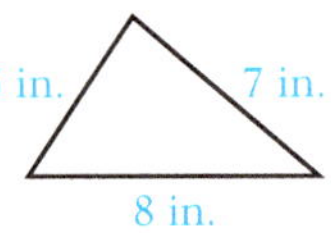

SOLUTION

$P = a + b + c$

$= 5 \text{ in.} + 7 \text{ in.} + 8 \text{ in.}$

$= 20 \text{ in.}$

The perimeter of the triangle is 20 in.

Check your understanding 1

Find the perimeter of a triangle with sides of lengths 12 cm, 15 cm, and 18 cm.

SOLUTION See page S-2.

Perimeter of a Rectangle

Let L be the length (usually the longer side) of a rectangle and W be the width (usually the shorter side) of a rectangle.

The perimeter of the rectangle is $P = 2L + 2W$.

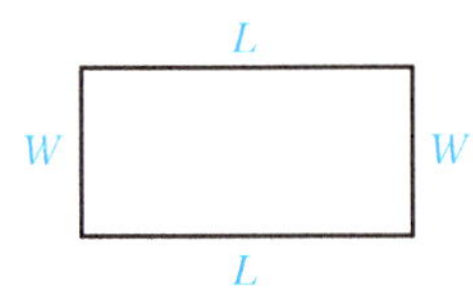

EXAMPLE

Find the perimeter of the rectangle shown at the right.

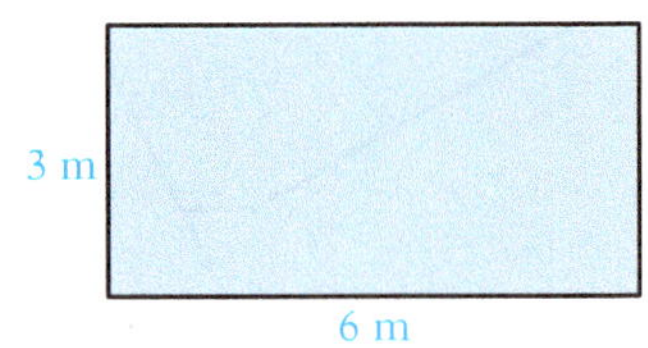

$P = 2L + 2W$

$= 2(6 \text{ m}) + 2(3 \text{ m})$ • $L = 6$ m; $W = 3$ m

$= 12 \text{ m} + 6 \text{ m}$

$= 18 \text{ m}$

The perimeter of the rectangle is 18 m.

Focus on finding the perimeter of a rectangle

Find the perimeter of a rectangle with a width of $\frac{2}{3}$ ft and a length of 2 ft.

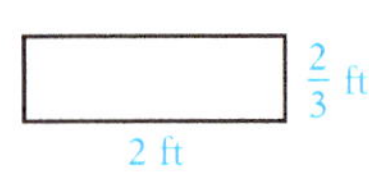

SOLUTION

$P = 2L + 2W$

$= 2(2 \text{ ft}) + 2\left(\frac{2}{3} \text{ ft}\right)$ • $L = 2$ ft; $W = \frac{2}{3}$ ft

$= 4 \text{ ft} + \frac{4}{3} \text{ ft}$

$= 5\frac{1}{3} \text{ ft}$

The perimeter of the rectangle is $5\frac{1}{3}$ ft.

Check your understanding 2

Find the perimeter of a rectangle with a length of 2 m and a width of 0.85 m.

SOLUTION See page S-2.

Apply the Concept

A building contractor must place a security fence around a rectangular construction lot that is 95 ft long and 72 ft wide. How many feet of fencing must the contractor buy?

SOLUTION

To find the amount of fencing needed, find the perimeter of the lot.

$P = 2L + 2W$

$= 2(95 \text{ ft}) + 2(72 \text{ ft})$ • $L = 95$ ft; $W = 72$ ft

$= 190 \text{ ft} + 144 \text{ ft}$

$= 334 \text{ ft}$

The contractor must buy 334 ft of fencing.

Take Note ➤ The perimeter of a square is the sum of the four sides:

$s + s + s + s = 4s$

Recall that a square is a rectangle in which all four sides are equal.

Perimeter of a Square

Let s be the length of a side of a square.

The perimeter of the square is $P = 4s$.

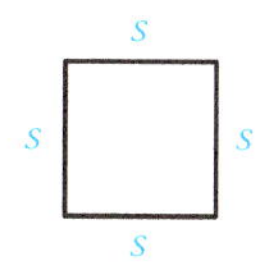

EXAMPLE

Find the perimeter of the square shown at the right.

$P = 4s$

$= 4(3 \text{ ft})$ • $s = 3$ ft

$= 12 \text{ ft}$

The perimeter of the square is 12 ft.

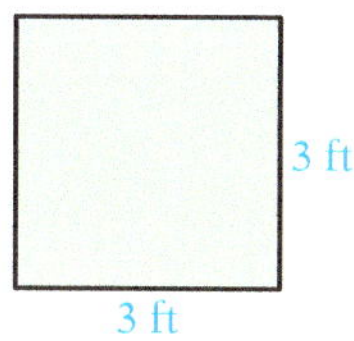

The perimeter of a circle is called its **circumference.** The circumference of a circle is equal to the product of pi (π) and the diameter. The value of π can be approximated as

$$\pi \approx 3.14 \text{ or } \pi \approx \frac{22}{7}$$

The π key on a calculator gives a more exact approximation of π.

Circumference of a Circle

Let d be the diameter of a circle.

The circumference of the circle is $C = \pi d$.

Because the diameter is twice the radius, the circumference is also given by $C = 2\pi r$.

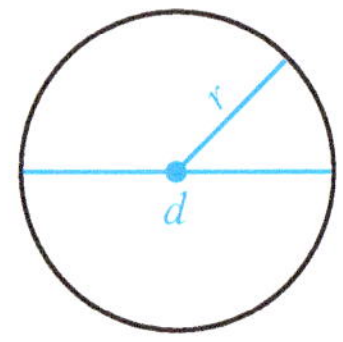

EXAMPLE

Find the circumference of the circle shown at the right.

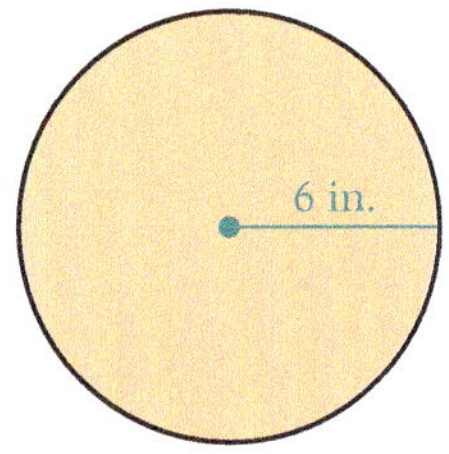

$C = 2\pi r$

$= 2\pi(6 \text{ in.})$ • $r = 6$ in.

$\approx 2(3.14)(6 \text{ in.})$

$= 37.68$ in.

The circumference of the circle is approximately 37.68 in.

Apply the Concept

The diameter of a car tire is 16 in. If the tire makes two complete revolutions, what distance has the tire traveled? Use 3.14 for π.

SOLUTION

To find the distance, first find the circumference of the tire. Because the tire makes two complete revolutions, multiply the circumference by 2.

$C = \pi d$

$\approx 3.14(16 \text{ in.})$ • $d = 16$ in.

$= 50.24$ in.

The circumference of the tire is approximately 50.24 in.

$2 \times 50.24 = 100.48$

The tire has traveled a distance of 100.48 in.

Focus on finding the circumference of a circle

Find the circumference of a circle with a radius of 18 cm. Use 3.14 for π.

SOLUTION

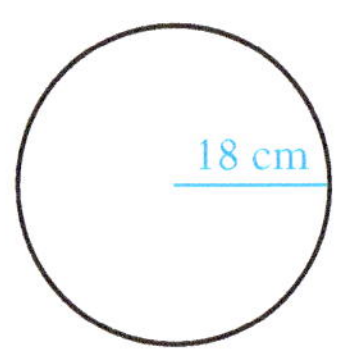

$C = 2\pi r$

$\approx 2 \cdot 3.14 \cdot 18$ cm

$= 113.04$ cm

The circumference is approximately 113.04 cm.

Check your understanding 3

Find the circumference of a circle with a diameter of 6 in. Use 3.14 for π.

SOLUTION See page S-2.

Objective 4.2A Practice

1. Find the perimeter of the triangle given below.

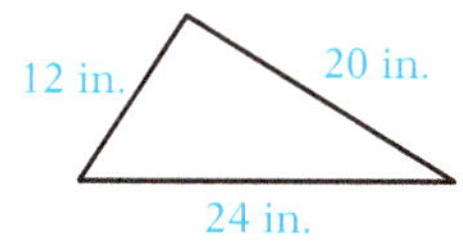

2. Find the perimeter of the square given below.

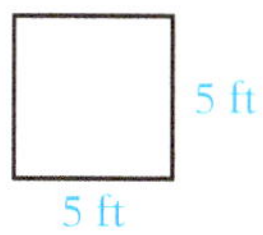

3. Find the circumference of the circle given below. Use 3.14 for π.

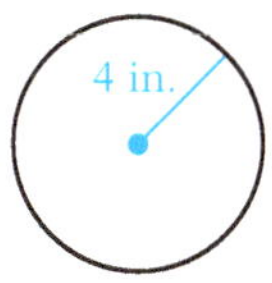

4. Find the perimeter of a rectangle with a length of 2 m and a width of 0.8 m.
5. A horse trainer wants to build a rectangular corral that is 60 ft wide and 75 ft long. How many feet of fencing will the trainer need to build the corral?
6. The diameter of a quarter dollar is 24.26 mm. What is the circumference of a quarter dollar? Use 3.14 for π. Round to the nearest hundredth.

Solutions on pp. S-9–S-10.

Objective 4.2B Find the perimeter of a composite geometric figure

A **composite geometric figure** is a figure made from two or more geometric figures. The following composite is made from part of a rectangle and part of a circle:

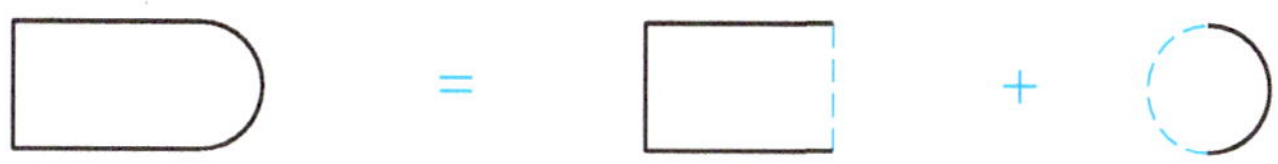

Perimeter of the composite figure $= 3$ sides of a rectangle $+ \frac{1}{2}$ the circumference of a circle

Perimeter of the composite figure $= 2L + W + \frac{1}{2}\pi d$

The perimeter of the composite figure below is found by adding twice the length plus the width plus one-half the circumference of the circle.

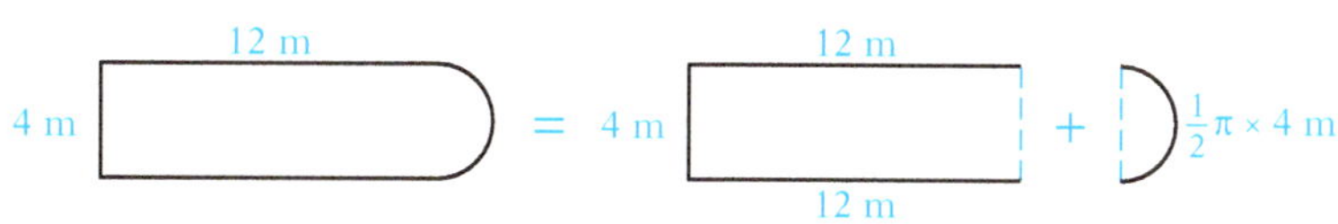

$$P = 2L + W + \frac{1}{2}\pi d$$

$$\approx 2(12 \text{ m}) + 4 \text{ m} + \frac{1}{2}(3.14)(4 \text{ m})$$

$$= 34.28 \text{ m}$$

• $L = 12$ m, $W = 4$ m, $d = 4$ m. *Note:* The diameter of the circle is equal to the width of the rectangle.

The perimeter is approximately 34.28 m.

Focus on finding the perimeter of a composite figure

Find the perimeter of the composite figure. Use $\frac{22}{7}$ for π.

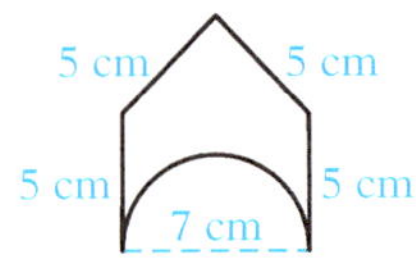

SOLUTION

$$\text{Perimeter of composite figure} = \underbrace{\text{sum of lengths of the four sides}} + \underbrace{\tfrac{1}{2}\text{ the circumference of the circle}}$$

$$P = 4s + \frac{1}{2}\pi d$$

$$\approx 4(5\text{ cm}) + \frac{1}{2}\left(\frac{22}{7}\right)(7\text{ cm})$$

$$= 20\text{ cm} + 11\text{ cm} = 31\text{ cm}$$

The perimeter is approximately 31 cm.

Check your understanding 4

Find the perimeter of the composite figure. Use 3.14 for π.

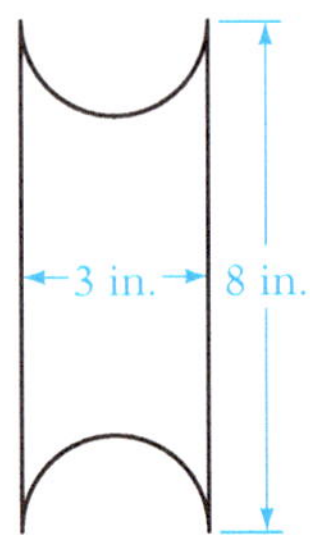

SOLUTION See page S-2.

Objective 4.2B Practice

For Exercises 1 to 4, find the perimeter of the given figure. If necessary, use 3.14 for π.

1.

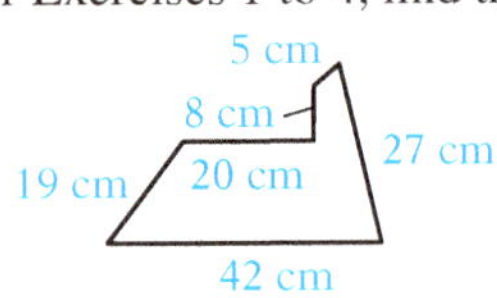

2.

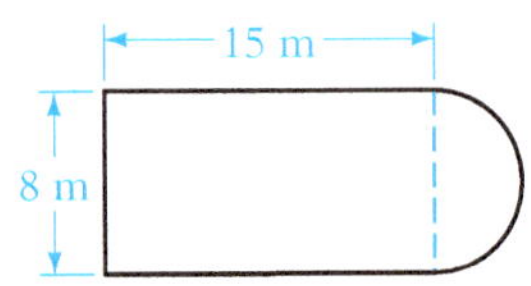

3.

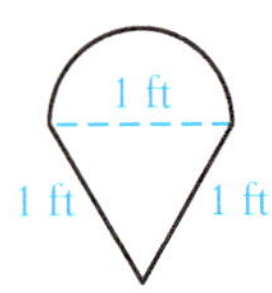

4.

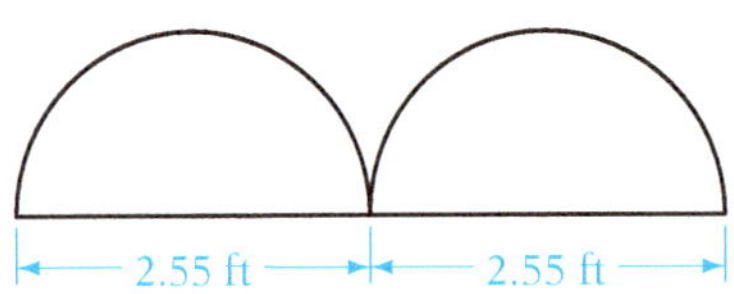

Solutions on p. S-10.

Objective 4.2C Solve application problems

Focus on solving an application

The dimensions of a triangular sail are 18 ft, 11 ft, and 15 ft. What is the perimeter of the sail?

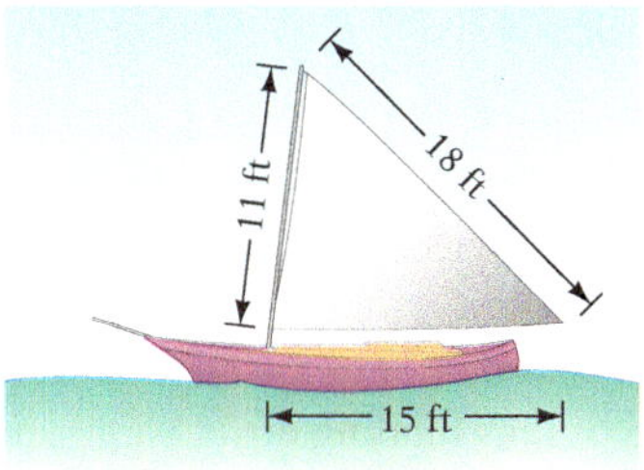

STRATEGY

To find the perimeter, use the formula for the perimeter of a triangle.

SOLUTION

$P = a + b + c$

$= 18 \text{ ft} + 11 \text{ ft} + 15 \text{ ft} = 44 \text{ ft}$

The perimeter of the sail is 44 ft.

Check your understanding 5

What is the perimeter of a standard piece of computer paper that measures $8\frac{1}{2}$ in. by 11 in.?

SOLUTION See page S-3.

Focus on solving an application

If fencing costs \$6.75 per foot, how much will it cost to fence a rectangular lot that is 108 ft wide and 240 ft long?

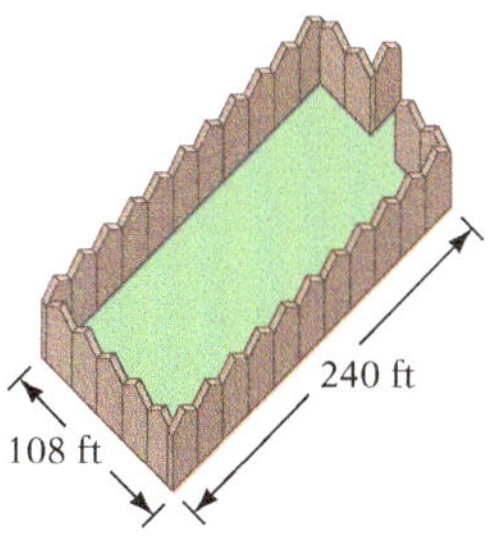

STRATEGY

To find the cost of the fence:

- Find the perimeter of the lot.
- Multiply the perimeter by the per-foot cost of the fencing.

SOLUTION

$P = 2L + 2W$

$= 2(240 \text{ ft}) + 2(108 \text{ ft})$

$= 480 \text{ ft} + 216 \text{ ft} = 696 \text{ ft}$

Cost $= 696 \times 6.75 = 4698$

The cost to fence the lot is \$4698.

Check your understanding 6

Metal stripping is being installed around a workbench that is 0.74 m wide and 3 m long. At $4.49 per meter, find the cost of the metal stripping. Round to the nearest cent.

SOLUTION See page S-3.

Objective 4.2C Practice

1. How many feet of fencing should be purchased to enclose a rectangular garden that is 18 ft long and 12 ft wide?
2. Find the length of molding needed to trim a circular table that is 3.8 ft in diameter. Use 3.14 for π.
3. Bias binding is to be sewed around the edge of a rectangular quilt measuring 72 in. by 45 in. Each package of bias binding costs $5.50 and contains 15 ft of binding. How many packages of bias binding are needed for the quilt?
4. A tricycle tire has a diameter of 12 in. How many feet does the tricycle travel when the wheel makes eight revolutions? Use 3.14 for π.

Solutions on p. S-11.

SECTION 4.3 Area of a Plane Geometric Figure

Objective 4.3A Find the area of a geometric figure

Area is a measure of the amount of surface in a region. Area can be used to describe, for example, the size of a rug, a parking lot, a farm, or a national park. Area is measured in square units.

A square that measures 1 in. on each side has an area of 1 square inch, which is written 1 in^2.

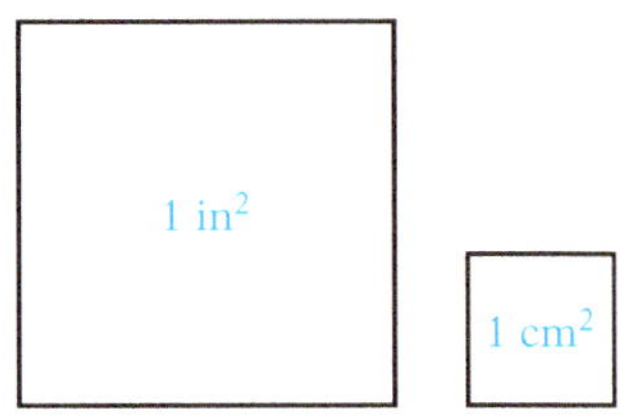

A square that measures 1 cm on each side has an area of 1 square centimeter, which is written 1 cm^2.

Larger areas can be measured in square feet (ft^2), square meters (m^2), square miles (mi^2), acres (43,560 ft^2), or any other square unit.

The area of a geometric figure is the number of squares that are necessary to cover the figure. In the figures below, two rectangles have been drawn and covered with squares. In the figure on the left, 12 squares, each of area 1 cm^2, were used to cover the rectangle. The area of the rectangle is 12 cm^2. In the figure on the right, 6 squares, each of area 1 in^2, were used to cover the rectangle. The area of the rectangle is 6 in^2.

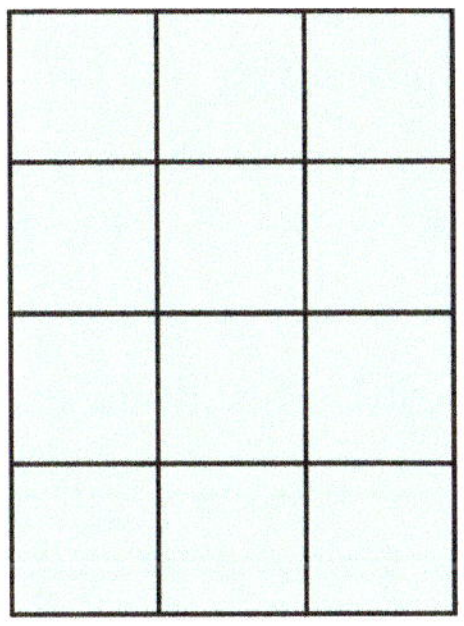

The area of the rectangle is 12 cm^2.

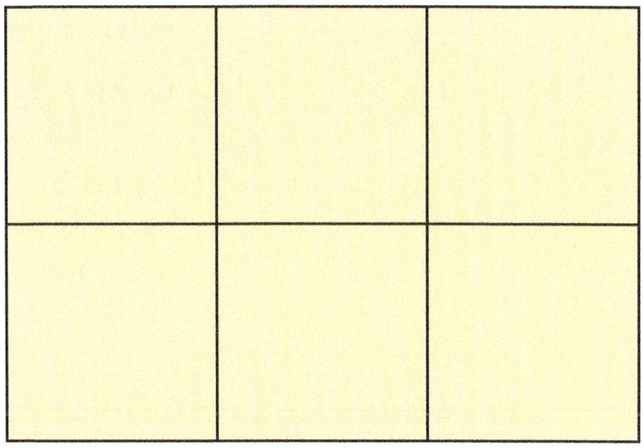

The area of the rectangle is 6 in^2.

Note from the above figures that the area of a rectangle can be found by multiplying the length of the rectangle by its width.

Area of a Rectangle

Let L be the length of a rectangle and W be the width of a rectangle.

The area of the rectangle is $A = LW$.

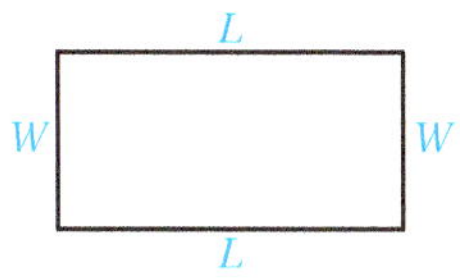

EXAMPLE

Find the area of the rectangle shown at the right.

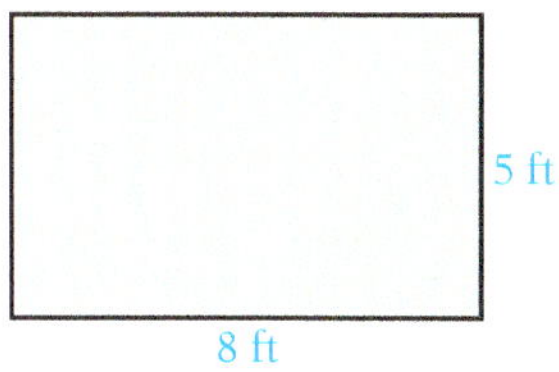

$A = LW$

$= (8\text{ ft})(5\text{ ft})$ • $L = 8$ ft; $W = 5$ ft

$= 40\text{ ft}^2$

The area of the rectangle is 40 ft^2.

Apply the Concept

A carpet installer charges $.30 per square foot to install wall-to-wall carpeting. How much would the installer charge to install wall-to-wall carpeting in a rectangular room that measures 12 ft by 14 ft?

SOLUTION

To find the cost, first find the area of the room. Then multiply the area by the cost per square foot.

$A = LW$

$= (14 \text{ ft})(12 \text{ ft})$ • $L = 14$ ft; $W = 12$ ft

$= 168 \text{ ft}^2$

The area of the room is 168 ft^2.

Cost $= 168 \cdot 0.30 = 50.40$

The installer would charge $50.40 to install the carpet.

Area of a Square

Let s be the length of one side of a square.

The area of the square is $A = s^2$.

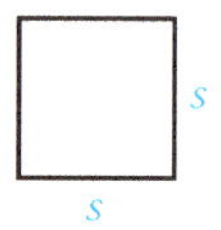

EXAMPLE

Find the area of the square shown at the right.

$A = s^2$

$= (14 \text{ cm})^2$ • $s = 14$ cm

$= 196 \text{ cm}^2$

The area of the square is 196 cm^2.

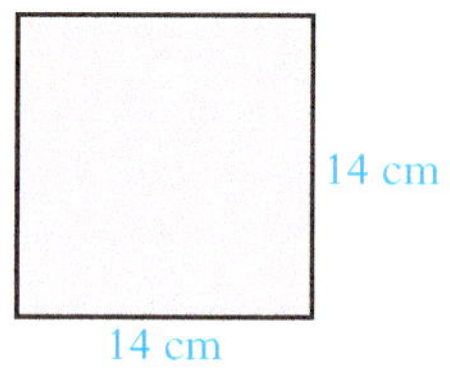

Area of a Triangle

In the figure at the right, $\overline{AB}$ is the base b of the triangle, and $\overline{CD}$, which is perpendicular to the base b, is the height h.

The area of a triangle is $A = \frac{1}{2}bh$.

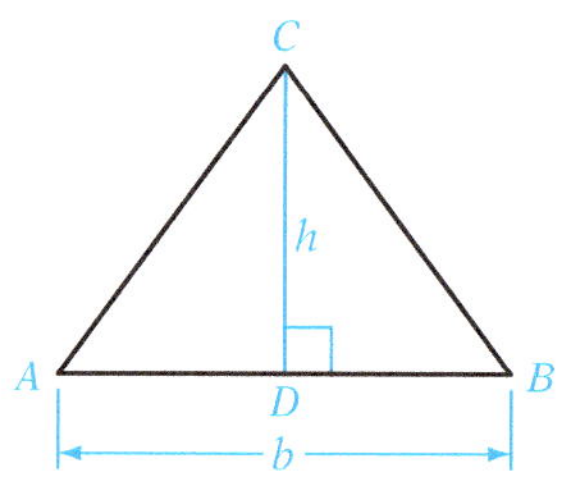

EXAMPLE

Find the area of the triangle shown at the right.

$A = \frac{1}{2}bh$

$= \frac{1}{2}(20 \text{ m})(5 \text{ m})$ • $b = 20$ m; $h = 5$ m

$= 50 \text{ m}^2$

The area of the triangle is 50 m^2.

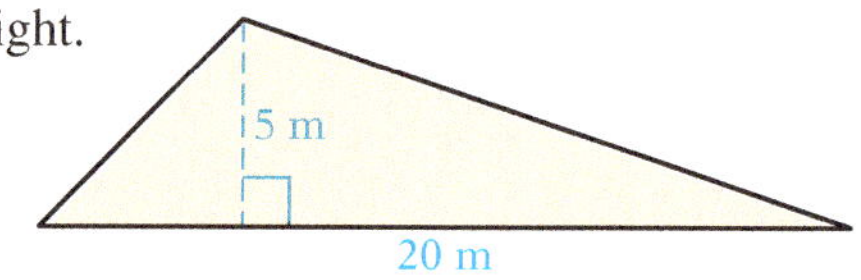

Area of a Circle

Let r be the radius of a circle.

The area of the circle is $A = \pi r^2$.

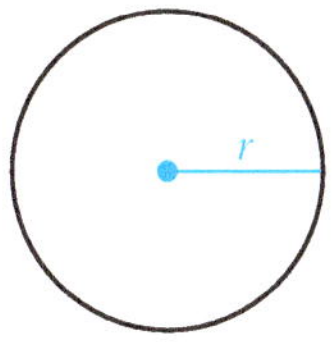

EXAMPLE

Find the area of the circle shown at the right.

$A = \pi r^2$

$= \pi(8 \text{ in.})^2 = 64\pi \text{ in}^2$

$\approx 64 \cdot 3.14 \text{ in}^2 = 200.96 \text{ in}^2$

The area is exactly 64π in^2.

The area is approximately 200.96 in^2.

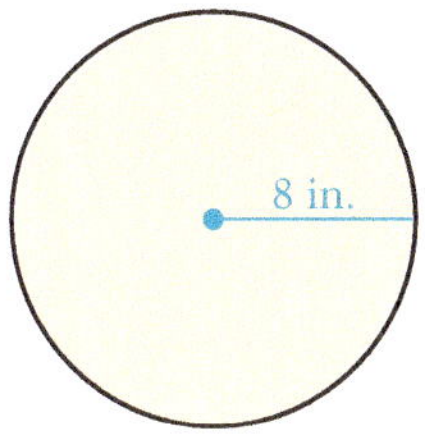

Take Note ➤
Note that we gave two answers for the area of the circle: the *exact* answer, which includes the symbol for π, and an *approximate* answer, which uses an approximation for π.

Apply the Concept

The bottom of a circular children's wading pool has a diameter of 5 ft. What is the area of the bottom of the pool?

SOLUTION

To find the area, first find the radius of the bottom of the pool. Then use the equation $A = \pi r^2$ to find the area. Use 3.14 for π.

$$r = \frac{\text{diameter}}{2} = \frac{5 \text{ ft}}{2} = 2.5 \text{ ft}$$

$A = \pi r^2$

$= \pi(2.5 \text{ ft})^2$

$\approx 3.14(2.5 \text{ ft})^2$

$= 19.625 \text{ ft}^2$

The area of the bottom of the pool is 19.625 ft^2.

Focus on finding the area of a circle

Find the area of a circle with a diameter of 9 cm. Use 3.14 for π.

SOLUTION

$$r = \frac{1}{2}d = \frac{1}{2}(9 \text{ cm}) = 4.5 \text{ cm}$$

$A = \pi r^2$

$\approx 3.14(4.5 \text{ cm})^2 = 63.585 \text{ cm}^2$

The area is approximately 63.585 cm^2.

Check your understanding 1

Find the area of a triangle with a base of 24 in. and a height of 14 in.

SOLUTION See page S-3.

Objective 4.3A Practice

For Exercises 1 to 4, find the area of the given figure. If necessary, use 3.14 for π.

1.

2.

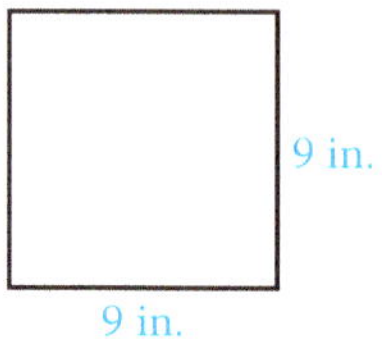

3.

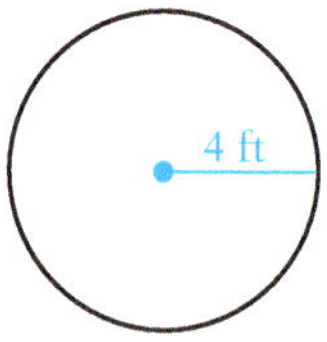

4.

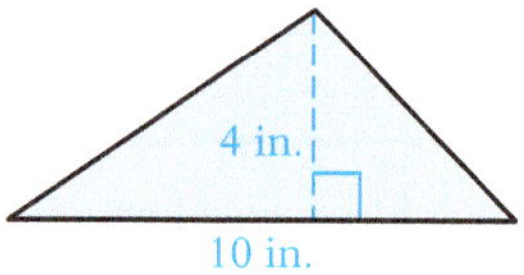

5. Find the area of a circle with a diameter of 40 cm. Use 3.14 for π.

Solutions on pp. S-11–S-12.

Objective 4.3B Find the area of a composite geometric figure

The area of the composite figure shown below is found by calculating the area of the rectangle and then subtracting the area of the triangle.

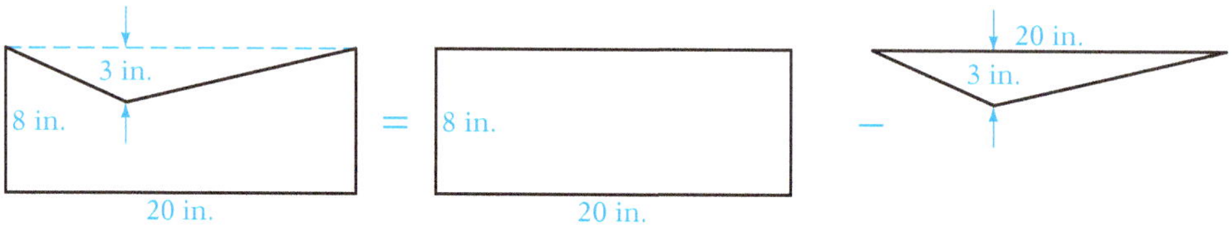

$$A = LW - \frac{1}{2}bh$$

$$= (20 \text{ in.})(8 \text{ in.}) - \frac{1}{2}(20 \text{ in.})(3 \text{ in.}) = 160 \text{ in}^2 - 30 \text{ in}^2 = 130 \text{ in}^2$$

Focus on finding the area of a composite figure

Find the area of the shaded portion of the figure. Use 3.14 for π.

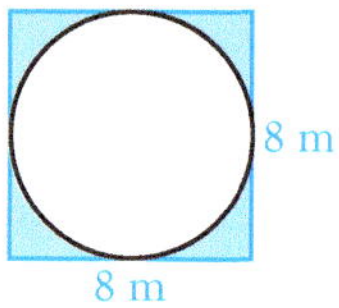

SOLUTION

Area of shaded portion = area of square − area of circle

$$\begin{aligned} A &= s^2 - \pi r^2 \\ &= (8\text{ m})^2 - \pi(4\text{ m})^2 \\ &\approx 64\text{ m}^2 - 3.14(16\text{ m}^2) \\ &= 64\text{ m}^2 - 50.24\text{ m}^2 = 13.76\text{ m}^2 \end{aligned}$$

The area is approximately 13.76 m^2.

Check your understanding 2

Find the area of the composite figure.

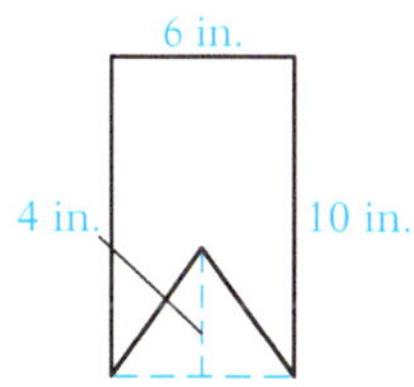

SOLUTION See page S-3.

Objective 4.3B Practice

For Exercises 1 to 4, find the area of the given figure. If necessary, use 3.14 for π.

1.

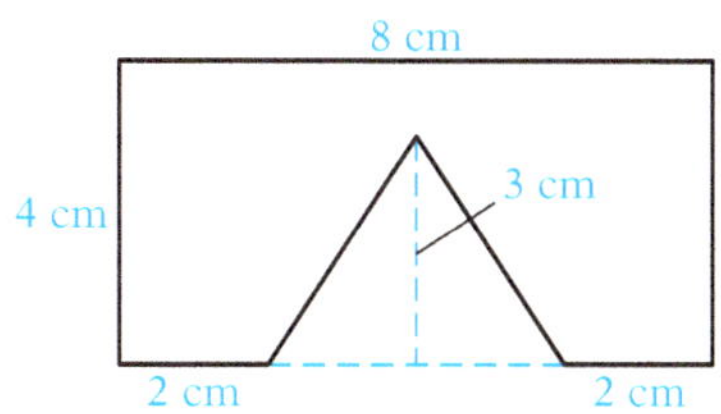

2.

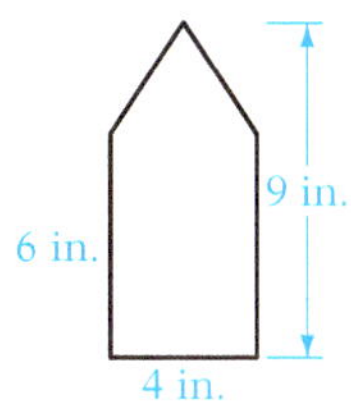

3.

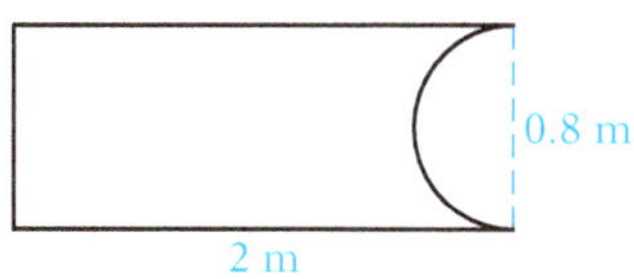

4.

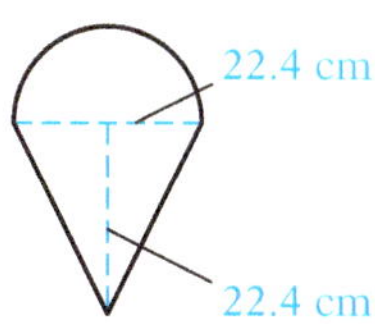

Solutions on p. S-12.

Objective 4.3C Solve application problems

Focus on solving an application

A walkway 2 m wide is built along the front and both sides of a building, as shown in the figure. Find the area of the walkway.

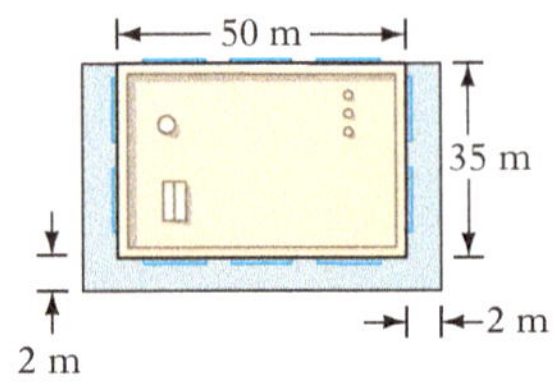

STRATEGY

To find the area of the walkway, add the area of the front section (54 m · 2 m) and the area of the two side sections (each 35 m · 2 m).

SOLUTION

Area of walkway = area of front section + 2(area of one side section)

$$
\begin{aligned}
A &= (54\text{ m})(2\text{ m}) + 2(35\text{ m})(2\text{ m}) \\
&= 108\text{ m}^2 + 140\text{ m}^2 \\
&= 248\text{ m}^2
\end{aligned}
$$

The area of the walkway is 248 m^2.

Check your understanding 3

New carpet is installed in a room measuring 9 ft by 12 ft. Find the area of the room in square yards. (9 ft^2 = 1 yd^2)

SOLUTION See pages S-3–S-4.

Objective 4.3C Practice

1. Artificial turf is being used to cover a playing field. The field is rectangular with a length of 100 yd and a width of 75 yd. How much artificial turf must be purchased to cover the field?
2. You plan to stain the wooden deck attached to your house. The deck measures 10 ft by 8 ft. A quart of stain will cost $11.87 and will cover 50 ft^2. How many quarts of stain should you buy?
3. A fabric wall hanging is to fill a space that measures 5 m by 3.5 m. Allowing for 0.1 m of the fabric to be folded back along each edge, how much fabric must be purchased for the wall hanging?
4. An irrigation system waters a circular field that has a 50-foot radius. Find the area watered by the irrigation system. Use 3.14 for π.
5. You want to tile your kitchen floor. The floor measures 12 ft by 9 ft. How many tiles, each a square with side $1\frac{1}{2}$ ft, should you purchase for the job?

Solutions on pp. S-13–S-14.

SECTION 4.4 Volume

Objective 4.4A Find the volume of a geometric solid

Volume is a measure of the amount of space inside a closed surface, or figure in space. Volume can be used to describe, for example, the amount of heating gas used for cooking, the amount of concrete delivered for the foundation of a house, or the amount of water in storage for a city's water supply.

A cube that is 1 ft on each side has a volume of 1 cubic foot, which is written 1 ft^3.

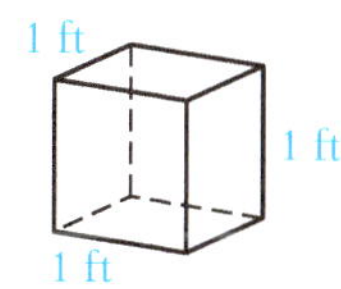

A cube that measures 1 cm on each side has a volume of 1 cubic centimeter, which is written 1 cm^3.

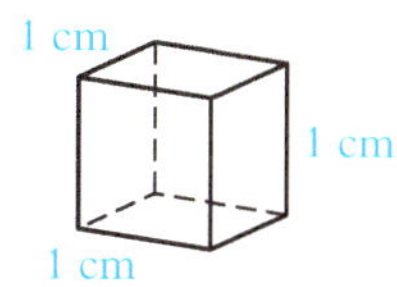

The volume of a solid is the number of cubes that are necessary to fill the solid exactly. The volume of the rectangular solid at the right is 24 cm^3 because it will hold exactly 24 cubes, each 1 cm on a side. Note that the volume can be found by multiplying the length times the width times the height.

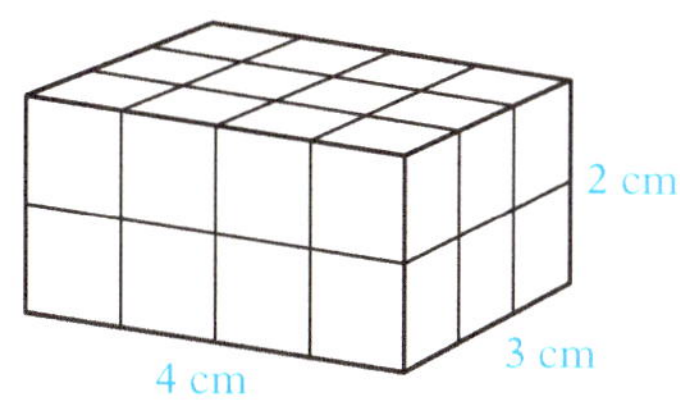

Volume of a Rectangular Solid

Let L be the length, W be the width, and H be the height of a rectangular solid.

The volume of the rectangular solid is $V = LWH$.

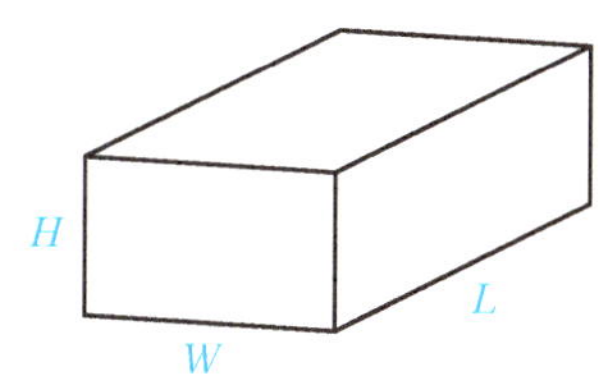

EXAMPLE

Find the volume of the rectangular solid shown at the right.

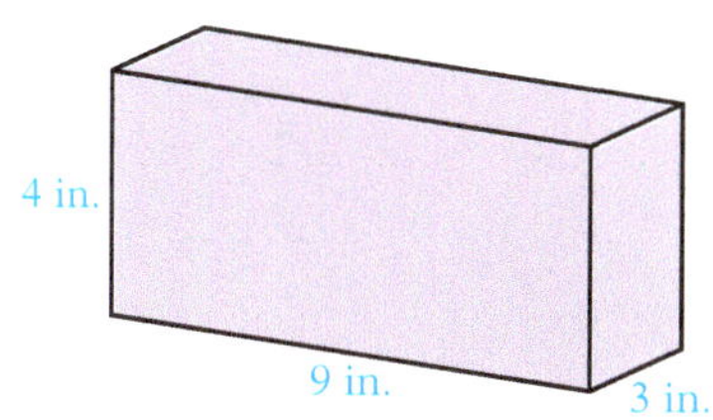

$V = LWH$

$= (9 \text{ in.})(3 \text{ in.})(4 \text{ in.})$ • $L = 9$ in., $W = 3$ in., $H = 4$ in.

$= 108 \text{ in}^3$

The volume of the rectangular solid is 108 in^3.

Focus on finding the volume of a rectangular solid

Find the volume of a rectangular solid with a length of 3 ft, a width of 1.5 ft, and a height of 2 ft.

SOLUTION

$$V = LWH$$
$$= (3 \text{ ft})(1.5 \text{ ft})(2 \text{ ft})$$
$$= 9 \text{ ft}^3$$

The volume is 9 ft^3.

Check your understanding 1

Find the volume of a rectangular solid with a length of 8 cm, a width of 3.5 cm, and a height of 4 cm.

SOLUTION See page S-4.

A **cube** is a rectangular solid for which the length, width, and height are all equal. The volume of a cube is found by multiplying the length of a side of the cube times itself three times ("side cubed").

Volume of a Cube

Let s be the length of one side of a cube.
The volume of the cube is $V = s^3$.

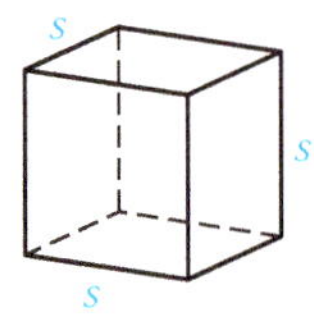

EXAMPLE

Find the volume of the cube shown at the right.

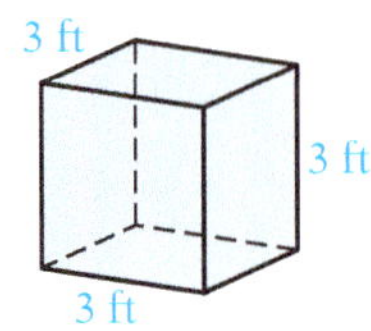

$$V = s^3$$
$$= (3 \text{ ft})^3 \quad \bullet\ s = 3 \text{ ft}$$
$$= 27 \text{ ft}^3$$

The volume of the cube is 27 ft^3.

Focus on finding the volume of a cube

Find the volume of a cube that has a side measuring 2.5 in.

SOLUTION

$$V = s^3$$
$$= (2.5 \text{ in.})^3$$
$$= 15.625 \text{ in}^3$$

The volume is 15.625 in^3.

Check your understanding 2

Find the volume of a cube with a side of length 5 cm.

SOLUTION See page S-4.

Volume of a Sphere

Let r be the radius of a sphere.

The volume of the sphere is $V = \frac{4}{3}\pi r^3$.

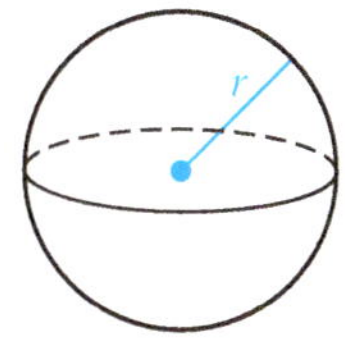

EXAMPLE

Find the volume of the sphere shown at the right. Use 3.14 for π. Round to the nearest hundredth.

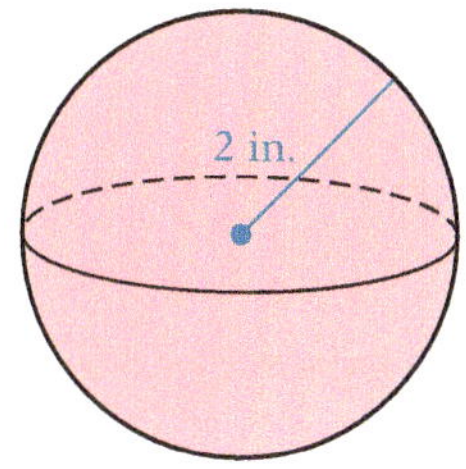

$$V = \frac{4}{3}\pi r^3$$
$$\approx \frac{4}{3}(3.14)(2 \text{ in.})^3 \quad \bullet\ r = 2 \text{ in.}$$
$$= \frac{4}{3}(3.14)(8 \text{ in}^3)$$
$$\approx 33.49 \text{ in}^3$$

The volume is approximately 33.49 in³.

Volume of a Cylinder

Let r be the radius of a cylinder.

The volume of the cylinder is $V = \pi r^2 h$.

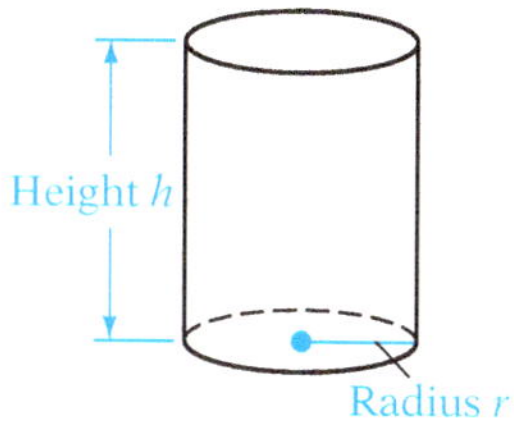

EXAMPLE

Find the volume of the cylinder shown at the right. Use 3.14 for π. Round to the nearest hundredth.

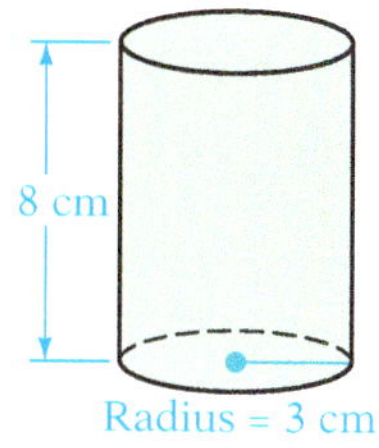

$$V = \pi r^2 h$$
$$\approx 3.14(3 \text{ cm})^2(8 \text{ cm}) \quad \bullet\ r = 3 \text{ cm}; h = 8 \text{ cm}$$
$$= 3.14(9 \text{ cm}^2)(8 \text{ cm})$$
$$= 226.08 \text{ cm}^3$$

The volume of the cylinder is approximately 226.08 cm³.

Focus on finding the volume of a cylinder

Find the volume of a cylinder with a radius of 12 cm and a height of 65 cm. Use 3.14 for π.

SOLUTION

$$V = \pi r^2 h$$
$$\approx 3.14(12 \text{ cm})^2(65 \text{ cm})$$
$$= 3.14(144 \text{ cm}^2)(65 \text{ cm})$$
$$= 29{,}390.4 \text{ cm}^3$$

The volume is approximately 29,390.4 cm³.

Check your understanding 3

Find the volume of a cylinder with a diameter of 14 in. and a height of 15 in. Use $\frac{22}{7}$ for π.

SOLUTION See page S-4.

Focus on finding the volume of a sphere

Find the volume of a sphere with a diameter of 12 in. Use 3.14 for π.

SOLUTION

$r = \frac{1}{2}d = \frac{1}{2}(12 \text{ in.}) = 6 \text{ in.}$ • Find the radius.

$V = \frac{4}{3}\pi r^3$ • Use the formula for the volume of a sphere

$\approx \frac{4}{3}(3.14)(6 \text{ in.})^3$

$= \frac{4}{3}(3.14)(216 \text{ in}^3)$

$= 904.32 \text{ in}^3$

The volume is approximately 904.32 in^3.

Check your understanding 4

Find the volume of a sphere with a radius of 3 m. Use 3.14 for π.

SOLUTION See page S-4.

Objective 4.4A Practice

1. Find the volume of the figure below.

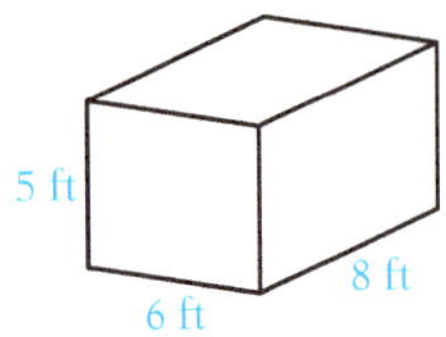

2. Find the volume of the figure below. Use 3.14 for π.

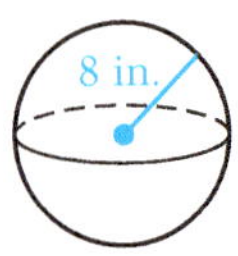

3. Find the volume of the figure below. Use 3.14 for π.

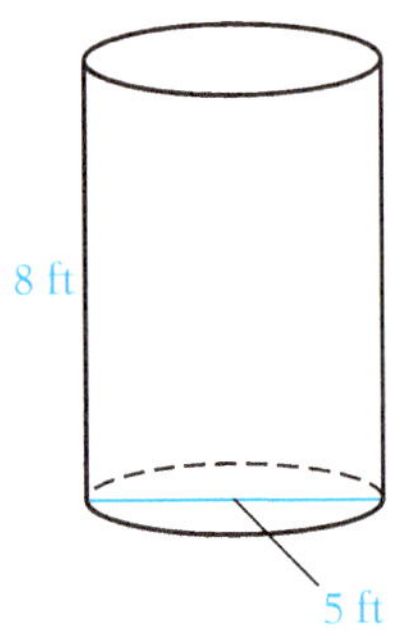

4. Find the volume of a cube with a side of length 2.14 m. Round to the nearest tenth.

5. Find the volume of a cylinder with a diameter of 12 ft and a height of 30 ft. Use 3.14 for π.

Solutions on p. S-14.

Objective 4.4B Find the volume of a composite geometric solid

A **composite geometric solid** is a solid made from two or more geometric solids. The solid shown is made from a cylinder and one-half of a sphere.

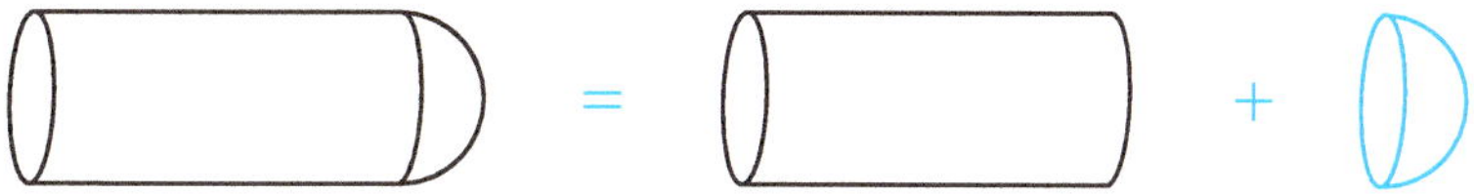

Volume of the composite solid = volume of the cylinder + $\frac{1}{2}$ the volume of the sphere

Find the volume of the composite solid shown above if the radius of the base of the cylinder is 3 in. and the height of the cylinder is 10 in. Use 3.14 for π.

The volume equals the volume of a cylinder plus one-half the volume of a sphere. The radius of the sphere equals the radius of the base of the cylinder.

$$V = \pi r^2 h + \frac{1}{2}\left(\frac{4}{3}\pi r^3\right)$$

$$\approx 3.14(3\text{ in.})^2(10\text{ in.}) + \frac{1}{2}\left(\frac{4}{3}\right)(3.14)(3\text{ in.})^3$$

$$= 3.14(9\text{ in}^2)(10\text{ in.}) + \frac{1}{2}\left(\frac{4}{3}\right)(3.14)(27\text{ in}^3)$$

$$= 282.6\text{ in}^3 + 56.52\text{ in}^3$$

$$= 339.12\text{ in}^3$$

The volume is approximately 339.12 in^3.

Focus on finding the volume of a composite figure

Find the volume of the solid shown in the figure. Use 3.14 for π.

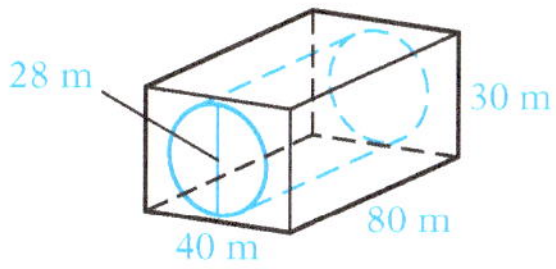

SOLUTION

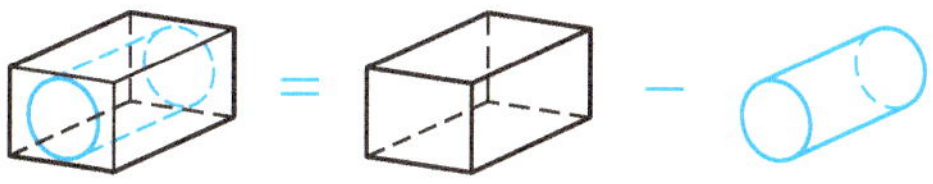

Volume of solid = volume of rectangular solid − volume of cylinder

$V = LWH - \pi r^2 h$

$\approx (80\text{ m})(40\text{ m})(30\text{ m}) - 3.14(14\text{ m})^2(80\text{ m})$

$= 96{,}000\text{ m}^3 - 49{,}235.2\text{ m}^3$

$= 46{,}764.8\text{ m}^3$

The volume is approximately 46,764.8 m^3.

Check your understanding 5

Find the volume of the solid shown in the figure. Use 3.14 for π.

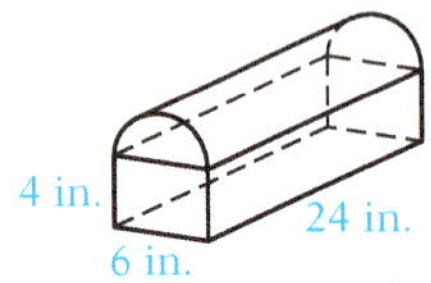

SOLUTION See page S-4.

Objective 4.4B Practice

For Exercises 1 to 4, find the volume of the given figure. If necessary, use 3.14 for π.

1.

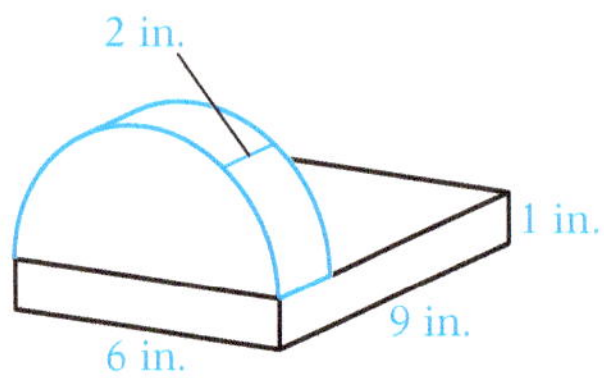

2.

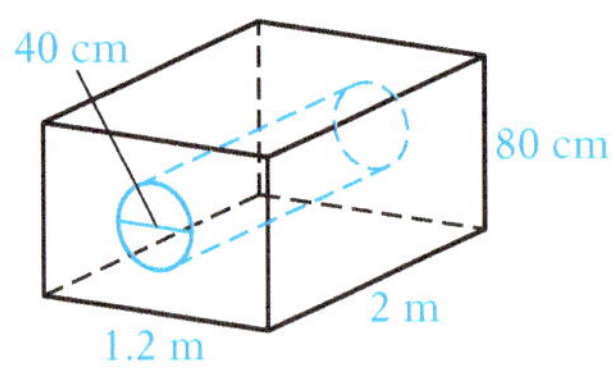

3.

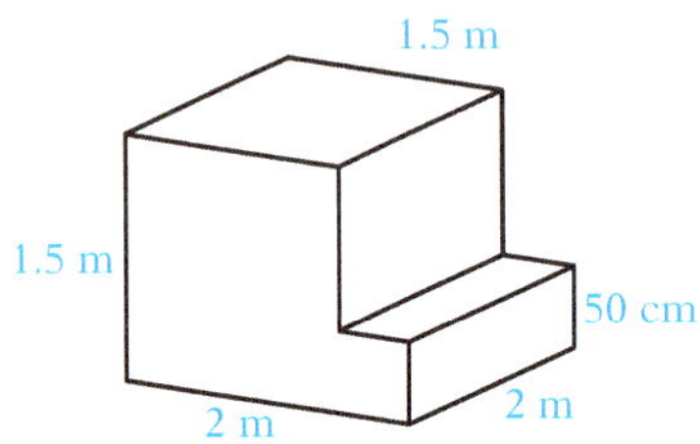

4.

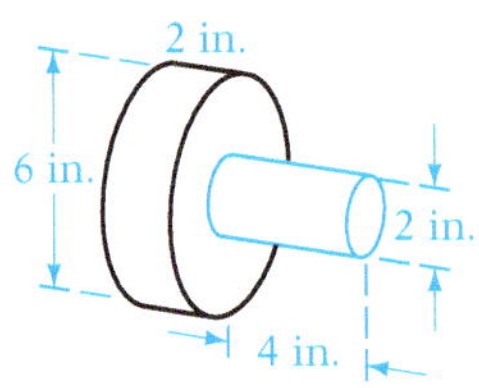

Solutions on pp. S-14–S-15.

Objective 4.4C Solve application problems

Focus on finding the volume of an aquarium

An aquarium is 28 in. long, 14 in. wide, and 16 in. high. Find the volume of the aquarium.

STRATEGY

To find the volume of the aquarium, use the formula for the volume of a rectangular solid.

SOLUTION

$$V = LWH$$
$$= (28 \text{ in.})(14 \text{ in.})(16 \text{ in.})$$
$$= 6272 \text{ in}^3$$

The volume of the aquarium is 6272 in^3.

Check your understanding 6

Find the volume of a freezer that is 7 ft long, 3 ft high, and 2.5 ft wide.

SOLUTION See page S-5.

Focus on finding the volume of a bushing

Find the volume of the bushing shown in the figure below. Use 3.14 for π.

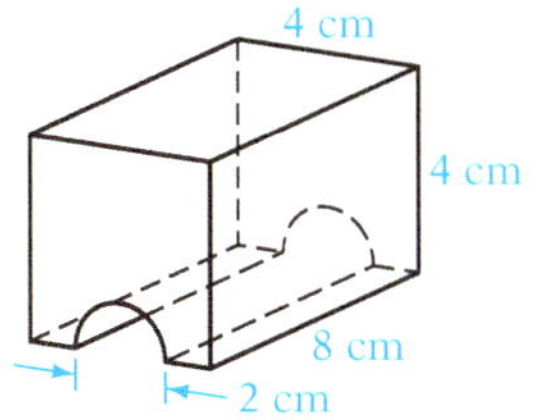

STRATEGY

To find the volume of the bushing, subtract the volume of the half-cylinder from the volume of the rectangular solid.

SOLUTION

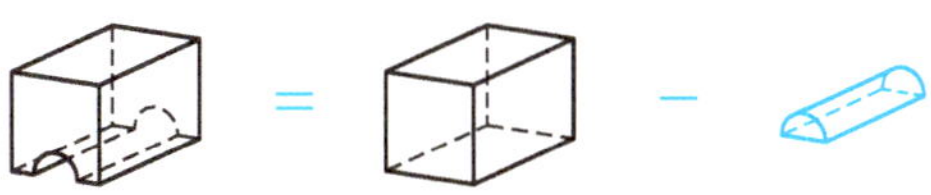

Volume of bushing = volume of rectangular solid − $\frac{1}{2}$ volume of cylinder

$$V = LWH - \frac{1}{2}\pi r^2 h$$
$$\approx (8 \text{ cm})(4 \text{ cm})(4 \text{ cm}) - \frac{1}{2}(3.14)(1 \text{ cm})^2(8 \text{ cm})$$
$$= 128 \text{ cm}^3 - 12.56 \text{ cm}^3$$
$$= 115.44 \text{ cm}^3$$

The volume of the bushing is approximately 115.44 cm^3.

Check your understanding 7

Find the volume of the channel iron shown in the figure below.

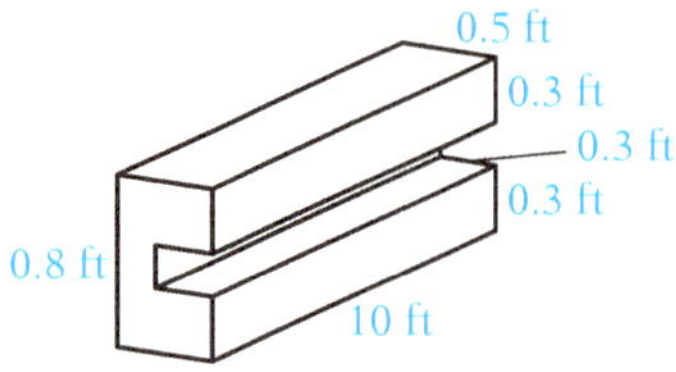

SOLUTION See page S-5.

Objective 4.4C Practice

1. A rectangular tank at a fish hatchery is 9 m long, 3 m wide, and 1.5 m deep. Find the volume of the water in the tank when the tank is full.
2. An oil tank, which is in the shape of a cylinder, is 4 m high and has a diameter of 6 m. The oil tank is two-thirds full. Find the number of cubic meters of oil in the tank. Use 3.14 for π. Round to the nearest hundredth.
3. A silo, which is in the shape of a cylinder, is 16 ft in diameter and has a height of 30 ft. The silo is three-fourths full. Find the volume of the portion of the silo that is not being used for storage. Use 3.14 for π.
4. An architect is designing the heating system for an auditorium and needs to know the volume of the structure. Find the volume of the auditorium with the measurements shown in the figure below. Use 3.14 for π.

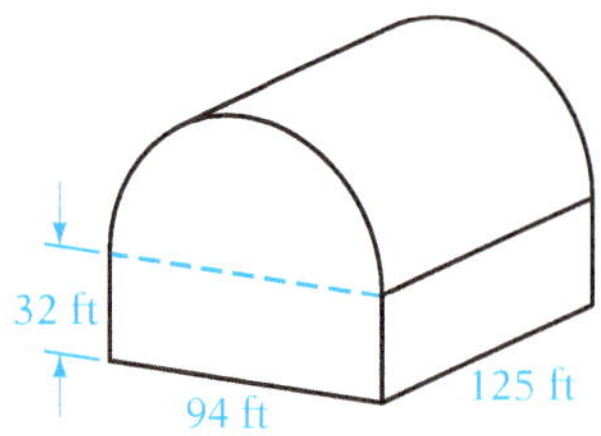

Solutions on pp. S-15–S-16.

SECTION 4.5 The Pythagorean Theorem

Objective 4.5A Apply the Pythagorean Theorem

The Greek mathematician Pythagoras is generally credited with the discovery that the square of the hypotenuse of a right triangle is equal to the sum of the squares of the two legs. This is called the **Pythagorean Theorem.**

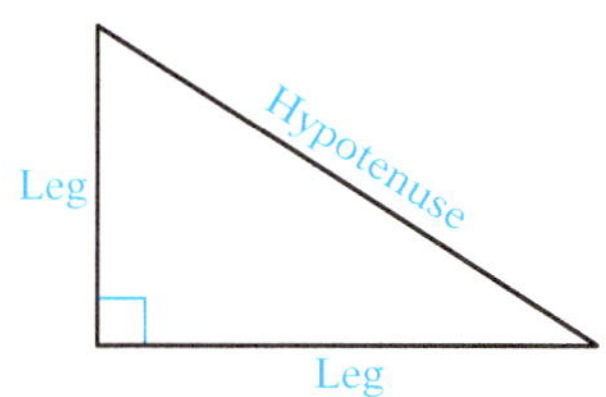

Pythagorean Theorem

If a and b are the lengths of the legs of a right triangle and c is the length of the hypotenuse, then $c^2 = a^2 + b^2$.

If the length of one side of a right triangle is unknown, then one of the following formulas can be used to find it.

If the length of the hypotenuse is unknown, use

$$\begin{aligned}\text{Hypotenuse} &= \sqrt{(\text{leg})^2 + (\text{leg})^2}\\ &= \sqrt{(3)^2 + (4)^2}\\ &= \sqrt{9 + 16}\\ &= \sqrt{25} = 5\end{aligned}$$

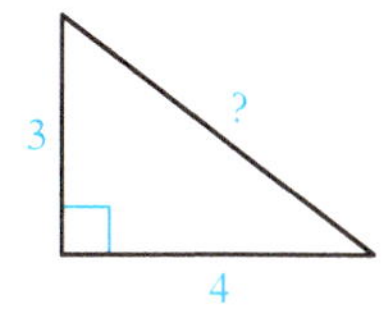

If the length of a leg is unknown, use

$$\begin{aligned}\text{Leg} &= \sqrt{(\text{hypotenuse})^2 - (\text{leg})^2}\\ &= \sqrt{(5)^2 - (4)^2}\\ &= \sqrt{25 - 16}\\ &= \sqrt{9} = 3\end{aligned}$$

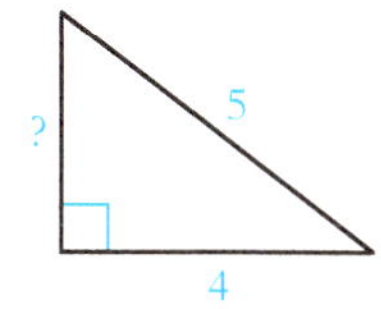

Focus on finding the hypotenuse of a right triangle

Find the hypotenuse of the triangle shown in the figure. Round to the nearest thousandth.

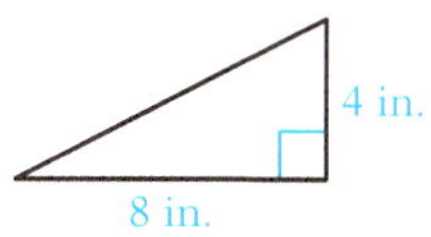

SOLUTION

$$\begin{aligned}\text{Hypotenuse} &= \sqrt{(\text{leg})^2 + (\text{leg})^2}\\ &= \sqrt{8^2 + 4^2}\\ &= \sqrt{64 + 16}\\ &= \sqrt{80} \approx 8.944\end{aligned}$$

The hypotenuse is approximately 8.944 in.

Check your understanding 1

Find the hypotenuse of the triangle shown in the figure. Round to the nearest thousandth.

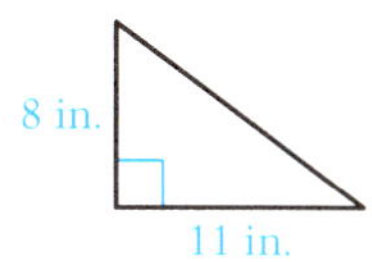

SOLUTION See page S-5.

Focus on finding the length of a leg of a right triangle

Find the length of the leg of the triangle shown in the figure. Round to the nearest thousandth.

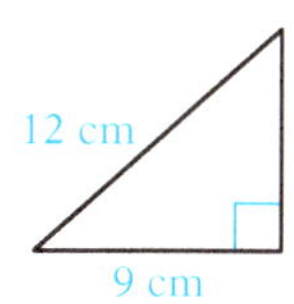

SOLUTION

$$\begin{aligned}\text{Leg} &= \sqrt{(\text{hypotenuse})^2 - (\text{leg})^2}\\ &= \sqrt{12^2 - 9^2}\\ &= \sqrt{144 - 81}\\ &= \sqrt{63} \approx 7.937\end{aligned}$$

The length of the leg is approximately 7.937 cm.

Check your understanding 2

Find the length of the leg of the triangle shown in the figure. Round to the nearest thousandth.

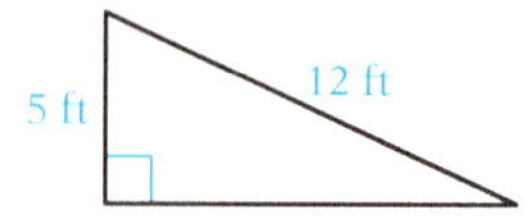

SOLUTION See page S-5.

Focus on solving an application

A 25-foot ladder is placed against a building at a point 21 ft above the ground, as shown in the figure. Find the distance from the base of the building to the base of the ladder. Round to the nearest thousandth.

STRATEGY

To find the distance from the base of the building to the base of the ladder, use the Pythagorean Theorem. The hypotenuse is the length of the ladder (25 ft). One leg is the distance along the building from the ground to the top of the ladder (21 ft). The distance from the base of the building to the base of the ladder is the unknown leg.

SOLUTION

$$\begin{aligned}\text{Leg} &= \sqrt{(\text{hypotenuse})^2 - (\text{leg})^2}\\ &= \sqrt{25^2 - 21^2}\\ &= \sqrt{625 - 441}\\ &= \sqrt{184} \approx 13.565\end{aligned}$$

The distance is approximately 13.565 ft.

Check your understanding 3

Find the distance between the centers of the holes in the metal plate shown in the figure. Round to the nearest thousandth.

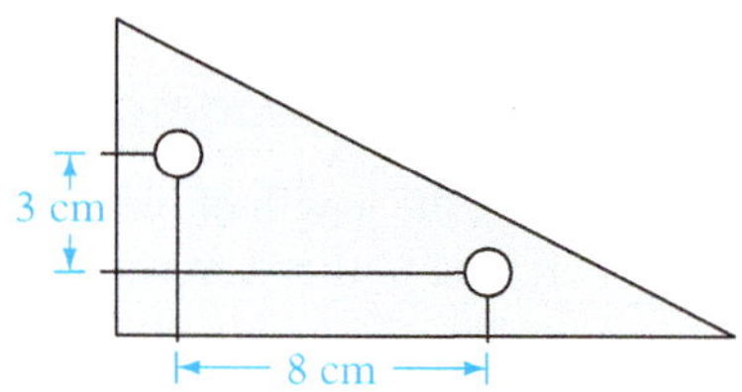

SOLUTION See page S-6.

Objective 4.5A Practice

For Exercises 1 to 3, find the unknown side of the triangle. Round to the nearest thousandth.

1.

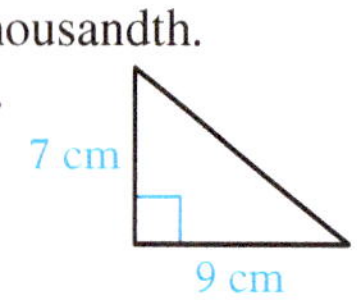

2.

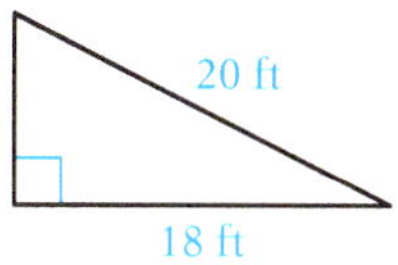

3.

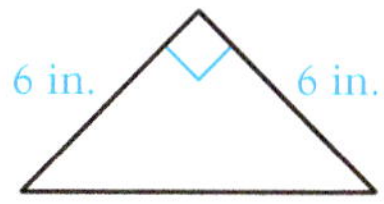

4. In the figure below, find the length of the ramp used to roll barrels up to the loading dock, which is 3.5 ft high. Round to the nearest hundredth.

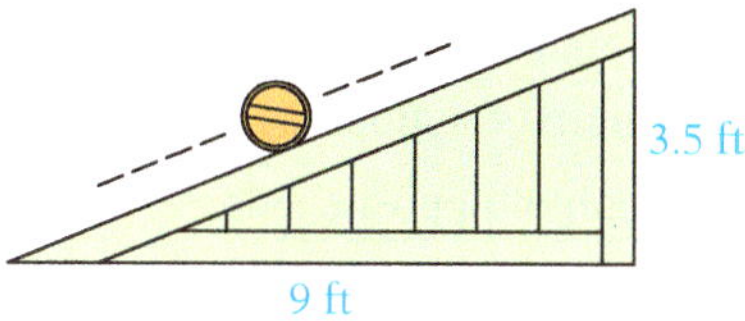

5. A ladder 8 m long is placed against a building in preparation for washing the windows. How high on the building does the ladder reach when the bottom of the ladder is 3 m from the base of the building? Round to the nearest tenth.

Solutions on pp. S-16–S-17.

SECTION 4.6 Right Triangle Trigonometry

Objective 4.6A Find the values of trigonometric ratios of a right triangle

The study of **trigonometry,** a term that comes from two Greek words meaning "triangle measurement," began about 2000 years ago, partially as a means of solving surveying problems. In this section, we will examine **right triangle trigonometry**—that is, trigonometry that applies only to right triangles.

When working with right triangles, it is convenient to refer to the side *opposite* an angle and to the side *adjacent to* (next to) an angle. The hypotenuse of a right triangle is defined to be neither adjacent to nor opposite either of the acute angles of the triangle.

Take Note ➤
In trigonometry, it is common practice to use Greek letters to name the angles of a triangle. Here are some frequently used letters: α (alpha), β (beta), and θ (theta). The word *alphabet* is derived from the first two letters of the Greek alphabet, α and β.

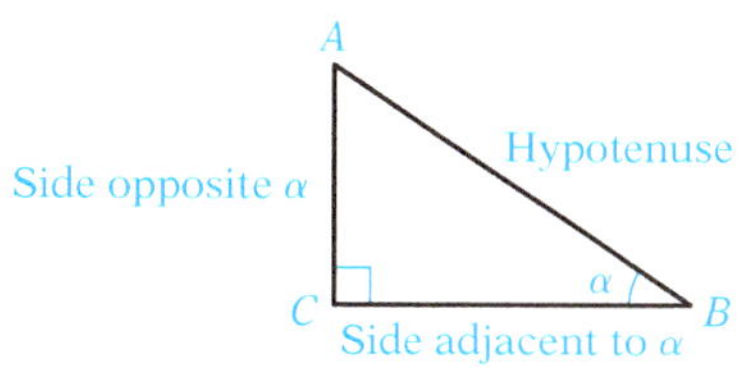

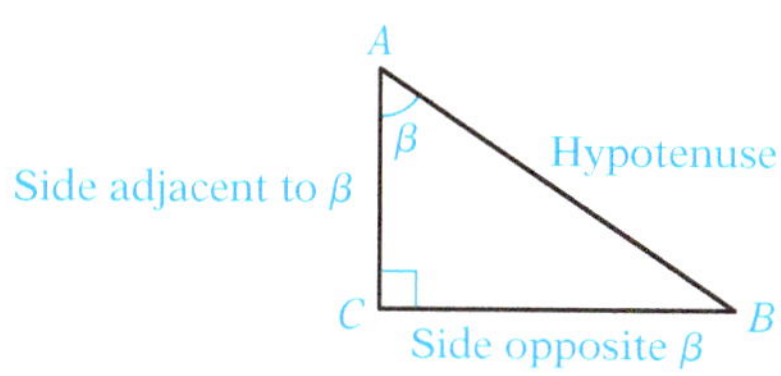

Consider the right triangle shown in Figure 4.1. Six possible ratios can be formed using the lengths of the sides of the triangle.

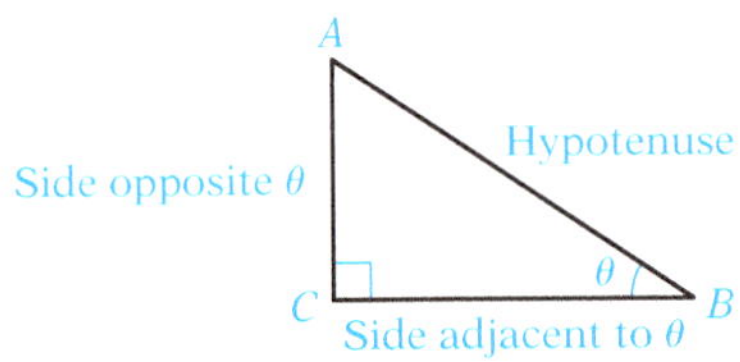

Figure 4.1

The Trigonometric Ratios of an Acute Angle of a Right Triangle

If θ is an acute angle of a right triangle ABC, then

$$\sin\theta = \frac{\text{length of opposite side}}{\text{length of hypotenuse}} \qquad \csc\theta = \frac{\text{length of hypotenuse}}{\text{length of opposite side}}$$

$$\cos\theta = \frac{\text{length of adjacent side}}{\text{length of hypotenuse}} \qquad \sec\theta = \frac{\text{length of hypotenuse}}{\text{length of adjacent side}}$$

$$\tan\theta = \frac{\text{length of opposite side}}{\text{length of adjacent side}} \qquad \cot\theta = \frac{\text{length of adjacent side}}{\text{length of opposite side}}$$

For convenience, we will write opp, adj, and hyp as abbreviations for *the length of the* opposite side, adjacent side, and hypotenuse, respectively. Using this convention, the definitions of the trigonometric ratios are written as follows:

$$\sin\theta = \frac{\text{opp}}{\text{hyp}} \qquad \csc\theta = \frac{\text{hyp}}{\text{opp}}$$

$$\cos\theta = \frac{\text{adj}}{\text{hyp}} \qquad \sec\theta = \frac{\text{hyp}}{\text{adj}}$$

$$\tan\theta = \frac{\text{opp}}{\text{adj}} \qquad \cot\theta = \frac{\text{adj}}{\text{opp}}$$

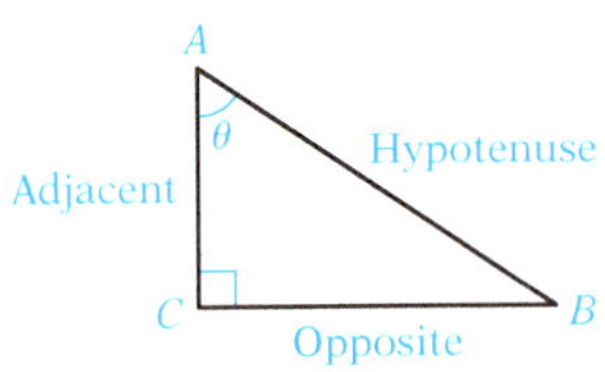

All of the trigonometric ratios have applications, but the sine, cosine, and tangent ratios are used most frequently. For the remainder of this section, we will focus on these ratios.

When working with trigonometric ratios, be sure to draw a diagram and label the adjacent and opposite sides of an angle. For instance, in the definition above, if we had placed θ at angle A, then the triangle would have been labeled as shown at the left. The definitions of the ratios remain the same.

$$\sin\theta = \frac{\text{opp}}{\text{hyp}} \qquad \cos\theta = \frac{\text{adj}}{\text{hyp}} \qquad \tan\theta = \frac{\text{opp}}{\text{adj}}$$

Focus on finding the values of trigonometric ratios

For the right triangle at the right, find the values of $\sin\theta$, $\cos\theta$, and $\tan\theta$. Round to the nearest ten-thousandth.

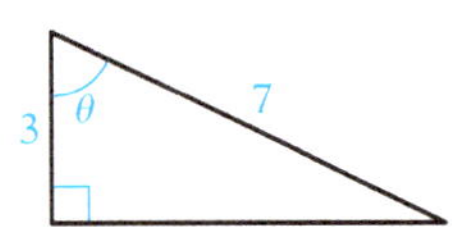

SOLUTION

Use the Pythagorean Theorem to find the length of the side opposite θ.

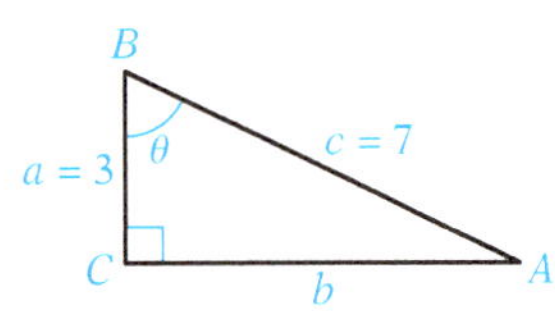

$$a^2 + b^2 = c^2$$
$$3^2 + b^2 = 7^2 \qquad \text{• } a = 3, c = 7$$
$$9 + b^2 = 49$$
$$b^2 = 40$$
$$b = \sqrt{40} \approx 6.3246$$

Using the definitions of the trigonometric ratios, we have

$$\sin\theta = \frac{\text{opp}}{\text{hyp}} \approx \frac{6.3246}{7} \approx 0.9035 \qquad \cos\theta = \frac{\text{adj}}{\text{hyp}} = \frac{3}{7} \approx 0.4286$$

$$\tan\theta = \frac{\text{opp}}{\text{adj}} \approx \frac{6.3246}{3} \approx 2.1082$$

Check your understanding 1

For the right triangle at the right, find the values of $\sin\theta$, $\cos\theta$, and $\tan\theta$.

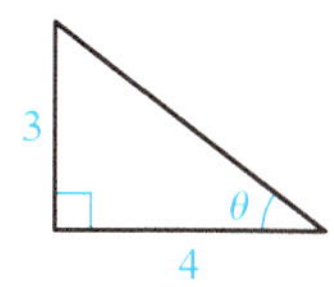

SOLUTION See page S-6.

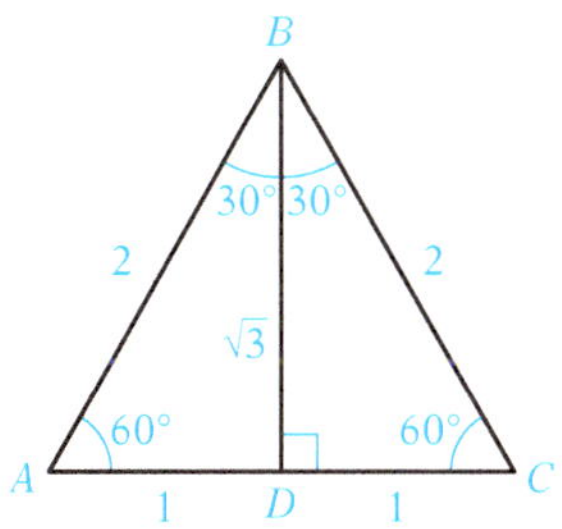

We will often want to know the value of a trigonometric ratio for a given angle. Triangle ABC at the left is an equilateral triangle with sides of length 2 units and angle bisector $\overline{BD}$. Because $\overline{BD}$ bisects $\angle ABC$, the measures of $\angle ABD$ and $\angle DBC$ are both 30°. The angle bisector $\overline{BD}$ also bisects $\overline{AC}$. Therefore, $AD = 1$ and $DC = 1$. Using the Pythagorean Theorem, we can find the measure of BD.

$$\begin{aligned}
(DC)^2 + (BD)^2 &= (BC)^2 \\
1^2 + (BD)^2 &= 2^2 && \text{• } DC = 1,\ BC = 2 \\
1 + (BD)^2 &= 4 && \text{• Solve for } BD. \\
(BD)^2 &= 3 \\
BD &= \sqrt{3}
\end{aligned}$$

Using the definitions of the trigonometric ratios and triangle BCD, we can find the values of the sine, cosine, and tangent of 30° and 60°.

$$\sin 30^\circ = \frac{\text{opp}}{\text{hyp}} = \frac{1}{2} = 0.5 \qquad \sin 60^\circ = \frac{\text{opp}}{\text{hyp}} = \frac{\sqrt{3}}{2} \approx 0.8660$$

$$\cos 30^\circ = \frac{\text{adj}}{\text{hyp}} = \frac{\sqrt{3}}{2} \approx 0.8660 \qquad \cos 60^\circ = \frac{\text{adj}}{\text{hyp}} = \frac{1}{2} = 0.5$$

$$\tan 30^\circ = \frac{\text{opp}}{\text{adj}} = \frac{1}{\sqrt{3}} \approx 0.5774 \qquad \tan 60^\circ = \frac{\text{opp}}{\text{adj}} = \sqrt{3} \approx 1.7320$$

Take Note ➤
Just as distances can be measured in feet, miles, meters, and other units, angles can be measured in various units: degrees, radians, and grads. In this section, we use only degree measurements for angles, so be sure your calculator is in degree mode.

On a TI-83/84, press the MODE key to determine whether the calculator is in degree mode.

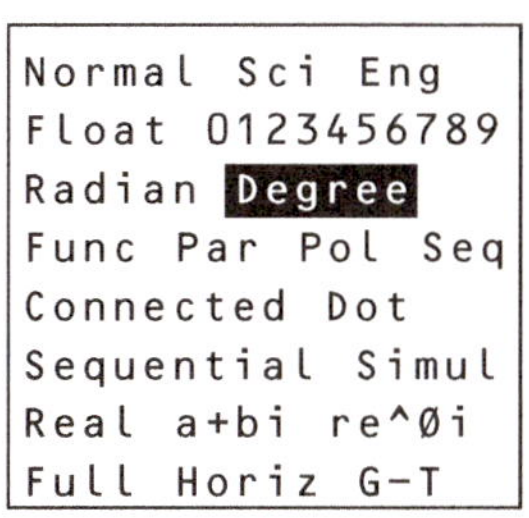

The properties of an equilateral triangle enabled us to calculate the values of the trigonometric ratios for 30° and 60°. Calculating values of the trigonometric ratios for most other angles, however, would be quite difficult. Fortunately, many calculators have been programmed to allow us to estimate these values.

To find tan 30° on a TI-83/84 calculator, confirm that your calculator is in the "degree" mode. Press the tan key and type in 30). Then press ENTER.

$$\tan 30^\circ \approx 0.5774$$

On a scientific calculator, type in 30, and then press the tan key.

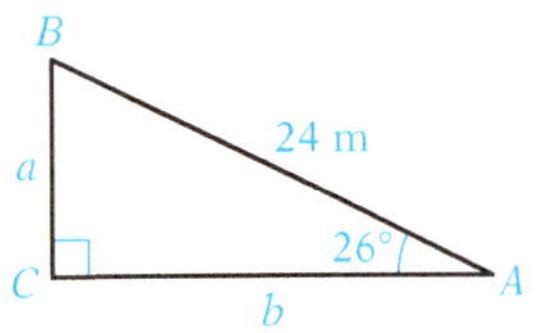

Focus on finding the length of a side of a triangle

For the right triangle shown at the left, find the length of side a. Round to the nearest tenth of a meter.

SOLUTION

We are given the measure of $\angle A$ and the hypotenuse. We want to find the length of side a. Side a is opposite $\angle A$. The sine function involves the side opposite an angle and the hypotenuse.

$$\sin A = \frac{\text{opp}}{\text{hyp}}$$

$$\sin 26^\circ = \frac{a}{24} \qquad \bullet\ A = 26^\circ,\ \text{hypotenuse} = 24\text{ m.}$$

$$24(\sin 26^\circ) = 24\left(\frac{a}{24}\right) \qquad \bullet\ \text{Multiply each side by 24.}$$

$$10.5 \approx a \qquad \bullet\ \text{Use a calculator to find } \sin 26^\circ\text{. Multiply the result in the display by 24.}$$

The length of side a is approximately 10.5 m.

Check your understanding 2

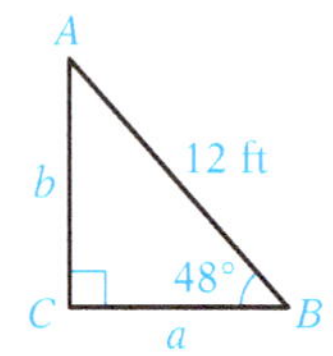

For the right triangle shown at the right, find the length of side a. Round to the nearest tenth of a foot.

SOLUTION See page S-6.

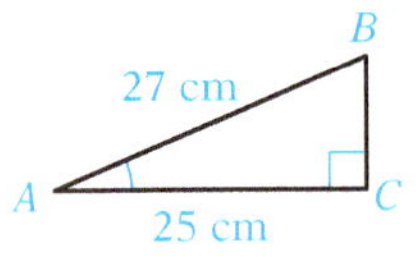

Sometimes it is necessary to find the measure of one of the acute angles of a right triangle. For instance, suppose it is necessary to find the measure of $\angle A$ in the figure at the left. Because the length of the side adjacent to $\angle A$ is known and the length of the hypotenuse is known, we can write

$$\cos A = \frac{\text{adj}}{\text{hyp}}$$

$$\cos A = \frac{25}{27}$$

The solution of this equation is the measure of the angle whose cosine is $\frac{25}{27}$. This angle can be found by using the $\boxed{\cos^{-1}}$ key on a calculator. The notation "$\cos^{-1}$" is read as "the inverse cosine of x."

$$\cos^{-1}\left(\frac{25}{27}\right) \approx 22.19160657$$

To the nearest tenth of a degree, the measure of $\angle A$ is 22.2°.

Take Note ➤
The expression $\sin^{-1}(x)$ is sometimes written $\arcsin(x)$. The two expressions are equivalent. The expressions $\cos^{-1}(x)$ and $\arccos(x)$ are equivalent, as are $\tan^{-1}(x)$ and $\arctan(x)$.

Inverse Sine, Inverse Cosine, and Inverse Tangent

$\sin^{-1}(x)$ is defined as the angle whose sine is x, $0 < x < 1$.

$\cos^{-1}(x)$ is defined as the angle whose cosine is x, $0 < x < 1$.

$\tan^{-1}(x)$ is defined as the angle whose tangent is x, $x > 0$.

Take Note ➤
To find an inverse on a calculator, we usually press the [INV] or [2nd] key prior to pushing the [sin], [cos], or [tan] key. Some calculators have [$\sin^{-1}$], [$\cos^{-1}$], and [$\tan^{-1}$] keys. Consult the instruction manual for your calculator.

Focus on evaluating an inverse sine expression

Use a calculator to find $\sin^{-1}(0.9171)$. Round to the nearest tenth of a degree.

SOLUTION

$\sin^{-1}(0.9171) \approx 66.5°$ • The calculator must be in degree mode. Press the keys for inverse sine, followed by .9171 [ )]. Press [ENTER].

Check your understanding 3

Use a calculator to find $\tan^{-1}(0.3165)$. Round to the nearest tenth of a degree.

SOLUTION See page S-6.

Take Note ➤
If
$\sin\theta = 0.7239$,
then
$\theta = \sin^{-1}(0.7239)$.

Focus on finding the measure of an angle using inverse sine

Given $\sin\theta = 0.7239$, find θ. Use a calculator. Round to the nearest tenth of a degree.

SOLUTION

This is equivalent to finding $\sin^{-1}(0.7239)$. The calculator must be in the degree mode.

$$\sin^{-1}(0.7239) \approx 46.4°$$
$$\theta \approx 46.4°$$

Check your understanding 4

Given $\tan\theta = 0.5681$, find θ. Use a calculator. Round to the nearest tenth of a degree.

SOLUTION See page S-6.

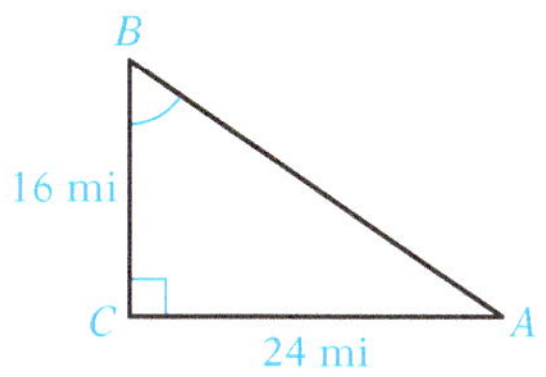

Focus on finding the measure of an angle in a right triangle

For the right triangle shown at the left, find the measure of $\angle B$. Round to the nearest tenth of a degree.

SOLUTION

We want to find the measure of $\angle B$, and we are given the lengths of the sides opposite $\angle B$ and adjacent to $\angle B$. The tangent ratio involves the side opposite an angle and the side adjacent to that angle.

$$\tan B = \frac{\text{opp}}{\text{adj}}$$
$$\tan B = \frac{24}{16}$$
$$B = \tan^{-1}\left(\frac{24}{16}\right)$$
$$B \approx 56.3°$$

• Use the [$\tan^{-1}$] key on a calculator.

The measure of $\angle B$ is approximately 56.3°.

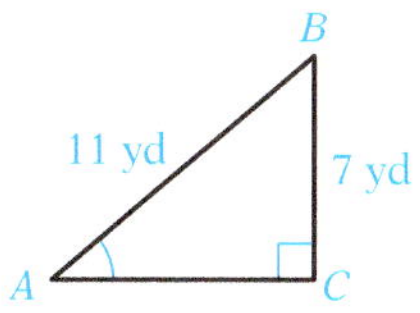

Check your understanding 5

For the right triangle shown at the left, find the measure of $\angle A$. Round to the nearest tenth of a degree.

SOLUTION See page S-7.

Objective 4.6A Practice

1. Find the values of $\sin\theta$, $\cos\theta$, and $\tan\theta$ for the given right triangle. Round to the nearest ten-thousandth.

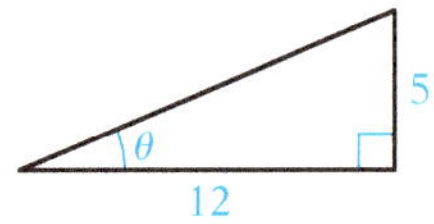

2. Find the values of $\sin\theta$, $\cos\theta$, and $\tan\theta$ for the given right triangle. Round to the nearest ten-thousandth.

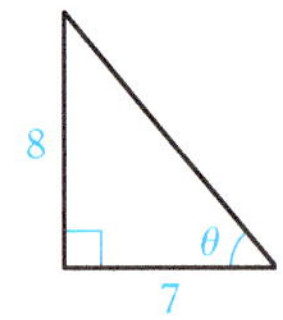

3. Given $\sin\theta = 0.6239$, find θ. Round to the nearest tenth.

4. For the given triangle, find the length of side a. Round to the nearest tenth.

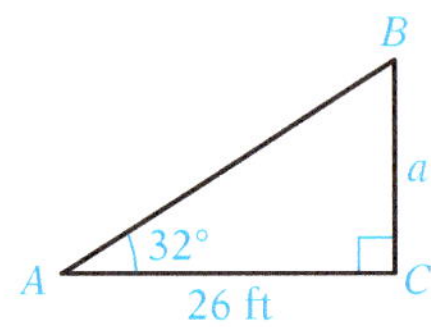

5. For the given triangle, find the value of θ. Round to the nearest tenth.

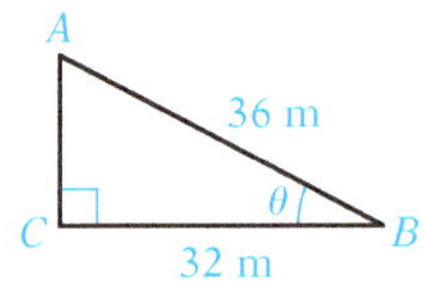

Solutions on pp. S-17–S-18.

Objective 4.6B Solve applications involving trigonometric ratios

Engineers can use trigonometry to find the distance across a ravine.

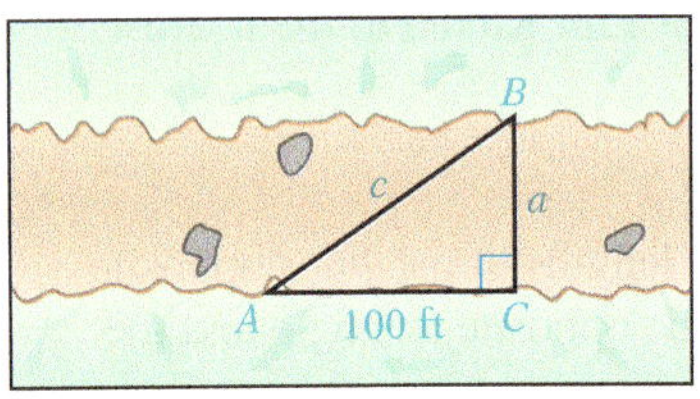

Focus on solving an application

Suppose, in the diagram at the left, engineers measure angle A as 33.8° and AC as 100 ft. Find the length a across the ravine. Round to the nearest tenth.

SOLUTION

The tangent involves the side opposite and the side adjacent to the known angle A.

$$\tan A = \frac{\text{opp}}{\text{adj}}$$

$$\tan 33.8^\circ = \frac{a}{100}$$

$$100(\tan 33.8^\circ) = 100\left(\frac{a}{100}\right)$$

• Multiply each side of the equation by 100.

$$66.9 \approx a$$

• Use a calculator to find tan 33.8°. Multiply the result in the display by 100.

The distance across the ravine is approximately 66.9 feet.

Check your understanding 6

A surveyor wants to determine the distance across a lake. The surveyor measures distance BC as 31 m and angle C as 52°. Find the distance AB across the lake. Round to the nearest tenth.

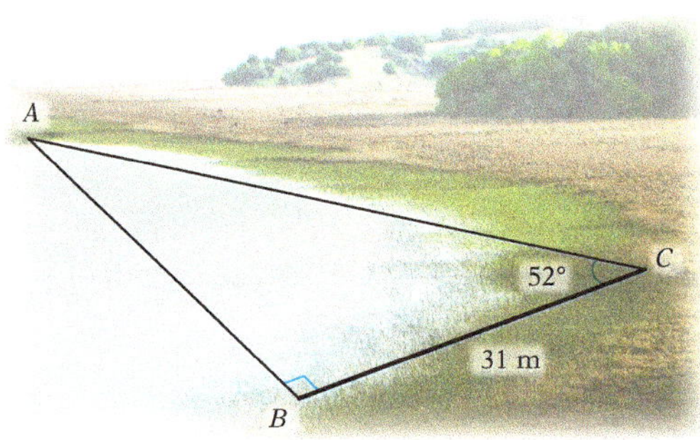

SOLUTION See page S-7.

Angles of elevation and depression are measured with respect to a horizontal line. If the object being observed is above the observer, the acute angle formed by the line of sight and the horizontal line is an **angle of elevation**. If the object being observed is below the observer, the acute angle formed by the line of sight and the horizontal line is an **angle of depression**.

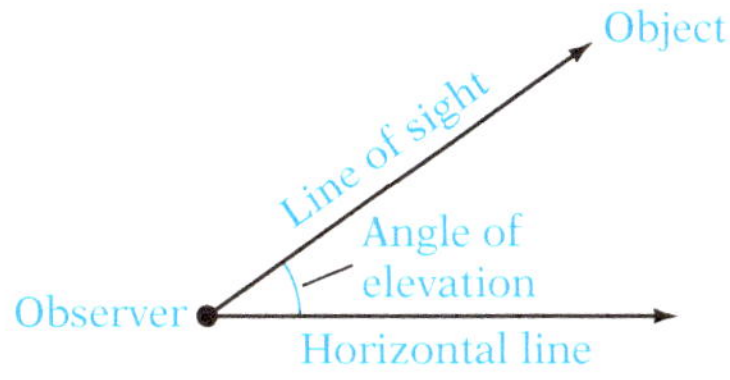

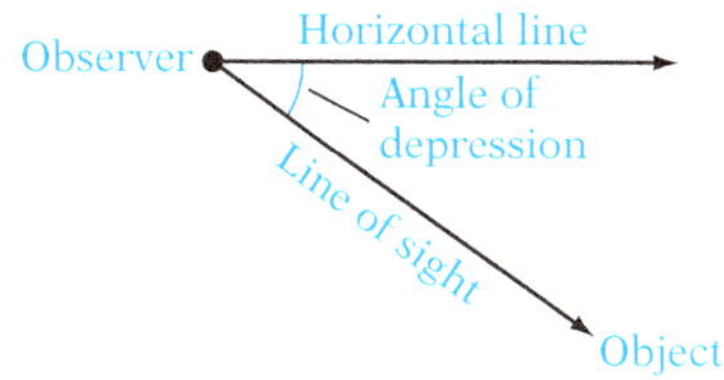

Focus on solving an angle of elevation problem

The angle of elevation from a point 62 ft away from the base of a flagpole to the top of the flagpole is 34°. Find the height of the flagpole. Round to the nearest tenth of a foot.

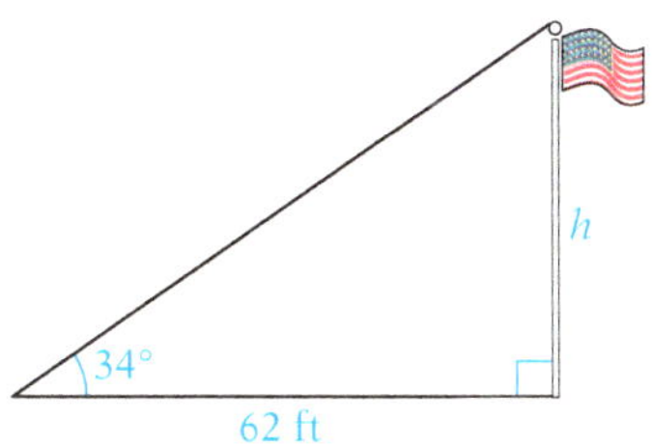

SOLUTION

Draw a diagram. To find the height h, write a trigonometric ratio that relates the given information and the unknown side of the triangle.

$$\tan 34° = \frac{h}{62}$$

$$62(\tan 34°) = 62\left(\frac{h}{62}\right)$$ • Solve for h.

$$41.8 \approx h$$ • Use a calculator to find tan 34°. Multiply the result in the display by 62.

The height of the flagpole is approximately 41.8 ft.

Check your understanding 7

The angle of depression from the top of a lighthouse that is 20 m high to a boat on the water is 25°. How far is the boat from the base of the lighthouse? Round to the nearest tenth of a meter.

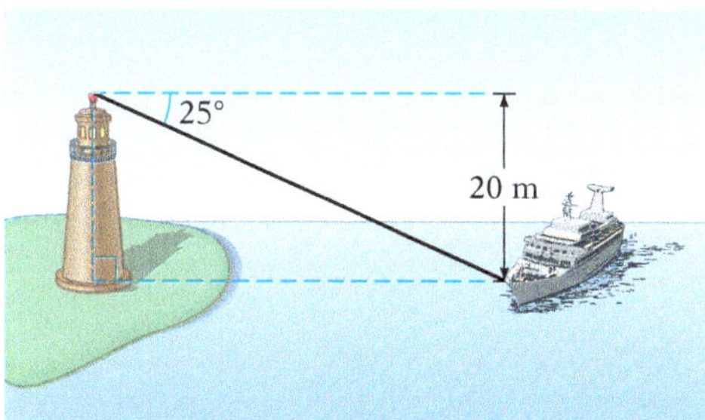

SOLUTION See page S-7.

Objective 4.6B Practice

1. A skateboarder wishes to build a jump ramp that is inclined at a 19.0° angle and that has a maximum height of 32.0 in. Find the horizontal width x of the ramp. Round to the nearest tenth.

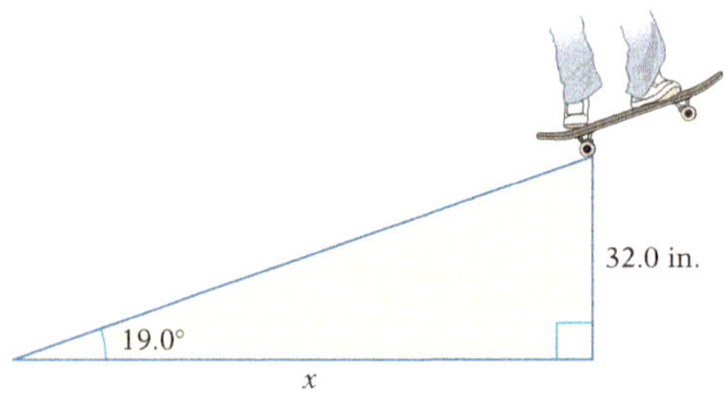

2. A 12-foot ladder is resting against a wall and makes an angle of 52° with the ground. Find the height to which the ladder reaches on the wall. Round to the nearest tenth.
3. Two buildings are 75 ft apart. The angle of depression from the top of one building to the top of the second building is 6°. How many feet shorter than the first building is the second building? Round to the nearest tenth.
4. A surveyor determines that the angle of elevation from a transit to the top of a building is 27.8°. The transit is positioned 5.5 ft above ground level and 131 ft from the building. Find the height of the building. Round to the nearest tenth of a foot.

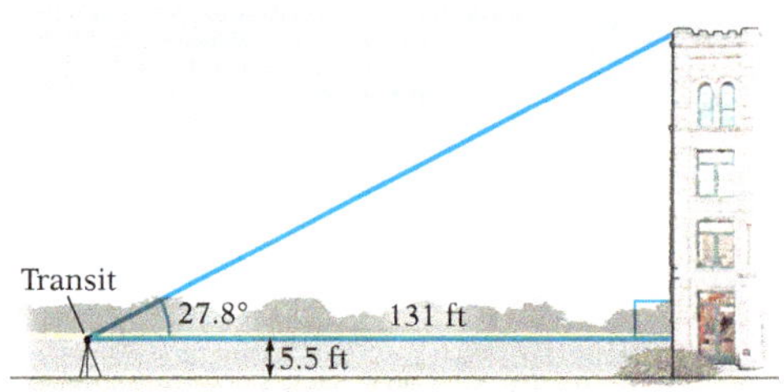

Solutions on pp. S-18–S-20.

SOLUTIONS TO MODULE 4

Solutions to Check Your Understanding

Section 4.1

Check your understanding 1

$$\begin{aligned} QT &= QR + RS + ST \\ 62 &= 24 + RS + 17 \\ 62 &= 41 + RS \\ 62 - 41 &= 41 - 41 + RS \\ 21 &= RS \end{aligned}$$

Check your understanding 2

Let x represent the supplement of a $32°$ angle. The sum of supplementary angles is $180°$.

$$\begin{aligned} x + 32° &= 180° \\ x + 32° - 32° &= 180° - 32° \\ x &= 148° \end{aligned}$$

$148°$ is the supplement of $32°$.

Check your understanding 3

$$\begin{aligned} \angle a + 68° &= 118° \\ \angle a + 68° - 68° &= 118° - 68° \\ \angle a &= 50° \end{aligned}$$

Check your understanding 4

The sum of the three angles of a triangle is $180°$.

$$\begin{aligned} \angle A + \angle B + \angle C &= 180° \\ \angle A + 62° + 45° &= 180° \\ \angle A + 107° &= 180° \\ \angle A + 107° - 107° &= 180° - 107° \\ \angle A &= 73° \end{aligned}$$

The measure of the third angle is $73°$.

Check your understanding 5

In a right triangle, one angle measures $90°$ and the two acute angles are complementary.

$$\begin{aligned} \angle A + \angle B &= 90° \\ \angle A + 7° &= 90° \\ \angle A + 7° - 7° &= 90° - 7° \\ \angle A &= 83° \end{aligned}$$

The other angles measure $90°$ and $83°$.

Check your understanding 6

$$r = \frac{1}{2}d$$

$$r = \frac{1}{2}(6 \text{ in.}) = 3 \text{ in.}$$

The radius is 3 in.

Check your understanding 7

Angles a and b are supplementary angles.

$$\begin{aligned} \angle a + \angle b &= 180^\circ \\ 125^\circ + \angle b &= 180^\circ \\ 125^\circ - 125^\circ + \angle b &= 180^\circ - 125^\circ \\ \angle b &= 55^\circ \end{aligned}$$

Check your understanding 8

$\angle c$ and $\angle a$ are corresponding angles. Corresponding angles are equal.

$\angle c = \angle a = 120^\circ$

$\angle b$ and $\angle c$ are supplementary angles.

$$\begin{aligned} \angle b + \angle c &= 180^\circ \\ \angle b + 120^\circ &= 180^\circ \\ \angle b + 120^\circ - 120^\circ &= 180^\circ - 120^\circ \\ \angle b &= 60^\circ \end{aligned}$$

Section 4.2

Check your understanding 1

$$\begin{aligned} P &= a + b + c \\ &= 12 \text{ cm} + 15 \text{ cm} + 18 \text{ cm} \\ &= 45 \text{ cm} \end{aligned}$$

The perimeter of the triangle is 45 cm.

Check your understanding 2

$$\begin{aligned} P &= 2L + 2W \\ &= 2(2 \text{ m}) + 2(0.85 \text{ m}) \\ &= 4 \text{ m} + 1.7 \text{ m} \\ &= 5.7 \text{ m} \end{aligned}$$

The perimeter of the rectangle is 5.7 m.

Check your understanding 3

$$\begin{aligned} C &= \pi d \\ &\approx 3.14 \cdot 6 \text{ in.} \\ &= 18.84 \text{ in.} \end{aligned}$$

The circumference is approximately 18.84 in.

Check your understanding 4

$$\begin{array}{lcccc} \text{Perimeter} & & \text{two} & & \text{the} \\ \text{of composite} & = & \text{lengths of a} & + & \text{circumference} \\ \text{figure} & & \text{rectangle} & & \text{of a circle} \end{array}$$

$$\begin{aligned} P &= 2L + \pi d \\ &\approx 2(8 \text{ in.}) + 3.14(3 \text{ in.}) \\ &= 16 \text{ in.} + 9.42 \text{ in.} \\ &= 25.42 \text{ in.} \end{aligned}$$

The perimeter is approximately 25.42 in.

Check your understanding 5

STRATEGY

To find the perimeter, use the formula for the perimeter of a rectangle.

SOLUTION

$$\begin{aligned} P &= 2L + 2W \\ &= 2(11 \text{ in.}) + 2\left(8\frac{1}{2} \text{ in.}\right) \\ &= 22 \text{ in.} + 17 \text{ in.} \\ &= 39 \text{ in.} \end{aligned}$$

The perimeter of the computer paper is 39 in.

Check your understanding 6

STRATEGY

To find the cost:

- Find the perimeter of the workbench.
- Multiply the perimeter by the per-meter cost of the stripping.

SOLUTION

$$\begin{aligned} P &= 2L + 2W \\ &= 2(3 \text{ m}) + 2(0.74 \text{ m}) \\ &= 6 \text{ m} + 1.48 \text{ m} \\ &= 7.48 \text{ m} \end{aligned}$$

$\$4.49 \times 7.48 = \$33.5852$

The cost is \$33.59.

Section 4.3

Check your understanding 1

$$A = \frac{1}{2}bh = \frac{1}{2}(24 \text{ in.})(14 \text{ in.}) = 168 \text{ in}^2$$

The area is 168 in^2.

Check your understanding 2

$$\begin{aligned} A &= \text{area of rectangle} - \text{area of triangle} \\ A &= LW - \frac{1}{2}bh \\ &= (10 \text{ in.} \times 6 \text{ in.}) - \left(\frac{1}{2} \times 6 \text{ in.} \times 4 \text{ in.}\right) \\ &= 60 \text{ in}^2 - 12 \text{ in}^2 \\ &= 48 \text{ in}^2 \end{aligned}$$

The area is 48 in^2.

Check your understanding 3

STRATEGY

To find the area of the room:

- Find the area in square feet.
- Convert to square yards.

SOLUTION

$$A = LW$$
$$= 12 \text{ ft} \cdot 9 \text{ ft}$$
$$= 108 \text{ ft}^2$$

$$108 \cancel{\text{ft}^2} \times \frac{1 \text{ yd}^2}{9 \cancel{\text{ft}^2}} = \frac{108}{9} \text{ yd}^2$$
$$= 12 \text{ yd}^2$$

The area of the room is 12 yd^2.

Section 4.4

Check your understanding 1

$$V = LWH$$
$$= (8 \text{ cm})(3.5 \text{ cm})(4 \text{ cm})$$
$$= 112 \text{ cm}^3$$

The volume is 112 cm^3.

Check your understanding 2

$$V = s^3$$
$$= (5 \text{ cm})^3$$
$$= 125 \text{ cm}^3$$

The volume is 125 cm^3.

Check your understanding 3

$$r = \frac{1}{2}d = \frac{1}{2}(14 \text{ in.}) = 7 \text{ in.}$$

$$V = \pi r^2 h$$
$$\approx \frac{22}{7}(7 \text{ in.})^2(15 \text{ in.})$$
$$= 2310 \text{ in}^3$$

The volume is approximately 2310 in^3.

Check your understanding 4

$$V = \frac{4}{3}\pi r^3$$
$$\approx \frac{4}{3}(3.14)(3 \text{ m})^3$$
$$= 113.04 \text{ m}^3$$

The volume is approximately 113.04 m^3.

Check your understanding 5

$$V = \text{volume of rectangular solid} + \frac{1}{2} \text{volume of cylinder}$$

$$V = LWH + \frac{1}{2}\pi r^2 h$$
$$\approx (24 \text{ in.})(6 \text{ in.})(4 \text{ in.}) + \frac{1}{2}(3.14)(3 \text{ in.})^2(24 \text{ in.})$$
$$= 576 \text{ in}^3 + 339.12 \text{ in}^3$$
$$= 915.12 \text{ in}^3$$

The volume is approximately 915.12 in^3.

Check your understanding 6

STRATEGY

To find the volume of the freezer, use the formula for the volume of a rectangular solid.

SOLUTION

$$\begin{aligned} V &= LWH \\ &= (7 \text{ ft})(2.5 \text{ ft})(3 \text{ ft}) \\ &= 52.5 \text{ ft}^3 \end{aligned}$$

The volume of the freezer is 52.5 ft^3.

Check your understanding 7

STRATEGY

To find the volume of the channel iron, subtract the volume of the cut-out rectangular solid from the volume of the large rectangular solid.

SOLUTION

$$\begin{aligned} V &= LWH - LWH \\ &= (10 \text{ ft})(0.5 \text{ ft})(0.8 \text{ ft}) - (10 \text{ ft})(0.3 \text{ ft})(0.2 \text{ ft}) \\ &= 4 \text{ ft}^3 - 0.6 \text{ ft}^3 \\ &= 3.4 \text{ ft}^3 \end{aligned}$$

The volume of the channel iron is 3.4 ft^3.

Section 4.5

Check your understanding 1

$$\begin{aligned} \text{Hypotenuse} &= \sqrt{(\text{leg})^2 + (\text{leg})^2} \\ &= \sqrt{8^2 + 11^2} \\ &= \sqrt{64 + 121} \\ &= \sqrt{185} \\ &\approx 13.601 \end{aligned}$$

The hypotenuse is approximately 13.601 in.

Check your understanding 2

$$\begin{aligned} \text{Leg} &= \sqrt{(\text{hypotenuse})^2 - (\text{leg})^2} \\ &= \sqrt{12^2 - 5^2} \\ &= \sqrt{144 - 25} \\ &= \sqrt{119} \\ &\approx 10.909 \end{aligned}$$

The length of the leg is approximately 10.909 ft.

Check your understanding 3

STRATEGY

To find the distance between the holes, use the Pythagorean Theorem. The hypotenuse is the distance between the holes. The length of each leg is given (3 cm and 8 cm).

SOLUTION

$$\begin{aligned}\text{Hypotenuse} &= \sqrt{(\text{leg})^2 + (\text{leg})^2}\\ &= \sqrt{3^2 + 8^2}\\ &= \sqrt{9 + 64}\\ &= \sqrt{73}\\ &\approx 8.544\end{aligned}$$

The distance is approximately 8.544 cm.

Section 4.6

Check your understanding 1

Use the Pythagorean Theorem to find the length of the hypotenuse.

$$\begin{aligned}a^2 + b^2 &= c^2\\ 3^2 + 4^2 &= c^2\\ 9 + 16 &= c^2\\ 25 &= c^2\\ \sqrt{25} &= \sqrt{c^2}\\ 5 &= c\end{aligned}$$

Using the definitions of the trigonometric ratios, we can write the following.

$$\sin\theta = \frac{\text{opp}}{\text{hyp}} = \frac{3}{5}, \quad \cos\theta = \frac{\text{adj}}{\text{hyp}} = \frac{4}{5}, \quad \tan\theta = \frac{\text{opp}}{\text{adj}} = \frac{3}{4}$$

Check your understanding 2

We are given the measure of $\angle B$ and the hypotenuse. We want to find the length of side a. The cosine function involves the side adjacent to an angle and the hypotenuse.

$$\begin{aligned}\cos B &= \frac{\text{adj}}{\text{hyp}}\\ \cos 48^\circ &= \frac{a}{12}\\ 12(\cos 48^\circ) &= 12\left(\frac{a}{12}\right)\\ 8.0 &\approx a\end{aligned}$$

The length of side a is approximately 8.0 ft.

Check your understanding 3

$\tan^{-1}(0.3165) \approx 17.6^\circ$

Check your understanding 4

$\theta \approx \tan^{-1}(0.5681)$

$\theta \approx 29.6^\circ$

Check your understanding 5

We want to find the measure of $\angle A$, and we are given the length of the side opposite $\angle A$ and the hypotenuse. The sine function involves the side opposite an angle and the hypotenuse.

$$\sin A = \frac{\text{opp}}{\text{hyp}}$$

$$\sin A = \frac{7}{11}$$

$$A = \sin^{-1}\frac{7}{11}$$

$$A \approx 39.5^\circ$$

The measure of $\angle A$ is approximately 39.5°.

Check your understanding 6

Side AB is opposite angle C and side BC is adjacent to angle C. The tangent involves the side opposite and the side adjacent to the known angle C.

$$\tan C = \frac{\text{opp}}{\text{adj}}$$

$$\tan 52^\circ = \frac{AB}{31}$$

$$31(\tan 52^\circ) = 31\left(\frac{AB}{31}\right)$$

$$39.7 \approx AB$$

The lake is approximately 39.7 m wide.

Check your understanding 7

Let d be the distance from the base of the lighthouse to the boat. To find the distance d, write a trigonometric ratio that relates the given information to the unknown side of the triangle.

$$\tan 25^\circ = \frac{20}{d}$$

$$d(\tan 25^\circ) = d\left(\frac{20}{d}\right)$$

$$d = \frac{20}{\tan 25^\circ}$$

$$d \approx 42.9$$

The boat is approximately 42.9 m from the base of the lighthouse.

Solutions to Objective Practice Exercises

Objective 4.1A **1.**

$$AD = AB + BC + CD$$

$$54 = 21 + 14 + CD$$

$$54 = 35 + CD$$

$$54 - 35 = 35 - 35 + CD$$

$$19 = CD$$

2. Let x represent the complement of 62°. The sum of complementary angles is 90°.

$$\begin{aligned} x + 62^\circ &= 90^\circ \\ x + 62^\circ - 62^\circ &= 90^\circ - 62^\circ \\ x &= 28^\circ \end{aligned}$$

28° is the complement of 62°.

3. Let x represent the supplement of 162°. The sum of supplementary angles is 180°.

$$\begin{aligned} x + 162^\circ &= 180^\circ \\ x + 162^\circ - 162^\circ &= 180^\circ - 162^\circ \\ x &= 18^\circ \end{aligned}$$

18° is the supplement of 162°.

4. $\angle AOB = 64^\circ + 72^\circ = 136^\circ$

5. $$\begin{aligned} \angle a + 13^\circ &= 90^\circ \\ \angle a + 13^\circ - 13^\circ &= 90^\circ - 13^\circ \\ \angle a &= 77^\circ \end{aligned}$$

Objective 4.1B

1. The sum of the three angles of a triangle is 180°.

$$\begin{aligned} \angle A + \angle B + \angle C &= 180^\circ \\ \angle A + 13^\circ + 65^\circ &= 180^\circ \\ \angle A + 78^\circ &= 180^\circ \\ \angle A + 78^\circ - 78^\circ &= 180^\circ - 78^\circ \\ \angle A &= 102^\circ \end{aligned}$$

The measure of the other angle is 102°.

2. In a right triangle, one angle measures 90° and the two acute angles are complementary.

$$\begin{aligned} \angle A + \angle B &= 90^\circ \\ \angle A + 62^\circ &= 90^\circ \\ \angle A + 62^\circ - 62^\circ &= 90^\circ - 62^\circ \\ \angle A &= 28^\circ \end{aligned}$$

The other angles measure 90° and 28°.

3. The sum of the three angles of a triangle is 180°.

$$\begin{aligned} \angle A + \angle B + \angle C &= 180^\circ \\ \angle A + 30^\circ + 45^\circ &= 180^\circ \\ \angle A + 75^\circ &= 180^\circ \\ \angle A + 75^\circ - 75^\circ &= 180^\circ - 75^\circ \\ \angle A &= 105^\circ \end{aligned}$$

The measure of the other angle is 105°.

4. $d = 2r$

$d = 2(24 \text{ cm}) = 48 \text{ cm}$

The diameter is 48 cm.

5. $r = \frac{1}{2}d$

$r = \frac{1}{2}(1.2 \text{ m}) = 0.6 \text{ m}$

The radius is 0.6 m.

Objective 4.1C

1. $\angle a + 74° = 180°$ • Supplementary angles

$\angle a + 74° - 74° = 180° - 74°$

$\angle a = 106°$

$\angle b = 74°$ • Vertical angle

2. $\angle a = 112°$ • Vertical angle

$\angle b + 112° = 180°$ • Supplementary angles

$\angle b + 112° - 112° = 180° - 112°$

$\angle b = 68°$

3. $\angle b + 136° = 180°$ • Supplementary angles

$\angle b + 136° - 136° = 180° - 136°$

$\angle b = 44°$

$\angle a = \angle b$ • Corresponding angles

$\angle a = 44°$

4. $\angle a = 130°$ • Alternate interior angles

$\angle b + 130° = 180°$ • Supplementary angles

$\angle b + 130° - 130° = 180° - 130°$

$\angle b = 50°$

Objective 4.2A

1. $P = a + b + c$

$= 12 \text{ in.} + 20 \text{ in.} + 24 \text{ in.}$

$= 56 \text{ in.}$

The perimeter of the triangle is 56 in.

2. $P = 4s$

$= 4(5 \text{ ft})$

$= 20 \text{ ft}$

The perimeter of the square is 20 ft.

3. $C = 2\pi r$

$\approx 2(3.14)(4 \text{ in.})$

$= 25.12 \text{ in.}$

The circumference of the circle is approximately 25.12 in.

4. $P = 2L + 2W$

$= 2(2 \text{ m}) + 2(0.8 \text{ m})$

$= 4 \text{ m} + 1.6 \text{ m}$

$= 5.6 \text{ m}$

The perimeter of the rectangle is 5.6 m.

5. STRATEGY

To find the amount of fencing, find the perimeter of the corral using the width (60 ft) and the length (75 ft).

SOLUTION

$$\begin{aligned} P &= 2L + 2W \\ &= 2(75 \text{ ft}) + 2(60 \text{ ft}) \\ &= 150 \text{ ft} + 120 \text{ ft} \\ &= 270 \text{ ft} \end{aligned}$$

The trainer will need 270 ft of fencing.

6. STRATEGY

To find the circumference, replace the variable d in the formula for circumference, $C = \pi d$, with the diameter (24.26 mm). Use 3.14 for π.

SOLUTION

$$\begin{aligned} C &= \pi d \\ &\approx 3.14(24.26 \text{ mm}) \\ &\approx 76.18 \text{ mm} \end{aligned}$$

The circumference is approximately 76.18 mm.

Objective 4.2B

1. $$\begin{aligned} \text{Perimeter} &= \text{sum of the lengths of the sides} \\ &= 19 \text{ cm} + 20 \text{ cm} + 8 \text{ cm} + 5 \text{ cm} + 27 \text{ cm} + 42 \text{ cm} \\ &= 121 \text{ cm} \end{aligned}$$

The perimeter is 121 cm.

2. $$\begin{aligned} \text{Perimeter of composite figure} &= \text{Sum of the lengths of the three sides of the rectangle} + \tfrac{1}{2} \text{ the circumference of the circle} \\ &= 2L + W + \frac{1}{2}\pi d \\ &\approx 2(15 \text{ m}) + 8 \text{ m} + \frac{1}{2}(3.14)(8 \text{ m}) \\ &= 30 \text{ m} + 8 \text{ m} + 12.56 \text{ m} \\ &= 50.56 \text{ m} \end{aligned}$$

The perimeter is 50.56 m.

3. $$\begin{aligned} \text{Perimeter} &= \text{sum of the lengths of the two sides} + \frac{1}{2} \text{ the circumference of the circle} \\ &\approx 2 \cdot 1 \text{ ft} + \frac{1}{2}(3.14 \cdot 1 \text{ ft}) \\ &= 2 \text{ ft} + 1.57 \text{ ft} \\ &= 3.57 \text{ ft} \end{aligned}$$

The perimeter is approximately 3.57 ft.

4. $$\begin{aligned} \text{Perimeter} &= \text{total length of the side} + 2 \text{ times } \frac{1}{2} \text{ the circumference of each circle} \\ &\approx (2.55 \text{ ft} + 2.55 \text{ ft}) + 2 \cdot \frac{1}{2}(3.14 \cdot 2.55 \text{ ft}) \\ &= 5.10 \text{ ft} + 8.007 \text{ ft} \\ &= 13.107 \text{ ft} \end{aligned}$$

The perimeter is approximately 13.107 ft.

Objective 4.2C

1. STRATEGY

To find the amount of fencing, use the formula for the perimeter of a rectangle.

SOLUTION

$$P = 2L + 2W = 2 \cdot 18 + 2 \cdot 12$$
$$= 36 + 24 = 60$$

The amount of fencing needed is 60 ft.

2. STRATEGY

To find how much molding is needed, use the formula for the circumference of a circle.

SOLUTION

$$C = \pi d$$
$$\approx 3.14 \cdot 3.8 \text{ ft}$$
$$= 11.932 \text{ ft}$$

The length of molding needed for the table is approximately 11.932 ft.

3. STRATEGY

To find the amount of bias binding:

- Use the formula for the perimeter of a rectangle to find the amount of binding needed.
- Convert the amount to feet.
- Divide the amount by 15 to find the number of packages needed.

SOLUTION

$$P = 2L + 2W = 2 \cdot 72 + 2 \cdot 45$$
$$= 144 + 90 = 234$$

$234 \text{ in.} = 19.5 \text{ ft}$

$19.5 \div 15 = 1.3$

Since 1.3 packages are needed, 2 packages must be ordered.

4. STRATEGY

To find the distance the tricycle travels:

- Convert the diameter (12 in.) to feet.
- Use the formula for circumference to find the distance traveled in 1 revolution.
- Multiply the distance traveled in 1 revolution by the number of revolutions (8).

SOLUTION

$$12 \not{\text{in.}} \times \frac{1 \text{ ft}}{12 \not{\text{in.}}} = 1 \text{ ft}$$

$$C = \pi d$$
$$\approx 3.14 \cdot 1 = 3.14 \text{ ft}$$

$3.14 \text{ ft} \times 8 = 25.12 \text{ ft}$

The tricycle travels approximately 25.12 ft.

Objective 4.3A

1. $A = LW = 24 \text{ ft} \cdot 6 \text{ ft} = 144 \text{ ft}^2$

2. $A = s^2 = (9 \text{ in.})^2 = 81 \text{ in}^2$

3. $A = \pi r^2$

$$\approx 3.14(4 \text{ ft})^2 = 50.24 \text{ ft}^2$$

4. $A = \frac{1}{2}bh$

$= \frac{1}{2} \cdot (10 \text{ in.})(4 \text{ in.}) = 20 \text{ in}^2$

5. $r = \frac{1}{2}d = \frac{1}{2} \cdot 40 \text{ cm} = 20 \text{ cm}$

$A = \pi r^2 \approx 3.14(20 \text{ cm})^2$

$= 1256 \text{ cm}^2$

Objective 4.3B

1. Area = area of rectangle − area of triangle

$= (LW) - \left(\frac{1}{2}bh\right)$

$= (8 \text{ cm} \cdot 4 \text{ cm}) - \left(\frac{1}{2} \cdot 4 \text{ cm} \cdot 3 \text{ cm}\right)$

$= 32 \text{ cm}^2 - 6 \text{ cm}^2$

$= 26 \text{ cm}^2$

2. Area = area of rectangle + area of triangle

$= (LW) + \left(\frac{1}{2}bh\right)$

$= (6 \text{ in.} \cdot 4 \text{ in.}) + \left(\frac{1}{2} \cdot 4 \text{ in.} \cdot 3 \text{ in.}\right)$

$= 24 \text{ in}^2 + 6 \text{ in}^2$

$= 30 \text{ in}^2$

3. Area = area of rectangle − $\frac{1}{2}$ area of circle

$= LW - \frac{1}{2}\pi r^2$

$\approx (2 \text{ m} \cdot 0.8 \text{ m}) - \frac{1}{2}(3.14)(0.4 \text{ m})^2$

$= 1.6 \text{ m}^2 - 0.2512 \text{ m}^2$

$= 1.3488 \text{ m}^2$

4. Area = area of triangle + $\frac{1}{2}$ area of circle

$= \left(\frac{1}{2}bh\right) + \frac{1}{2} \cdot \pi r^2$

$\approx \left(\frac{1}{2} \cdot 22.4 \text{ cm} \cdot 22.4 \text{ cm}\right) + \frac{1}{2} \cdot 3.14(11.2 \text{ cm})^2$

$= 250.88 \text{ cm}^2 + 196.9408 \text{ cm}^2$

$= 447.8208 \text{ cm}^2$

Objective 4.3C

1. STRATEGY

To find the amount of artificial turf needed, find the area of a rectangle with length 100 yd and width 75 yd.

SOLUTION

$A = LW$

$= 100 \text{ yd} \cdot 75 \text{ yd}$

$= 7500 \text{ yd}^2$

The amount of artificial turf needed is 7500 yd^2.

2. STRATEGY

To find the amount of stain:

- Find the area of a rectangle that measures 10 ft by 8 ft.
- Divide the area by the area that one quart of stain will cover (50 ft^2).

SOLUTION

$A = LW$

$= 10 \text{ ft} \cdot 8 \text{ ft}$

$= 80 \text{ ft}^2$

$80 \text{ ft}^2 \div 50 \text{ ft}^2 = 1.6$

You will need 1.6 quarts of stain, so you should buy 2 qt.

3. STRATEGY

To find the amount of fabric:

- Find the length and width of the fabric by adding 0.1 m to each side of the wall hanging.
- Find the area of a rectangle with the new length and width.

SOLUTION

$5 \text{ m} + 0.1 \text{ m} + 0.1 \text{ m} = 5.2 \text{ m}$

$3.5 \text{ m} + 0.1 \text{ m} + 0.1 \text{ m} = 3.7 \text{ m}$

$A = LW$

$= 5.2 \text{ m} \cdot 3.7 \text{ m}$

$= 19.24 \text{ m}^2$

The amount of fabric needed is 19.24 m^2.

4. STRATEGY

To find the area of the field, find the area of a circle with a radius of 50 ft.

SOLUTION

$A = \pi r^2 \approx 3.14 \cdot 50 \text{ ft} \cdot 50 \text{ ft}$

$= 7850 \text{ ft}^2$

The area watered by the irrigation system is approximately 7850 ft^2.

5. STRATEGY

To find the number of tiles to be purchased:

- Use the formula for the area of a rectangle to find the area of the kitchen floor.
- Divide the area of the kitchen floor by the area of one tile $\left(1\frac{1}{2}\text{ft}\right)^2$.

SOLUTION

$$\begin{aligned} A &= LW \\ &= 12 \text{ ft} \cdot 9 \text{ ft} \\ &= 108 \text{ ft}^2 \end{aligned}$$

$$\begin{aligned} 108 \text{ ft}^2 &\div \left(1\frac{1}{2} \text{ ft}\right)^2 \\ &= 108 \div \frac{9}{4} \\ &= 108 \cdot \frac{4}{9} = 48 \end{aligned}$$

You should purchase 48 tiles for your kitchen floor.

Objective 4.4A

1. $V = LWH$
$= 6 \text{ ft} \cdot 8 \text{ ft} \cdot 5 \text{ ft} = 240 \text{ ft}^3$

2. $V = \frac{4}{3}\pi r^3 \approx \frac{4}{3}(3.14)(8 \text{ in.})^3$
$\approx 2143.57 \text{ in}^3$

3. $r = \frac{1}{2}d = \frac{1}{2} \cdot 5 \text{ ft} = 2.5 \text{ ft}$
$V = \pi r^2 h$
$\approx 3.14(2.5 \text{ ft})^2 \cdot 8 \text{ ft} = 157 \text{ ft}^3$

4. $V = s^3$
$\approx 2.14 \text{ m} \cdot 2.14 \text{ m} \cdot 2.14 \text{ m}$
$\approx 9.8 \text{ m}^3$

5. $r = \frac{1}{2}d = \frac{1}{2} \cdot 12 \text{ ft} = 6 \text{ ft}$
$V = \pi r^2 h$
$\approx 3.14(6 \text{ ft})^2(30 \text{ ft})$
$= 3391.2 \text{ ft}^3$

Objective 4.4B

1. Volume $= \frac{1}{2}$volume of cylinder + volume of rectangular solid
$= \left[\frac{1}{2} \cdot \pi(\text{radius})^2 \cdot \text{height}\right] + (\text{length} \cdot \text{width} \cdot \text{height})$
$\approx \left(\frac{1}{2} \cdot 3.14 \cdot 3 \text{ in.} \cdot 3 \text{ in.} \cdot 2 \text{ in.}\right) + (6 \text{ in.} \cdot 9 \text{ in.} \cdot 1 \text{ in.})$
$= 28.26 \text{ in}^3 + 54 \text{ in}^3 = 82.26 \text{ in}^3$

2. Volume = volume of rectangular solid − volume of cylinder
$= (\text{length} \cdot \text{width} \cdot \text{height}) - [\pi(\text{radius})^2 \cdot \text{height}]$
$\approx (1.20 \text{ m} \cdot 2 \text{ m} \cdot 0.80 \text{ m}) - (3.14 \cdot 0.20 \text{ m} \cdot 0.20 \text{ m} \cdot 2 \text{ m})$
$= 1.92 \text{ m}^3 - 0.2512 \text{ m}^3$
$= 1.6688 \text{ m}^3$

3. Volume = volume of rectangular solid + volume of rectangular solid
$= (\text{length} \cdot \text{width} \cdot \text{height}) + (\text{length} \cdot \text{width} \cdot \text{height})$
$= (2\text{ m} \cdot 1.5\text{ m} \cdot 1.5\text{ m}) + (2\text{ m} \cdot 0.5\text{ m} \cdot 0.5\text{ m})$
$= 4.50\text{ m}^3 + 0.50\text{ m}^3 = 5\text{ m}^3$

4. Volume = volume of cylinder + volume of cylinder
$= [\pi(\text{radius})^2 \cdot \text{height}] + [\pi(\text{radius})^2 \cdot \text{height}]$
$\approx (3.14 \cdot 3\text{ in.} \cdot 3\text{ in.} \cdot 2\text{ in.}) + (3.14 \cdot 1\text{ in.} \cdot 1\text{ in.} \cdot 4\text{ in.})$
$= 56.52\text{ in}^3 + 12.56\text{ in}^3$
$= 69.08\text{ in}^3$

Objective 4.4C

1. STRATEGY
To find the volume of the tank, use the formula for the volume of a rectangular solid.

SOLUTION
$V = LWH$
$= 9\text{ m} \cdot 3\text{ m} \cdot 1.5\text{ m}$
$= 40.5\text{ m}^3$

The volume of the water in the tank is 40.5 m^3.

2. STRATEGY
To find the amount of oil:

- Use the formula for the volume of a cylinder to find the volume of the tank.
- Multiply the volume by $\frac{2}{3}$ to find the amount of oil.

SOLUTION
$V = \pi r^2 h$
$\approx 3.14(3\text{ m})^2(4\text{ m})$
$= 113.04\text{ m}^3$

Amount of oil $= \frac{2}{3} \cdot 113.04\text{ m}^3$
$= 75.36\text{ m}^3$

The amount of oil is approximately 75.36 m^3.

3. STRATEGY
To find the volume not used for storage:

- Use the formula for the volume of a cylinder to find the volume of the silo.
- Multiply the volume of the silo by $\frac{1}{4}$ to find the volume not used for storage.

SOLUTION
$V = \pi r^2 h$
$\approx 3.14(8\text{ ft})^2(30\text{ ft})$
$= 6028.80\text{ ft}^3$

Amount not used $= \frac{1}{4} \cdot 6028.80\text{ ft}^3$
$\approx 1507.2\text{ ft}^3$

Approximately 1507.2 ft^3 of the silo is not used for storage.

4. STRATEGY

To find the volume of the auditorium, find the volume of the rectangular solid and add half the volume of the cylinder.

SOLUTION

$$
\begin{aligned}
V &= LWH + \frac{1}{2}[\pi r^2 h] \\
&\approx 125 \text{ ft} \cdot 94 \text{ ft} \cdot 32 \text{ ft} \\
&\quad + \frac{1}{2}[3.14(47 \text{ ft})^2 \cdot 125 \text{ ft}] \\
&= 376{,}000 \text{ ft}^3 + 433{,}516.25 \text{ ft}^3 \\
&= 809{,}516.25 \text{ ft}^3
\end{aligned}
$$

The volume of the auditorium is approximately 809,516.25 ft^3.

Objective 4.5A

1.

$$
\begin{aligned}
\text{Hypotenuse} &= \sqrt{(\text{leg})^2 + (\text{leg})^2} \\
&= \sqrt{(7 \text{ cm})^2 + (9 \text{ cm})^2} \\
&= \sqrt{49 \text{ cm}^2 + 81 \text{ cm}^2} \\
&= \sqrt{130 \text{ cm}^2} \\
&\approx 11.402 \text{ cm}
\end{aligned}
$$

2.

$$
\begin{aligned}
\text{Leg} &= \sqrt{(\text{hypotenuse})^2 - (\text{leg})^2} \\
&= \sqrt{(20 \text{ ft})^2 - (18 \text{ ft})^2} \\
&= \sqrt{400 \text{ ft}^2 - 324 \text{ ft}^2} \\
&= \sqrt{76 \text{ ft}^2} \\
&\approx 8.718 \text{ ft}
\end{aligned}
$$

3.

$$
\begin{aligned}
\text{Hypotenuse} &= \sqrt{(\text{leg})^2 + (\text{leg})^2} \\
&= \sqrt{(6 \text{ in.})^2 + (6 \text{ in.})^2} \\
&= \sqrt{36 \text{ in}^2 + 36 \text{ in}^2} \\
&= \sqrt{72 \text{ in}^2} \\
&\approx 8.485 \text{ in.}
\end{aligned}
$$

4. STRATEGY

To find the length of the ramp, use the Pythagorean Theorem. The ramp is the hypotenuse of a right triangle.

SOLUTION

$$
\begin{aligned}
\text{Hypotenuse} &= \sqrt{(\text{leg})^2 + (\text{leg})^2} \\
&= \sqrt{(9 \text{ ft})^2 + (3.5 \text{ ft})^2} \\
&= \sqrt{81 \text{ ft}^2 + 12.25 \text{ ft}^2} \\
&= \sqrt{93.25 \text{ ft}^2} \\
&\approx 9.66 \text{ ft}
\end{aligned}
$$

The ramp is approximately 9.66 ft long.

5. STRATEGY

To find how high on the building the ladder reaches, use the Pythagorean Theorem. The hypotenuse is the length of the ladder (8 m). One leg is the distance from the base of the building to the base of the ladder (3 m). The distance along the building from the ground to the top of the ladder is the unknown leg.

SOLUTION

$$\begin{aligned} \text{Leg} &= \sqrt{(\text{hypotenuse})^2 - (\text{leg})^2} \\ &= \sqrt{(8 \text{ m})^2 - (3 \text{ m})^2} \\ &= \sqrt{64 \text{ m}^2 - 9 \text{ m}^2} \\ &= \sqrt{55 \text{ m}^2} \\ &\approx 7.4 \text{ m} \end{aligned}$$

The ladder reaches approximately 7.4 m high on the building.

Objective 4.6A

1. Use the Pythagorean Theorem to find the hypotenuse.

$$\begin{aligned} c^2 &= a^2 + b^2 \\ c^2 &= 12^2 + 5^2 \\ c^2 &= 144 + 25 \\ c^2 &= 169 \\ \sqrt{c^2} &= \sqrt{169} \\ c &= 13 \end{aligned}$$

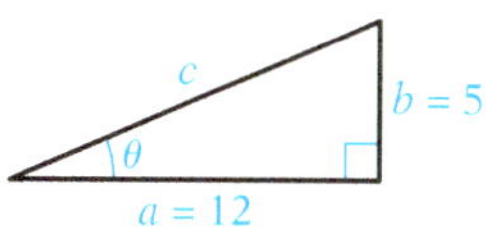

Use the definitions of the trigonometric ratios.

$$\sin\theta = \frac{\text{opp}}{\text{hyp}} = \frac{5}{13} \approx 0.3846$$

$$\cos\theta = \frac{\text{adj}}{\text{hyp}} = \frac{12}{13} \approx 0.9231$$

$$\tan\theta = \frac{\text{opp}}{\text{adj}} = \frac{5}{12} \approx 0.4167$$

2. Use the Pythagorean Theorem to find the hypotenuse.

$$\begin{aligned} c^2 &= a^2 + b^2 \\ c^2 &= 8^2 + 7^2 \\ c^2 &= 64 + 49 \\ c^2 &= 113 \\ \sqrt{c^2} &= \sqrt{113} \\ c &= \sqrt{113} \end{aligned}$$

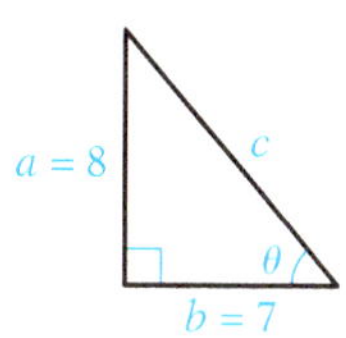

Use the definitions of the trigonometric ratios.

$$\sin\theta = \frac{8}{\sqrt{113}} \approx 0.7526$$

$$\cos\theta = \frac{7}{\sqrt{113}} \approx 0.6585$$

$$\tan\theta = \frac{8}{7} \approx 1.1429$$

3. Use a calculator set to degree mode.

$$\sin^{-1}(0.6239) \approx 38.6°$$
$$\theta \approx 38.6°$$

4. We want to find the length of side a, and we are given the measure of $\angle A$ and the length of the side adjacent to $\angle A$. Use the tangent ratio.

$$\tan A = \frac{\text{opp}}{\text{adj}}$$
$$\tan 32° = \frac{a}{26}$$
$$26(\tan 32°) = 26\left(\frac{a}{26}\right)$$
$$16.2 \approx a$$

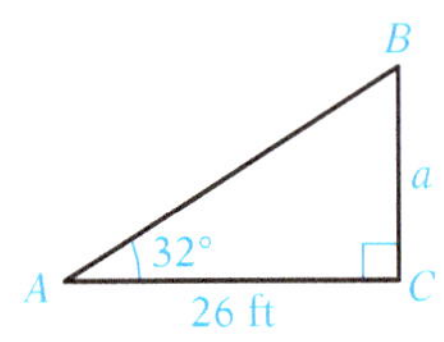

Side a measures approximately 16.2 ft.

5. We want to find the value of θ, and we are given the lengths of the hypotenuse and the side adjacent to θ. Use the cosine ratio.

$$\cos\theta = \frac{\text{adj}}{\text{hyp}}$$
$$\cos\theta = \frac{32}{36}$$
$$\theta = \cos^{-1}\left(\frac{32}{36}\right)$$
$$\theta \approx 27.3°$$

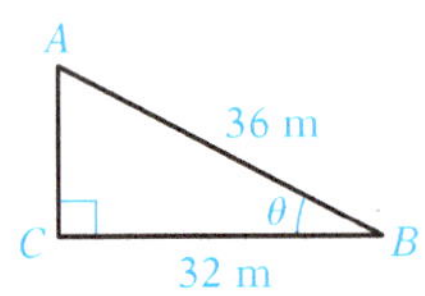

Objective 4.6B **1.** STRATEGY

To find the horizontal width x of the ramp, write a trigonometric ratio that relates the given information and the unknown side of the triangle. We are given the measure of an angle and the length of the side opposite that angle. To find the length of the adjacent side, use the tangent ratio.

SOLUTION

$$\tan\theta = \frac{\text{opp}}{\text{adj}}$$
$$\tan 19° = \frac{32}{x}$$
$$x(\tan 19°) = x\left(\frac{32}{x}\right)$$
$$x(\tan 19°) = 32$$
$$x = \frac{32}{\tan 19°}$$
$$x \approx 92.9$$

The horizontal width of the ramp is approximately 92.9 in.

2. STRATEGY

Draw a sketch.

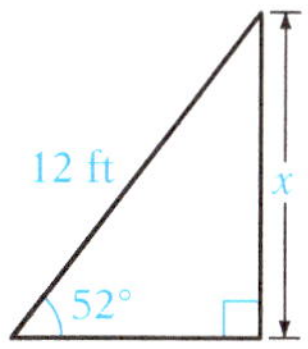

To find the height to which the ladder reaches on the wall, write a trigonometric ratio that relates the given information and the unknown side of the triangle. We are given the measure of an angle and the length of the hypotenuse. To find the length of the opposite side, use the sine ratio.

SOLUTION

$$\sin \theta = \frac{\text{opp}}{\text{hyp}}$$

$$\sin 52° = \frac{x}{12}$$

$$12(\sin 52°) = 12\left(\frac{x}{12}\right)$$

$$9.5 \approx x$$

The ladder reaches to a height of approximately 9.5 ft on the wall.

3. STRATEGY

Draw a sketch.

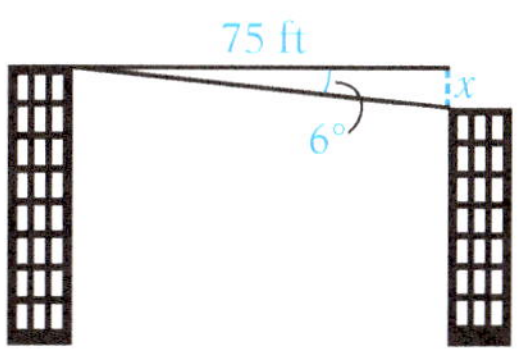

To find the difference in the building heights, find the length of side x. We are given the measure of an angle and the length of the side adjacent to that angle. To find the length of the opposite side, use the tangent ratio.

SOLUTION

$$\tan \theta = \frac{\text{opp}}{\text{adj}}$$

$$\tan 6° = \frac{x}{75}$$

$$75(\tan 6°) = 75\left(\frac{x}{75}\right)$$

$$7.9 \approx x$$

The second building is approximately 7.9 ft shorter than the first building.

4. STRATEGY

To find the height of the building:

- Use the given information to find the unknown side of the right triangle. We are given the measure of an angle and the length of the side adjacent to that angle. To find the length of the opposite side, use the tangent ratio.
- Add the height of the transit, 5.5 ft, to the length of the opposite side.

SOLUTION

$$\begin{aligned}\tan\theta &= \frac{\text{opp}}{\text{adj}}\\ \tan 27.8^\circ &= \frac{x}{131}\\ 131(\tan 27.8^\circ) &= 131\left(\frac{x}{131}\right)\\ 69.1 &\approx x\end{aligned}$$

$$69.1 + 5.5 = 74.6$$

The height of the building is approximately 74.6 ft.

MODULE

5 Applications of Functions

SECTION

5.1 The Rectangular Coordinate System

Objective 5.1A Graph points in a rectangular coordinate system

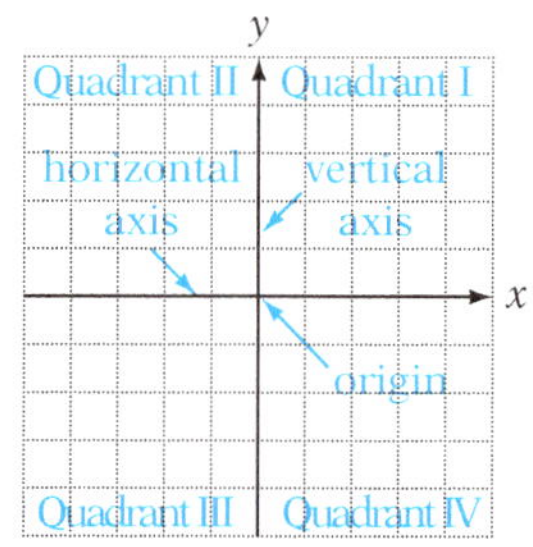

A **rectangular coordinate system** is formed by two number lines, one horizontal and one vertical, that intersect at the zero point of each line. The point of intersection is called the **origin.** The two lines are called **coordinate axes,** or simply **axes.**

The axes determine a **plane,** which can be thought of as a large, flat sheet of paper. The two axes divide the plane into four regions called **quadrants,** numbered counterclockwise from I to IV.

Each point in the plane can be identified by a pair of numbers called an **ordered pair.** The first number of the pair measures a horizontal distance and is called the **abscissa.** The second number of the pair measures a vertical distance and is called the **ordinate.** The **coordinates of a point** are the numbers in the ordered pair associated with the point. The abscissa is also called the **first coordinate** of the ordered pair, and the ordinate is also called the **second coordinate** of the ordered pair.

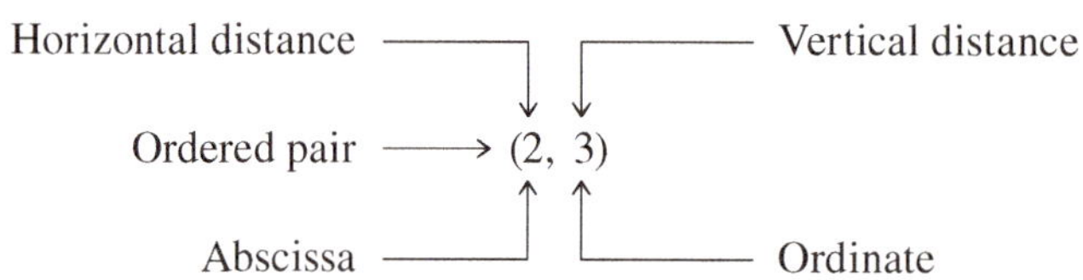

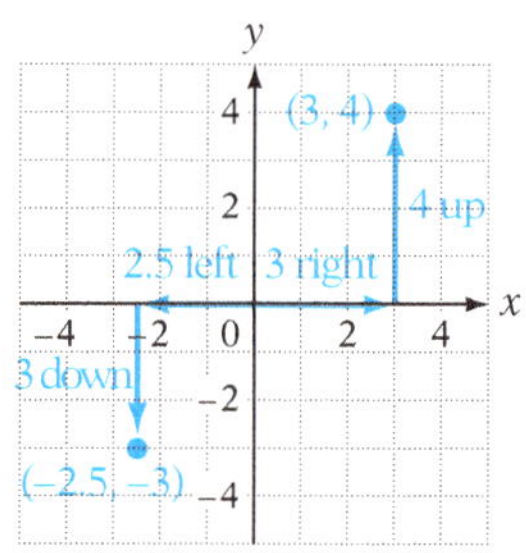

Graphing, or plotting, an ordered pair in the plane means placing a dot at the location given by the ordered pair. The **graph of an ordered pair** is the dot drawn at the coordinates of the point in the plane. The points whose coordinates are (3, 4) and (−2.5, −3) are graphed in the figure at the right.

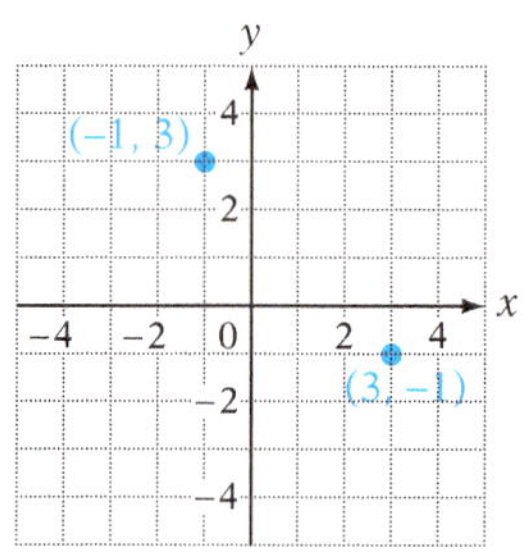

The points whose coordinates are (3, −1) and (−1, 3) are graphed at the right. Note that the graphed points are in different locations. The *order* of the coordinates of an ordered pair is important.

Take Note ➤ The concept of *ordered pair* is an important concept. Remember: There are two numbers (a *pair*), and the *order* in which they are listed is important.

Instead of saying "the point whose coordinates are (3, −1)," we will frequently use the notation $P(3, -1)$. For instance, $P(4, 3)$ is read "the point P whose coordinates are (4, 3)."

When drawing a rectangular coordinate system, we often label the horizontal axis x and the vertical axis y. In this case, the coordinate system is called an ***xy*-coordinate system.** The coordinates of the points are given by ordered pairs (x, y), where the abscissa is called the ***x*-coordinate** and the ordinate is called the ***y*-coordinate.**

Each point in the plane is associated with an ordered pair, and each ordered pair is associated with a point in the plane. Although only the labels for integers are given on a coordinate grid, the graph of any ordered pair can be approximated. For example, the points whose coordinates are $(-2.3, 4.1)$ and $(\sqrt{2}, -\sqrt{3})$ are shown in the graph at the right.

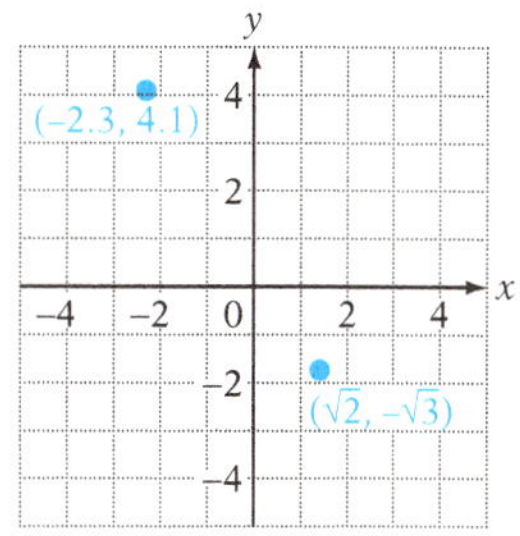

Focus on graphing ordered pairs

Graph the ordered pairs $(-2, -3)$, $(3, -2)$, $(1, 3)$, and $(4, 1)$.

SOLUTION

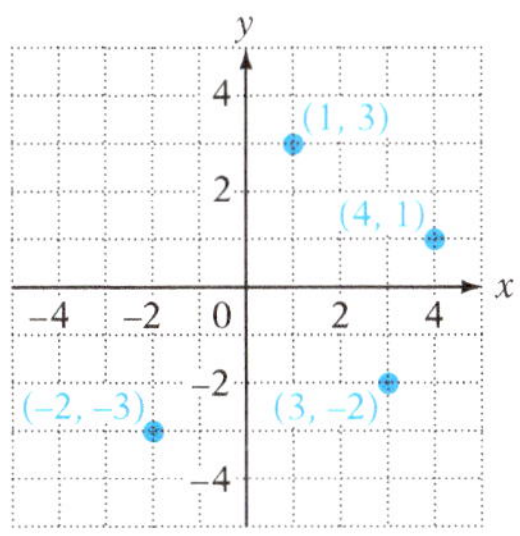

Check your understanding 1

Graph the ordered pairs $(-1, 3)$, $(1, 4)$, $(-4, 0)$, and $(-2, -1)$.

SOLUTION See page S-1.

Focus on finding the coordinates of a point

Find the coordinates of each point.

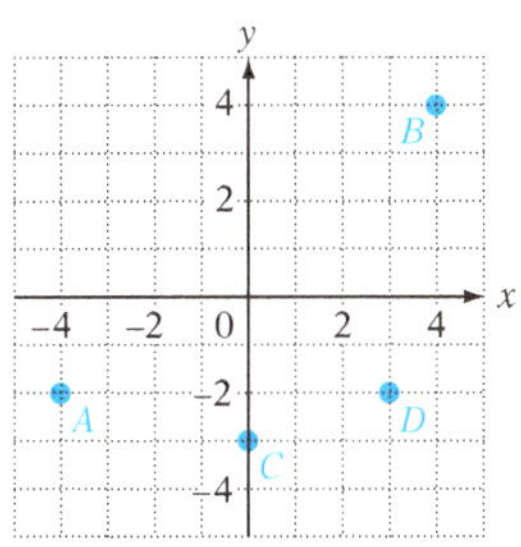

SOLUTION

$A(-4, -2)$ $B(4, 4)$ $C(0, -3)$ $D(3, -2)$

Check your understanding 2

Find the coordinates of each point.

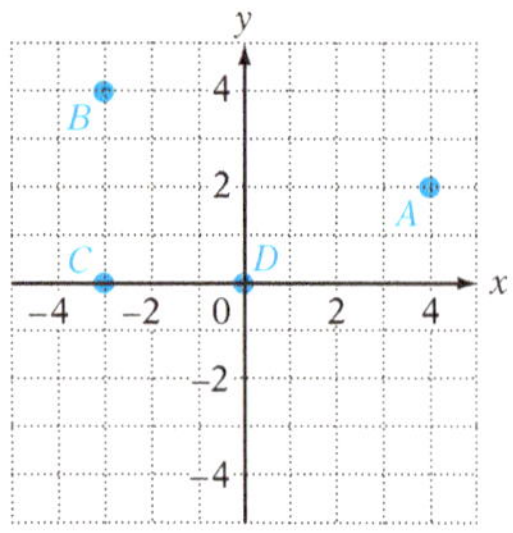

SOLUTION See page S-1.

Objective 5.1A Practice

For Exercises 1 to 3, graph the ordered pairs.

1. (5, 2), (3, −5), (−2, 1), and (0, 3)

2. (−3, −3), (5, −1), (−2, 4), and (0, −5)

3. (−2, −3), (1, −1), (−4, 5), and (−1, 0)

For Exercises 4 and 5, find the coordinates of each point.

4.

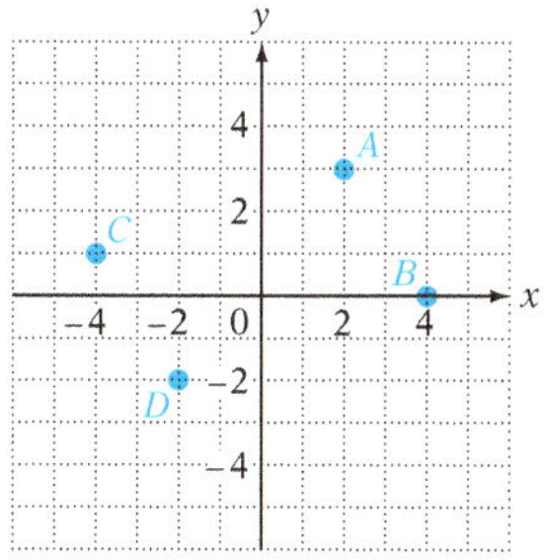

5.

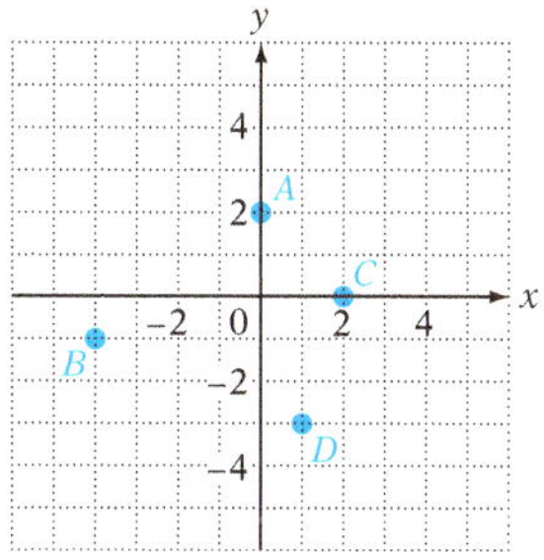

Solutions on p. S-11.

Objective 5.1B Graph a scatter diagram

Discovering a relationship between two variables is an important task in the study of mathematics. These relationships occur in many forms and in a wide variety of applications. Here are some examples:

> A botanist wants to know the relationship between the number of bushels of wheat yielded per acre and the amount of watering per acre.
>
> An environmental scientist wants to know the relationship between the incidence of skin cancer and the amount of ozone in the atmosphere.
>
> A business analyst wants to know the relationship between the price of a product and the number of products that are sold at that price.

A researcher may investigate the relationship between two variables by means of *regression analysis,* which is a branch of statistics. The study of the relationship between two variables may begin with a **scatter diagram,** which is a graph of the ordered pairs of the known data.

The following table gives data collected by a university registrar comparing the grade point averages of graduating high school seniors and their scores on a national test.

GPA, x	3.50	3.50	3.25	3.00	3.00	2.75	2.50	2.50	2.00	2.00	1.50
Test Score, y	1500	1100	1200	1200	1000	1000	1000	900	800	900	700

The scatter diagram for these data is shown below.

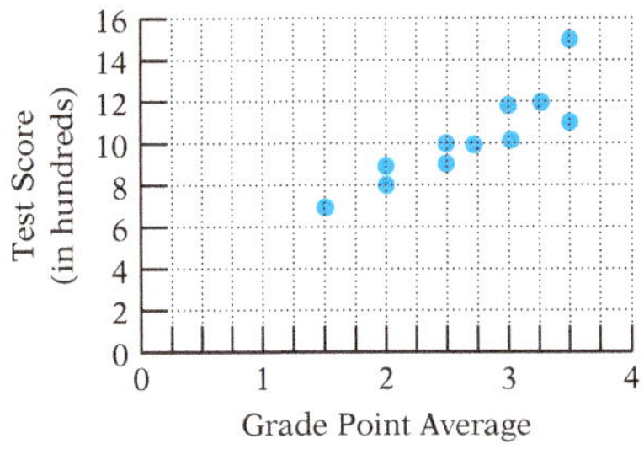

Each ordered pair represents the GPA and test score for a student. For example, the ordered pair (2.75, 1000) indicates a student with a GPA of 2.75 who had a test score of 1000.

The dot on the scatter diagram at (3, 12) represents the student with a GPA of 3.00 and a test score of 1200.

Focus on creating a scatter diagram

To test a heart medicine, a doctor measured the heart rates, in beats per minute, of five patients before and after they took the medication. The results are recorded in the scatter diagram. One patient's heart rate before taking the medication was 75 beats per minute. What was this patient's heart rate after taking the medication?

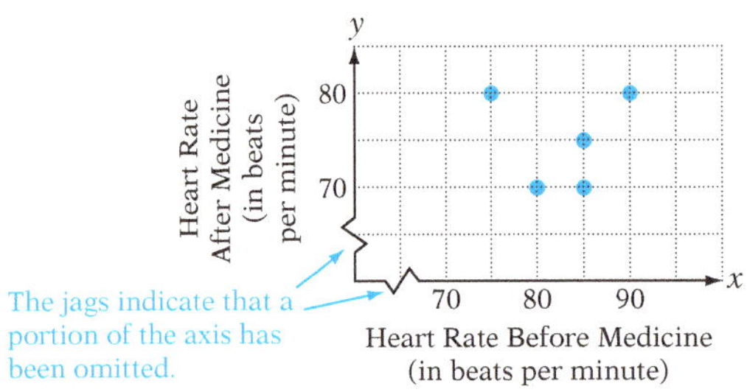

SOLUTION

Locate 75 beats per minute on the x-axis. Follow the vertical line from 75 to the point plotted in the diagram. Follow a horizontal line from that point to the y-axis. Read the number where the horizontal line intersects the y-axis.

The ordered pair is (75, 80), which indicates that the patient's heart rate before taking the medication was 75 and the heart rate after taking the medication was 80.

Check your understanding 3

A sports statistician collected data on the total number of yards gained and the total number of points scored by a college football team for six games. The data are recorded in the following table. Graph the scatter diagram for the data.

Yards, x	300	400	350	400	300	450
Points, y	18	24	14	21	21	30

SOLUTION See page S-1.

Objective 5.1B Practice

1. The numbers of years of previous work experience and the monthly salaries of six people who completed a bachelor's degree in marketing are recorded in the following table. Graph the scatter diagram for these data.

Years of Experience, *x*	2	0	5	2	3	1
Salary (in Hundreds), *y*	30	25	45	35	30	35

2. An exercise physiologist measured the time, in minutes, a person spent on a treadmill at a fast walk and the heart rate of that person. The results are recorded in the following table. Draw a scatter diagram for these data.

Time on Treadmill, *x*	2	10	5	8	6	5
Heart Rate, *y*	75	90	80	90	85	85

3. Sherlock Holmes solved a crime by recognizing a relationship between the length, in inches, of a person's stride and the height of that person in inches. The data for six people are recorded in the table below. Graph the scatter diagram for these data.

Length of Stride, *x*	15	25	20	25	15	30
Height, *y*	60	70	65	65	65	75

4. The scatter diagram at the right shows the record times for races of different lengths at a junior high school track meet. What was the record time for the 800-meter race?

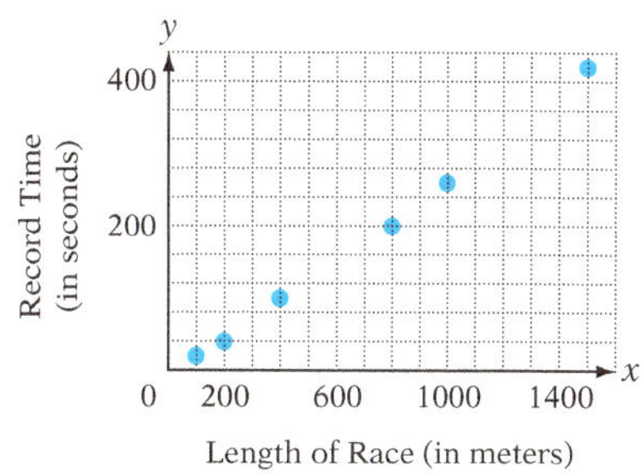

Solutions on p. S-11.

Objective 5.1C Graph an equation in two variables

The *xy*-coordinate system is used to graph equations in *two variables.* Examples of equations in two variables are shown at the right.

$$y = 3x + 7$$
$$y = x^2 - 4x + 3$$
$$x^2 + y^2 = 25$$
$$x = \frac{y}{y^2 + 4}$$

A **solution of an equation in two variables** is an ordered pair (x, y) whose coordinates make the equation a true statement.

Is the ordered pair $(-3, 7)$ a solution of the equation $y = -2x + 1$?

$$\begin{array}{c|l} y & = -2x + 1 \\ \hline 7 & -2(-3) + 1 \\ 7 & 6 + 1 \end{array}$$

$7 = 7$

- Replace x by -3 and y by 7.
- Simplify.
- Compare the results. If the resulting equation is a true statement, the ordered pair is a solution of the equation. If the resulting equation is not a true statement, the ordered pair is not a solution of the equation.

Yes, the ordered pair $(-3, 7)$ is a solution of the equation.

In addition to the ordered pair $(-3, 7)$, there are many other ordered-pair solutions of the equation $y = -2x + 1$. For example, $(-5, 11)$, $(0, 1)$, $\left(-\frac{3}{2}, 4\right)$, and $(4, -7)$ are also solutions of the equation.

In general, an equation in two variables has an infinite number of solutions. By choosing any value of x and substituting that value into the equation, we can calculate a corresponding value for y. The resulting ordered-pair solution (x, y) of the equation can be graphed in a rectangular coordinate system.

Graph the solutions (x, y) of $y = x^2 - 1$ when x equals $-2, -1, 0, 1,$ and 2.

Substitute each value of x into the equation and solve for y. Then graph the resulting ordered pairs by placing a dot at the coordinates of each point. It is convenient to record the ordered-pair solutions in a table similar to the one shown below. The graph of the ordered pairs is shown to the right of the table.

x	$y = x^2 - 1$	y	(x, y)
-2	$y = (-2)^2 - 1$	3	$(-2, 3)$
-1	$y = (-1)^2 - 1$	0	$(-1, 0)$
0	$y = 0^2 - 1$	-1	$(0, -1)$
1	$y = 1^2 - 1$	0	$(1, 0)$
2	$y = 2^2 - 1$	3	$(2, 3)$

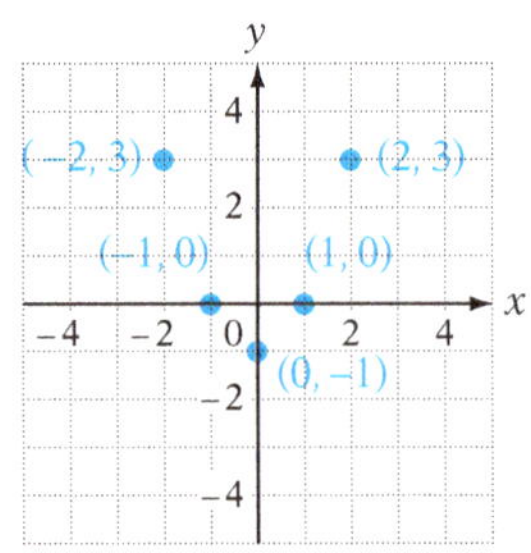

Generally, when we graph an equation in two variables, we include *all* the solutions, not just some selected ones as we did above. The graph of an equation is the graph of all of the ordered-pair solutions of the equation.

Consider $y = -2x + 1$. We can find ordered-pair solutions when $x = -2, -1, 0, 1, 2,$ and 3. The results are shown in the table below. The graph of these solutions is shown in Figure 1.

x	$y = -2x + 1$	y	(x, y)
-2	$y = -2(-2) + 1$	5	$(-2, 5)$
-1	$y = -2(-1) + 1$	3	$(-1, 3)$
0	$y = -2(0) + 1$	1	$(0, 1)$
1	$y = -2(1) + 1$	-1	$(1, -1)$
2	$y = -2(2) + 1$	-3	$(2, -3)$
3	$y = -2(3) + 1$	-5	$(3, -5)$

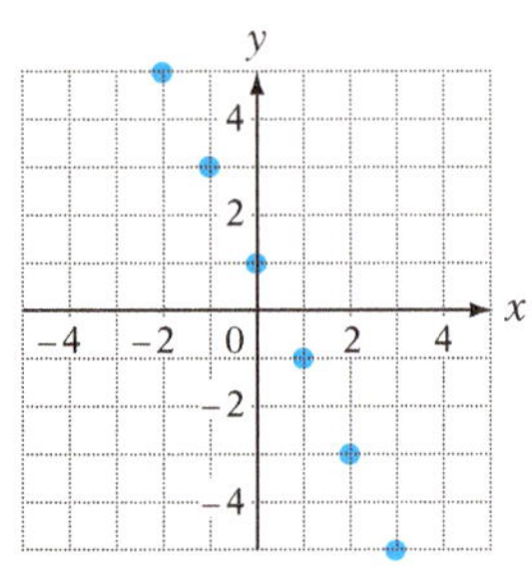

Figure 1

If we find additional solutions, such as when $x = -1.5, -0.5, 0.5, 1.5$, and 2.5, we get more points, as shown in Figure 2. If we continued to add more and more points, there would be so many dots that the graph would look like the straight line in Figure 3, which is the graph of $y = -2x + 1$. The arrowheads indicate that the graph extends forever in both directions.

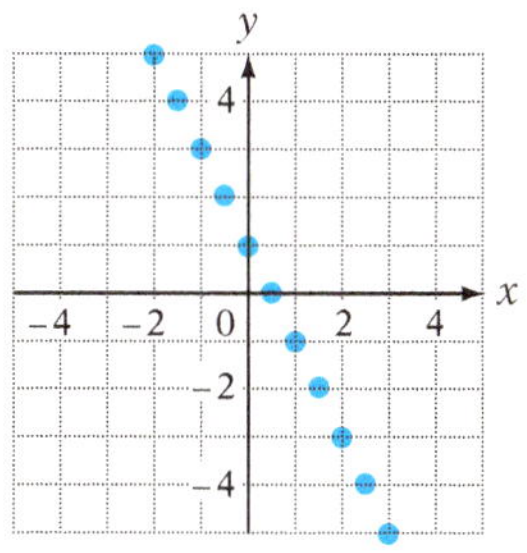

Figure 2

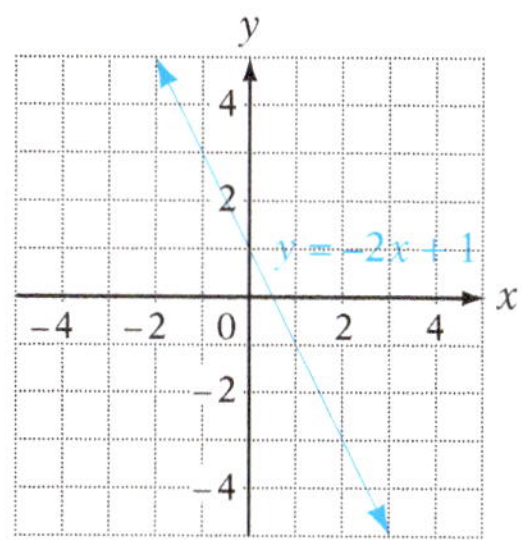

Figure 3

Focus on graphing an equation in two variables

Graph $y = \frac{1}{2}x + 1$ by first plotting the solutions of the equation when $x = -4, -2, 0, 2$, and 4.

SOLUTION

Determine the ordered-pair solutions (x, y) for the given values of x. Plot the points, and then connect the points to form the graph.

x	$y = \frac{1}{2}x + 1$	y
-4	$y = \frac{1}{2}(-4) + 1$	-1
-2	$y = \frac{1}{2}(-2) + 1$	0
0	$y = \frac{1}{2}(0) + 1$	1
2	$y = \frac{1}{2}(2) + 1$	2
4	$y = \frac{1}{2}(4) + 1$	3

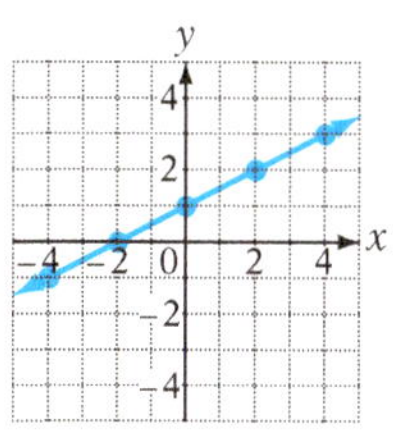

Check your understanding 4

Graph $y = -\frac{2}{3}x + 3$ by first plotting the solutions of the equation when $x = -6, -3, 0, 3$, and 6.

SOLUTION See page S-1.

It may be necessary to plot a number of ordered pairs in order to create an accurate graph of an equation in two variables.

Focus on graphing an equation in two variables

Graph $y = x^2 - 2x - 3$ by first plotting the solutions of the equation when $x = -2, -1, 0, 1, 2, 3$, and 4.

SOLUTION

Determine the ordered-pair solutions (x, y) for the given values of x. Plot the points, and then connect the points to form the graph.

x	$y = x^2 - 2x - 3$	y
-2	$y = (-2)^2 - 2(-2) - 3$	5
-1	$y = (-1)^2 - 2(-1) - 3$	0
0	$y = (0)^2 - 2(0) - 3$	-3
1	$y = (1)^2 - 2(1) - 3$	-4
2	$y = (2)^2 - 2(2) - 3$	-3
3	$y = (3)^2 - 2(3) - 3$	0
4	$y = (4)^2 - 2(4) - 3$	5

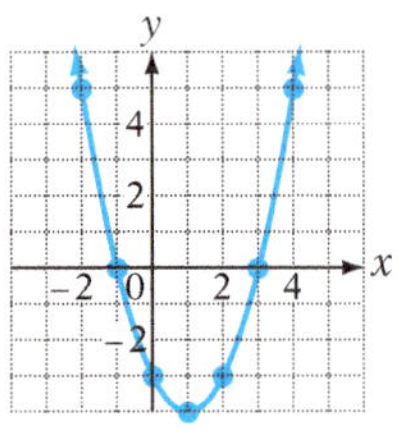

Check your understanding 5

Graph $y = -x^2 + 4$ by first plotting the solutions of the equation when $x = -3, -2, -1, 0, 1, 2$, and 3.

SOLUTION See page S-2.

Objective 5.1C Practice

For Exercises 1 to 5, graph the equation. First plot the solutions of the equation for the given values of x, and then connect the points with a smooth graph.

1. $y = 2x - 3$
$x = -1, 0, 1, 2, 3, 4$

2. $y = -\frac{2}{3}x + 1$
$x = -6, -3, 0, 3, 6$

3. $y = x^2 - 4$
$x = -3, -2, -1, 0, 1, 2, 3$

4. $y = -x^2 + 2x + 3$
$x = -2, -1, 0, 1, 2, 3, 4$

5. $y = |x + 2|$
$x = -5, -4, -3, -2, -1, 0, 1$

Solutions on pp. S-12–S-13.

SECTION 5.2 Introduction to Functions

Objective 5.2A Evaluate a function

Tips for Success ➤
Have you considered joining a study group? Getting together regularly with other students in the class to go over material and quiz each other can be very beneficial. See *AIM for Success*.

In mathematics and its applications, we must often investigate a relationship between two quantities. Here is a financial application: Consider a person who is planning to finance the purchase of a car. If the current interest rate for a 5-year loan is 5%, the equation that describes the relationship between the amount that is borrowed B and the monthly payment P is $P = 0.018871B$.

For each amount the purchaser may borrow (B), there is a certain monthly payment (P). The relationship between the amount borrowed and the payment can be recorded as a set of ordered pairs, where the first coordinate of each pair is the amount borrowed and the second coordinate is the monthly payment. Some of these ordered pairs are shown at the right.

$0.018871B = P$

(6000, 113.23)
(7000, 132.10)
(8000, 150.97)
(9000, 169.84)

A relationship between two quantities is not always given by an equation. The table at the right describes a grading scale that defines a relationship between a score on a test and a letter grade. For each score, the table assigns only one letter grade. The ordered pair (84, B) indicates that a score of 84 receives a letter grade of B.

Score	Grade
90–100	A
80–89	B
70–79	C
60–69	D
0–59	F

The bar graph at the right shows the number of people who watched the Super Bowl for the years 2007 to 2012. The jagged line between 0 and 90 on the vertical axis indicates that a portion of the vertical axis has been omitted. The data in the graph can be written as a set of ordered pairs.

{(2007, 93.2), (2008, 97.4), (2009, 98.7), (2010, 106.5), (2011, 111.01), (2012, 111.35)}

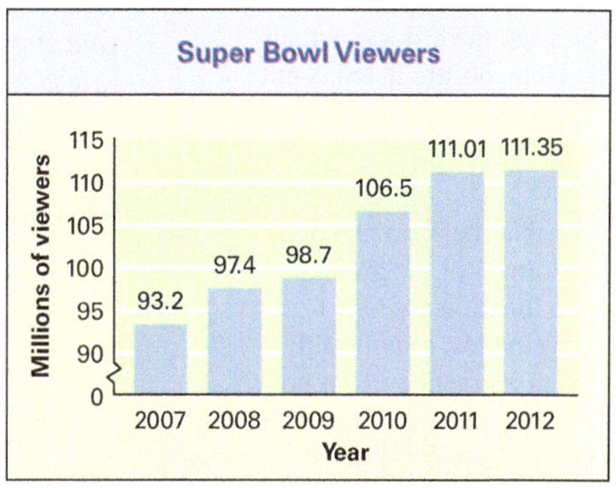

This set is a function. There are no two ordered pairs with the same first coordinate. The ordered pair (2010, 106.5) means that in 2010, the number of people who watched the Super Bowl was 106.5 million.

In each of the above examples, there is a rule (an equation, a table, or a graph) that determines a certain set of ordered pairs.

Definition of a Function

A **function** is a set of ordered pairs in which no two ordered pairs have the same first coordinate. The **domain** of a function is the set of first coordinates of the ordered pairs; the **range** of a function is the set of second coordinates of the ordered pairs.

EXAMPLES

1. $\{(1, 2), (2, 4), (3, 6), (4, 8)\}$
Domain $= \{1, 2, 3, 4\}$ Range $= \{2, 4, 6, 8\}$

2. $\{(-1, 0), (0, 0), (1, 0), (2, 0), (3, 0)\}$
Domain $= \{-1, 0, 1, 2, 3\}$ Range $= \{0\}$

Now consider the set of ordered pairs $\{(1, 2), (4, 5), (7, 8), (4, 6)\}$. This set of ordered pairs is *not* a function. There are two ordered pairs, (4, 5) and (4, 6), with the same first coordinate. This set of ordered pairs is called a *relation*. A **relation** is any set of ordered pairs. A function is a special type of relation. The concepts of domain and range apply to relations as well as functions.

Determine whether each set of ordered pairs is a function. State the domain and range.

A. $\{(2, 3), (4, 6), (6, 8), (10, 6)\}$

B. $\{(2, 2), (1, 1), (0, 0), (2, -2), (1, -1)\}$

A. No two ordered pairs have the same first element. The set of ordered pairs is a function. The domain is $\{2, 4, 6, 10\}$. The range is $\{3, 6, 8\}$.

B. The ordered pairs (2, 2) and $(2, -2)$ have the same first coordinate. The set of ordered pairs is not a function. The domain is $\{0, 1, 2\}$. The range is $\{-2, -1, 0, 1, 2\}$.

Take Note ➤
For a set, the order in which the elements are listed is not important. For instance,

$$\{a, b, c\} = \{b, a, c\}$$

Note that the elements of the domain of the grading-scale function are listed from smallest to largest. It is common practice to list both domain and range elements in order from smallest to largest.

Consider again the three examples of functions given at the beginning of this objective. For the equation $0.018871B = P$, the domain is the possible amounts a consumer might borrow to purchase a car. Let's assume that the most a person would borrow is \$50,000. Then the domain is $\{B \mid 0 \le B \le 50{,}000\}$. The range is all possible monthly payments. The largest monthly payment is $P = 0.018871(50{,}000) \approx 943.55$, so the range is $\{P \mid 0 \le P \le 943.55\}$.

For the grading-scale function, the domain is all possible test scores. The domain is $\{0, 1, 2, 3, ..., 97, 98, 99, 100\}$. The range is all possible grades. The range is $\{A, B, C, D, F\}$.

For a graph, the domain is represented on the horizontal axis and the range is represented on the vertical axis. For the graph of the Super Bowl data, the domain is the set of years. The domain is $\{2007, 2008, 2009, 2010, 2011, 2012\}$. The range is the number of people watching each year. The range is $\{93.2, 97.4, 98.7, 106.5, 111.01, 111.35\}$.

Focus on finding the domain and range of a function

Find the domain and range of the function $\{(5, 3), (9, 7), (13, 7), (17, 3)\}$.

SOLUTION

The domain is the set of first coordinates. The range is the set of second coordinates.

The domain is $\{5, 9, 13, 17\}$.

The range is $\{3, 7\}$.

Check your understanding 1

Find the domain and range of the function $\{(-1, 5), (3, 5), (4, 5), (6, 5)\}$.

SOLUTION See page S-2.

The **square function,** which pairs each real number with its square, can be defined by the equation

$$y = x^2$$

This equation states that for a given value of x in the domain, the corresponding value of y in the range is the square of x. For instance, if $x = 6$, then $y = 36$ and if $x = -7$, then $y = 49$. Because the value of y *depends* on the value of x, y is called the **dependent variable** and x is called the **independent variable.**

Take Note ➤ A pictorial representation of the square function is shown at the right. The function acts as a machine that changes a number from the domain into the square of the number.

A function can be thought of as a rule that pairs one number with another number. For instance, the square function pairs a number with its square. The ordered pairs for the values shown at the right are $(-5, 25)$, $\left(\frac{3}{5}, \frac{9}{25}\right)$, $(0, 0)$, and $(3, 9)$. For this function, the second coordinate is the square of the first coordinate. If x represents the first coordinate, then the second coordinate is x^2 and the ordered pair is (x, x^2).

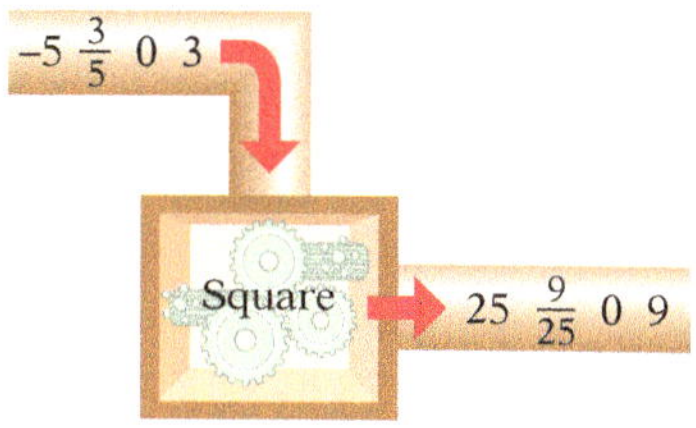

A function cannot have two ordered pairs with *different* second coordinates and the same first coordinate. However, a function may contain ordered pairs with the *same* second coordinate. For instance, the square function has the ordered pairs $(-3, 9)$ and $(3, 9)$; the second coordinates are the same but the first coordinates are different.

Not every equation in two variables defines a function. For instance, consider the equation

$$y^2 = x^2 + 9$$

Because

$$5^2 = 4^2 + 9 \qquad \text{and} \qquad (-5)^2 = 4^2 + 9$$

the ordered pairs $(4, 5)$ and $(4, -5)$ are both solutions of the equation. Consequently, there are two ordered pairs that have the same first coordinate (4) but *different* second coordinates (5 and -5). Therefore, the equation does not define a function.

The phrase "y is a function of x," or the same phrase with different variables, is used to describe an equation in two variables that defines a function. To emphasize that the equation represents a function, **function notation** is used.

Just as the variable x is commonly used to represent a number, the letter f is commonly used to name a function. The square function is written in function notation as follows:

Take Note ➤
The dependent variable y and the notation $f(x)$ can be used interchangeably.

This is the value of the function. It is the number that is paired with x.

$$f(x) = x^2$$

The name of the function is f.

This is an algebraic expression that defines the relationship between the dependent and independent variables.

The symbol $f(x)$ is read "the value of f at x" or "f of x."

It is important to note that $f(x)$ does *not* mean f times x. The symbol $f(x)$ is the **value of the function** and represents the value of the dependent variable for a given value of the independent variable. We often write $y = f(x)$ to emphasize the relationship between the independent variable x and the dependent variable y. Remember that y and $f(x)$ are different symbols for the same number.

The letters used to represent a function are somewhat arbitrary. All of the following equations represent the same function.

$$\left.\begin{aligned} f(x) &= x^2 \\ s(t) &= t^2 \\ P(v) &= v^2 \end{aligned}\right\} \text{Each equation represents the square function.}$$

The process of determining $f(x)$ for a given value of x is called **evaluating a function.** For instance, to evaluate $f(x) = x^2$ when $x = 4$, replace x by 4 and simplify.

$$\begin{aligned} f(x) &= x^2 \\ f(4) &= 4^2 = 16 \end{aligned}$$

The *value* of the function is 16 when $x = 4$. An ordered pair of the function is $(4, 16)$.

Focus on evaluating a function

Evaluate $g(t) = 3t^2 - 5t + 1$ when $t = -2$.

SOLUTION

$$\begin{aligned} g(t) &= 3t^2 - 5t + 1 \\ g(-2) &= 3(-2)^2 - 5(-2) + 1 \\ &= 3(4) - 5(-2) + 1 \\ &= 12 + 10 + 1 = 23 \end{aligned}$$

• Replace t by -2 and then simplify.

When t is -2, the value of the function is 23.

Take Note ➤
Because $g(-2) = 23$, $(-2, 23)$ is an ordered pair of the function.

Check your understanding 2

Evaluate $G(x) = \frac{3x}{x + 2}$ when $x = -4$.

SOLUTION See page S-2.

Apply the Concept

The height $s(t)$, in feet, of a ball above the ground t seconds after it is thrown upward at a velocity of 64 ft/s is given by $s(t) = -16t^2 + 64t + 4$. Find the height of the ball 1.5 s after it is released.

SOLUTION

To find the height, evaluate the function when $t = 1.5$.

$$\begin{aligned} s(t) &= -16t^2 + 64t + 4 \\ s(1.5) &= -16(1.5)^2 + 64(1.5) + 4 && \text{• Evaluate the function when } t = 1.5. \\ &= -16(2.25) + 64(1.5) + 4 \\ &= -36 + 96 + 4 \\ &= 64 \end{aligned}$$

The ball is 64 ft above the ground 1.5 s after it is released.

When a function is represented by an equation, the domain of the function is all real numbers for which the value of the function is a real number. For instance:

- The domain of $f(x) = x^2$ is all real numbers, because the square of every real number is a real number. In set-builder notation, the domain is $\{x \mid -\infty < x < \infty\}$.
- The domain of $g(x) = \frac{1}{x-2}$ is all real numbers except 2, because when $x = 2$, $g(2) = \frac{1}{2-2} = \frac{1}{0}$, which is not a real number. The domain is $\{x \mid x \neq 2\}$.

Find the domain of $f(x) = 2x^2 - 7x + 1$.

Because the value of $2x^2 - 7x + 1$ is a real number for any value of x, no values are excluded from the domain of $f(x) = 2x^2 - 7x + 1$. The domain of the function is all real numbers, or $\{x \mid -\infty < x < \infty\}$.

Focus on finding the domain of a function given by an equation

What is the domain of $f(x) = \frac{2}{x-5}$?

SOLUTION

For $x = 5$, $f(x) = f(5) = \frac{2}{5-5} = \frac{2}{0}$, which is undefined. So, 5 is excluded from the domain of f. The domain is $\{x \mid x \neq 5\}$.

Check your understanding 3

What is the domain of $f(x) = 3x^2 - 5x + 2$?

SOLUTION See page S-2.

Objective 5.2A Practice

1. Determine whether $\{(0, 0), (1, 0), (2, 0), (3, 0), (4, 0)\}$ is a function. State the domain and range.
2. Given $f(x) = 5x - 4$, evaluate $f(-2)$.
3. Given $q(r) = r^2 - 4$, evaluate $q(-5)$.
4. Given $s(t) = t^3 - 3t + 4$, evaluate $s(2)$.
5. What is the domain of the function $H(x) = \frac{1}{2}x^2$? Write the answer in set-builder notation.
6. What is the domain of the function $f(x) = 3x^2 + x + 4$? Write the answer in set-builder notation.

Solutions on p. S-13.

Objective 5.2B Graph a function

The **graph of a function** is a graph of all of the ordered pairs of the function. Graphing a function is similar to graphing an equation in two variables. Evaluate the function for selected values of x. Plot the corresponding ordered pairs and then connect the points to form the graph.

Focus on graphing a function and estimating its domain and range

Graph $f(x) = x^2 + 2x - 3$. Begin by plotting the ordered pairs (x, y) of the function when x equals $-4, -3, -2, -1, 0, 1$, and 2. Then connect the points to form the graph. From the graph, estimate the domain and range of the function.

SOLUTION

Create a table. Remember that the dependent variable is y, and $y = f(x)$.

x	$f(x) = x^2 + 2x - 3$	y	(x, y)
-4	$f(-4) = (-4)^2 + 2(-4) - 3$	5	$(-4, 5)$
-3	$f(-3) = (-3)^2 + 2(-3) - 3$	0	$(-3, 0)$
-2	$f(-2) = (-2)^2 + 2(-2) - 3$	-3	$(-2, -3)$
-1	$f(-1) = (-1)^2 + 2(-1) - 3$	-4	$(-1, -4)$
0	$f(0) = (0)^2 + 2(0) - 3$	-3	$(0, -3)$
1	$f(1) = (1)^2 + 2(1) - 3$	0	$(1, 0)$
2	$f(2) = (2)^2 + 2(2) - 3$	5	$(2, 5)$

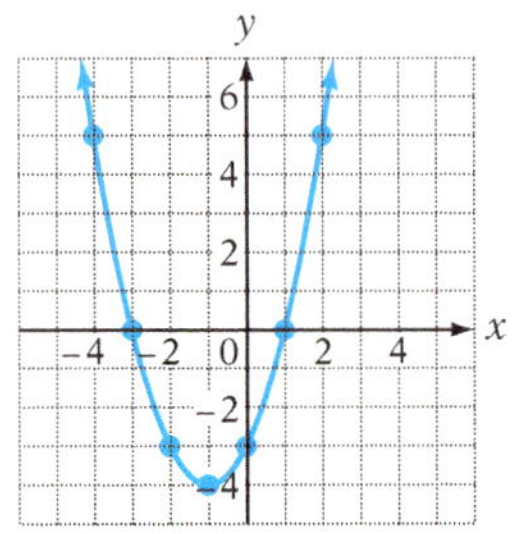

Because $x^2 + 2x - 3$ is a real number for all values of x, the domain of $f(x) = x^2 + 2x - 3$ is all real numbers. We can write the domain in set-builder notation as $\{x | -\infty < x < \infty\}$ or in interval notation as $(-\infty, \infty)$. Because all values of y are greater than or equal to -4, the range is $\{y | y \geq -4\}$ or, in interval notation, $[-4, \infty)$.

Check your understanding 4

Graph $g(x) = \frac{3}{2}x - 2$. Begin by plotting the ordered pairs (x, y) of the function when $x = -2, 0, 2$, and 4. Estimate the domain and range from the graph. Write the answers using interval notation.

SOLUTION See page S-2.

The graph of $f(x) = x^2 + 2x - 3$ is shown again at the right. As we can see from the graph, the point with coordinates $(-1, 3)$ is not on the graph. Evaluating the function at -1, we have

$$f(x) = x^2 + 2x - 3$$
$$f(-1) = (-1)^2 + 2(-1) - 3 = -4 \neq 3$$

Because $f(-1) \neq 3$, the ordered pair $(-1, 3)$ does not belong to the function, and the point with coordinates $(-1, 3)$ is not on the graph of f.

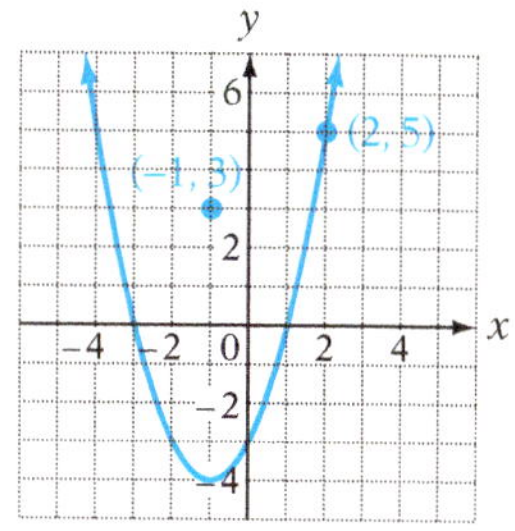

Every ordered pair (x, y) on the graph of a function f satisfies $y = f(x)$, and every ordered pair that satisfies $y = f(x)$ is a point on the graph of the function.

Take Note ➤
We chose values of t that are multiples of 2 to make it easier to plot the points. For instance, if we had selected $t = 3$, then

$$s = g(3) = -\frac{3}{2}(3) + 1$$
$$= -\frac{9}{2} + 1 = -\frac{7}{2}$$

Now we would need to estimate $-\frac{7}{2}$ on the coordinate grid—not impossible, but not as easy as plotting an integer value.

Focus on graphing a function

Graph $s = g(t) = -\frac{3}{2}t + 1$. Begin by plotting the ordered pairs (t, s) of the function when $t = -2, 0$, and 2. Then connect the points to form the graph.

SOLUTION

The variables in a function can be different from x and y. The key point is that the independent variable, t, is represented on the horizontal axis. We can choose any other variable for the dependent variable. In this case, we chose s. The dependent variable is represented on the vertical axis.

t	$g(t) = -\frac{3}{2}t + 1$	s	(t, s)
-2	$g(-2) = -\frac{3}{2}(-2) + 1$	4	$(-2, 4)$
0	$g(0) = -\frac{3}{2}(0) + 1$	1	$(0, 1)$
2	$g(2) = -\frac{3}{2}(2) + 1$	-2	$(2, -2)$

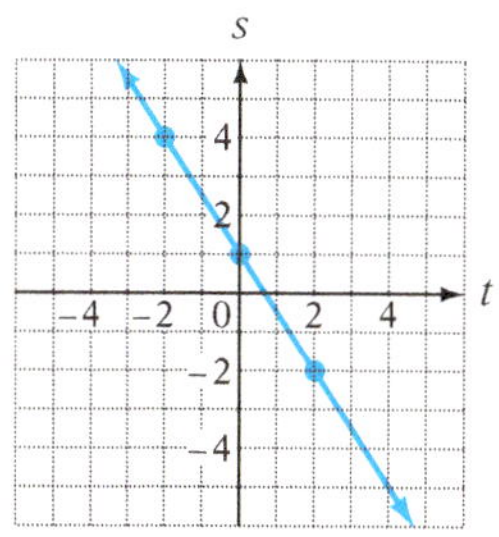

Because $-\frac{3}{2}t + 1$ is a real number for all values of t, the domain of $s = g(t) = -\frac{3}{2}t + 1$ is $\{t \mid -\infty < t < \infty\}$. From the graph, the range is $\{s \mid -\infty < s < \infty\}$.

Check your understanding 5

Graph $h(u) = -u^2 + 4u$. Begin by plotting the ordered pairs (u, v) of the function when $u = -1, 0, 1, 2, 3, 4$, and 5. Estimate the domain and range from the graph. Write the answers using set-builder notation.

SOLUTION See page S-2.

Focus on graphing an absolute value function

Graph $f(x) = |2x + 4| - 1$. Begin by plotting the ordered pairs (x, y) of the function when $x = -5, -4, -3, -2, -1, 0$, and 1. Estimate the domain and range from the graph. Write the answers using interval notation.

SOLUTION

The independent variable is x. The dependent variable is y.

x	$f(x) = \|2x + 4\| - 1$	y
-5	$f(-5) = \|2(-5) + 4\| - 1$	5
-4	$f(-4) = \|2(-4) + 4\| - 1$	3
-3	$f(-3) = \|2(-3) + 4\| - 1$	1
-2	$f(-2) = \|2(-2) + 4\| - 1$	-1
-1	$f(-1) = \|2(-1) + 4\| - 1$	1
0	$f(0) = \|2(0) + 4\| - 1$	3
1	$f(1) = \|2(1) + 4\| - 1$	5

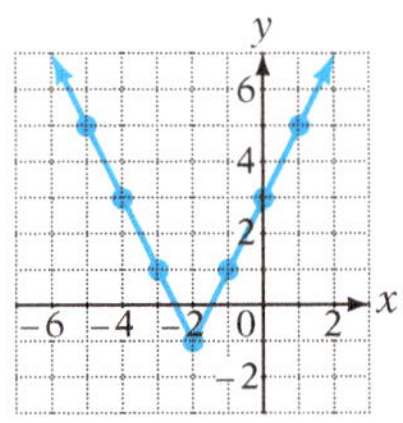

The domain is $(-\infty, \infty)$. The range is $[-1, \infty)$.

Check your understanding 6

Graph $g(t) = -|t + 2| + 3$. Begin by plotting the ordered pairs (t, s) of the function when $t = -6, -4, -2, 0, 2$, and 4. Estimate the domain and range from the graph. Write the answers using set-builder notation.

SOLUTION See page S-3.

Focus on estimating the domain and range of a function

Estimate the domain and range of the function graphed at the right. Write the answers in interval notation.

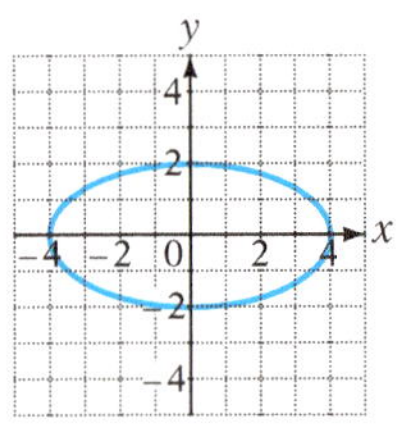

SOLUTION

The points on the graph have x-values between -4 and 4, and y-values between -2 and 2. The domain is $[-4, 4]$. The range is $[-2, 2]$.

Check your understanding 7

Estimate the domain and range of the function graphed at the right. Write the answers in set-builder notation.

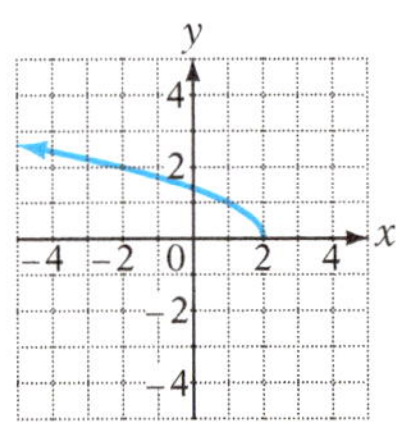

SOLUTION See page S-3.

Objective 5.2B Practice

For Exercises 1 to 4, graph the function. First plot the ordered pairs of the function for the given values of the independent variable. Then connect the points to form the graph. Estimate the domain and range from the graph. Write the answers in set-builder notation.

1. $f(x) = -x^2 - 4x + 1$
$x = -5, -4, -3, -2, -1, 0, 1$

2. $f(x) = 2|x + 1| - 1$
$x = -4, -3, -2, -1, 0, 1, 2$

3. $h(u) = |2u| - 3$
$u = -4, -3, -2, -1, 0, 1, 2, 3, 4$

4. $f(t) = -|4 - 2t| + 3$
$t = -2, -1, 0, 1, 2, 3, 4, 5$

Solutions on pp. S-13–S-14.

Objective 5.2C Apply the vertical line test

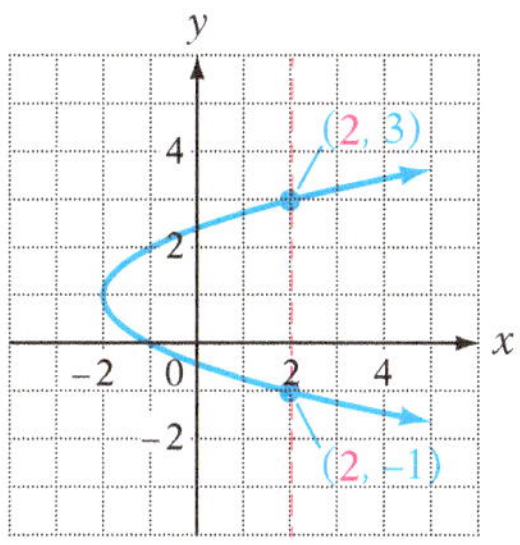

Consider the graph at the left. Note that $(2, 3)$ and $(2, -1)$ are two ordered pairs on the graph. Because there are two ordered pairs with the same first coordinate, the graph is not the graph of a function. Note also that the vertical line through $(2, 3)$ and $(2, -1)$ intersects the graph at more than one point. As long as there is one vertical line that intersects a graph more than once, the graph is not the graph of a function.

The vertical line test is used to determine whether a graph is the graph of a function.

Vertical Line Test

A graph defines a function if any vertical line intersects the graph at no more than one point.

Take Note ➤
For the graph at the far right, note that there are values of x for which there is only one value of y. For instance, when $x = -5$, $y = 4$. For the relation to be a function, however, *every* value of x in the domain of the function must pair with exactly one value of y. If there is even one value of x that pairs with two or more values of y, the condition for a function is not met. This is what the vertical line test states: If there is any place on the graph where a vertical line intersects the graph more than once, the graph is not the graph of a function.

The vertical line test is a graphical interpretation of the fact that each value in the domain of a function is paired with *exactly one* value in the range of the function.

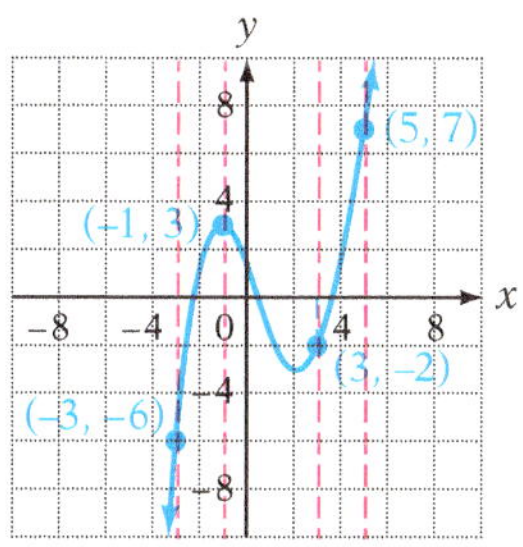

For each x, there is *exactly one* value of y. For instance, when $x = -3$, $y = -6$. This is the graph of a function.

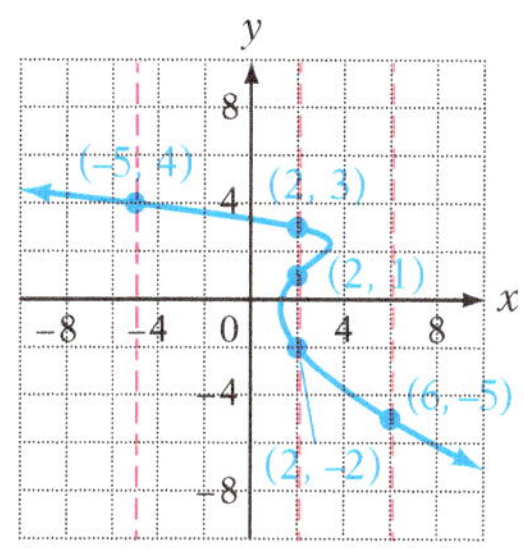

Some values of x can be paired with more than one value of y. For instance, 2 can be paired with -2, 1, and 3. This is not the graph of a function.

Determine whether the graph is the graph of a function.

A.

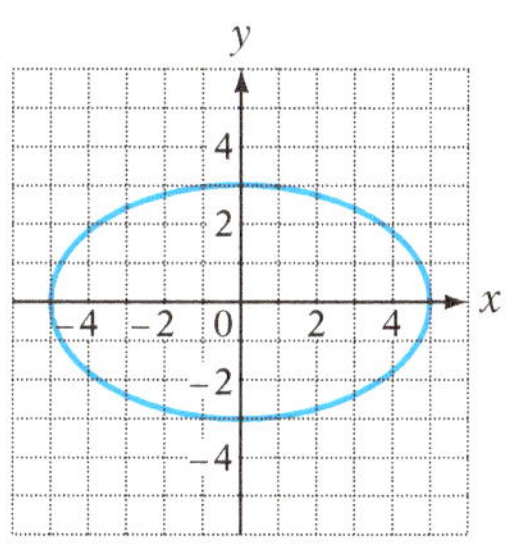

B.

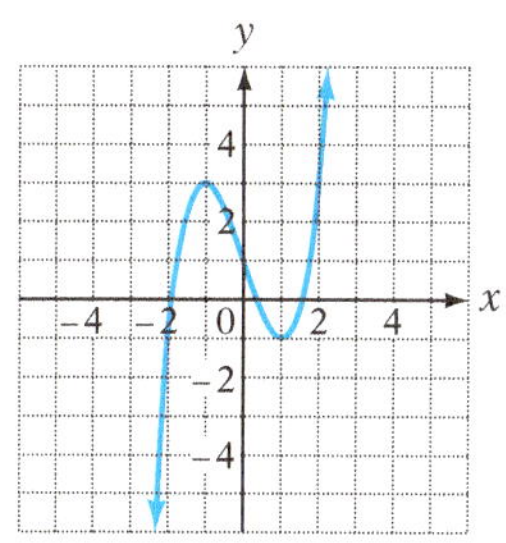

A. There are vertical lines that intersect the graph at more than one point.

B. Every vertical line intersects the graph at most once.

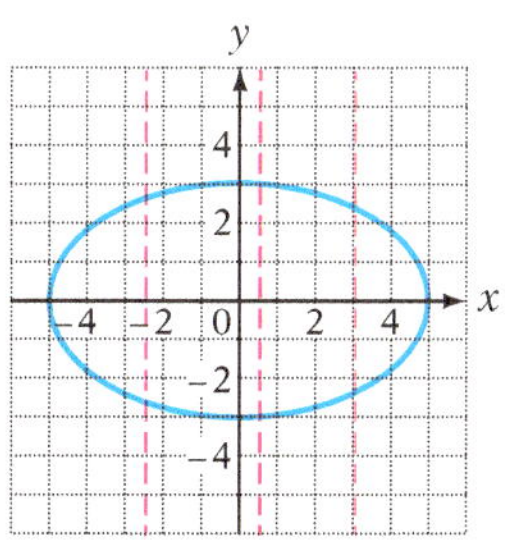

The graph is not the graph of a function.

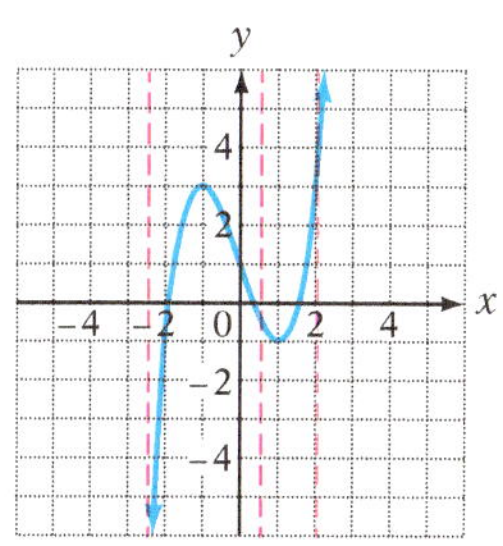

The graph is the graph of a function.

Focus on applying the vertical line test

Use the vertical line test to determine whether the graph shown below is the graph of a function.

SOLUTION

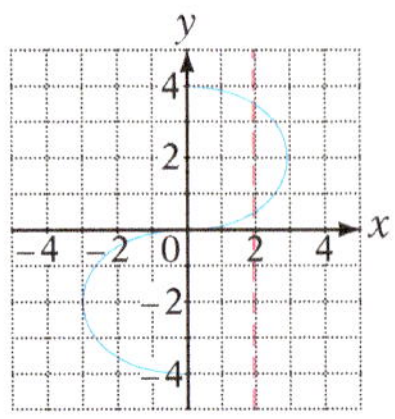

The dashed vertical line intersects the graph more than once. The graph is not the graph of a function.

Check your understanding 8

Use the vertical line test to determine whether the graph shown at the right is the graph of a function.

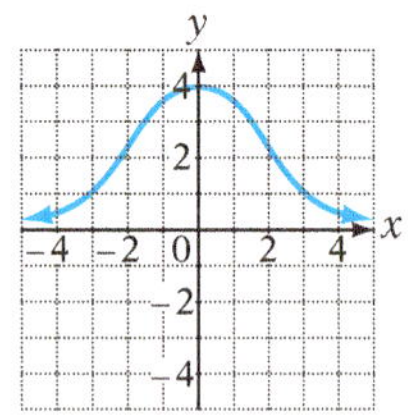

SOLUTION See page S-3.

Objective 5.2C Practice

For Exercises 1 to 4, use the vertical line test to determine whether the graph is the graph of a function.

1.

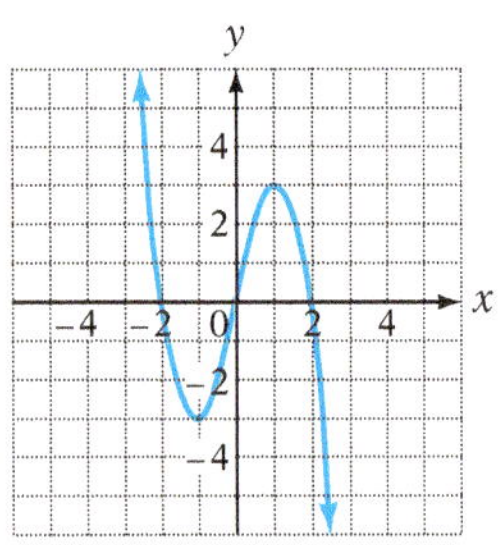

2.

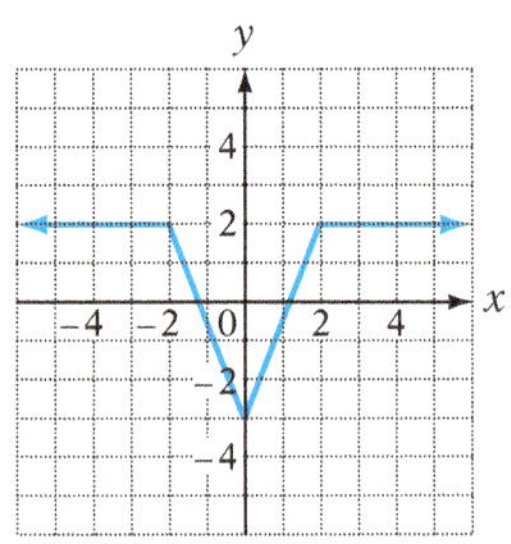

3.

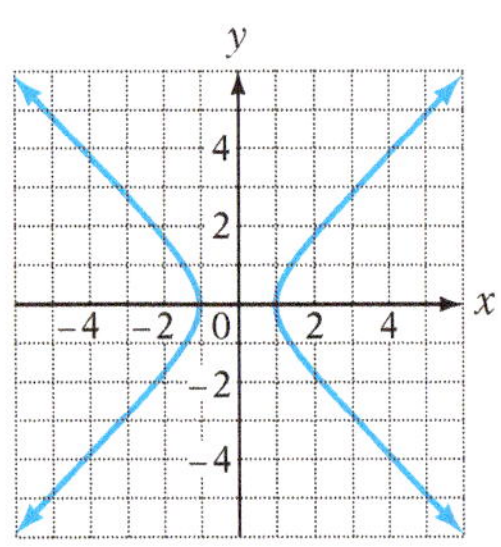

4.

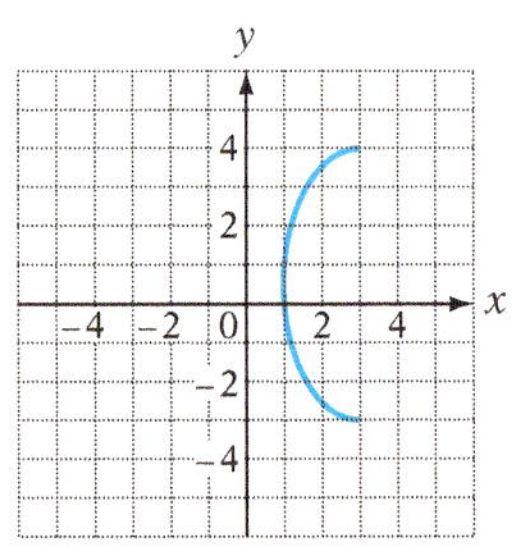

Solutions on p. S-14.

SECTION 5.3

Linear Functions

Objective 5.3A Graph a linear function

Recall that the ordered pairs of a function can be written as $(x, f(x))$ or as (x, y). The graph of a function is a graph of the ordered pairs (x, y) that belong to the function. Certain functions have characteristic graphs. A function that can be written in the form $f(x) = mx + b$ (or $y = mx + b$) is called a **linear function** because its graph is a straight line.

Examples of linear functions are shown at the right. Note that the exponent on each variable is 1.

$f(x) = 2x + 5$ $\quad (m = 2, b = 5)$

$P(t) = 3t - 2$ $\quad (m = 3, b = -2)$

$y = -2x$ $\quad (m = -2, b = 0)$

$y = -\frac{2}{3}x + 1$ $\quad \left(m = -\frac{2}{3}, b = 1\right)$

$g(z) = z - 2$ $\quad (m = 1, b = -2)$

The equation $y = x^2 + 4x + 3$ is not a linear function because it includes a term with a variable squared. The equation $f(x) = \frac{3}{x - 2}$ is not a linear function because a variable occurs in the denominator. Another example of an equation that is not a linear function is $y = \sqrt{x} + 4$; this equation contains a variable within a radical and so is not a linear function.

There are many applications of linear functions.

Apply the Concept

The linear function $F = T(C) = \frac{9}{5}C + 32$ is used to convert a Celsius temperature C to a Fahrenheit temperature F. Find the Fahrenheit temperature when the Celsius temperature is 20°C.

SOLUTION

$$F = T(C) = \frac{9}{5}C + 32$$

$$= T(20) = \frac{9}{5}(20) + 32 \quad \text{• Evaluate the function when } C = 20.$$

$$= 36 + 32 = 68$$

When the Celsius temperature is 20°C, the Fahrenheit temperature is 68°F.

Whether an equation is written as $f(x) = mx + b$ or as $y = mx + b$, the equation represents a linear function, and the graph of the equation is a straight line.

Because the graph of a linear function is a straight line, and a straight line is determined by two points, the graph of a linear function can be drawn by finding only two of the ordered pairs of the function. However, it is recommended that you find at least *three* ordered pairs to ensure accuracy.

Take Note ➤
When the coefficient of x is a fraction, choose values of x that are multiples of the denominator of the fraction. This will result in coordinates that are integers.

Focus on graphing a linear function

Graph: $f(x) = -\frac{1}{2}x + 3$

SOLUTION

x	$y = f(x)$
−4	5
0	3
2	2

• Find at least three ordered pairs. Because the coefficient of x is a fraction with denominator 2, choosing values of x that are divisible by 2 simplifies the calculations. The ordered pairs can be displayed in a table.

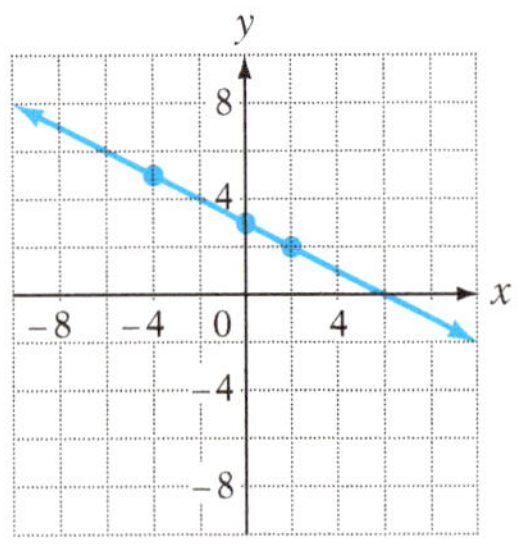

• Graph the ordered pairs and draw a line through the points.

Check your understanding 1

Graph: $f(x) = \frac{3}{5}x - 4$

SOLUTION See page S-3.

Focus on graphing a linear function

Graph: $y = \frac{2}{3}x$

SOLUTION

x	y
0	0
3	2
−3	−2

• Find at least three ordered pairs. Because the coefficient of x is a fraction with denominator 3, choosing values of x that are divisible by 3 simplifies the calculations. The ordered pairs can be displayed in a table.

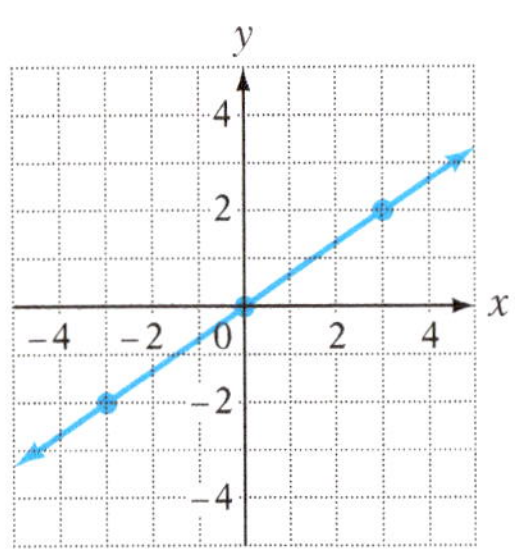

• Graph the ordered pairs and draw a line through the points.

Check your understanding 2

Graph: $y = -\frac{3}{4}x$

SOLUTION See page S-3.

There are a variety of applications of linear functions.

Take Note ➤
In many applications, the domain of the variable is such that the equation makes sense. For this example, it would not be sensible to have values of t that are less than zero. This would indicate negative time! The number 10 is somewhat arbitrary, but after 10 min most people's heart rates would level off, and a linear function would no longer apply.

Focus on solving an application using a linear function

The heart rate R after t minutes for a person taking a brisk walk can be approximated by the equation $R = 2t + 72$. Graph this equation for $0 \le t \le 10$. The point with coordinates (5, 82) is on the graph. Write a sentence that describes the meaning of this ordered pair.

SOLUTION

Find the values of R for $t = 0$ and $t = 10$. When $t = 0$, $R = 72$. When $t = 10$, $R = 84$.

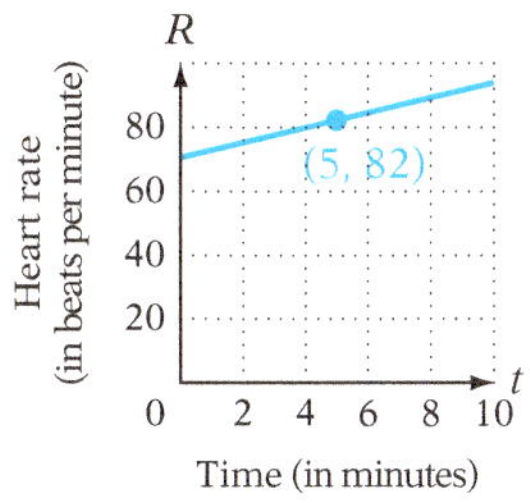

• Graph (0, 72), (5, 82), and (10, 84). Draw the line segment that contains the three points.

The ordered pair (5, 82) means that after 5 min, the person's heart rate is 82 beats per minute.

Check your understanding 3

The height h (in inches) of a person and the length L (in inches) of that person's stride while walking are related. The equation $h = \frac{3}{4}L + 50$ approximates this relationship. Graph this equation for $15 \le L \le 40$. The point with coordinates (32, 74) is on the graph. Write a sentence that describes the meaning of this ordered pair.

SOLUTION See page S-3.

Objective 5.3A Practice

1. Graph: $y = -2x + 3$

2. Graph: $f(x) = \frac{3}{2}x$

3. Graph: $f(x) = \frac{3}{4}x + 2$

4. Graph: $y = -\frac{3}{2}x - 3$

5. The Large Hadron Collider, or LHC, is a machine that is capable of accelerating a proton to a velocity that is 99.99% the speed of light. At this speed, a proton will travel approximately 0.98 ft in a billionth of a second (one nanosecond). (*Source:* news.yahoo.com) The equation $d = 0.98t$ gives the distance d, in feet, traveled by a proton in t nanoseconds. Graph this equation for $0 \le t \le 10$. The point with coordinates (4, 3.92) is on the graph. Write a sentence that explains the meaning of this ordered pair.

Solutions on pp. S-14–S-15.

Objective 5.3B Graph an equation of the form $Ax + By = C$

The equation $Ax + By = C$, where A and B are coefficients and C is a constant, is also a *linear equation in two variables.* This equation can be written in the form $y = mx + b$.

Write $4x - 3y = 6$ in the form $y = mx + b$.

$4x - 3y = 6$

$-3y = -4x + 6$ • Subtract $4x$ from each side of the equation.

$y = \frac{4}{3}x - 2$ • Divide each side of the equation by -3. This is the form $y = mx + b$, with $m = \frac{4}{3}$ and $b = -2$.

To graph an equation of the form $Ax + By = C$, first solve the equation for y. Then follow the same procedure used to graph an equation of the form $y = mx + b$.

Focus on graphing a linear equation of the form $Ax + By = C$

Graph: $3x + 2y = 6$

SOLUTION

$3x + 2y = 6$

$2y = -3x + 6$

$y = -\frac{3}{2}x + 3$ • Solve the equation for y.

x	y
0	3
2	0
4	−3

• Find at least three solutions. Choose multiples of 2 for x.

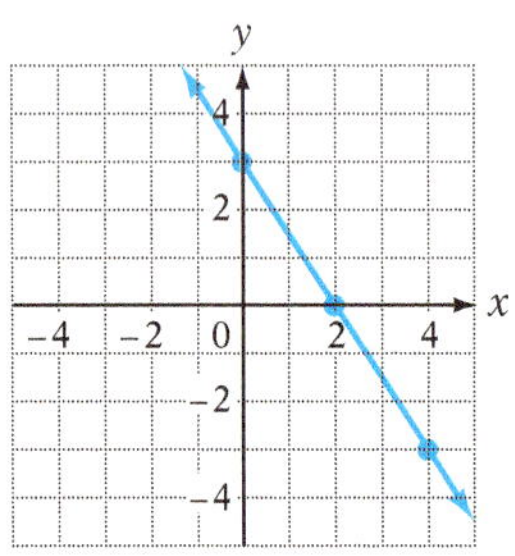

• Graph the ordered pairs and draw a line through the points.

Check your understanding 4

Graph: $-3x + 2y = 4$

SOLUTION See page S-4.

If one of the coefficients A or B is zero, the graph of $Ax + By = C$ is a horizontal or vertical line. Consider the equation $y = -2$, where A, the coefficient of x, is 0. We can write this equation in *two* variables as $0 \cdot x + y = -2$. No matter what value of x is selected, $0 \cdot x = 0$. Therefore, y equals -2 for all values of x. The table below shows some ordered-pair solutions of $y = -2$. The graph is shown to the right of the table.

x	$0 \cdot x + y = -2$	y	(x, y)
-1	$0 \cdot (-1) + y = -2$	-2	$(-1, -2)$
0	$0 \cdot 0 + y = -2$	-2	$(0, -2)$
3	$0 \cdot 3 + y = -2$	-2	$(3, -2)$

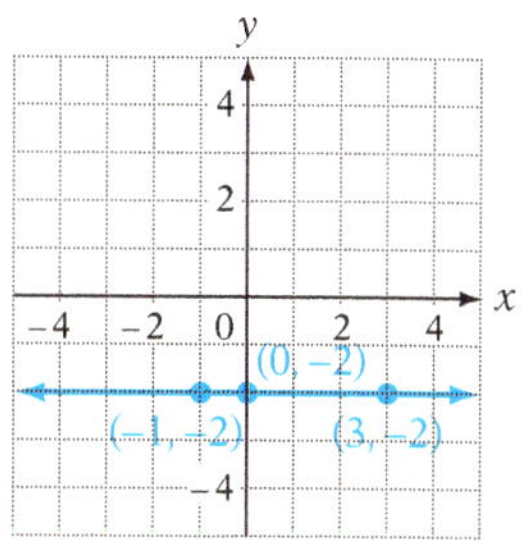

Graph of $y = b$

The **graph of $y = b$** is a horizontal line passing through the point with coordinates $(0, b)$.

Focus on graphing a horizontal line

Graph: $y + 4 = 0$

SOLUTION

Solve for y.

$y + 4 = 0$

$y = -4$

The graph of $y = -4$ is a horizontal line passing through the point with coordinates $(0, -4)$.

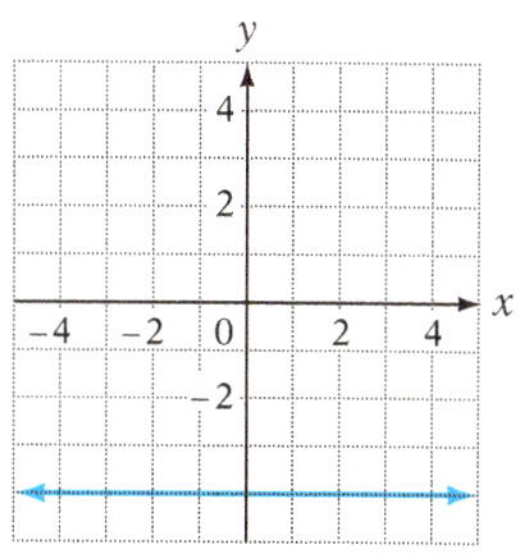

Check your understanding 5

Graph: $y - 3 = 0$

SOLUTION See page S-4.

The equation $y = -4$ represents a function. Some of the ordered pairs of this function are $(-1, -4)$, $(0, -4)$, and $(3, -4)$. In function notation, we write $f(x) = -4$. This function is an example of a *constant function*. For every value of x, the value of the function is the constant -4. For instance, we have $f(17) = -4$, $f(\sqrt{2}) = -4$, and $f(\pi) = -4$.

Constant Function

A function given by $f(x) = b$, where b is a constant, is a **constant function.** The graph of a constant function is a horizontal line passing through the point $P(0, b)$.

Now consider the equation $x = 2$, where B, the coefficient of y, is zero. We write this equation in two variables by writing $x + 0 \cdot y = 2$. No matter what value of y is selected, $0 \cdot y = 0$. Therefore, x equals 2 for all values of y.

Graph of $x = a$

The **graph of $x = a$** is a vertical line passing through the point with coordinates $(a, 0)$.

Focus on graphing a vertical line

Graph: $x = 2$

SOLUTION

For any value of y, the value of x is 2. Some ordered-pair solutions of $x = 2$ are shown in the table below. Graph the ordered-pair solutions and then draw a line through the points.

y	$x + 0 \cdot y = 2$	x	(x, y)
6	$x + 0 \cdot 6 = 2$	2	(2, 6)
1	$x + 0 \cdot 1 = 2$	2	(2, 1)
−4	$x + 0 \cdot (-4) = 2$	2	(2, −4)

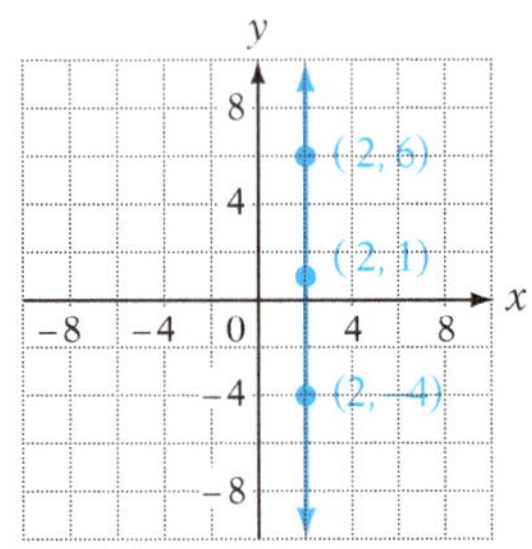

Check your understanding 6

Graph: $x = -4$

SOLUTION See page S-4.

Take Note ➤
The equation $y = b$ represents a function. The equation $x = a$ does *not* represent a function. Remember, not all equations represent functions.

Recall that a function is a set of ordered pairs in which no two ordered pairs have the same first coordinate and different second coordinates. Because (2, 6), (2, 1), and (2, −4) are ordered pairs belonging to the equation $x = 2$, this equation does not represent a function, and its graph is not the graph of a function.

The graph of the equation $x - 2y = 4$ is shown at the right. The graph crosses the x-axis at the point with coordinates (4, 0). This point is called the ***x*-intercept.** The graph crosses the y-axis at the point with coordinates (0, −2). This point is called the ***y*-intercept.**

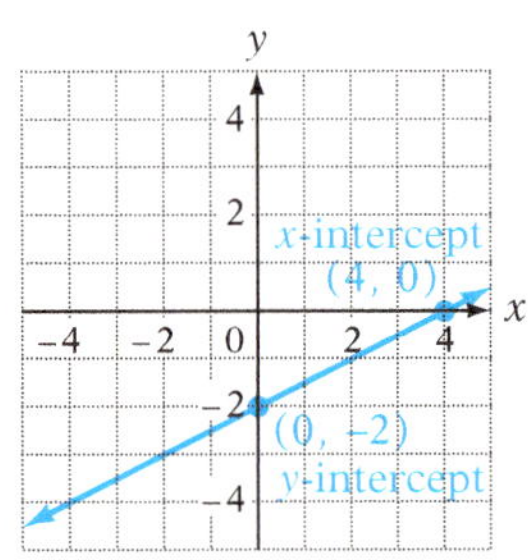

Take Note ➤
The x-intercept occurs when $y = 0$.
The y-intercept occurs when $x = 0$.

Find the coordinates of the x- and y-intercepts of the graph of the equation $3x + 4y = -12$.

To find the x-intercept, let $y = 0$. (Any point on the x-axis has y-coordinate 0.)

$$\begin{aligned} 3x + 4y &= -12 \\ 3x + 4(0) &= -12 \\ 3x &= -12 \\ x &= -4 \end{aligned}$$

The x-intercept has coordinates $(-4, 0)$.

To find the y-intercept, let $x = 0$. (Any point on the y-axis has x-coordinate 0.)

$$\begin{aligned} 3x + 4y &= -12 \\ 3(0) + 4y &= -12 \\ 4y &= -12 \\ y &= -3 \end{aligned}$$

The y-intercept has coordinates $(0, -3)$.

Focus on graphing by using the x- and y-intercepts

Graph $3x - 2y = 6$ by using the x- and y-intercepts.

SOLUTION

$$\begin{aligned} 3x - 2y &= 6 \\ 3x - 2(0) &= 6 \\ 3x &= 6 \\ x &= 2 \end{aligned}$$

• To find the x-intercept, let $y = 0$. Then solve for x.

The x-intercept has coordinates $(2, 0)$.

$$\begin{aligned} 3x - 2y &= 6 \\ 3(0) - 2y &= 6 \\ -2y &= 6 \\ y &= -3 \end{aligned}$$

• To find the y-intercept, let $x = 0$. Then solve for y.

The y-intercept has coordinates $(0, -3)$.

Graph the x- and y-intercepts. Draw a line through the two points.

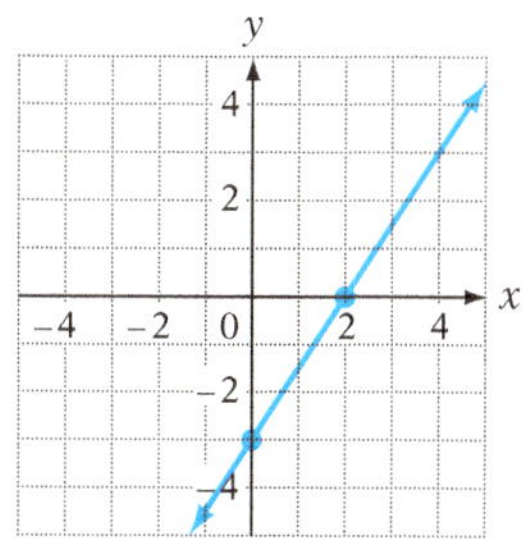

Check your understanding 7

Graph $3x - y = 2$ by using the x- and y-intercepts.

SOLUTION See page S-4.

Objective 5.3B Practice

1. Graph $2x - 5y = 10$ by first writing the equation in the form $y = mx + b$.
2. Graph $y = -2$
3. Graph $x = -3$

For Exercises 4 to 6, find the x- and y-intercepts and graph.

4. $x - 2y = -4$
5. $2x - 3y = 9$
6. $3x + 2y = 4$

Solutions on pp. S-15–S-16.

Objective 5.3C Find the slope of a line given two points

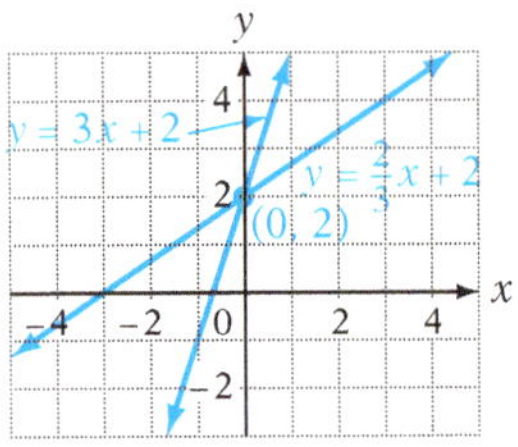

The graphs of $y = 3x + 2$ and $y = \frac{2}{3}x + 2$ are shown at the left. Each graph crosses the y-axis at the point $P(0, 2)$, but the graphs have different slants. The **slope** of a line is a measure of the slant of the line. The symbol for slope is m.

The slope of a line containing two points is the ratio of the change in the y values of the two points to the change in the x values. The line containing the points whose coordinates are $(-1, -3)$ and $(5, 2)$ is shown below.

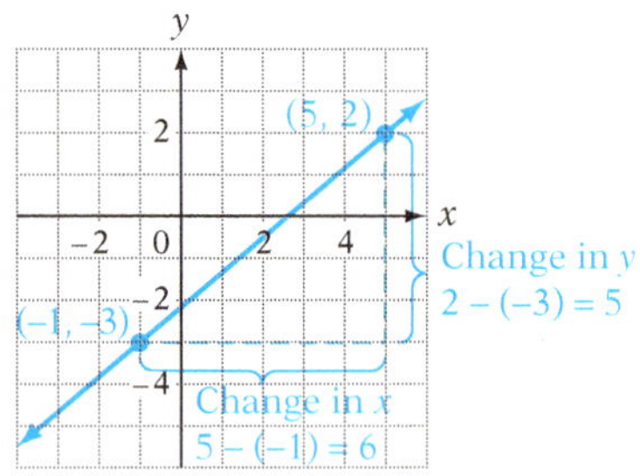

The change in the y values is the difference between the y-coordinates of the two points.

Change in $y = 2 - (-3) = 5$

The change in the x values is the difference between the x-coordinates of the two points.

Change in $x = 5 - (-1) = 6$

The slope of the line between the two points is the ratio of the change in y to the change in x.

$$\text{Slope} = m = \frac{\text{change in } y}{\text{change in } x} = \frac{5}{6} \qquad m = \frac{2 - (-3)}{5 - (-1)} = \frac{5}{6}$$

In general, if $P_1(x_1, y_1)$ and $P_2(x_2, y_2)$ are two points on a line, then

Change in $y = y_2 - y_1$ Change in $x = x_2 - x_1$

Using these ideas, we can state a formula for slope.

Slope Formula

The slope of the line containing the two points $P_1(x_1, y_1)$ and $P_2(x_2, y_2)$ is given by

$$m = \frac{y_2 - y_1}{x_2 - x_1}, \; x_1 \neq x_2$$

Frequently, the Greek letter Δ (delta) is used to designate the change in a variable. Using this notation, we can write equations for the change in y and the change in x as follows:

Change in $y = \Delta y = y_2 - y_1$ Change in $x = \Delta x = x_2 - x_1$

Using this notation, the slope formula is written $m = \frac{\Delta y}{\Delta x}$.

Take Note ➤
When finding the slope of the line between two points, it does not matter which point is called P_1 and which is called P_2. We could have labeled the points $P_1(4, 5)$ and $P_2(-2, 0)$, reversing the names of P_1 and P_2. Then

$$m = \frac{y_2 - y_1}{x_2 - x_1}$$

$$m = \frac{0 - 5}{-2 - 4} = \frac{-5}{-6} = \frac{5}{6}$$

The result is the same.

Focus on finding the slope of a line: positive slope

Find the slope of the line passing through the points $P_1(-2, 0)$ and $P_2(4, 5)$.

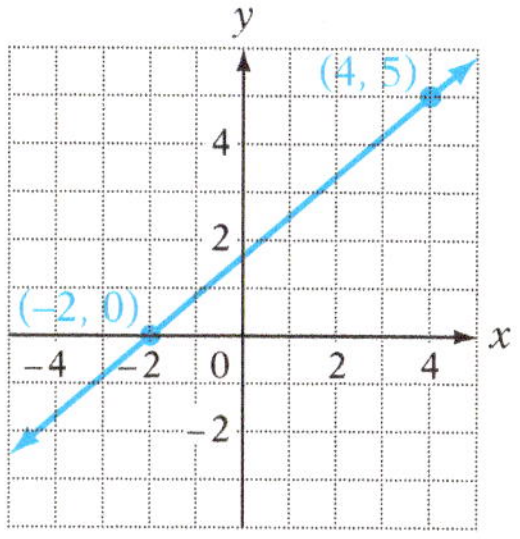

Positive slope

SOLUTION

From $P_1(-2, 0)$, we have $x_1 = -2, y_1 = 0$. From $P_2(4, 5)$, we have $x_2 = 4, y_2 = 5$. Now use the slope formula.

$$m = \frac{y_2 - y_1}{x_2 - x_1} = \frac{5 - 0}{4 - (-2)} = \frac{5}{6}$$

The slope of the line is $\frac{5}{6}$.

A line that slants upward to the right has a **positive slope.**

Check your understanding 8

Find the slope of the line containing $P_1(2, 5)$ and $P_2(-4, 2)$.

SOLUTION See page S-5.

Focus on finding the slope of a line: negative slope

Find the slope of the line passing through the points $P_1(-3, 4)$ and $P_2(4, 2)$.

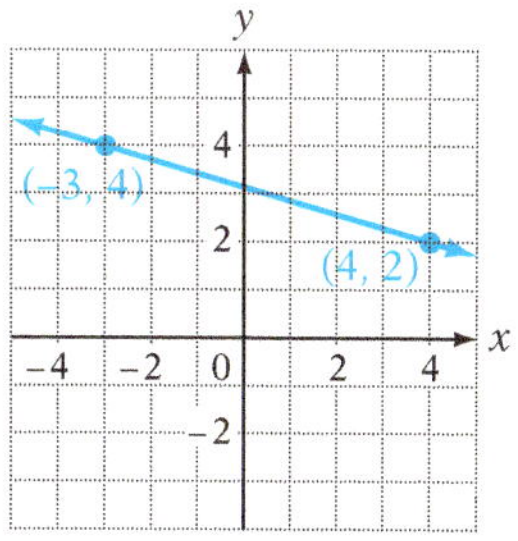

Negative slope

SOLUTION

From $P_1(-3, 4)$, we have $x_1 = -3, y_1 = 4$. From $P_2(4, 2)$, we have $x_2 = 4, y_2 = 2$. Now use the slope formula.

$$m = \frac{y_2 - y_1}{x_2 - x_1} = \frac{2 - 4}{4 - (-3)} = \frac{-2}{7} = -\frac{2}{7}$$

The slope of the line is $-\frac{2}{7}$.

A line that slants downward to the right has a **negative slope.**

Check your understanding 9

Find the slope of the line containing $P_1(4, -3)$ and $P_2(2, 7)$.

SOLUTION See page S-5.

Focus on finding the slope of a line: zero slope

Find the slope of the line passing through the points $P_1(-2, 2)$ and $P_2(4, 2)$.

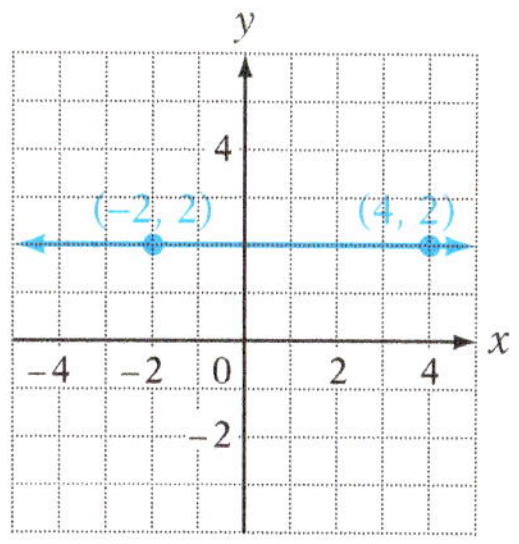

Zero slope

SOLUTION

From $P_1(-2, 2)$, we have $x_1 = -2, y_1 = 2$. From $P_2(4, 2)$, we have $x_2 = 4, y_2 = 2$. Now use the slope formula.

$$m = \frac{y_2 - y_1}{x_2 - x_1} = \frac{2 - 2}{4 - (-2)} = \frac{0}{6} = 0$$

The slope of the line is 0.

A horizontal line has **zero slope.**

Check your understanding 10

Find the slope of the line containing $P_1(-3, 4)$ and $P_2(5, 4)$.

SOLUTION See page S-5.

Focus on finding the slope of a line: undefined slope

Find the slope of the line passing through the points $P_1(1, -2)$ and $P_2(1, 3)$.

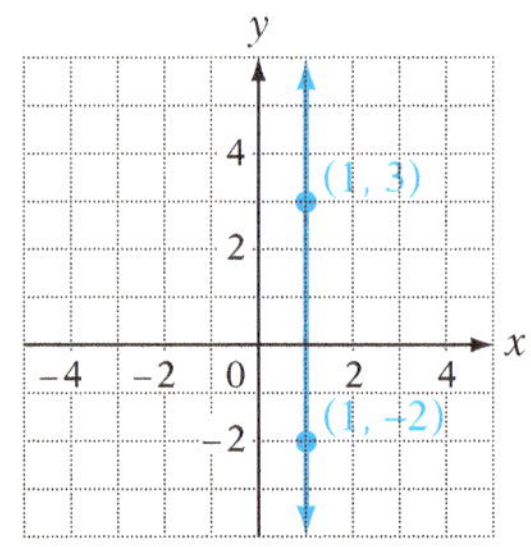

Undefined

SOLUTION

From $P_1(1, -2)$, we have $x_1 = 1, y_1 = -2$. From $P_2(1, 3)$, we have $x_2 = 1, y_2 = 3$. Now use the slope formula.

$$m = \frac{y_2 - y_1}{x_2 - x_1} = \frac{3 - (-2)}{1 - 1} = \frac{5}{0}$$

$\frac{5}{0}$ is not a real number, because division by zero is undefined.

The slope of the line is undefined.

A vertical line has **undefined slope.** A vertical line is also said to have **no slope.**

Check your understanding 11

Find the slope of the line containing $P_1(6, -1)$ and $P_2(6, 7)$.

SOLUTION See page S-5.

Tips for Success ➤
To learn mathematics, you must be an active participant. Listening to and watching your professor do mathematics is not enough. Take notes in class, mentally think through every question your instructor asks, and try to answer it even if you are not called on to do so. Ask questions when you have them. See *AIM for Success*.

There are many applications of slope. Here are two examples.

The first record time for the 1-mile run was recorded in 1865 in England. Richard Webster ran the mile in 4 min 36.5 s. His average speed was approximately 19 ft/s.

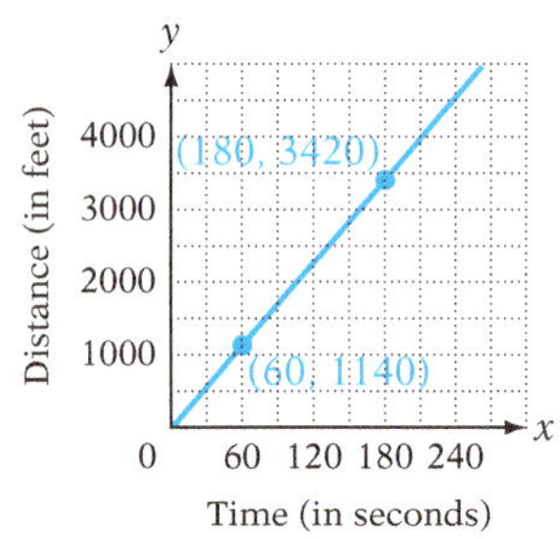

The graph at the right shows the distance Webster ran during that run. From the graph, note that after 60 s (1 min) he had traveled 1140 ft, and after 180 s (3 min) he had traveled 3420 ft.

We will choose $P_1(60, 1140)$ and $P_2(180, 3420)$. The slope of the line between these two points is

$$m = \frac{y_2 - y_1}{x_2 - x_1} = \frac{3420 - 1140}{180 - 60} = \frac{2280}{120} = 19$$

Note that the slope of the line is the same as Webster's average speed, 19 ft/s. Average speed is related to slope.

The resale value of a 2010 Chevrolet Corvette declines as the number of miles the car is driven increases, as shown in the graph at the right. From the graph, note that after the car is driven 25,000 mi, its value is \$34,400, and after the car is driven 50,000 mi, its value is \$33,000. (*Source*: Edmunds.com, December 2011)

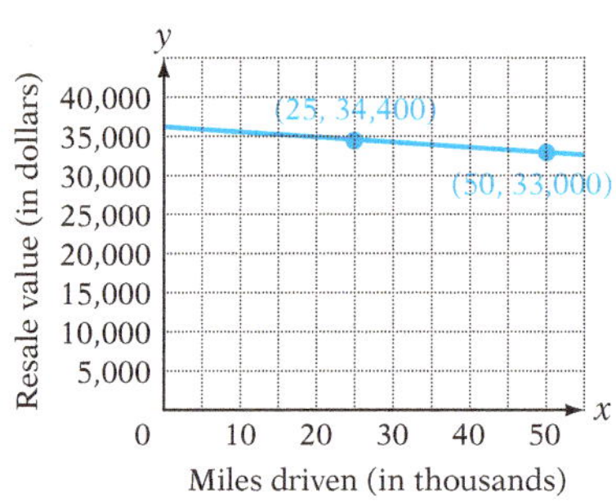

We will choose $P_1(25, 34{,}400)$ and $P_2(50, 33{,}000)$. The slope of the line between the two points is

$$m = \frac{y_2 - y_1}{x_2 - x_1} = \frac{33{,}000 - 34{,}400}{50 - 25} = \frac{-1400}{25} \approx -56$$

If we interpret negative slope as decreasing value, then the slope of the line represents the dollar decline in the value of the car for each 1000 mi driven. Thus the value of the car decreases by \$56 for each 1000 mi driven.

In general, any quantity that is expressed using the word *per* is represented mathematically as slope. In the first example, the slope represented the average speed, 19 ft/s. In the second example, the slope represented the rate at which the value of the car was decreasing, \$56 for each 1000 mi driven.

Focus on solving an application involving slope

The graph at the right shows the relationship between the cost of an item and the sales tax. Find the slope of the line between the two points shown on the graph. Write a sentence that states the meaning of the slope.

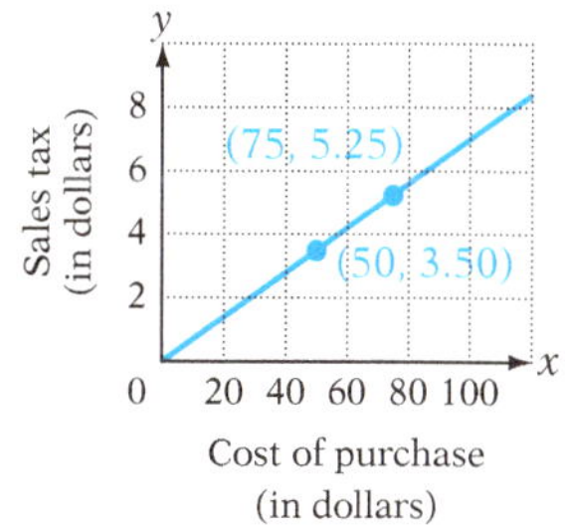

SOLUTION

$$m = \frac{5.25 - 3.50}{75 - 50}$$

• Choose $P_1(50, 3.50)$ and $P_2(75, 5.25)$.

$$= \frac{1.75}{25} = 0.07$$

A slope of 0.07 means that the sales tax is \$.07 per dollar.

Check your understanding 12

The graph at the right shows the decrease in the value of a recycling truck over 6 years. Find the slope of the line between the two points shown on the graph. Write a sentence that states the meaning of the slope.

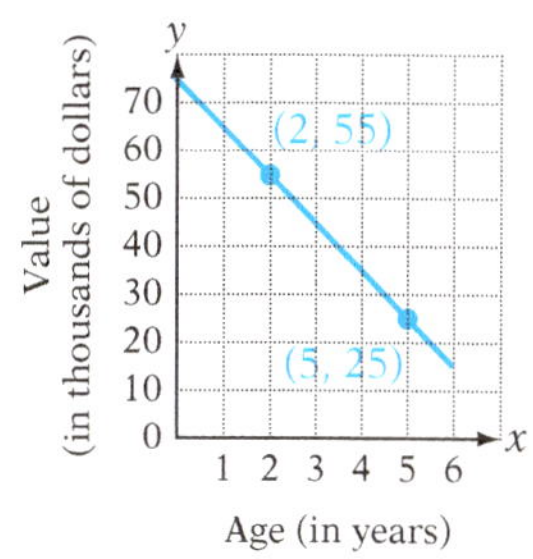

SOLUTION See page S-5.

Objective 5.3C Practice

For Exercises 1 to 5, find the slope of the line containing the given points.

1. $P_1(-1, 4)$ and $P_2(2, 5)$
2. $P_1(-1, 3)$ and $P_2(-4, 5)$
3. $P_1(2, 4)$ and $P_2(2, -2)$
4. $P_1(2, 3)$ and $P_2(-1, 3)$
5. $P_1(0, 4)$ and $P_2(-2, 5)$

6. The graph below shows how the amount of gas in the tank of a car decreases as the car is driven. Find the slope of the line. Write a sentence that states the meaning of the slope.

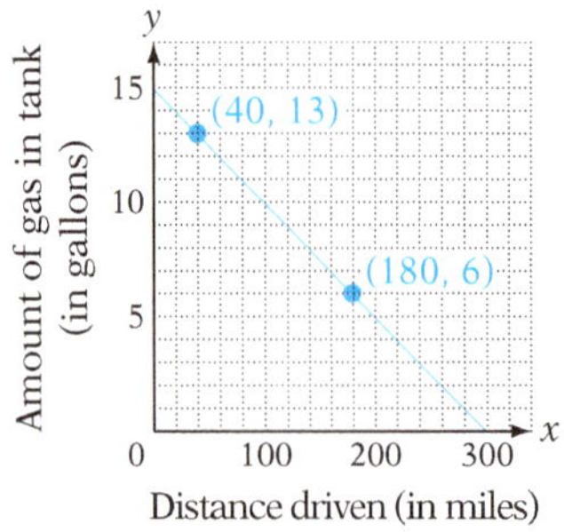

Solutions on p. S-17.

Objective 5.3D Graph a line given a point and the slope

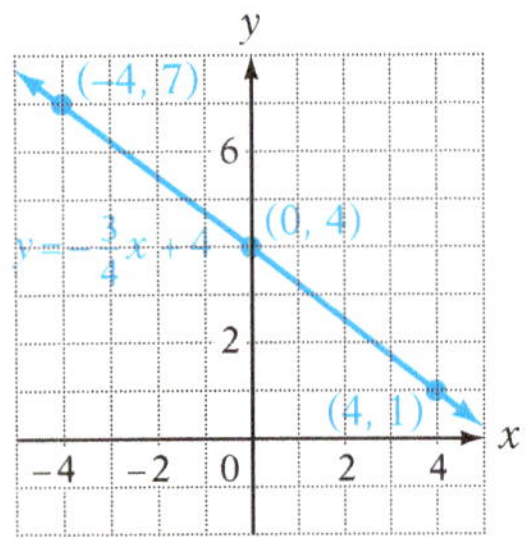

The graph of the equation $y = -\frac{3}{4}x + 4$ is shown at the left. The points with coordinates $(-4, 7)$ and $(4, 1)$ are on the graph. The slope of the line is

$$m = \frac{7 - 1}{-4 - 4} = \frac{6}{-8} = -\frac{3}{4}$$

Note that the slope of the line is the coefficient of x in the equation of the line.

Recall that the y-intercept is found by replacing x by zero and solving for y.

$$y = -\frac{3}{4}x + 4 \qquad y = -\frac{3}{4}(0) + 4 = 4$$

The coordinates of the y-intercept are $(0, 4)$. Note that the y-coordinate of the y-intercept is the constant term of the equation of the line.

Slope-Intercept Form of a Straight Line

The equation $y = mx + b$ is called the **slope-intercept form** of a straight line. The slope of the line is m, the coefficient of x. The coordinates of the y-intercept are $(0, b)$.

When the equation of a straight line is in the form $y = mx + b$, its graph can be drawn by using the slope and y-intercept. First locate the y-intercept. Use the slope to find a second point on the line. Then draw a line through the two points.

Take Note ➤
When graphing a line by using its slope and y-intercept, *always* start at the y-intercept.

Focus on graphing a line using the slope and y-intercept

Graph $y = \frac{5}{3}x - 4$ by using the slope and y-intercept.

SOLUTION

The slope is the coefficient of x: $m = \frac{5}{3} = \frac{\text{change in } y}{\text{change in } x}$.

The coordinates of the y-intercept are $(0, -4)$.

Beginning at the y-intercept, which has coordinates $(0, -4)$, move up 5 units (change in y) and then right 3 units (change in x).

The point with coordinates $(3, 1)$ is a second point on the graph. Draw a line through the points with coordinates $(0, -4)$ and $(3, 1)$.

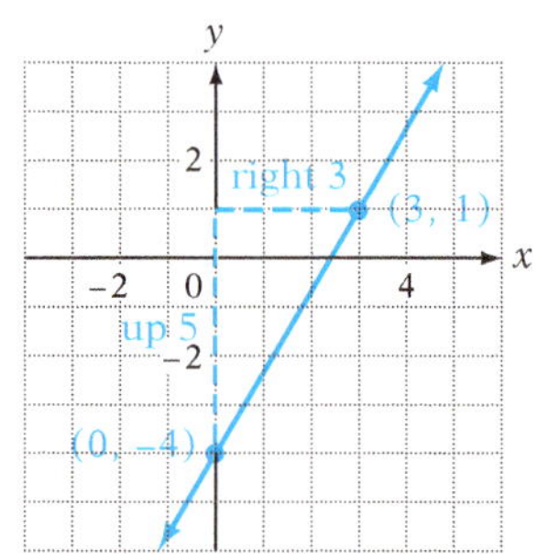

Check your understanding 13

Graph $y = -\frac{3}{2}x + 4$ by using the slope and y-intercept.

SOLUTION See page S-5.

Focus on graphing a line using the slope and y-intercept

Graph $x + 2y = 4$ by using the slope and y-intercept.

SOLUTION

Write $x + 2y = 4$ in the form $y = mx + b$ by solving the equation for y.

$$x + 2y = 4$$
$$2y = -x + 4$$
$$y = -\frac{1}{2}x + 2 \qquad \bullet\ m = -\frac{1}{2} = \frac{-1}{2},\ y\text{-intercept} = (0, 2)$$

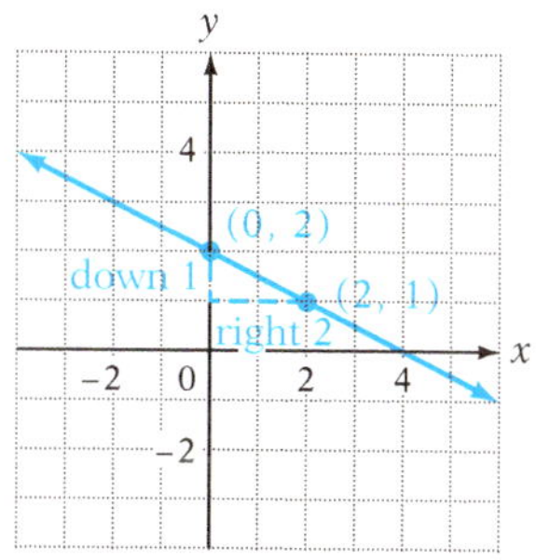

Beginning at the y-intercept, which has coordinates (0, 2), move down 1 unit (change in y) and then right 2 units (change in x). The point with coordinates (2, 1) is a second point on the graph. Draw a line through the points with coordinates (0, 2) and (2, 1).

Check your understanding 14

Graph $2x + 3y = 6$ by using the slope and y-intercept.

SOLUTION See page S-6.

The graph of a line can be drawn when any point on the line and the slope of the line are given.

Take Note ➤ This example differs from the preceding two in that a point other than the y-intercept is given. In this case, start at the given point.

Focus on graphing a line using the slope and a point on the line

Graph the line that passes through $P(-4, -4)$ and has slope 2.

SOLUTION

When the slope is an integer, write it as a fraction with denominator 1.

$$m = 2 = \frac{2}{1} = \frac{\text{change in } y}{\text{change in } x}$$

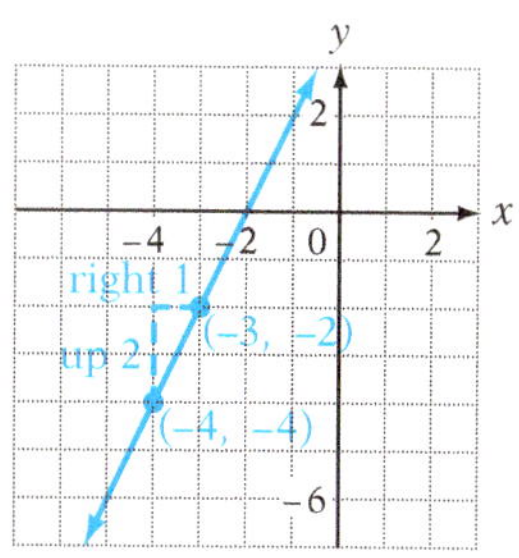

Beginning at the point $P(-4, -4)$, move up 2 units (change in y) and then right 1 unit (change in x). The point with coordinates $(-3, -2)$ is a second point on the graph. Draw a line through the points with coordinates $(-4, -4)$ and $(-3, -2)$.

Check your understanding 15

Graph the line that passes through $P(-2, 3)$ and has slope $-\frac{4}{3}$.

SOLUTION See page S-6.

Objective 5.3D Practice

1. Graph $y = \frac{2}{3}x - 3$ by using the slope and y-intercept.
2. Graph $y = -\frac{3}{2}x$ by using the slope and y-intercept.
3. Graph $y = 2x - 4$ by using the slope and y-intercept.
4. Graph the line that passes through the given point and has the given slope: $P(-1, -3)$; slope $\frac{4}{3}$
5. Graph the line that passes through the given point and has the given slope: $P(2, 1)$; slope -4

Solutions on pp. S-17–S-18.

SECTION 5.4 Linear Models

Objective 5.4A Find the equation of a line given a point and the slope

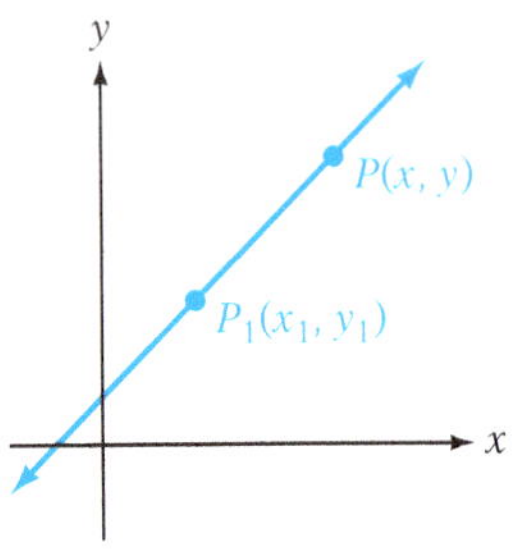

One method of finding the equation of a line when the slope and *any* point on the line are known involves using the *point-slope formula.* This formula is derived from the formula for the slope of a line as follows.

Let $P_1(x_1, y_1)$ be the given point on the line, and let $P(x, y)$ be any other point on the line. See the graph at the left.

$$\frac{y - y_1}{x - x_1} = m \qquad \text{• Use the formula for the slope of a line.}$$

$$\frac{y - y_1}{x - x_1}(x - x_1) = m(x - x_1) \qquad \text{• Multiply each side by } (x - x_1).$$

$$y - y_1 = m(x - x_1) \qquad \text{• Simplify.}$$

Point-Slope Formula

Let m be the slope of a line, and let $P_1(x_1, y_1)$ be a point on the line. The equation of the line can be found from the **point-slope formula:**

$$y - y_1 = m(x - x_1)$$

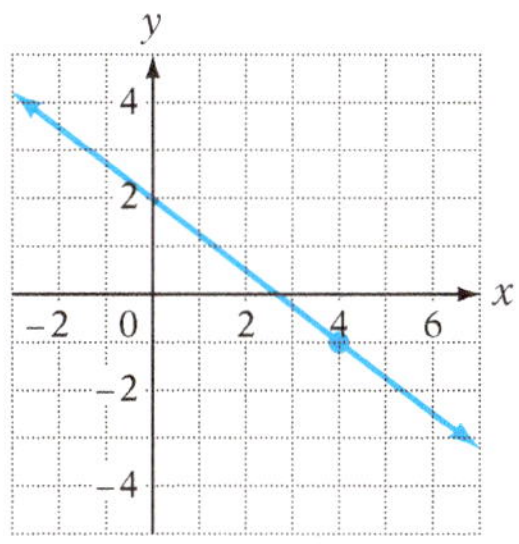

$y = -\frac{3}{4}x + 2$

Find the equation of the line that contains the point $P(4, -1)$ and has slope $-\frac{3}{4}$.

$y - y_1 = m(x - x_1)$ • Use the point-slope formula.

$y - (-1) = \left(-\frac{3}{4}\right)(x - 4)$ • $m = -\frac{3}{4}$, $(x_1, y_1) = (4, -1)$

$y + 1 = -\frac{3}{4}x + 3$ • Simplify.

$y = -\frac{3}{4}x + 2$ • Write the equation in the form $y = mx + b$.

When the slope of a line is undefined, the point-slope formula cannot be used.

Find the equation of the line that passes through the point $P(4, 3)$ and whose slope is undefined.

Because the slope is undefined, the point-slope formula cannot be used to find the equation. Instead, recall that when the slope of a line is undefined, the line is vertical. The equation of a vertical line is $x = a$, where a is the x-coordinate of the x-intercept. Because the line is vertical and passes through $P(4, 3)$, the x-intercept has coordinates $P(4, 0)$.

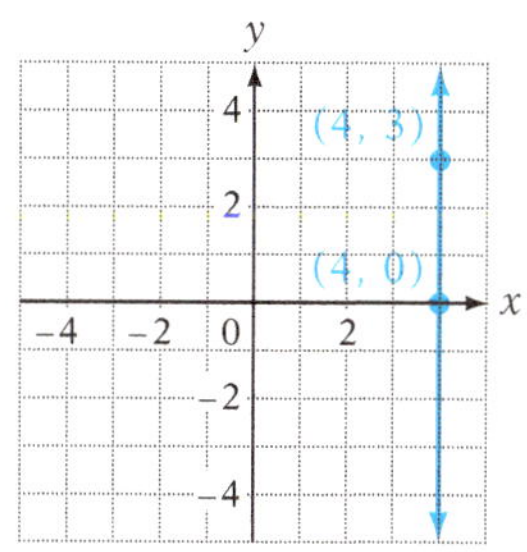

The equation of the line is $x = 4$.

Focus on finding the equation of a line given a point and the slope

Find the equation of the line that contains the point $P(-2, 4)$ and has slope 2.

SOLUTION

$m = 2 \qquad (x_1, y_1) = (-2, 4)$

$$y - y_1 = m(x - x_1)$$
$$y - 4 = 2[x - (-2)]$$
$$y - 4 = 2(x + 2)$$
$$y - 4 = 2x + 4$$
$$y = 2x + 8$$

The equation of the line is $y = 2x + 8$.

Check your understanding 1

Find the equation of the line that contains the point $P(-3, -2)$ and has slope $-\frac{1}{3}$.

SOLUTION See page S-6.

Objective 5.4A Practice

For Exercises 1 to 5, find the equation of the line that contains the given point and has the given slope.

1. $P(2, 3)$, $m = \frac{1}{2}$

2. $P(-1, 7)$, $m = -3$

3. $P(3, 5)$, $m = -\frac{2}{3}$

4. $P(3, -4)$, slope is undefined

5. $P(-2, -3)$, $m = 0$

Solutions on p. S-18.

Objective 5.4B Find the equation of a line given two points

The point-slope formula and the formula for slope are used to find the equation of a line when two points are known.

Focus on finding the equation of a line given two points

Find the equation of the line containing $P_1(3, 2)$ and $P_2(-5, 6)$.

SOLUTION

To use the point-slope formula, we must know the slope. Use the formula for slope to determine the slope of the line between the two given points.

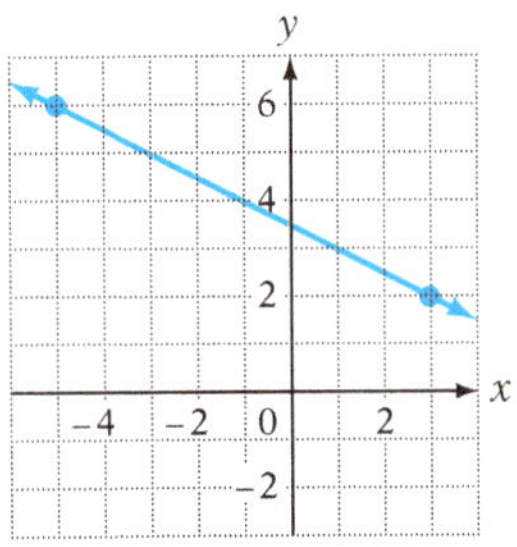

$y = -\frac{1}{2}x + \frac{7}{2}$

$$m = \frac{y_2 - y_1}{x_2 - x_1} = \frac{6 - 2}{-5 - 3} = \frac{4}{-8} = -\frac{1}{2}$$

Now use the point-slope formula with $m = -\frac{1}{2}$ and $(x_1, y_1) = (3, 2)$.

$$y - y_1 = m(x - x_1)$$ • Use the point-slope formula.

$$y - 2 = \left(-\frac{1}{2}\right)(x - 3)$$ • $m = -\frac{1}{2}$, $(x_1, y_1) = (3, 2)$

$$y - 2 = -\frac{1}{2}x + \frac{3}{2}$$ • Simplify.

$$y = -\frac{1}{2}x + \frac{7}{2}$$ • Solve for y.

The equation of the line is $y = -\frac{1}{2}x + \frac{7}{2}$.

Check your understanding 2

Find the equation of the line passing through the points $P_1(2, 0)$ and $P_2(5, 3)$.

SOLUTION See page S-6.

Focus on finding the equation of a line given two points

Find the equation of the line containing $P_1(2, -3)$ and $P_2(2, 5)$.

SOLUTION

$$m = \frac{y_2 - y_1}{x_2 - x_1} = \frac{5 - (-3)}{2 - 2} = \frac{8}{0}$$

The slope is undefined. The graph is a vertical line.

The equation of the line is $x = 2$.

Check your understanding 3

Find the equation of the line containing $P_1(2, 3)$ and $P_2(-5, 3)$.

SOLUTION See page S-7.

Objective 5.4B Practice

For Exercises 1 to 5, find the equation of the line that contains the given points.

1. $P_1(0, -3), P_2(-4, 5)$

2. $P_1(-1, -2), P_2(3, 4)$

3. $P_1(-2, 3), P_2(2, -1)$

4. $P_1(3, -4), P_2(-2, -4)$

5. $P_1(-2, 5), P_2(-2, -5)$

Solutions on p. S-19.

Objective 5.4C Solve application problems by using linear models

Linear functions can be used to model a variety of applications in science and business. For each application, data are collected and the independent and dependent variables are selected. Then a linear function that models the data is determined.

Focus on solving an application

Suppose a manufacturer has determined that at a price of \$150, consumers will purchase 1 million portable music players, and at a price of \$125, consumers will purchase 1.25 million portable music players. Describe this situation with a linear function. Use the function to predict how many portable music players consumers will purchase if the price is \$80.

STRATEGY

- Select the independent and dependent variables. Because we are trying to determine the number of portable music players, that quantity is the *dependent* variable, y. The price of a portable music player is the *independent* variable, x.
- From the given data, two ordered pairs are (150, 1) and (125, 1.25). (The ordinates are in millions of units.) Use these ordered pairs to determine the linear function.
- Evaluate the function when $x = 80$ to predict how many portable music players consumers will purchase if the price is \$80.

SOLUTION

Choose $P_1(150, 1)$ and $P_2(125, 1.25)$.

$$m = \frac{y_2 - y_1}{x_2 - x_1} = \frac{1.25 - 1}{125 - 150} = \frac{0.25}{-25} = -0.01$$

$$y - y_1 = m(x - x_1)$$
$$y - 1 = -0.01(x - 150)$$
$$y = -0.01x + 2.50$$

The linear function is $f(x) = -0.01x + 2.50$.

$$f(80) = -0.01(80) + 2.50 = 1.7$$

Consumers will purchase 1.7 million portable music players at a price of \$80.

Check your understanding 4

Gabriel Daniel Fahrenheit invented the mercury thermometer in 1717. In terms of readings on this thermometer, water freezes at 32°F and boils at 212°F. In 1742, Anders Celsius invented the Celsius temperature scale. On this scale, water freezes at 0°C and boils at 100°C. Determine a linear function that can be used to predict the Celsius temperature when the Fahrenheit temperature is known.

SOLUTION See page S-7.

The table below shows the recorded maximum exercise heart rates of selected individuals of various ages who exercise regularly.

Age x, in Years	20	25	30	32	43	55	28	42	50	55	62
Maximum Heart Rate y, in Beats Per Minute	160	150	148	145	140	130	155	140	132	125	125

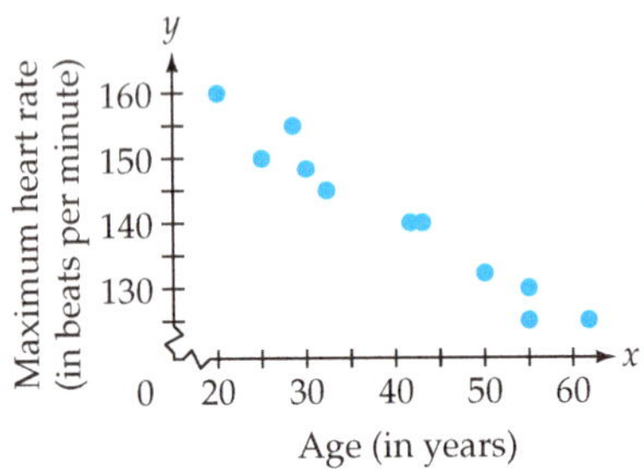

The graph at the left shows the ordered pairs of the table. These ordered pairs suggest that the maximum exercise heart rate of an individual decreases as the person's age increases.

Although these points do not lie on a straight line, it is possible to find a line that *approximately fits* the data. Select two data points and find the equation of the line that passes through the two points. Suppose we choose (20, 160) as P_1 and (62, 125) as P_2. Then the slope of the line between P_1 and P_2 is

$$m = \frac{y_2 - y_1}{x_2 - x_1} = \frac{125 - 160}{62 - 20} = -\frac{35}{42} = -\frac{5}{6}$$

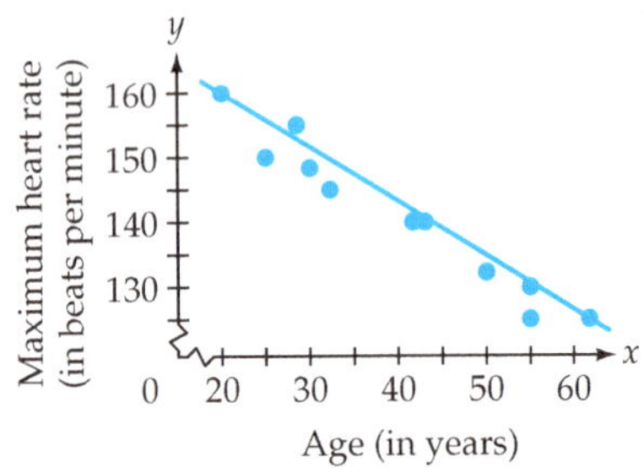

The graph of $y = -\frac{5}{6}x + \frac{530}{3}$

Now use the point–slope formula.

$$y - y_1 = m(x - x_1)$$

$$y - 160 = -\frac{5}{6}(x - 20)$$ • $m = -\frac{5}{6}$, $x_1 = 20$, $y_1 = 160$

$$y - 160 = -\frac{5}{6}x + \frac{50}{3}$$ • Use the distributive property to remove the parentheses on the right side of the equation.

$$y = -\frac{5}{6}x + \frac{530}{3}$$ • Add 160 to each side of the equation.

The graph of $y = -\frac{5}{6}x + \frac{530}{3}$ is shown at the left. This line *approximates* the data.

In the above example, if we had chosen different points, the result would have been a different equation. Among all the lines that might be chosen, statisticians generally pick the *least-squares line*, which is also called the *regression line*, as the line that most closely approximates the data. Regression lines are discussed in detail in a statistics course.

Focus on finding the equation of a line that approximates data

Sodium thiosulfate is used by photographers to develop some types of film. The amount of this chemical that will dissolve in water depends on the temperature of the water. The table below gives the number of grams of sodium thiosulfate that will dissolve in 100 ml of water at various temperatures.

Temperature x, in Degrees Celsius	20	35	50	60	75	90	100
Amount of Sodium Thiosulfate Dissolved y, in Grams	50	80	120	145	175	205	230

Use the points $P_1(20, 50)$ and $P_2(100, 230)$ to find the equation of a line that approximates the data. Use the equation to approximate the number of grams of sodium thiosulfate that will dissolve in 100 ml of water kept at a temperature of 70°C.

STRATEGY

- Find the slope of the line that passes through the given points.
- Use the point-slope formula to find the equation of the line.
- Substitute the given temperature into the equation, and sove for y.

SOLUTION

$$m = \frac{y_2 - y_1}{x_2 - x_1} = \frac{230 - 50}{100 - 20} = \frac{180}{80} = \frac{9}{4}$$

$$y - y_1 = m(x - x_1)$$
$$y - 50 = \frac{9}{4}(x - 20)$$
$$y - 50 = \frac{9}{4}x - 45$$
$$y = \frac{9}{4}x + 5$$

The equation of a line that approximates the data is $y = \frac{9}{4}x + 5$.

Replace x by 70 and solve for y.

$$y = \frac{9}{4}x + 5$$
$$y = \frac{9}{4}(70) + 5$$
$$= 157.5 + 5$$
$$= 162.5$$

The equation predicts that approximately 163 g of sodium thiosulfate will dissolve at a temperature of 70°C.

Check your understanding 5

The table below shows the data collected by a chemistry student who is trying to determine a relationship between the temperature T, in degrees Celsius, and the volume V, in liters, of 1 g of oxygen at a constant pressure. Chemists refer to this relationship as *Charles's law.*

Temperature *T*, in Degrees Celsius	−100	−75	−50	−25	0	25	50
Volume *V*, in Liters	0.43	0.5	0.57	0.62	0.7	0.75	0.81

Use the points $P_1(-100, 0.43)$ and $P_2(50, 0.81)$ to find the equation of a line that approximates the data. Use the equation to find the volume of 1 g of oxygen when the temperature is −30°C. Round to the nearest hundredth.

SOLUTION See page S-7.

Objective 5.4C Practice

1. A jogger running at 9 mph burns approximately 14 Calories per minute.

a. Write a linear function for the number of Calories burned by the jogger in terms of the number of minutes run.

b. Use your function to find the number of Calories the jogger has burned after jogging for 32 min.

2. A cellular phone company offers several different service options. One option, for people who plan on using the phone only in emergencies, costs the user \$4.95 per month plus \$.59 per minute for each minute the phone is used.
   a. Write a linear function for the monthly cost of the phone in terms of the number of minutes the phone is used.
   b. Use your function to find the cost of using the cellular phone for 13 min in one month.
3. At sea level, the boiling point of water is 100°C. At an altitude of 2 km, the boiling point of water is 93°C.
   a. Write a linear function for the boiling point of water in terms of the altitude above sea level.
   b. Use your function to predict the boiling point of water on top of Mount Everest, which is approximately 8.85 km above sea level. Round to the nearest degree.
4. A research hospital did a study on the relationship between stress and diastolic blood pressure. The results from eight patients in the study are given in the table below. Blood pressure measurements are given in milliliters of mercury.

Stress Test Score, x	55	62	58	78	92	88	75	80
Blood Pressure, y	70	85	72	85	96	90	82	85

   Use the points $P_1(55, 70)$ and $P_2(92, 96)$ to find the equation of a line that approximates the data. Use the equation to predict the blood pressure of a patient whose stress test score was 70.

Solutions on pp. S-20–S-21.

SECTION

5.5 Quadratic Functions

Objective 5.5A Graph a quadratic function

Quadratic Function

A **quadratic function** is a function that can be expressed by the equation $f(x) = ax^2 + bx + c, a \neq 0$.

EXAMPLES

1. $f(x) = 2x^2 - 3x + 4$ $\quad a = 2, b = -3, c = 4$
2. $g(x) = x^2 + 4x$ $\quad a = 1, b = 4, c = 0$
3. $h(x) = 6 - x^2$ $\quad a = -1, b = 0, c = 6$

The graph of a quadratic function can be drawn by finding ordered pairs that belong to the function.

Graph: $f(x) = x^2 - 2x - 3$

Evaluate the function for various values of x. Find enough ordered pairs to determine the shape of the graph.

x	$f(x) = x^2 - 2x - 3$	$f(x)$	(x, y)
−2	$f(-2) = (-2)^2 - 2(-2) - 3$	5	(−2, 5)
−1	$f(-1) = (-1)^2 - 2(-1) - 3$	0	(−1, 0)
0	$f(0) = (0)^2 - 2(0) - 3$	−3	(0, −3)
1	$f(1) = (1)^2 - 2(1) - 3$	−4	(1, −4)
2	$f(2) = (2)^2 - 2(2) - 3$	−3	(2, −3)
3	$f(3) = (3)^2 - 2(3) - 3$	0	(3, 0)
4	$f(4) = (4)^2 - 2(4) - 3$	5	(4, 5)

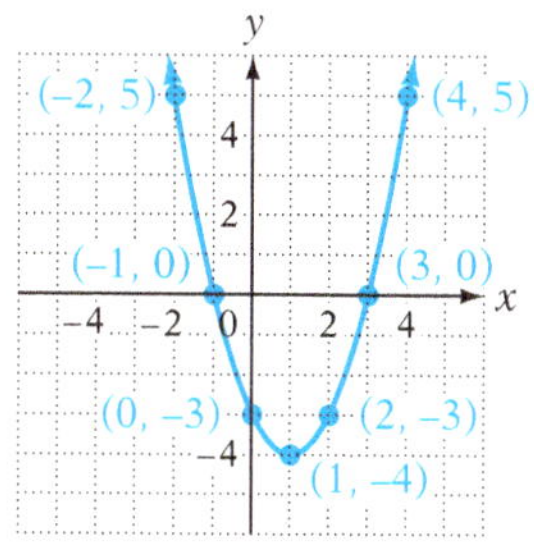

$f(x) = x^2 - 2x - 3$

The graph of a quadratic function, called a **parabola,** is "cup" shaped, as shown above. The cup can open up or down. The parabola opens up when $a > 0$ and opens down when $a < 0$.

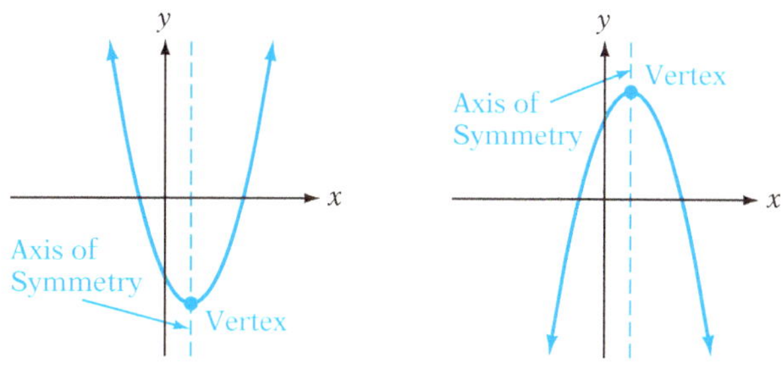

The **vertex** of a parabola is the point with the least y-coordinate when $a > 0$ and the point with the greatest y-coordinate when $a < 0$. The line that passes through the vertex and is parallel to the y-axis is called the **axis of symmetry.** To understand the axis of symmetry, think of folding the graph along that line. The two halves of the graph would match up.

Find the coordinates of the vertex and the equation of the axis of symmetry of the graph of $F(x) = x^2 + 4x + 3$.

$F(x) = x^2 + 4x + 3$

$F(x) = (x^2 + 4x) + 3$

• To find the coordinates of the vertex, complete the square. Group the variable terms.

$F(x) = (x^2 + 4x + 4) - 4 + 3$

• Complete the square of $x^2 + 4x$. Add and subtract $\left[\frac{1}{2}(4)\right]^2 = 4$ to and from $x^2 + 4x$.

$F(x) = (x + 2)^2 - 1$

• Factor and combine like terms.

Because a, the coefficient of x^2, is positive $(a = 1)$, the parabola opens up and the vertex is the point with the least y-coordinate. Because $(x + 2)^2 \geq 0$ for all values of x, the least y-coordinate occurs when $(x + 2)^2 = 0$. When $x = -2$, this expression equals zero.

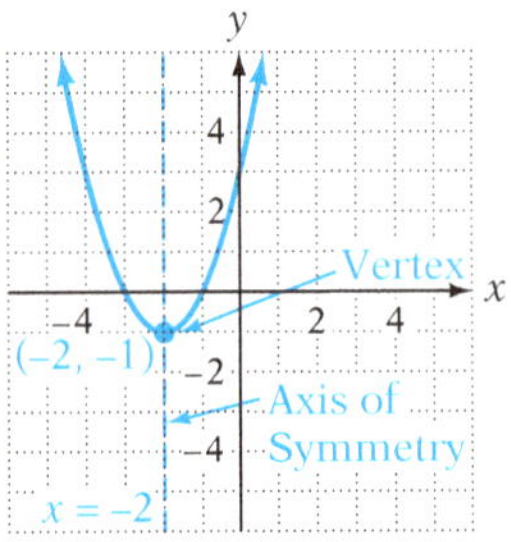

The x-coordinate of the vertex is -2.

$$F(x) = (x + 2)^2 - 1$$

$$F(-2) = (-2 + 2)^2 - 1 = -1$$

• To find the y-coordinate of the vertex, evaluate the function at $x = -2$.

The y-coordinate of the vertex is -1.

From these results, the coordinates of the vertex are $(-2, -1)$. The equation of the axis of symmetry is $x = -2$.

By following the process illustrated in the above example and completing the square of $f(x) = ax^2 + bx + c$, we can find a formula for the coordinates of the vertex of a parabola.

Vertex and Axis of Symmetry of a Parabola

Let $f(x) = ax^2 + bx + c$, $a \neq 0$, be the equation of a parabola. The coordinates of the vertex are $\left(-\frac{b}{2a}, f\left(-\frac{b}{2a}\right)\right)$. The equation of the axis of symmetry is $x = -\frac{b}{2a}$.

EXAMPLE

Let $f(x) = -2x^2 + 8x + 5$. For this equation, $a = -2$, $b = 8$, and $c = 5$.

x-coordinate of the vertex: $-\frac{b}{2a} = -\frac{8}{2(-2)} = 2$

y-coordinate of the vertex: $f\left(-\frac{b}{2a}\right) = f(2) = -2(2)^2 + 8(2) + 5 = 13$

The coordinates of the vertex are $(2, 13)$.

The equation of the axis of symmetry is $x = 2$.

Focus on finding the vertex and axis of symmetry of a parabola

Find the coordinates of the vertex and the equation of the axis of symmetry for the parabola with equation $y = x^2 + 2x - 3$. Then sketch its graph.

SOLUTION

$x = -\frac{b}{2a} = -\frac{2}{2(1)} = -1$ • Find the x-coordinate of the vertex. $a = 1$ and $b = 2$.

$y = x^2 + 2x - 3$ • Find the y-coordinate of the vertex by replacing x with -1 and solving for y.
$y = (-1)^2 + 2(-1) - 3$
$y = 1 - 2 - 3 = -4$

The coordinates of the vertex are $(-1, -4)$. The equation of the axis of symmetry is $x = -1$.

Find some ordered-pair solutions of the equation and record these in a table. Because the graph is symmetric with respect to the line with equation $x = -1$, choose values of x greater than -1.

x	$y = x^2 + 2x - 3$
0	-3
1	0
2	5

Graph the ordered-pair solutions on a rectangular coordinate system. Use symmetry to locate points of the graph on the other side of the axis of symmetry. Remember that corresponding points on the graph are the same distance from the axis of symmetry.

Draw a parabola through the points.

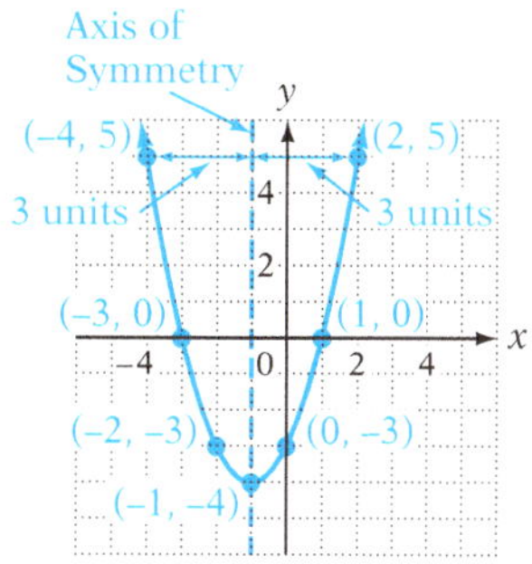

Check your understanding 1

Find the coordinates of the vertex and the equation of the axis of symmetry for the parabola with equation $y = x^2 - 2$. Then sketch its graph.

SOLUTION See page S-8.

Because $f(x) = ax^2 + bx + c$ is a real number for all real numbers x, the domain of a quadratic function is all real numbers. The range of a quadratic function can be determined from the y-coordinate of the vertex.

Focus on finding the domain and range of a quadratic function

Graph $f(x) = -2x^2 - 4x + 3$. State the domain and range of f.

SOLUTION

Because a is negative $(a = -2)$, the graph of f will open down.
The x-coordinate of the vertex is

$$x = -\frac{b}{2a} = -\frac{-4}{2(-2)} = -1.$$

The y-coordinate of the vertex is
$f(-1) = -2(-1)^2 - 4(-1) + 3 = 5.$
The coordinates of the vertex are $(-1, 5)$.

Evaluate $f(x)$ for various values of x, and use symmetry to draw the graph.

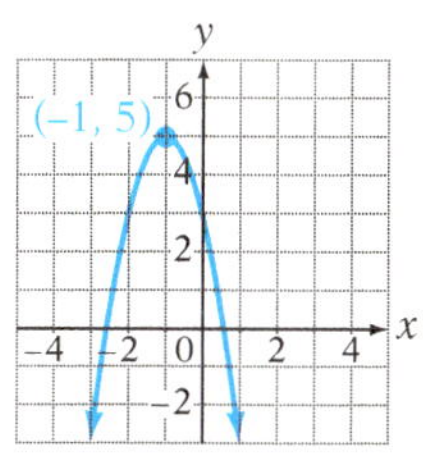

Because $f(x) = -2x^2 - 4x + 3$ is a real number for all values of x, the domain of f is $\{x|x \in \text{real numbers}\}$. The vertex of the parabola is the highest point on the graph. Because the y-coordinate at that point is 5, the range of f is $\{y|y \leq 5\}$.

Check your understanding 2

Graph $g(x) = x^2 + 4x - 2$. State the domain and range of the function.

SOLUTION See page S-8.

Objective 5.5A Practice

For Exercises 1 to 3, find the coordinates of the vertex and the equation of the axis of symmetry for the parabola with the given equation. Then sketch its graph.

1. $y = x^2 - 2$

2. $y = -\frac{1}{2}x^2 + 2$

3. $y = x^2 - 3x + 2$

4. Graph $f(x) = -2x^2 - 3x + 2$. State the domain and range of the function.

5. Graph $f(x) = x^2 + 4x - 3$. State the domain and range of the function.

Solutions on pp. S-21–S-22.

Objective 5.5B Find the x-intercepts of a quadratic function

Recall that a point at which a graph crosses the x- or y-axis is called an *intercept* of the graph. The x-intercepts of the graph of an equation occur when $y = 0$; the y-intercepts occur when $x = 0$.

The graph of $y = x^2 + 3x - 4$ is shown at the right. The points whose coordinates are $(-4, 0)$ and $(1, 0)$ are x-intercepts of the graph.

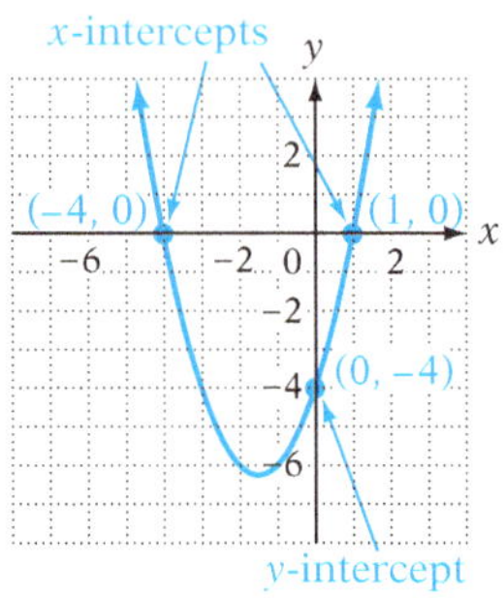

Focus on finding the x-intercepts of a quadratic function

Find the coordinates of the x-intercepts of the parabola whose equation is $y = 2x^2 - x - 6$.

SOLUTION

$y = 2x^2 - x - 6$

$0 = 2x^2 - x - 6$ • To find the x-intercepts, let $y = 0$.

$0 = (2x + 3)(x - 2)$ • Factor.

$2x + 3 = 0 \qquad x - 2 = 0$ • Use the Principle of Zero Products.

$x = -\frac{3}{2} \qquad x = 2$

The coordinates of the x-intercepts are $(-\frac{3}{2}, 0)$ and $(2, 0)$. See the graph at the left.

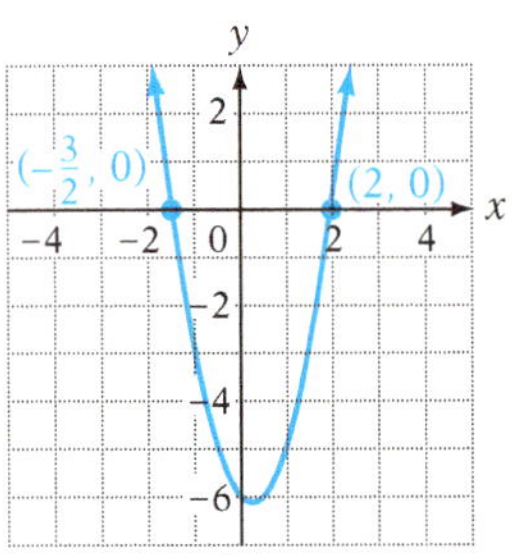

$y = 2x^2 - x - 6$

Check your understanding 3

Find the coordinates of the x-intercepts of the parabola whose equation is $f(x) = x^2 + 2x - 8$.

SOLUTION See page S-8.

If $ax^2 + bx + c = 0$ has a double root, then the graph of $y = ax^2 + bx + c$ intersects the x-axis at one point. In that case, the graph is said to be **tangent** to the x-axis.

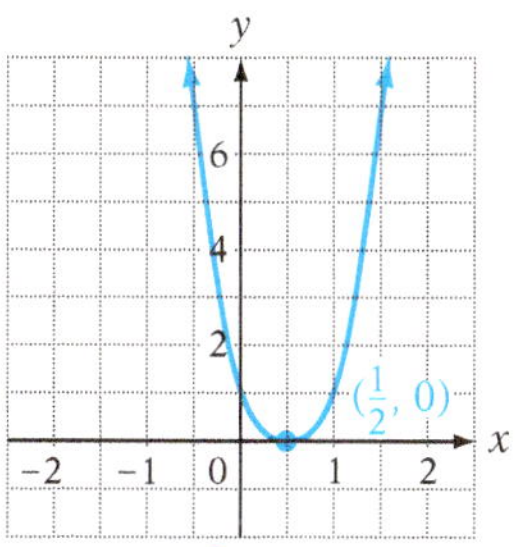

$y = 4x^2 - 4x + 1$

Find the coordinates of the x-intercept of the parabola whose equation is $y = 4x^2 - 4x + 1$.

$y = 4x^2 - 4x + 1$

$0 = 4x^2 - 4x + 1$ • To find the x-intercept, let $y = 0$.

$0 = (2x - 1)(2x - 1)$ • Solve for x by factoring and using the Principle of Zero Products.

$2x - 1 = 0 \qquad 2x - 1 = 0$

$2x = 1 \qquad 2x = 1$

$x = \frac{1}{2} \qquad x = \frac{1}{2}$

The coordinates of the x-intercept are $(\frac{1}{2}, 0)$. See the graph at the left.

Objective 5.5B Practice

For Exercises 1 to 5, find the coordinates of the x-intercepts of the parabola given by the equation.

1. $y = x^2 - 4$
2. $y = 3x^2 + 6x$
3. $y = 3x^2 - 19x - 14$
4. $y = x^2 + 4x - 3$
5. $y = -x^2 - 4x - 5$

Solutions on pp. S-22–S-23.

Objective 5.5C Find the minimum or maximum of a quadratic function

The graph of $f(x) = x^2 - 2x + 3$ is shown at the right. Because a is positive, the parabola opens up. The vertex of the parabola is the lowest point on the parabola. It is the point that has the minimum y-coordinate. Therefore, the value of the function at this point is a **minimum.**

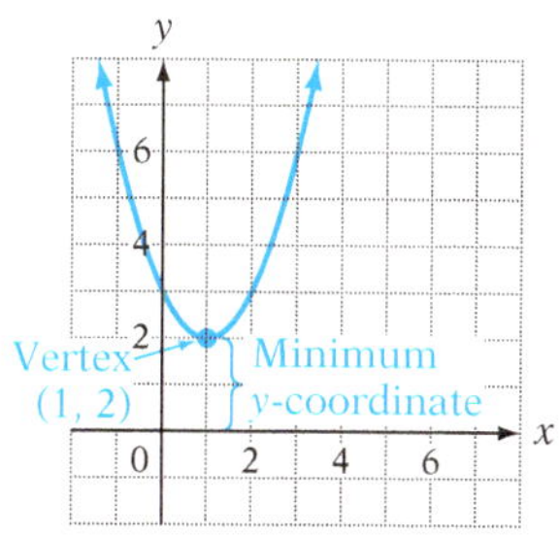

The graph of $f(x) = -x^2 + 2x + 1$ is shown at the right. Because a is negative, the parabola opens down. The vertex of the parabola is the highest point on the parabola. It is the point that has the maximum y-coordinate. Therefore, the value of the function at this point is a **maximum.**

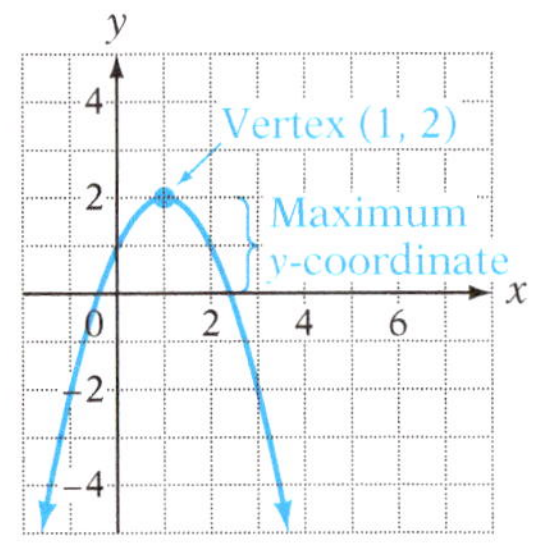

To find the minimum or maximum value of a quadratic function, first find the x-coordinate of the vertex. Then evaluate the function at that value.

Focus on finding the minimum or maximum value of a function

Find the maximum or minimum value of the function $f(x) = -2x^2 + 4x + 3$.

SOLUTION

$x = -\frac{b}{2a} = -\frac{4}{2(-2)} = 1$ • Find the x-coordinate of the vertex. $a = -2$, $b = 4$

$f(x) = -2x^2 + 4x + 3$

$f(1) = -2(1)^2 + 4(1) + 3$ • Evaluate the function at $x = 1$.

$f(1) = 5$

Because $a < 0$, the graph of f opens down. Therefore, the function has a maximum value.

The maximum value of the function is 5. See the graph at the left.

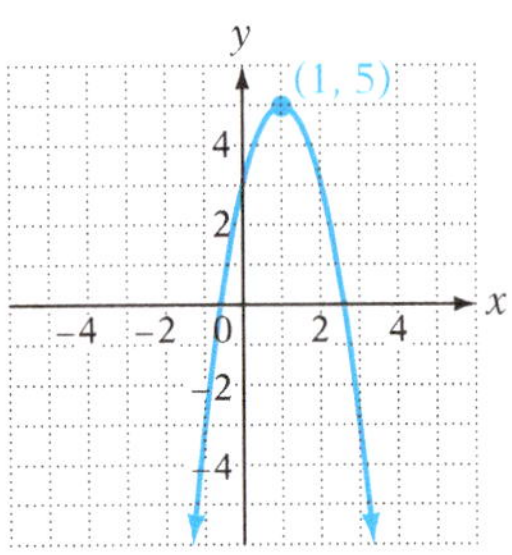

$f(x) = -2x^2 + 4x + 3$

Check your understanding 4

Find the maximum or minimum value of the function $f(x) = 2x^2 - 3x + 1$.

SOLUTION See page S-8.

Objective 5.5C Practice

For Exercises 1 to 4, find the maximum or minimum value of the function.

1. $f(x) = x^2 - 2x + 3$

2. $f(x) = -2x^2 + 4x - 3$

3. $f(x) = -2x^2 + 4x - 5$

4. $f(x) = 2x^2 + 3x - 8$

Solutions on p. S-24.

Objective 5.5D Solve applications of quadratic functions

A mason is forming a rectangular floor for a storage shed. The perimeter of the rectangle is 44 ft. What dimensions of the rectangle will give the floor a maximum area? What is the maximum area?

We are given the perimeter of the rectangle, and we want to find the dimensions of the rectangle that will yield the maximum area for the floor. Use the equation for the perimeter of a rectangle.

$$P = 2L + 2W$$
$$44 = 2L + 2W \quad \bullet\ P = 44$$
$$22 = L + W \quad \bullet\ \text{Divide both sides of the equation by 2.}$$
$$22 - L = W \quad \bullet\ \text{Solve for } W.$$

Now use the equation for the area of a rectangle. Use substitution to express the area in terms of L.

$$A = LW$$
$$A = L(22 - L) \quad \bullet\ \text{From the equation above, } W = 22 - L.$$
$$A = 22L - L^2$$

The area of the rectangle is $22L - L^2$. To find the length of the rectangle, find the L-coordinate of the vertex of the function $f(L) = -L^2 + 22L$.

$$L = -\frac{b}{2a} = -\frac{22}{2(-1)} = 11 \quad \bullet\ \text{For the function } f(L) = -L^2 + 22L,\ a = -1 \text{ and } b = 22.$$

The length of the rectangle is 11 ft.

$$W = 22 - L$$
$$W = 22 - 11 = 11$$

• To find the width, replace L in $22 - L$ by 11, the L-coordinate of the vertex, and evaluate.

The width of the rectangle is 11 ft.

The dimensions of the rectangle that would give the floor a maximum area are 11 ft by 11 ft.

$$f(L) = -L^2 + 22L$$
$$f(11) = -(11)^2 + 22(11)$$
$$= -121 + 242 = 121$$

• To find the maximum area of the floor, evaluate the function $f(L) = -L^2 + 22L$ at 11, the L-coordinate of the vertex.

The maximum area of the floor is 121 ft^2.

The graph of the function $f(L) = -L^2 + 22L$ is shown at the right. Note that the vertex of the parabola is (11, 121). For any value of L less than 11, the area of the floor will be less than 121 ft^2. For any value of L greater than 11, the area of the floor will be less than 121 ft^2. 121 is the maximum value of the function, and the maximum value occurs when $L = 11$.

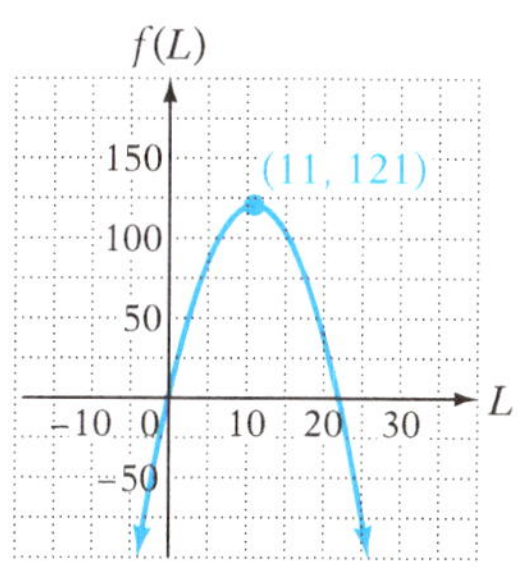

Focus on solving an application

A mining company has determined that the cost c, in dollars per ton, of mining a mineral is given by

$$c(x) = 0.2x^2 - 2x + 12$$

where x is the number of tons of the mineral that are mined. Find the number of tons of the mineral that should be mined to minimize the cost. What is the minimum cost?

STRATEGY

- To find the number of tons that will minimize the cost, find the x-coordinate of the vertex.
- To find the minimum cost, evaluate the function at the x-coordinate of the vertex.

SOLUTION

$$x = -\frac{b}{2a} = -\frac{-2}{2(0.2)} = 5$$

To minimize the cost, 5 tons should be mined.

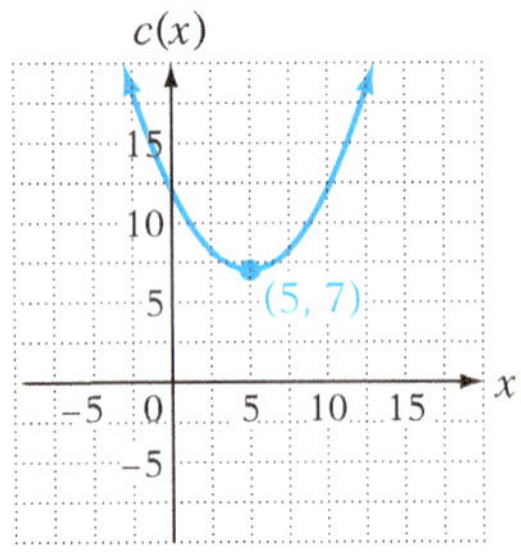

$c(x) = 0.2x^2 - 2x + 12$

$c(5) = 0.2(5)^2 - 2(5) + 12 = 5 - 10 + 12 = 7$

The minimum cost per ton is \$7.

Note: The graph of the function $c(x) = 0.2x^2 - 2x + 12$ is shown at the left. The vertex of the parabola is $(5, 7)$. For any value of x less than 5, the cost per ton is greater than \$7. For any value of x greater than 5, the cost per ton is greater than \$7. 7 is the minimum value of the function, and the minimum value occurs when $x = 5$.

Check your understanding 5

The height s, in feet, of a ball thrown straight up is given by $s(t) = -16t^2 + 64t$, where t is the time in seconds. Find the time it takes the ball to reach its maximum height. What is the maximum height?

SOLUTION See page S-9.

Focus on solving a problem by writing a quadratic equation

Find two numbers whose difference is 10 and whose product is a minimum.

STRATEGY

- Let x and y represent the two numbers.
- Express y in terms of x.

$y - x = 10$ • The difference of the numbers is 10.

$y = x + 10$ • Solve for y.

- Express the product of the numbers in terms of x.

$xy = x(x + 10)$ • $y = x + 10$

$f(x) = x^2 + 10x$ • The function f represents the product of the two numbers.

- To find one of the two numbers, find the x-coordinate of the vertex of $f(x) = x^2 + 10x$.
- To find the other number, replace x in $x + 10$ by the x-coordinate of the vertex and evaluate.

SOLUTION

$$x = -\frac{b}{2a} = -\frac{10}{2(1)} = -5$$

$x + 10 = -5 + 10 = 5$

The numbers are -5 and 5.

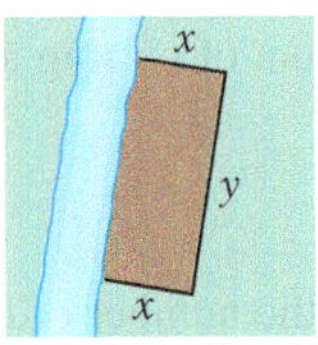

Check your understanding 6

A rectangular area is being fenced along a stream to enclose a picnic area. If 100 ft of fencing is available, what dimensions of the rectangle will produce the maximum area for picnicking? See the figure at the left.

SOLUTION See page S-9.

Objective 5.5D Practice

1. An event in the Summer Olympics is 10-meter springboard diving. In this event, the height s, in meters, of a diver above the water t seconds after jumping is given by $s(t) = -4.9t^2 + 7.8t + 10$. What is the maximum height that the diver will reach? Round to the nearest tenth.
2. A manufacturer of microwave ovens believes that the revenue R, in dollars, the company receives is related to the price P, in dollars, of an oven by $R(P) = 125P - 0.25P^2$. What price will give the maximum revenue?
3. The Buckingham Fountain in Chicago shoots water from a nozzle at the base of the fountain. The height h, in feet, of the water above the ground t seconds after it leaves the nozzle is given by $h(t) = -16t^2 + 90t + 15$. What is the maximum height of the water? Round to the nearest tenth.
4. The suspension cable that supports a small footbridge hangs in the shape of a parabola. The height h, in feet, of the cable above the bridge is given by $h(x) = 0.25x^2 - 0.8x + 25$, where x is the distance in feet from one end of the bridge. What is the minimum height of the cable above the bridge?

Solutions on pp. S-25–S-26.

SECTION 5.6 Exponential Functions

Objective 5.6A Graph an exponential function

Data suggest that since the year 2000, the number of Internet users worldwide has been increasing at a rate of approximately 17% per year. The graph at the right shows this growth. This graph depicts an example of **exponential growth.**

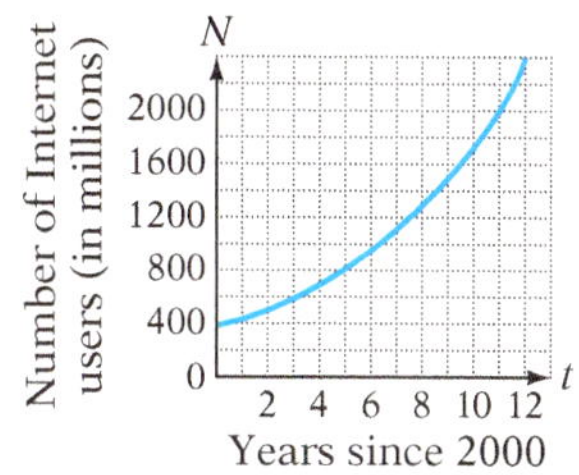

Nuclear medicine physicians use radioisotopes for the diagnosis and treatment of certain diseases. One of the most widely used isotopes is technetium-99m. One use of this isotope is in the diagnosis of cardiovascular disease. The graph at the right shows the amount of technetium-99m in a patient after its injection into the patient. This graph depicts an example of **exponential decay.**

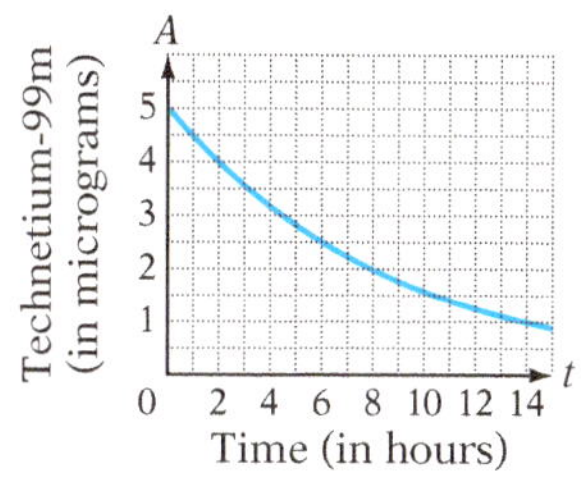

Definition of an Exponential Function

The **exponential function** with **base *b*** is defined by

$$f(x) = b^x$$

where $b > 0$, $b \neq 1$, and x is any real number.

In the definition of an exponential function, b, the base, is required to be positive. If the base were a negative number, the value of the function would be a complex number for some values of x. For instance, the value of $f(x) = (-4)^x$ when $x = \frac{1}{2}$ is $f\left(\frac{1}{2}\right) = (-4)^{\frac{1}{2}} = \sqrt{-4} = 2i$. To avoid complex number values of a function, the base of the exponential function is always a positive number.

Evaluate $f(x) = 2^x$ at $x = 3$ and $x = -2$.

$f(3) = 2^3 = 8$ • Substitute 3 for x and simplify.

$f(-2) = 2^{-2} = \dfrac{1}{2^2} = \dfrac{1}{4}$ • Substitute -2 for x and simplify.

To evaluate an exponential expression at an irrational number such as $\sqrt{2}$, we obtain an approximation to the value of the function by approximating the irrational number. For instance, the value of $f(x) = 4^x$ when $x = \sqrt{2}$ can be approximated by using an approximation of $\sqrt{2}$.

$$f(\sqrt{2}) = 4^{\sqrt{2}} \approx 4^{1.4142} \approx 7.1029$$

Because $f(x) = b^x$ $(b > 0, b \neq 1)$ can be evaluated at both rational and irrational numbers, the domain of f is all real numbers. And because $b^x > 0$ for all values of x, the range of f is the positive real numbers.

Focus on evaluating an exponential function

Evaluate $f(x) = 2^{3x-1}$ at $x = 1$ and $x = -1$.

SOLUTION

$$f(x) = 2^{3x-1}$$
$$f(1) = 2^{3(1)-1} = 2^2 = 4$$
$$f(-1) = 2^{3(-1)-1} = 2^{-4} = \frac{1}{2^4} = \frac{1}{16}$$

Check your understanding 1

Evaluate $f(x) = 2^{2x+1}$ at $x = 0$ and $x = -2$.

SOLUTION See page S-9.

A frequently used base in applications of exponential functions is an irrational number designated by e. The number e is approximately 2.71828183. It is an irrational number, so it has a nonterminating, nonrepeating decimal representation.

Natural Exponential Function
The function defined by $f(x) = e^x$ is called the **natural exponential function.**

The $\boxed{e^x}$ key on a calculator can be used to evaluate the natural exponential function.

Focus on evaluating the natural exponential function

Evaluate $f(x) = e^x$ at $x = 2$, $x = -3$, and $x = \pi$. Round to the nearest ten-thousandth.

SOLUTION

$$f(x) = e^x$$
$$f(2) = e^2 \approx 7.3891$$
$$f(-3) = e^{-3} \approx 0.0498$$
$$f(\pi) = e^{\pi} \approx 23.1407$$

Check your understanding 2

Evaluate $f(x) = e^x$ at $x = 1.2$, $x = -2.5$, and $x = e$. Round to the nearest ten-thousandth.

SOLUTION See page S-10.

Some of the properties of an exponential function can be seen by considering its graph.

Graph: $f(x) = 2^x$

To graph $f(x) = 2^x$, think of the function as the equation $y = 2^x$. Choose values of x, and find the corresponding values of y. The results can be recorded in a table, as shown below.

Graph the ordered pairs on a rectangular coordinate system. Connect the points with a smooth curve.

x	$f(x) = 2^x$	y
-2	$f(-2) = 2^{-2} = \frac{1}{2^2} = \frac{1}{4}$	$\frac{1}{4}$
-1	$f(-1) = 2^{-1} = \frac{1}{2}$	$\frac{1}{2}$
0	$f(0) = 2^0 = 1$	1
1	$f(1) = 2^1 = 2$	2
2	$f(2) = 2^2 = 4$	4
3	$f(3) = 2^3 = 8$	8

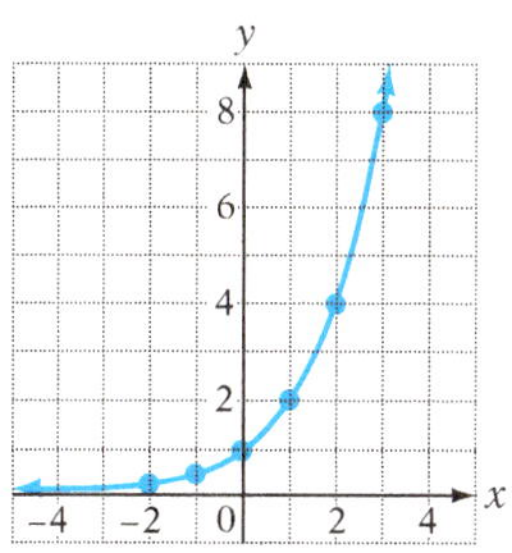

Note that a vertical line would intersect the graph in the example above at only one point. Therefore, by the vertical line test, the graph of $f(x) = 2^x$ is the graph of a function. Also note that a horizontal line would intersect the graph at only one point. Therefore, the graph of $f(x) = 2^x$ is the graph of a 1–1 function.

Graph: $f(x) = \left(\frac{1}{2}\right)^x$

Think of the function as the equation $y = \left(\frac{1}{2}\right)^x$. Choose values of x, and find the corresponding values of y.

Graph the ordered pairs on a rectangular coordinate system. Connect the points with a smooth curve.

x	$f(x) = \left(\frac{1}{2}\right)^x$	y
-3	$f(-3) = \left(\frac{1}{2}\right)^{-3} = 2^3 = 8$	8
-2	$f(-2) = \left(\frac{1}{2}\right)^{-2} = 2^2 = 4$	4
-1	$f(-1) = \left(\frac{1}{2}\right)^{-1} = 2^1 = 2$	2
0	$f(0) = \left(\frac{1}{2}\right)^{0} = 1$	1
1	$f(1) = \left(\frac{1}{2}\right)^{1} = \frac{1}{2}$	$\frac{1}{2}$
2	$f(2) = \left(\frac{1}{2}\right)^{2} = \frac{1}{4}$	$\frac{1}{4}$

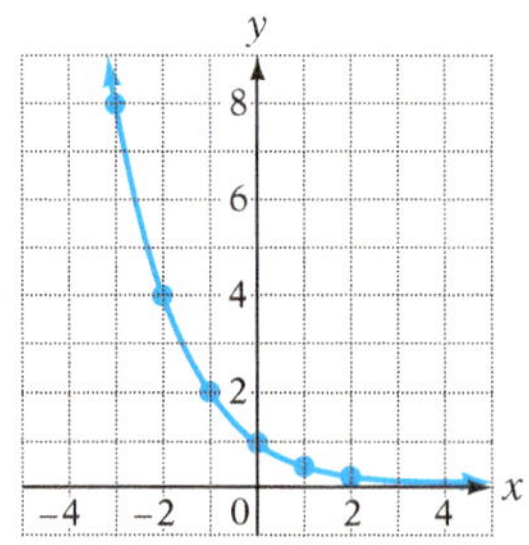

Focus on graphing an exponential function

Graph. **a.** $f(x) = 3^{\frac{1}{2}x-1}$ **b.** $f(x) = 2^x - 1$

SOLUTION

a.

x	$y = 3^{\frac{1}{2}x-1}$
-2	$\frac{1}{9}$
0	$\frac{1}{3}$
2	1
4	3

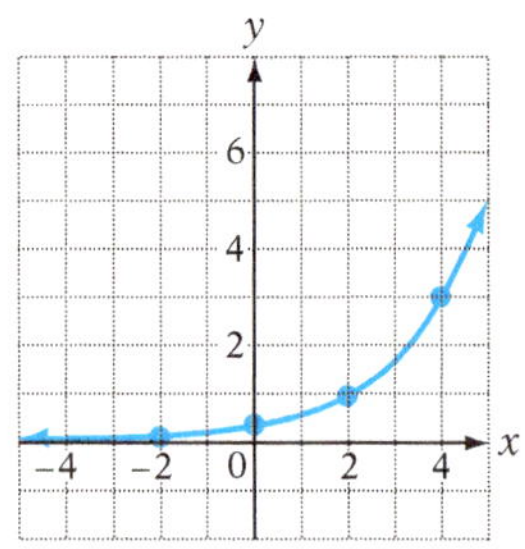

b.

x	$y = 2^x - 1$
-2	$-\frac{3}{4}$
-1	$-\frac{1}{2}$
0	0
1	1
2	3
3	7

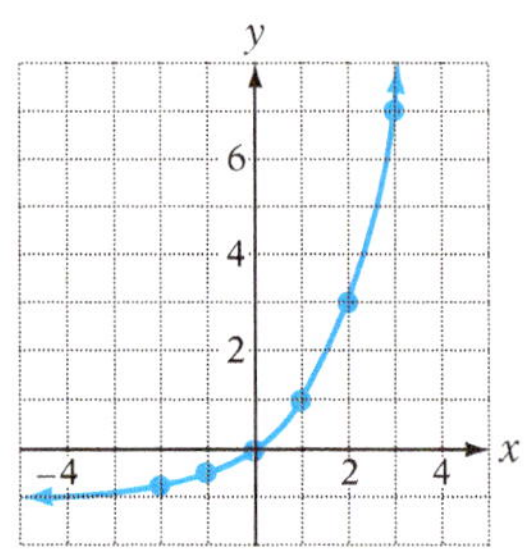

Check your understanding 3

Graph. **a.** $f(x) = 2^{-\frac{1}{2}x}$ **b.** $f(x) = 2^x + 1$

SOLUTION See page S-10.

Focus on graphing a natural exponential function

Graph. **a.** $f(x) = e^{x-2}$ **b.** $f(x) = e^x - 1$

SOLUTION

a.

x	$y = e^{x-2}$
-3	0.007
-2	0.02
-1	0.05
0	0.14
1	0.37
2	1
3	2.72
4	7.39

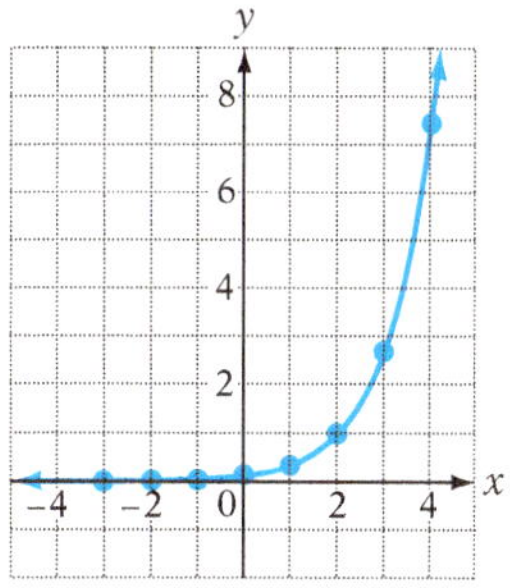

b.

x	$y = e^x - 1$
-2	-0.86
-1	-0.63
0	0
1	1.72
2	6.39

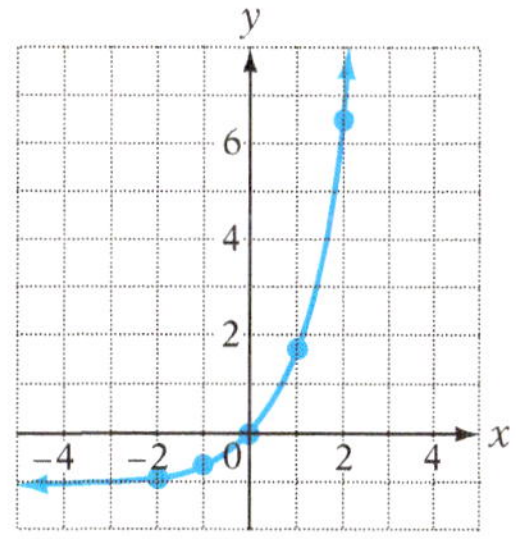

Check your understanding 4

Graph. **a.** $f(x) = 2^{-x} + 2$ **b.** $f(x) = e^{-2x} - 4$

SOLUTION See page S-10.

Objective 5.6A Practice

1. Given $f(x) = 2^{-2x}$, evaluate:

a. $f(-3)$ **b.** $f(2)$ **c.** $f(0)$

2. Given $P(t) = e^{-\frac{1}{2}t}$, evaluate the following. Round to the nearest ten-thousandth.

a. $P(-3)$ **b.** $P(4)$ **c.** $P\left(\frac{1}{2}\right)$

3. Graph: $f(x) = 2^{x-1}$

4. Graph: $f(x) = \left(\frac{2}{3}\right)^x$

5. Graph: $f(x) = 2^x - 3$

6. Graph: $f(x) = e^{x-1}$

Solutions on p. S-26.

Objective 5.6B Solve applications of exponential functions

Focus on solving an application involving an exponential function

When an amount of money P is placed in an account that earns compound interest, the value A of the money after t years is given by the compound interest formula

$$A = P\left(1 + \frac{r}{n}\right)^{nt}$$

where r is the annual interest rate as a decimal and n is the number of compounding periods per year. Suppose \$500 is placed in an account that earns 8% interest compounded daily. Find the value of the investment after 5 years.

SOLUTION

Use the compound interest formula. Because interest is compounded daily, $n = 365$.

$$A = P\left(1 + \frac{r}{n}\right)^{nt}$$

$$= 500\left(1 + \frac{0.08}{365}\right)^{365(5)} \quad \text{• } P = 500, r = 0.08, n = 365, t = 5$$

$$\approx 500(1.491759) \approx 745.88$$

After 5 years, the value of the investment is \$745.88.

Check your understanding 5

The radioactive isotope iodine-131 is used to monitor thyroid activity. The number of grams N of iodine-131 in the body t hours after an injection is given by $N(t) = 1.5\left(\frac{1}{2}\right)^{t/193.7}$. Find the number of grams of the isotope in the body 24 h after an injection. Round to the nearest ten-thousandth of a gram.

SOLUTION See page S-11.

The next example is based on Newton's law of cooling. This exponential function can be used to model the temperature of something that is being cooled.

Focus on solving an application involving an exponential function

A cup of coffee is heated to 160°F and placed in a room that maintains a temperature of 70°F. The temperature of the coffee after t minutes is given by $T(t) = 70 + 90e^{-0.0485t}$. Find the temperature of the coffee 20 min after it is placed in the room. Round to the nearest degree.

SOLUTION

Evaluate the function $T(t) = 70 + 90e^{-0.0485t}$ at $t = 20$.

$$T(t) = 70 + 90e^{-0.0485t}$$

$$T(20) = 70 + 90e^{-0.0485(20)} \quad \text{• Substitute 20 for } t.$$

$$\approx 70 + 34.1$$

$$\approx 104.1$$

After 20 min, the temperature of the coffee is about 104°F.

Check your understanding 6

The function $A(t) = 200e^{-0.014t}$ gives the amount of aspirin, in milligrams, in a patient's bloodstream t minutes after the aspirin has been administered. Find the amount of aspirin in the patient's bloodstream after 45 min. Round to the nearest milligram.

SOLUTION See page S-11.

Objective 5.6B Practice

In Exercises 1 and 2, use the compound interest formula $A = P\left(1 + \frac{r}{n}\right)^{nt}$, where P is the amount deposited, A is the value of the money after t years, r is the annual interest rate as a decimal, and n is the number of compounding periods per year.

1. A computer network specialist deposits \$2500 into a retirement account that earns 7.5% annual interest, compounded daily. What is the value of the investment after 20 years?

2. A \$10,000 certificate of deposit (CD) earns 5% annual interest, compounded daily. What is the value of the investment after 20 years?

3. An isotope of technetium is used to prepare images of internal body organs. This isotope has a half-life (time required for half the material to erode) of approximately 6 h. If a patient is injected with 30 mg of this isotope, what will be the technetium level in the patient after 3 h? Use the function $A(t) = 30\left(\frac{1}{2}\right)^{t/6}$, where A is the technetium level, in milligrams, in the patient after t hours. Round to the nearest tenth of a milligram.

Solutions on p. S-27.

SOLUTIONS TO MODULE 5

Solutions to Check Your Understanding

Section 5.1 Check your understanding 1

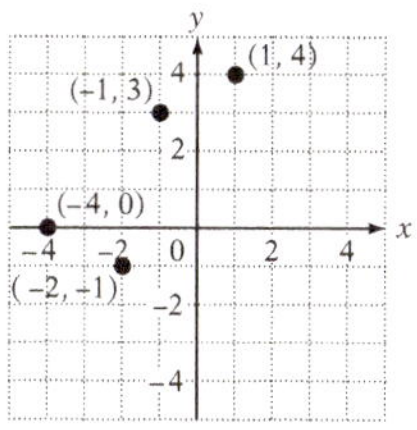

Check your understanding 2

$A(4, 2)$
$B(-3, 4)$
$C(-3, 0)$
$D(0, 0)$

Check your understanding 3

SOLUTION

Graph the ordered pairs on a rectangular coordinate system. The horizontal axis represents the number of yards gained, and the vertical axis represents the number of points scored.

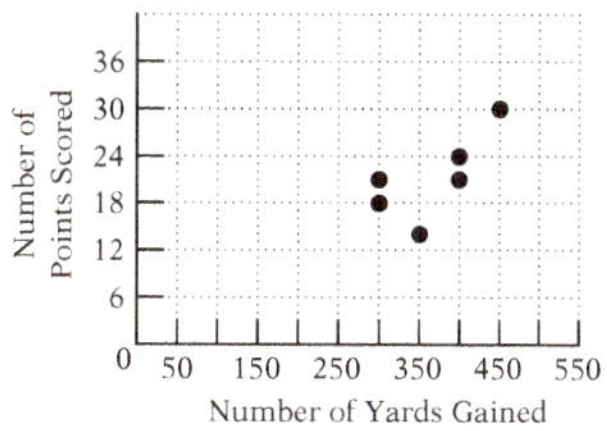

Check your understanding 4

x	$y = -\frac{2}{3}x + 3$
−6	7
−3	5
0	3
3	1
6	−1

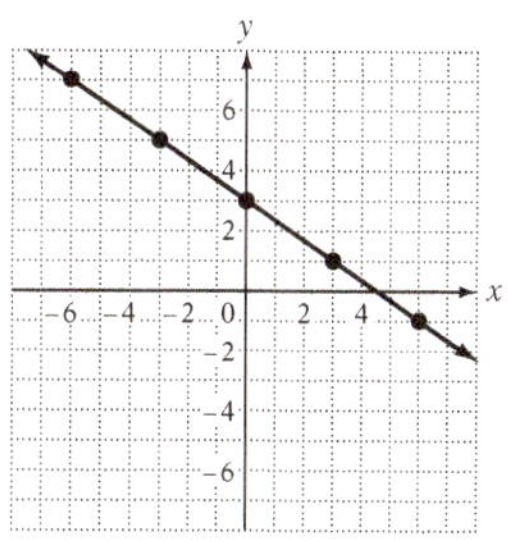

Check your understanding 5

x	$y = -x^2 + 4$
−3	−5
−2	0
−1	3
0	4
1	3
2	0
3	−5

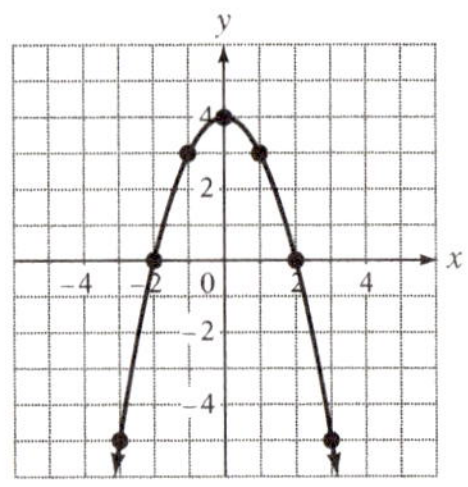

Section 5.2

Check your understanding 1

The domain is the set of first coordinates. The range is the set of second coordinates.
Domain: $\{-1, 3, 4, 6\}$
Range: $\{5\}$

Check your understanding 2

$$G(x) = \frac{3x}{x+2}$$

$$G(-4) = \frac{3(-4)}{-4+2} = \frac{-12}{-2} = 6$$

Check your understanding 3

Because $3x^2 - 5x + 2$ evaluates to a real number for any value of x, the domain of $f(x) = 3x^2 - 5x + 2$ is all real numbers, or $\{x|-\infty < x < \infty\}$.

Check your understanding 4

x	$g(x) = \frac{3}{2}x - 2$
−2	−5
0	−2
2	1
4	4

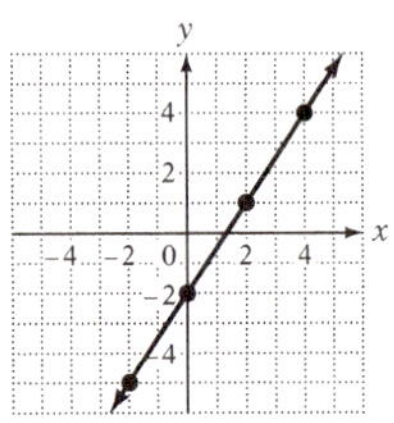

Domain: $(-\infty, \infty)$; Range: $(-\infty, \infty)$

Check your understanding 5

u	$h(u) = -u^2 + 4u$
−1	−5
0	0
1	3
2	4
3	3
4	0
5	−5

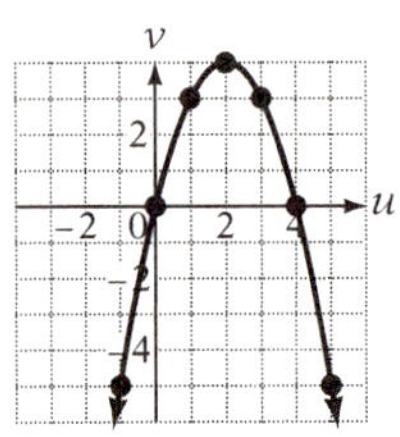

Domain: $\{u|-\infty < u < \infty\}$; Range: $\{v|v \leq 4\}$

Check your understanding 6

t	$g(t) = -\lvert t + 2\rvert + 3$
-6	-1
-4	1
-2	3
0	1
2	-1
4	-3

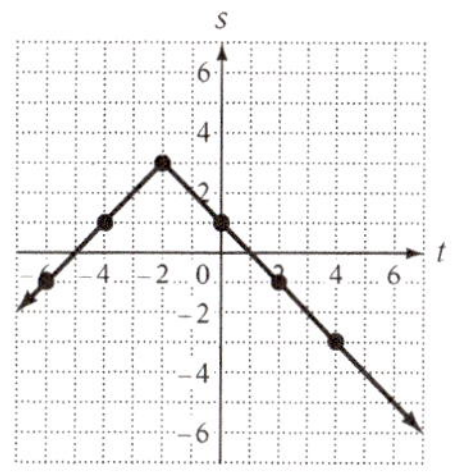

Domain: $\{t \mid -\infty < t < \infty\}$; Range: $\{s \mid s \leq 3\}$

Check your understanding 7

Domain: $\{x \mid x \leq 2\}$; Range: $\{y \mid y \geq 0\}$

Check your understanding 8

Every vertical line intersects the graph at most once. The graph is the graph of a function.

Section 5.3

Check your understanding 1

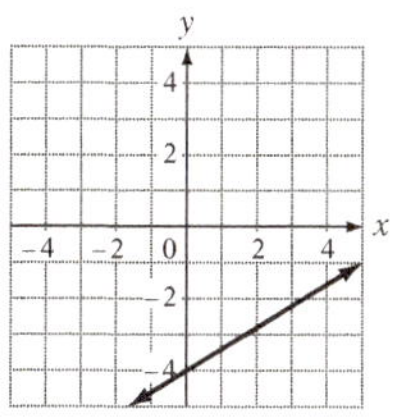

Check your understanding 2

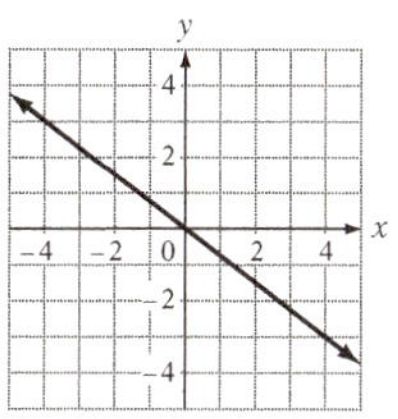

Check your understanding 3

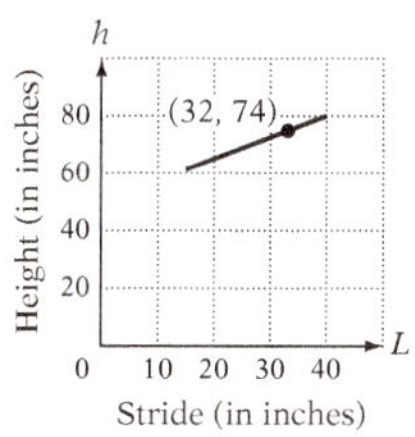

- Graph $h = \frac{3}{4}L + 50$.

  When $L = 20$, $h = 65$.
  When $L = 40$, $h = 80$.

The ordered pair $(32, 74)$ means that a person with a stride of 32 in. is 74 in. tall.

Check your understanding 4

$$-3x + 2y = 4$$
$$2y = 3x + 4$$
$$y = \frac{3}{2}x + 2$$

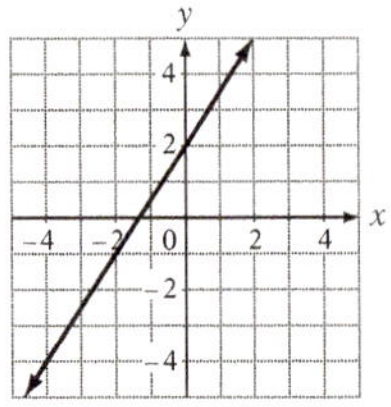

Check your understanding 5

$$y - 3 = 0$$
$$y = 3$$

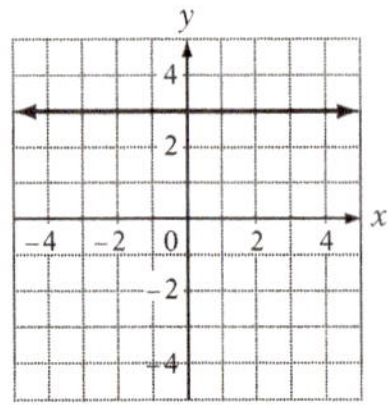

• The graph of $y = 3$ goes through the point with coordinates $(0, 3)$.

Check your understanding 6

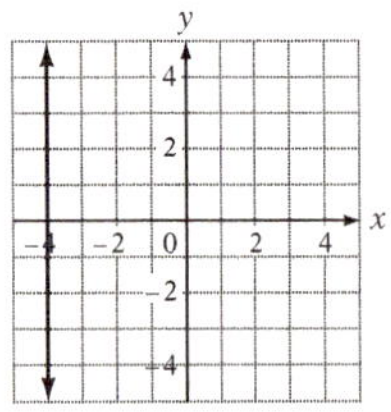

• The graph of $x = -4$ goes through the point with coordinates $(-4, 0)$.

Check your understanding 7

x-intercept:

$$3x - y = 2$$
$$3x - 0 = 2 \quad \text{• Let } y = 0.$$
$$3x = 2$$
$$x = \frac{2}{3}$$

x-intercept: $\left(\frac{2}{3}, 0\right)$

y-intercept:

$$3x - y = 2$$
$$3(0) - y = 2 \quad \text{• Let } x = 0.$$
$$-y = 2$$
$$y = -2$$

y-intercept: $(0, -2)$

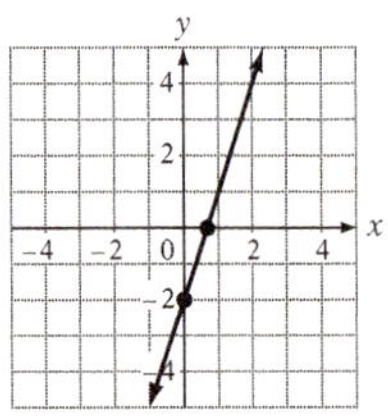

Check your understanding 8

Use $P_1(2, 5)$ and $P_2(-4, 2)$.

$$m = \frac{y_2 - y_1}{x_2 - x_1} = \frac{2 - 5}{-4 - 2} = \frac{-3}{-6} = \frac{1}{2}$$

The slope is $\frac{1}{2}$.

Check your understanding 9

Use $P_1(4, -3)$ and $P_2(2, 7)$.

$$m = \frac{y_2 - y_1}{x_2 - x_1} = \frac{7 - (-3)}{2 - 4} = \frac{10}{-2} = -5$$

The slope is -5.

Check your understanding 10

Use $P_1(-3, 4)$ and $P_2(5, 4)$.

$$m = \frac{y_2 - y_1}{x_2 - x_1} = \frac{4 - 4}{5 - (-3)} = \frac{0}{8} = 0$$

The slope of the line is zero.

Check your understanding 11

Use $P_1(6, -1)$ and $P_2(6, 7)$.

$$m = \frac{y_2 - y_1}{x_2 - x_1} = \frac{7 - (-1)}{6 - 6} = \frac{8}{0}$$

Division by zero is not defined.
The slope of the line is undefined.

Check your understanding 12

Choose $P_1(5, 25{,}000)$ and $P_2(2, 55{,}000)$.

$$\begin{aligned} m &= \frac{y_2 - y_1}{x_2 - x_1} \\ &= \frac{55{,}000 - 25{,}000}{2 - 5} \\ &= \frac{30{,}000}{-3} \\ &= -10{,}000 \end{aligned}$$

• $(x_1, y_1) = (5, 25{,}000)$, $(x_2, y_2) = (2, 55{,}000)$

A slope of $-10{,}000$ means that the value of the recycling truck is decreasing by \$10,000 per year.

Check your understanding 13

$$m = -\frac{3}{2} = \frac{-3}{2}$$

y-intercept $= (0, 4)$

Beginning at $(0, 4)$, move down 3 (the change in y) and to the right 2 (the change in x). Draw a line between the two points.

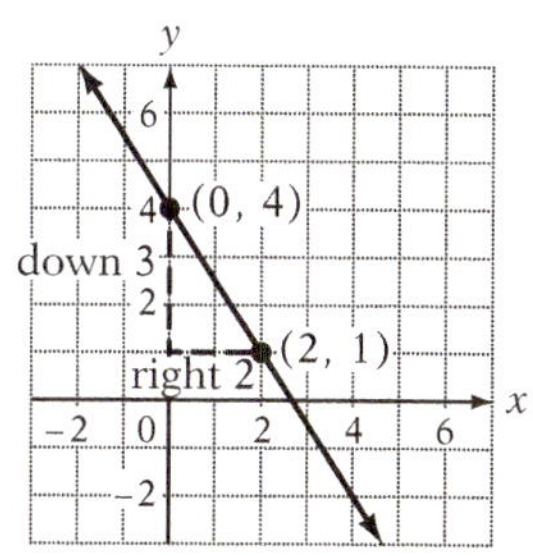

Check your understanding 14

$$2x + 3y = 6$$
$$3y = -2x + 6$$
$$y = -\frac{2}{3}x + 2$$
$$m = -\frac{2}{3} = \frac{-2}{3}$$

y-intercept $= (0, 2)$

Beginning at $(0, 2)$, move down 2 (the change in y) and to the right 3 (the change in x). Draw a line between the two points.

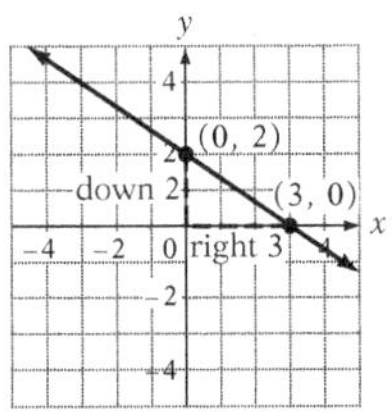

Check your understanding 15

Locate $P(-2, 3)$.

$$m = -\frac{4}{3} = \frac{-4}{3}$$

Beginning at $(-2, 3)$, move down 4 (the change in y) and to the right 3 (the change in x). Draw a line between the two points.

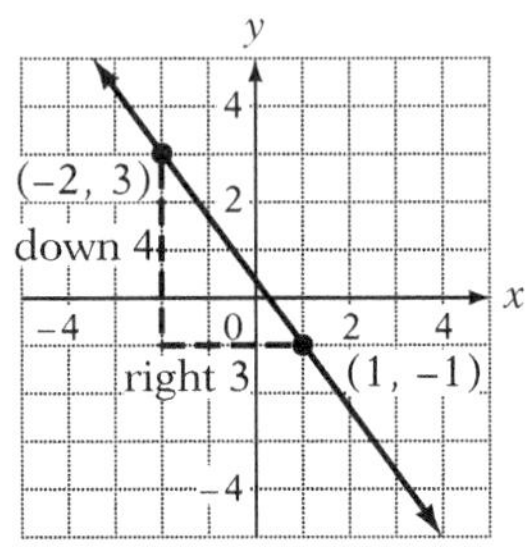

Section 5.4

Check your understanding 1

$$m = -\frac{1}{3}; \quad (x_1, y_1) = (-3, -2)$$
$$y - y_1 = m(x - x_1)$$
$$y - (-2) = -\frac{1}{3}[x - (-3)]$$
$$y + 2 = -\frac{1}{3}(x + 3)$$
$$y + 2 = -\frac{1}{3}x - 1$$
$$y = -\frac{1}{3}x - 3$$

The equation of the line is $y = -\frac{1}{3}x - 3$.

Check your understanding 2

Use $P_1(2, 0)$ and $P_2(5, 3)$.

$$m = \frac{y_2 - y_1}{x_2 - x_1} = \frac{3 - 0}{5 - 2} = \frac{3}{3} = 1$$
$$y - y_1 = m(x - x_1)$$
$$y - 0 = 1(x - 2)$$
$$y = 1(x - 2)$$
$$y = x - 2$$

The equation of the line is $y = x - 2$.

Check your understanding 3

Use $P_1(2, 3)$ and $P_2(-5, 3)$.

$$m = \frac{y_2 - y_1}{x_2 - x_1} = \frac{3 - 3}{-5 - 2} = \frac{0}{-7} = 0$$

The line has zero slope, so the line is a horizontal line.
All points on the line have an ordinate of 3. The equation of the line is $y = 3$.

Check your understanding 4

STRATEGY

- Select the independent and dependent variables. The function will predict the Celsius temperature, so that quantity is the dependent variable, y. The Fahrenheit temperature is the independent variable, x.
- From the given data, two ordered pairs are (212, 100) and (32, 0). Use these ordered pairs to determine the linear function.

SOLUTION

Choose $P_1(32, 0)$ and $P_2(212, 100)$.

$$m = \frac{y_2 - y_1}{x_2 - x_1} = \frac{100 - 0}{212 - 32} = \frac{100}{180} = \frac{5}{9}$$

$$y - y_1 = m(x - x_1)$$

$$y - 0 = \frac{5}{9}(x - 32)$$

$$y = \frac{5}{9}(x - 32), \text{ or } C = \frac{5}{9}(F - 32)$$

The linear function is $f(F) = \frac{5}{9}(F - 32)$.

Check your understanding 5

STRATEGY

- Find the slope of the line between the two given points.
- Use the point-slope formula to find the equation of the line.
- Evaluate the equation at $x = -30$.

SOLUTION

$$m = \frac{y_2 - y_1}{x_2 - x_1} = \frac{0.81 - 0.43}{50 - (-100)} = \frac{0.38}{150} \approx 0.00253$$

$$y - y_1 = m(x - x_1)$$

$$y - 0.43 = 0.00253[x - (-100)]$$

$$y - 0.43 = 0.00253(x + 100)$$

$$y - 0.43 = 0.00253x + 0.253$$

$$y = 0.00253x + 0.683$$

The equation of a line that approximates the data is $y = 0.00253x + 0.683$.

Replace x by -30 and solve for y.

$$y = 0.00253x + 0.683$$

$$y = 0.00253(-30) + 0.683$$

$$= 0.6071$$

The equation predicts that the volume will be approximately 0.61 L when the temperature is $-30°$C.

Section 5.5 Check your understanding 1

x-coordinate: $-\frac{b}{2a} = -\frac{0}{2(1)} = 0$ • $a = 1, b = 0$

$y = x^2 - 2$ • Find y when $x = 0$.
$\quad = 0^2 - 2$
$\quad = -2$

The coordinates of the vertex are $(0, -2)$.
The equation of the axis of symmetry is $x = 0$.

x	$y = x^2 - 2$
1	−1
2	2

• Find ordered-pair solutions using values of x greater than 0.

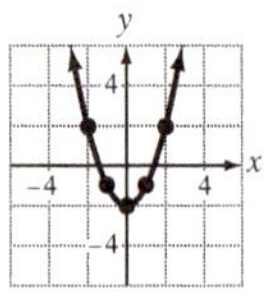

• Use symmetry to find two more points on the other side of the axis of symmetry.

Check your understanding 2

Because $a > 0$, the graph of g will open up.
The coordinates of the vertex are

$$x = -\frac{b}{2a} = \frac{-4}{2(1)} = -2$$

$$y = g(-2) = (-2)^2 + 4(-2) - 2 = -6$$

Find several ordered-pair solutions and then use symmetry to draw the graph.
The domain is $\{x|x \in \text{real numbers}\}$.
The range is $\{y|y \geq -6\}$.

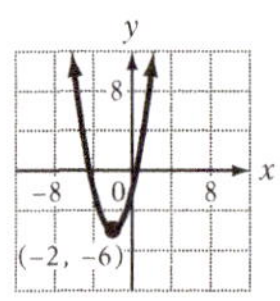

Check your understanding 3

$f(x) = x^2 + 2x - 8$
$0 = x^2 + 2x - 8$ • Replace $f(x)$ by 0.
$0 = (x + 4)(x - 2)$ • Factor.

$x + 4 = 0 \quad x - 2 = 0$ • Solve for x.
$x = -4 \quad x = 2$

The coordinates of the x-intercepts are $(-4, 0)$ and $(2, 0)$.

Check your understanding 4

$x = -\frac{b}{2a} = -\frac{-3}{2(2)} = \frac{3}{4}$ • Find the x-coordinate of the vertex.

$f(x) = 2x^2 - 3x + 1$

$f\left(\frac{3}{4}\right) = 2\left(\frac{3}{4}\right)^2 - 3\left(\frac{3}{4}\right) + 1$ • Evaluate f at $x = \frac{3}{4}$.

$= \frac{9}{8} - \frac{9}{4} + 1$

$= -\frac{1}{8}$

Because a is positive, the function has a minimum value.
The minimum value of the function is $-\frac{1}{8}$.

Check your understanding 5

STRATEGY

- To find the time it takes for the ball to reach its maximum height, find the t-coordinate of the vertex.
- To find the maximum height, evaluate the function at the t-coordinate of the vertex.

SOLUTION

$$t = -\frac{b}{2a} = -\frac{64}{2(-16)} = 2$$

The ball reaches its maximum height in 2 s.

$$s(t) = -16t^2 + 64t$$
$$s(2) = -16(2)^2 + 64(2) = -64 + 128 = 64$$

The maximum height is 64 ft.

Check your understanding 6

STRATEGY

- Width of rectangle: x

  Length of rectangle: y

  Amount of fencing, F: 100 ft

- Express the length of the rectangle in terms of x.

  $F = 2x + y$

  $100 = 2x + y$ • $F = 100$

  $100 - 2x = y$ • Solve for y.

  Express the area of the rectangle in terms of x.

  $A = xy$

  $A = x(100 - 2x)$ • $y = 100 - 2x$

  $A = -2x^2 + 100x$

- To find the width, find the x-coordinate of the vertex of $f(x) = -2x^2 + 100x$.
- To find the length, replace x in $y = 100 - 2x$ by the x-coordinate of the vertex.

SOLUTION

$$x = -\frac{b}{2a} = -\frac{100}{2(-2)} = 25$$

The width is 25 ft.

$$100 - 2x = 100 - 2(25)$$
$$= 100 - 50 = 50$$

The length is 50 ft.

Section 5.6

Check your understanding 1

$$f(x) = 2^{2x+1}$$
$$f(0) = 2^{2(0)+1} = 2^1 = 2$$
$$f(-2) = 2^{2(-2)+1} = 2^{-3} = \frac{1}{2^3} = \frac{1}{8}$$

Check your understanding 2

$$
\begin{aligned}
f(x) &= e^x \\
f(1.2) &= e^{1.2} \\
&\approx 3.3201 \\
f(-2.5) &= e^{-2.5} \\
&\approx 0.0821 \\
f(e) &= e^e \\
&\approx 15.1543
\end{aligned}
$$

Check your understanding 3

a.

x	$y = 2^{-\frac{1}{2}x}$
-4	4
-2	2
0	1
2	$\frac{1}{2}$

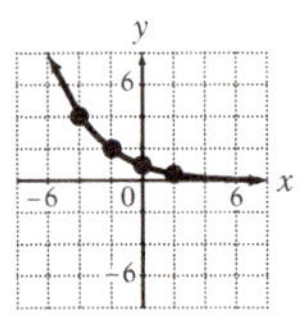

b.

x	$y = 2^x + 1$
-4	$1\frac{1}{16}$
-2	$1\frac{1}{4}$
0	2
2	5

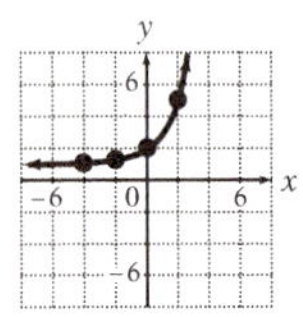

Check your understanding 4

a.

x	$y = 2^{-x} + 2$
-2	6
-1	4
0	3
1	$2\frac{1}{2}$
2	$2\frac{1}{4}$

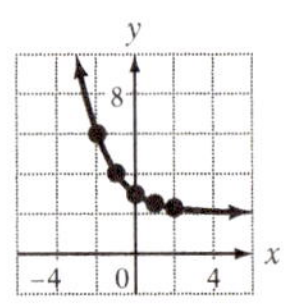

b.

x	$y = e^{-2x} - 4$
-2	50.60
-1	3.39
0	-3
2	-3.98
4	-4.00

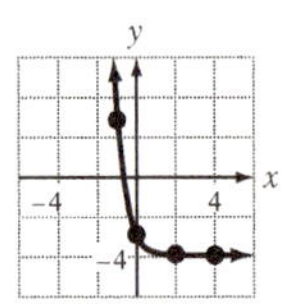

Check your understanding 5

$$N(t) = 1.5\left(\tfrac{1}{2}\right)^{t/193.7}$$

$$N(24) = 1.5\left(\tfrac{1}{2}\right)^{24/193.7}$$

$$\approx 1.5(0.9177) \approx 1.3766$$

After 24 h, there are approximately 1.3766 g of the isotope in the body.

Check your understanding 6

$$A(t) = 200e^{-0.014t}$$

$$A(45) = 200e^{-0.014(45)}$$

$$\approx 107$$

After 45 min, there are approximately 107 mg of aspirin in the patient's bloodstream.

Solutions to Objective Practice Exercises

Objective 5.1A

1.

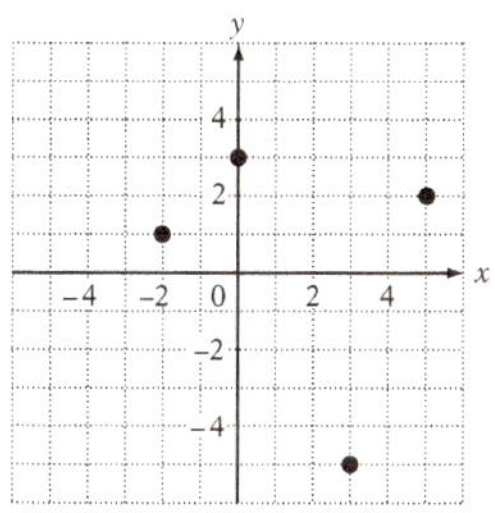

2.

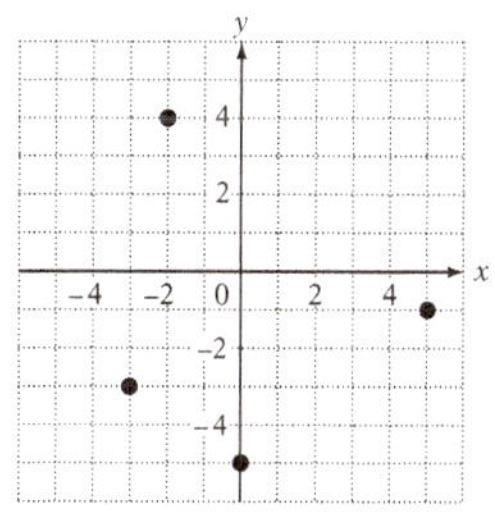

3.

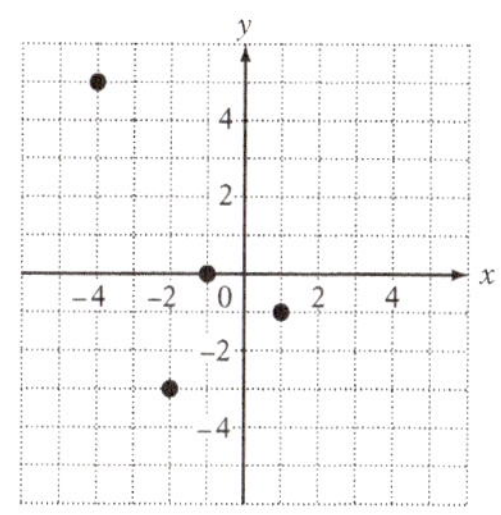

4. $A(2, 3)$

$B(4, 0)$

$C(-4, 1)$

$D(-2, -2)$

5. $A(0, 2)$

$B(-4, -1)$

$C(2, 0)$

$D(1, -3)$

Objective 5.1B

1.

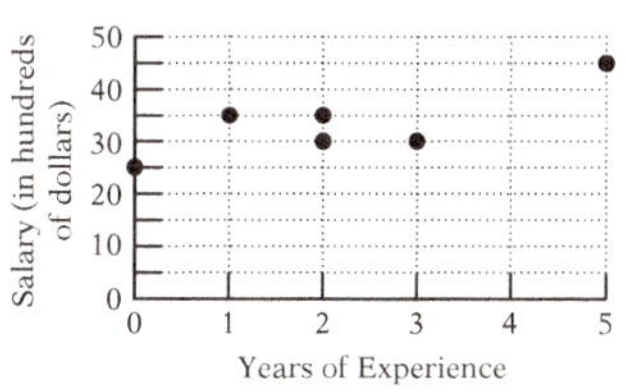

2.

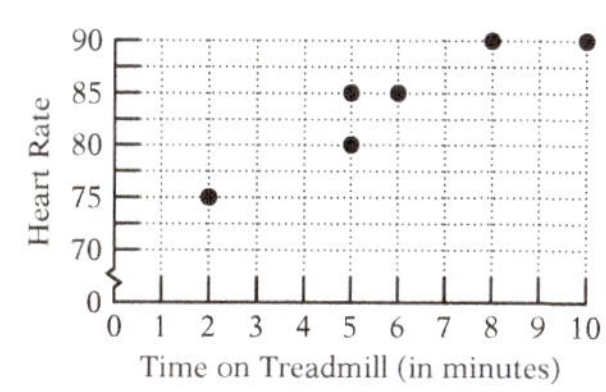

3.

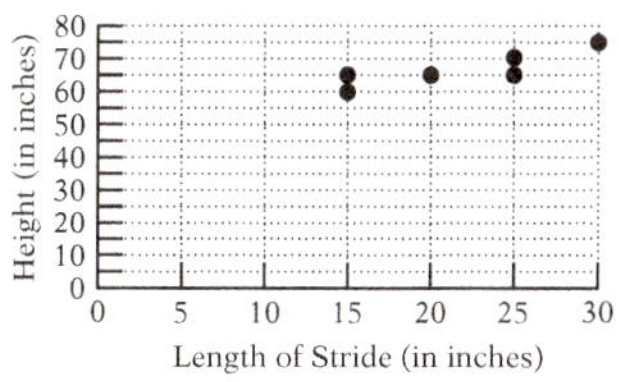

4. The record time for the 800-meter race was 200 s.

Objective 5.1C

1. $y = 2x - 3$

x	y
−1	−5
0	−3
1	−1
2	1
3	3
4	5

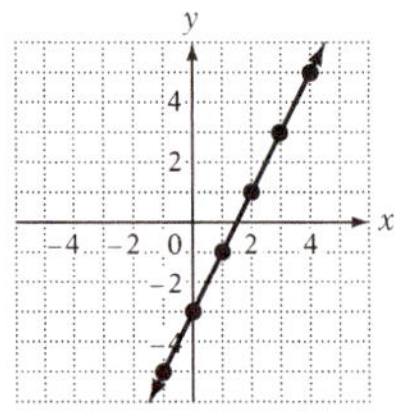

2. $y = -\frac{2}{3}x + 1$

x	y
−6	5
−3	3
0	1
3	−1
6	−3

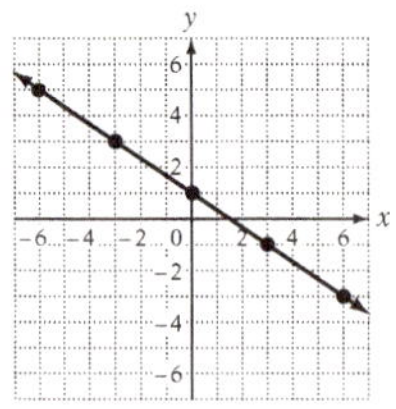

3. $y = x^2 - 4$

x	y
−3	5
−2	0
−1	−3
0	−4
1	−3
2	0
3	5

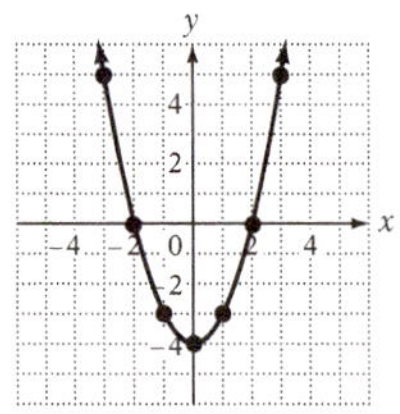

4. $y = -x^2 + 2x + 3$

x	y
−2	−5
−1	0
0	3
1	4
2	3
3	0
4	−5

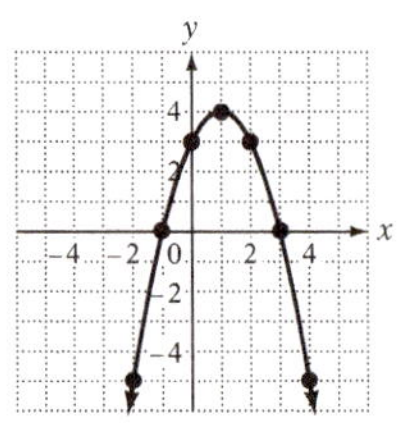

5. $y = |x + 2|$

x	y
−5	3
−4	2
−3	1
−2	0
−1	1
0	2
1	3

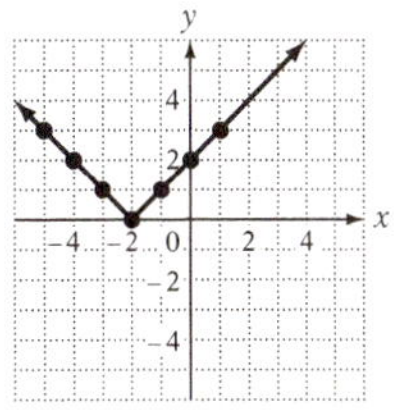

Objective 5.2A

1. This set represents a function because each x-coordinate is paired with only one y-coordinate.

$D = \{0, 1, 2, 3, 4\}$; $R = \{0\}$

2.
$f(x) = 5x - 4$
$f(-2) = 5(-2) - 4$
$f(-2) = -10 - 4$
$f(-2) = -14$

3.
$q(r) = r^2 - 4$
$q(-5) = (-5)^2 - 4$
$q(-5) = 25 - 4$
$q(-5) = 21$

4.
$s(t) = t^3 - 3t + 4$
$s(2) = (2)^3 - 3(2) + 4$
$s(2) = 8 - 6 + 4$
$s(2) = 6$

5. $\{x \mid -\infty < x < \infty\}$

6. $\{x \mid -\infty < x < \infty\}$

Objective 5.2B

1.

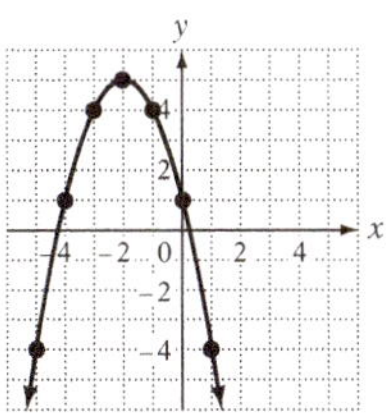

Domain $= \{x \mid -\infty < x < \infty\}$
Range $= \{y \mid y \le 5\}$

2.

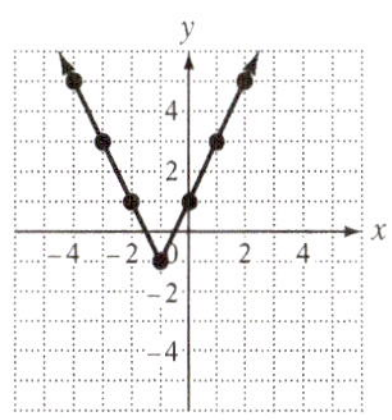

Domain $= \{x \mid -\infty < x < \infty\}$
Range $= \{y \mid y \ge -1\}$

3. 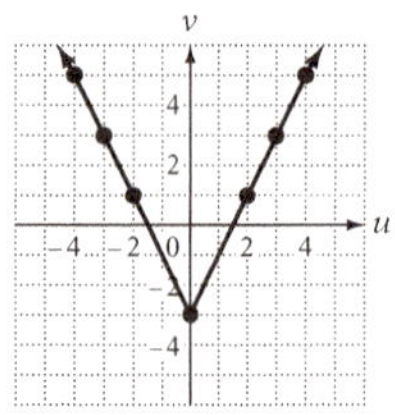

Domain $= \{u \mid -\infty < u < \infty\}$
Range $= \{v \mid v \geq -3\}$

4. 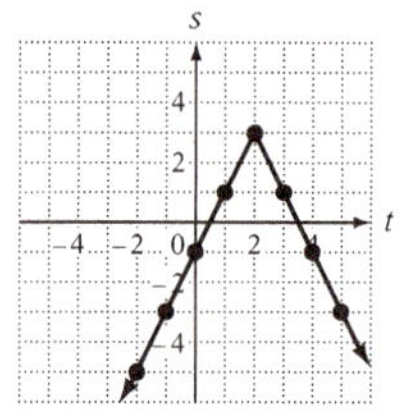

Domain $= \{t \mid -\infty < t < \infty\}$
Range $= \{s \mid s \leq 3\}$

Objective 5.2C

1. Yes
2. Yes
3. No
4. No

Objective 5.3A

1. $y = -2x + 3$

x	y
0	3
1	1
2	−1

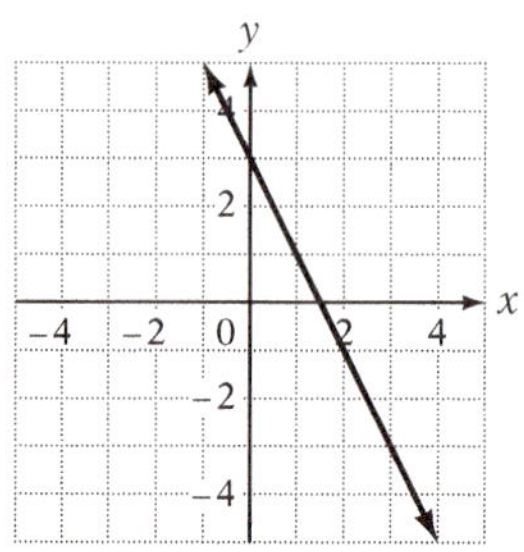

2. $f(x) = \frac{3}{2}x$

x	$f(x)$
−2	−3
0	0
2	3

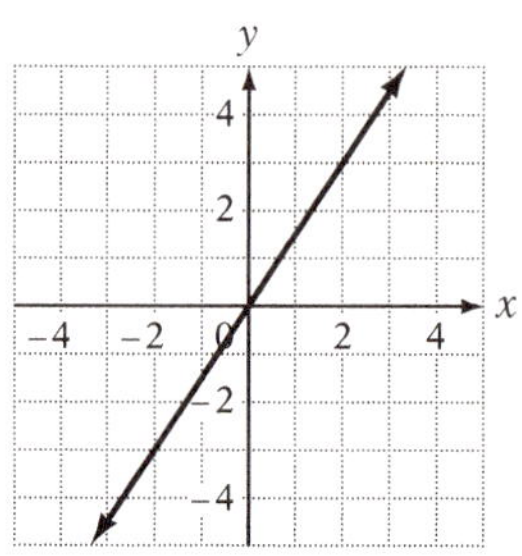

3. $f(x) = \frac{3}{4}x + 2$

x	$f(x)$
−4	−1
0	2
4	5

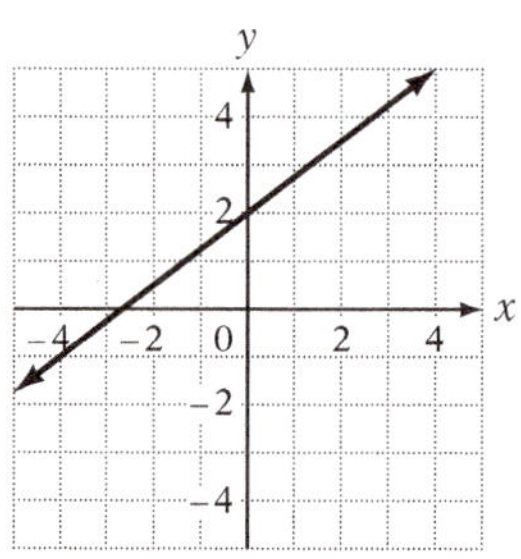

4. $y = -\frac{3}{2}x - 3$

x	y
−4	3
−2	0
0	−3

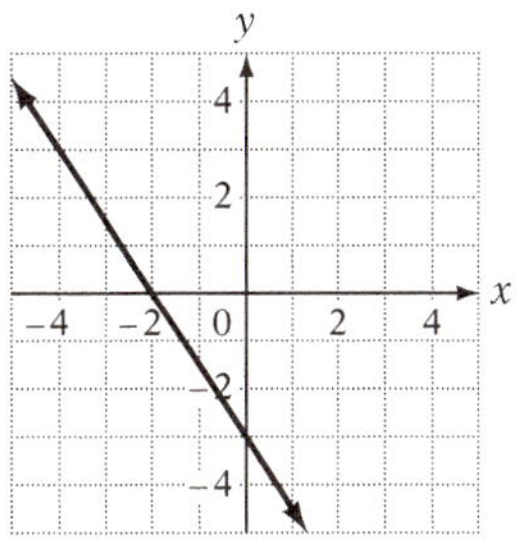

5. $d = 0.98t$

t	d
0	0
4	3.92

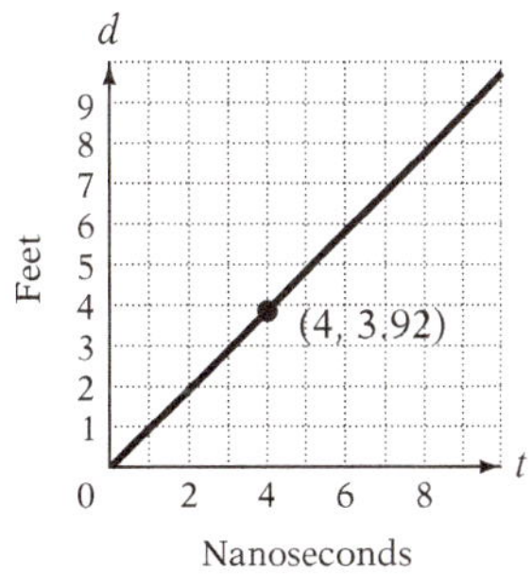

The proton travels 3.92 ft in 4 nanoseconds.

Objective 5.3B **1.** $2x - 5y = 10$

$$-5y = -2x + 10$$

$$y = \frac{2}{5}x - 2$$

x	y
0	−2
5	0
−5	−4

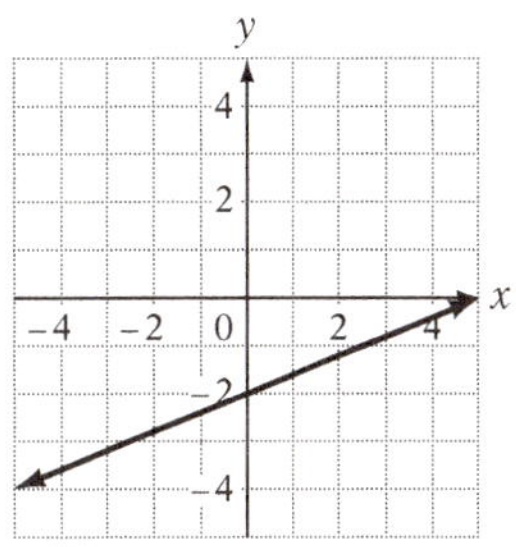

2. $y = -2$

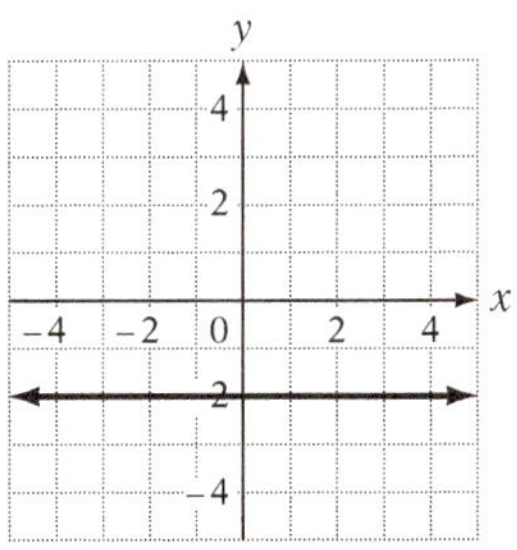

3. $x = -3$

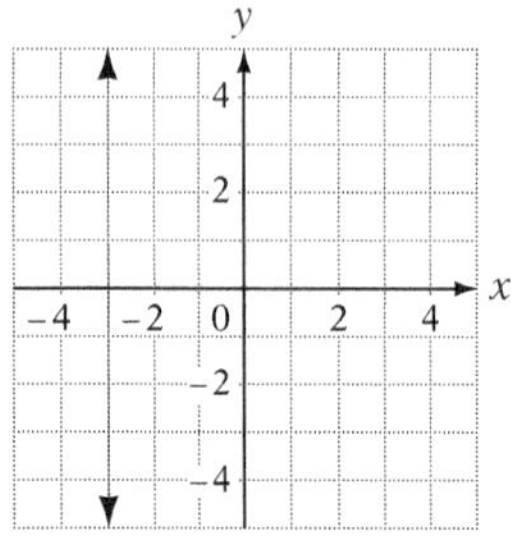

4. x-intercept:

$$\begin{aligned} x - 2y &= -4 \\ x - 2(0) &= -4 \\ x &= -4 \end{aligned}$$

$(-4, 0)$

y-intercept:

$$\begin{aligned} x - 2y &= -4 \\ 0 - 2y &= -4 \\ -2y &= -4 \\ y &= 2 \end{aligned}$$

$(0, 2)$

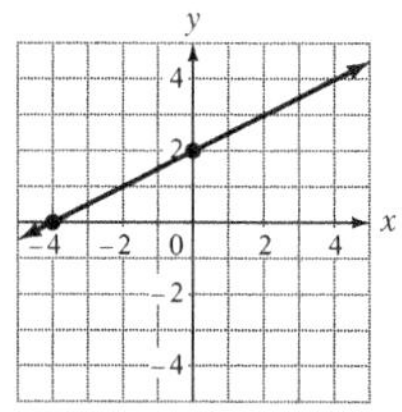

5. x-intercept:

$$\begin{aligned} 2x - 3y &= 9 \\ 2x - 3(0) &= 9 \\ 2x &= 9 \\ x &= \frac{9}{2} \end{aligned}$$

$\left(\frac{9}{2}, 0\right)$

y-intercept:

$$\begin{aligned} 2x - 3y &= 9 \\ 2(0) - 3y &= 9 \\ -3y &= 9 \\ y &= -3 \end{aligned}$$

$(0, -3)$

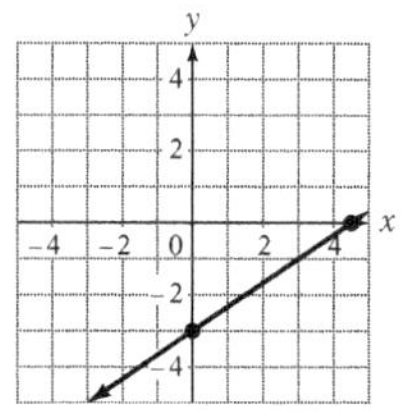

6. x-intercept:

$$\begin{aligned} 3x + 2y &= 4 \\ 3x + 2(0) &= 4 \\ 3x &= 4 \\ x &= \frac{4}{3} \end{aligned}$$

$\left(\frac{4}{3}, 0\right)$

y-intercept:

$$\begin{aligned} 3x + 2y &= 4 \\ 3(0) + 2y &= 4 \\ 2y &= 4 \\ y &= 2 \end{aligned}$$

$(0, 2)$

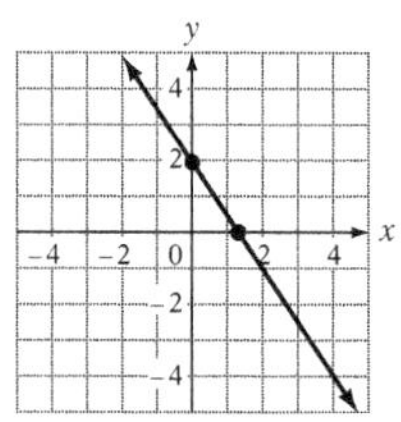

Objective 5.3C

1. $P_1(-1, 4), P_2(2, 5)$

$$m = \frac{y_2 - y_1}{x_2 - x_1} = \frac{5 - 4}{2 - (-1)} = \frac{1}{3}$$

The slope is $\frac{1}{3}$.

2. $P_1(-1, 3), P_2(-4, 5)$

$$m = \frac{y_2 - y_1}{x_2 - x_1} = \frac{5 - 3}{-4 - (-1)} = \frac{2}{-3} = -\frac{2}{3}$$

The slope is $-\frac{2}{3}$.

3. $P_1(2, 4), P_2(2, -2)$

$$m = \frac{y_2 - y_1}{x_2 - x_1} = \frac{4 - (-2)}{2 - 2} = \frac{6}{0}$$

The slope is undefined.

4. $P_1(2, 3), P_2(-1, 3)$

$$m = \frac{y_2 - y_1}{x_2 - x_1} = \frac{3 - 3}{2 - (-1)} = \frac{0}{3} = 0$$

The slope is zero.

5. $P_1(0, 4), P_2(-2, 5)$

$$m = \frac{y_2 - y_1}{x_2 - x_1} = \frac{5 - 4}{-2 - 0} = \frac{1}{-2} = -\frac{1}{2}$$

The slope is $-\frac{1}{2}$.

6. $m = \frac{13 - 6}{40 - 180} = \frac{7}{-140} = -0.05$

Approximately 0.05 gal of fuel is used for each mile the car is driven.

Objective 5.3D

1. $y = \frac{2}{3}x - 3$

$m = \frac{2}{3}$

y-intercept: $(0, -3)$

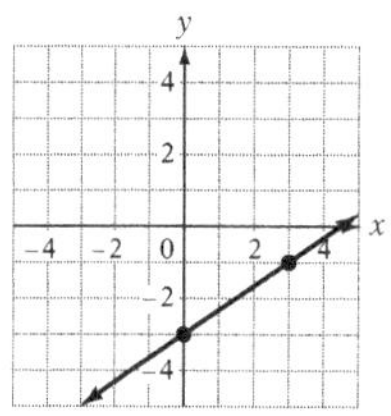

2. $y = -\frac{3}{2}x$

$m = -\frac{3}{2}$

y-intercept: $(0, 0)$

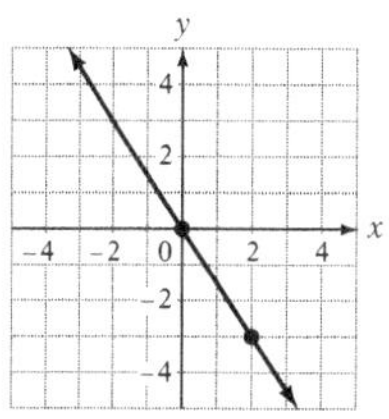

3. $y = 2x - 4$

$m = 2$

y-intercept: $(0, -4)$

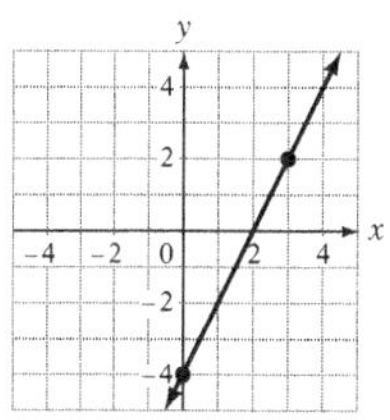

4. 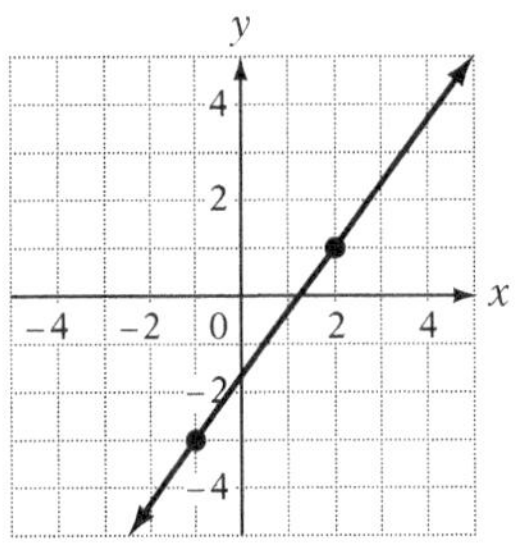

5. 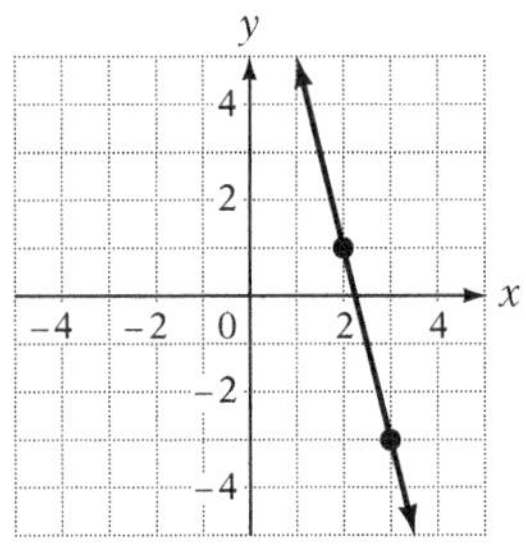

Objective 5.4A

1. $m = \frac{1}{2}, (x_1, y_1) = (2, 3)$

$$y - y_1 = m(x - x_1)$$
$$y - 3 = \frac{1}{2}(x - 2)$$
$$y - 3 = \frac{1}{2}x - 1$$
$$y = \frac{1}{2}x + 2$$

The equation of the line is $y = \frac{1}{2}x + 2$.

2. $m = -3, (x_1, y_1) = (-1, 7)$

$$y - y_1 = m(x - x_1)$$
$$y - 7 = -3[x - (-1)]$$
$$y - 7 = -3(x + 1)$$
$$y - 7 = -3x - 3$$
$$y = -3x + 4$$

The equation of the line is $y = -3x + 4$.

3. $m = -\frac{2}{3}, (x_1, y_1) = (3, 5)$

$$y - y_1 = m(x - x_1)$$
$$y - 5 = -\frac{2}{3}(x - 3)$$
$$y - 5 = -\frac{2}{3}x + 2$$
$$y = -\frac{2}{3}x + 7$$

The equation of the line is $y = -\frac{2}{3}x + 7$.

4. Slope is undefined, $(x_1, y_1) = (3, -4)$

The line is a vertical line. All points on the line have an abscissa of 3.

The equation of the line is $x = 3$.

5. $m = 0, (x_1, y_1) = (-2, -3)$

$$y - y_1 = m(x - x_1)$$
$$y - (-3) = 0[x - (-2)]$$
$$y + 3 = 0$$
$$y = -3$$

The equation of the line is $y = -3$.

Objective 5.4B

1. $P_1(0, -3), P_2(-4, 5)$

$$m = \frac{y_2 - y_1}{x_2 - x_1} = \frac{5 - (-3)}{-4 - 0} = \frac{8}{-4} = -2$$

$$\begin{aligned} y - y_1 &= m(x - x_1) \\ y - (-3) &= -2(x - 0) \\ y + 3 &= -2x \\ y &= -2x - 3 \end{aligned}$$

The equation of the line is $y = -2x - 3$.

2. $P_1(-1, -2), P_2(3, 4)$

$$m = \frac{y_2 - y_1}{x_2 - x_1} = \frac{4 - (-2)}{3 - (-1)} = \frac{6}{4} = \frac{3}{2}$$

$$\begin{aligned} y - y_1 &= m(x - x_1) \\ y - 4 &= \frac{3}{2}(x - 3) \\ y - 4 &= \frac{3}{2}x - \frac{9}{2} \\ y &= \frac{3}{2}x - \frac{1}{2} \end{aligned}$$

The equation of the line is $y = \frac{3}{2}x - \frac{1}{2}$.

3. $P_1(-2, 3), P_2(2, -1)$

$$m = \frac{y_2 - y_1}{x_2 - x_1} = \frac{3 - (-1)}{-2 - 2} = \frac{4}{-4} = -1$$

$$\begin{aligned} y - y_1 &= m(x - x_1) \\ y - (-1) &= -1(x - 2) \\ y + 1 &= -x + 2 \\ y &= -x + 1 \end{aligned}$$

The equation of the line is $y = -x + 1$.

4. $P_1(3, -4), P_2(-2, -4)$

$$m = \frac{y_2 - y_1}{x_2 - x_1} = \frac{-4 - (-4)}{3 - (-2)} = \frac{0}{5} = 0$$

$$\begin{aligned} y - y_1 &= m(x - x_1) \\ y - (-4) &= 0(x - 3) \\ y + 4 &= 0 \\ y &= -4 \end{aligned}$$

The equation of the line is $y = -4$.

5. $P_1(-2, 5), P_2(-2, -5)$

$$m = \frac{y_2 - y_1}{x_2 - x_1} = \frac{5 - (-5)}{-2 - (-2)} = \frac{10}{0}$$

The slope is undefined. The line is a vertical line. All points on the line have an abscissa of -2.

The equation of the line is $x = -2$.

Objective 5.4C

1. STRATEGY

Let x represent the number of minutes the jogger runs.

Let y represent the number of Calories burned.

Use the slope-intercept form to find the equation of the line.

SOLUTION

a. y-intercept $= (0, 0)$; slope $= 14$

$y = 14x + 0$

$y = 14x$

The linear function is $f(x) = 14x$.

b. $y = 14x = 14(32) = 448$

After jogging for 32 min, the jogger has burned 448 Calories.

2. STRATEGY

Let x represent the number of minutes the cell phone is used in a month.

Let y represent the monthly cost.

Use the slope-intercept form to find the equation of the line.

SOLUTION

a. y-intercept $= (0, 4.95)$; slope $= 0.59$

$y = 0.59x + 4.95$

The linear function is $f(x) = 0.59x + 4.95$.

b. $y = 0.59x + 4.95 = 0.59(13) + 4.95 = 12.62$

It will cost \$12.62 to use the phone for 13 min in one month.

3. STRATEGY

Let x represent the altitude above sea level in kilometers.

Let y represent the boiling point of water in degrees Celsius.

Use the point-slope formula to find the equation of the line.

SOLUTION

a. $P_1(0, 100)$, $P_2(2, 93)$

$$m = \frac{100 - 93}{0 - 2} = \frac{7}{-2} = -3.5$$

$$y - y_1 = m(x - x_1)$$
$$y - 100 = -3.5(x - 0)$$
$$y = -3.5x + 100$$

The linear function is $f(x) = -3.5x + 100$.

b. $y = -3.5x + 100 = -3.5(8.85) + 100 \approx 69$

Water will boil at approximately 69°C on top of Mount Everest.

4. STRATEGY

Find the slope of the line between the two given points.

Use the point-slope formula to find the equation of the line.

Evaluate the equation at $x = 70$.

SOLUTION

$$m = \frac{y_2 - y_1}{x_2 - x_1} = \frac{96 - 70}{92 - 55} = \frac{26}{37}$$

$$y - y_1 = m(x - x_1)$$

$$y - 70 = \frac{26}{37}(x - 55)$$

$$y - 70 \approx 0.7027(x - 55)$$

$$y - 70 = 0.7027x - 38.6485$$

$$y = 0.7027x + 31.3515$$

The equation of a line that approximates the data is $y = 0.7027x + 31.3515$.

Replace x by 70 and solve for y.

$$y = 0.7027x + 31.3515$$

$$y = 0.7027(70) + 31.3515$$

$$= 80.5405$$

The equation predicts that a patient with a stress test score of 70 will have a diastolic pressure of approximately 81 mm of mercury.

Objective 5.5A **1.** $a = 1, b = 0$

$$-\frac{b}{2a} = -\frac{0}{2(1)} = 0$$

$$y = 0^2 - 2 = -2$$

Vertex: $(0, -2)$

Axis of symmetry: $x = 0$

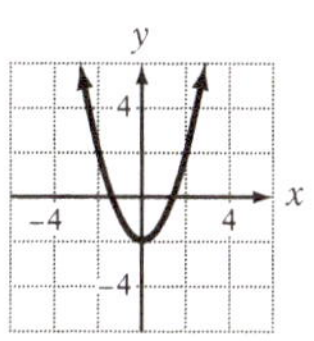

2. $a = -\frac{1}{2}, b = 0$

$$-\frac{b}{2a} = -\frac{0}{2\left(-\frac{1}{2}\right)} = 0$$

$$y = -\frac{1}{2}(0)^2 + 2 = 2$$

Vertex: $(0, 2)$

Axis of symmetry: $x = 0$

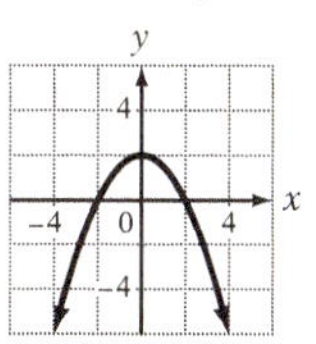

3. $a = 1, b = -3$

$$-\frac{b}{2a} = -\frac{-3}{2(1)} = \frac{3}{2}$$

$$y = \left(\frac{3}{2}\right)^2 - 3\left(\frac{3}{2}\right) + 2 = -\frac{1}{4}$$

Vertex: $\left(\frac{3}{2}, -\frac{1}{4}\right)$

Axis of symmetry: $x = \frac{3}{2}$

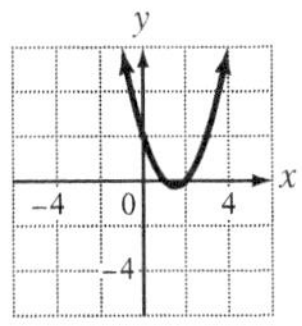

4.

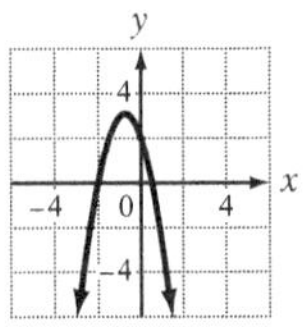

Domain: $\{x | x \in \text{real numbers}\}$

Range: $\left\{y \middle| y \leq \frac{25}{8}\right\}$

5.

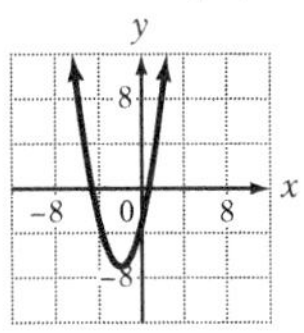

Domain: $\{x | x \in \text{real numbers}\}$

Range: $\{y | y \geq -7\}$

Objective 5.5B **1.** $y = x^2 - 4$

$0 = x^2 - 4$

$0 = (x - 2)(x + 2)$

$x - 2 = 0 \qquad x + 2 = 0$

$x = 2 \qquad x = -2$

The x-intercepts are $(2, 0)$ and $(-2, 0)$.

2. $y = 3x^2 + 6x$

$0 = 3x^2 + 6x$

$0 = 3x(x + 2)$

$3x = 0 \qquad x + 2 = 0$

$x = 0 \qquad x = -2$

The x-intercepts are $(0, 0)$ and $(-2, 0)$.

3. $y = 3x^2 - 19x - 14$

$0 = 3x^2 - 19x - 14$

$0 = (3x + 2)(x - 7)$

$3x + 2 = 0 \qquad x - 7 = 0$

$3x = -2 \qquad x = 7$

$x = -\frac{2}{3}$

The x-intercepts are $\left(-\frac{2}{3}, 0\right)$ and $(7, 0)$.

4. $y = x^2 + 4x - 3$

$0 = x^2 + 4x - 3$

$a = 1, b = 4, c = -3$

$$x = \frac{-b \pm \sqrt{b^2 - 4ac}}{2a}$$

$$= \frac{-4 \pm \sqrt{4^2 - 4(1)(-3)}}{2(1)}$$

$$= \frac{-4 \pm \sqrt{16 + 12}}{2}$$

$$= \frac{-4 \pm \sqrt{28}}{2}$$

$$= \frac{-4 \pm 2\sqrt{7}}{2}$$

$$= -2 \pm \sqrt{7}$$

The x-intercepts are $(-2 + \sqrt{7}, 0)$ and $(-2 - \sqrt{7}, 0)$.

5. $y = -x^2 - 4x - 5$

$0 = -x^2 - 4x - 5$

$a = -1, b = -4, c = -5$

$$x = \frac{-b \pm \sqrt{b^2 - 4ac}}{2a}$$

$$= \frac{-(-4) \pm \sqrt{(-4)^2 - 4(-1)(-5)}}{2(-1)}$$

$$= \frac{4 \pm \sqrt{16 - 20}}{-2}$$

$$= \frac{4 \pm \sqrt{-4}}{-2}$$

$$= \frac{4 \pm 2i}{-2}$$

$$= -2 \pm i$$

The equation has no real solutions.
The parabola has no x-intercepts.

Objective 5.5C

1. $f(x) = x^2 - 2x + 3$

$$x = -\frac{b}{2a} = -\frac{-2}{2(1)} = 1$$

$$f(x) = x^2 - 2x + 3$$

$$f(1) = (1)^2 - 2(1) + 3 = 2$$

Since a is positive, the function has a minimum value. The minimum value of the function is 2.

2. $f(x) = -2x^2 + 4x - 3$

$$x = -\frac{b}{2a} = -\frac{4}{2(-2)} = 1$$

$$f(x) = -2x^2 + 4x - 3$$

$$f(1) = -2(1)^2 + 4(1) - 3$$

$$= -2 + 4 - 3 = -1$$

Since a is negative, the function has a maximum value. The maximum value of the function is -1.

3. $f(x) = -2x^2 + 4x - 5$

$$x = -\frac{b}{2a} = -\frac{4}{2(-2)} = 1$$

$$f(x) = -2x^2 + 4x - 5$$

$$f(1) = -2(1)^2 + 4(1) - 5 = -2 + 4 - 5 = -3$$

Since a is negative, the function has a maximum value. The maximum value of the function is -3.

4. $f(x) = 2x^2 + 3x - 8$

$$x = -\frac{b}{2a} = -\frac{3}{2(2)} = -\frac{3}{4}$$

$$f(x) = 2x^2 + 3x - 8$$

$$f\left(-\frac{3}{4}\right) = 2\left(-\frac{3}{4}\right)^2 + 3\left(-\frac{3}{4}\right) - 8$$

$$= \frac{9}{8} - \frac{9}{4} - 8$$

$$= -\frac{73}{8}$$

Since a is positive, the function has a minimum value. The minimum value of the function is $-\frac{73}{8}$.

Objective 5.5D

1. STRATEGY

- To find the time it takes for the diver to reach the maximum height, find the t-coordinate of the vertex.
- To find the maximum height, evaluate the function at the t-coordinate of the vertex.

SOLUTION

$$t = -\frac{b}{2a} = -\frac{7.8}{2(-4.9)} \approx 0.80$$

The diver reaches the maximum height in about 0.8 s.

$$\begin{aligned} s(t) &= -4.9t^2 + 7.8t + 10 \\ s(0.8) &= -4.9(0.8)^2 + 7.8(0.8) + 10 \\ &= -3.136 + 6.24 + 10 \\ &\approx 13.1 \end{aligned}$$

The maximum height is approximately 13.1 m.

2. STRATEGY

To find the price that will give the maximum revenue, find the P-coordinate of the vertex.

SOLUTION

$$P = -\frac{b}{2a} = -\frac{125}{2\left(-\frac{1}{4}\right)} = 250$$

A price of \$250 will give the maximum revenue.

3. STRATEGY

- To find the time when the water reaches its maximum height, find the t-coordinate of the vertex.
- To find the maximum height, evaluate the function at the t-coordinate of the vertex.

SOLUTION

$$t = -\frac{b}{2a} = -\frac{90}{2(-16)} = 2.8125$$

The water reaches its maximum height in 2.8125 s.

$$\begin{aligned} h(t) &= -16t^2 + 90t + 15 \\ h(2.8125) &= -16(2.8125)^2 + 90(2.8125) + 15 \\ &\approx 141.6 \end{aligned}$$

The maximum height of the waterspout is approximately 141.6 ft.

4. STRATEGY

- To find the distance from one end of the bridge to the point at which the cable is at its minimum height, find the x-coordinate of the vertex.
- To find the minimum height, evaluate the function at the x-coordinate of the vertex.

SOLUTION

$$x = -\frac{b}{2a} = -\frac{-0.8}{2(0.25)} = 1.6$$

The cable is at its minimum height 1.6 ft from one end of the bridge.

$$\begin{aligned} h(x) &= 0.25x^2 - 0.8x + 25 \\ h(1.6) &= 0.25(1.6)^2 - 0.8(1.6) + 25 \\ &= 0.64 - 1.28 + 25 \\ &= 24.36 \end{aligned}$$

The minimum height is 24.36 ft.

Objective 5.6A

1. $f(x) = 2^{-2x}$

a. $f(-3) = 2^{-2(-3)} = 2^6 = 64$

b. $f(2) = 2^{-2(2)} = 2^{-4} = \frac{1}{2^4} = \frac{1}{16}$

c. $f(0) = 2^{-2(0)} = 2^0 = 1$

2. $P(t) = e^{-\frac{1}{2}t}$

a. $P(-3) = e^{-\frac{1}{2}(-3)} = e^{\frac{3}{2}} \approx 4.4817$

b. $P(4) = e^{-\frac{1}{2}(4)} = e^{-2} = \frac{1}{e^2} \approx 0.1353$

c. $P\left(\frac{1}{2}\right) = e^{-\frac{1}{2}\left(\frac{1}{2}\right)} = e^{-\frac{1}{4}} = \frac{1}{e^{0.25}} \approx 0.7788$

3.

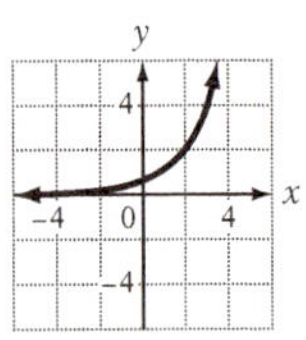

4.

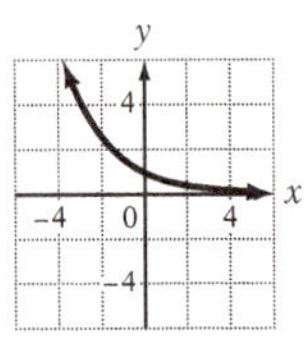

5.

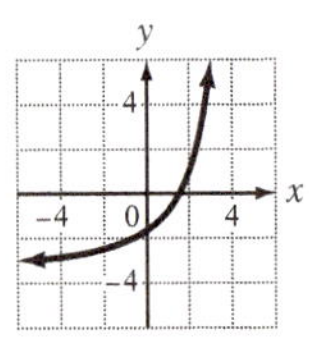

6.

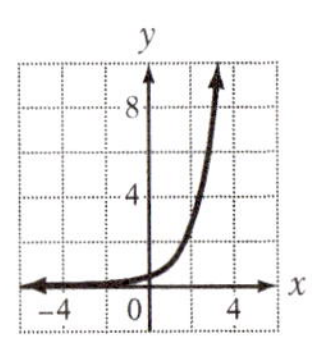

Objective 5.6B

1. STRATEGY

Use the compound interest formula with $P = 2500$, $r = 0.075$, $n = 365$ (compounded daily), and $t = 20$.

SOLUTION

$$A = P\left(1 + \frac{r}{n}\right)^{nt}$$

$$= 2500\left(1 + \frac{0.075}{365}\right)^{365\cdot 20}$$

$$\approx 2500(4.480998548)$$

$$\approx 11202.496$$

The value of the investment after 20 years will be \$11,202.50.

2. STRATEGY

Use the compound interest formula with $P = 10{,}000$, $r = 0.05$, $n = 365$ (compounded daily), and $t = 20$.

SOLUTION

$$A = P\left(1 + \frac{r}{n}\right)^{nt}$$

$$= 10{,}000\left(1 + \frac{0.05}{365}\right)^{365\cdot 20}$$

$$\approx 10{,}000(2.718095669)$$

$$\approx 27180.957$$

The value of the investment after 20 years will be \$27,180.96.

3. STRATEGY

Evaluate the function when $t = 3$.

SOLUTION

$$A(t) = 30\left(\frac{1}{2}\right)^{t/6}$$

$$A(3) = 30\left(\frac{1}{2}\right)^{3/6}$$

$$= 30\left(\frac{1}{2}\right)^{1/2}$$

$$\approx 30(0.7071068)$$

$$\approx 21.213$$

After 3 h, the level of technetium will be approximately 21.2 mg.

MODULE

6 Statistics and Probability

SECTION 6.1 **Organizing Data**

Objective **6.1A** Create frequency distributions

Objective **6.1B** Read histograms

Objective **6.1C** Read frequency polygons

SECTION 6.2 **Statistical Measures**

Objective **6.2A** Find the mean, median, and mode of a distribution

Objective **6.2B** Find the standard deviation of a distribution

SECTION 6.3 **Measures of Relative Position**

Objective **6.3A** Calculate percentiles and quartiles

Objective **6.3B** Create a box-and-whiskers plot

SECTION 6.4 **Introduction to Probability**

Objective **6.4A** Calculate the probability of simple events

Objective **6.4B** Calculate the probability of compound events

Objective **6.4C** Calculate the odds of an event

SECTION

6.1 Organizing Data

Objective 6.1A Create frequency distributions

Tips for Success ➤
A great many new vocabulary words are introduced in this chapter. There are six new terms on this page alone: statistics, population, frequency distribution, classes, range, and class width. All of these terms are in **bold type.** The bold type indicates that these are concepts you must know to learn the material. Be sure to study each new term as it is presented.

Statistics is the study of collecting, organizing, and interpreting data. Data are collected from a **population,** which is the set of all observations of interest. Here are two examples of populations.

A medical researcher wants to determine the effectiveness of a new drug in controlling blood pressure. The population for the researcher is the amount of change in blood pressure for each patient receiving the medication.

The quality control inspector of a precision instrument company wants to determine the diameters of ball bearings. The population for the inspector is the measure of the diameter of each ball bearing.

A **frequency distribution** is one method of organizing the data collected from a population. A frequency distribution is constructed by dividing the data gathered from the population into **classes.** Here is an example:

A ski association surveys 40 of its members, asking them to report the percent of their ski terrain that is rated expert. The results of the survey follow.

Percent of Expert Terrain at 40 Ski Resorts

14	24	8	31	27	9	12	32	24	27
12	21	24	23	12	31	30	31	26	34
13	18	29	33	34	21	28	23	11	10
25	20	14	18	15	11	17	29	21	25

To organize these data into a frequency distribution:

1. Find the smallest number (8) and the largest number (34) in the table. The difference between these two numbers is the **range** of the data.

   Range = 34 − 8 = 26

2. Decide how many classes the frequency distribution will contain. Frequency distributions usually have from 6 to 12 classes. The frequency distribution for this example will contain 6 classes.

3. Divide the range by the number of classes. If necessary, round the quotient to a whole number. This number is called the **class width.**

   $\frac{26}{6} \approx 4$. The class width is 4.

4. Form the classes of the frequency distribution.

Classes	
8–12	Add 4 to the smallest number.
13–17	Add 4 again.
18–22	Continue until a class contains
23–27	the largest number in the set
28–32	of data.
33–37	

These are the **lower class limits.** (left numbers) These are the **upper class limits.** (right numbers)

5. Complete the table by tabulating the data for each class. For each data value, place a slash next to the class that contains the value. Count the number of tallies in each class. This is the **class frequency.**

Frequency Distribution for Ski Resort Data

Classes	Tally	Frequency
8–12	////////	8
13–17	/////	5
18–22	//////	6
23–27	//////////	10
28–32	////////	8
33–37	///	3

Organizing data into a frequency distribution enables us to make statements about the data. For example, 27 (6 + 10 + 8 + 3) of the ski resorts reported that 18% or more of their terrain was rated expert.

An insurance agency tabulated the dollar amounts of 50 auto insurance premiums. The results are given in the following table.

Dollar Amounts of 50 Auto Insurance Premiums

475	224	722	721	815	351	596	625	981	748
993	881	361	560	574	742	703	998	435	873
882	278	455	803	985	305	522	900	638	810
677	688	410	505	890	186	829	631	882	991
484	339	950	579	539	422	326	793	453	118

Focus on creating a frequency distribution

For the table of auto insurance premiums given above, make a frequency distribution that has 6 classes.

STRATEGY

To make the frequency distribution:

- Find the range.
- Divide the range by 6, the number of classes. Round the quotient to the nearest whole number. This is the class width.
- Tabulate the data for each class.

SOLUTION

Range = 998 − 118 = 880

$$\text{Class width} = \frac{880}{6} \approx 147$$

Dollar Amount of Insurance Premiums

Classes	Tally	Frequency
118–265	///	3
266–413	///////	7
414–561	//////////	10
562–709	/////////	9
710–857	/////////	9
858–1005	////////////	12

Check your understanding 1

For the table of auto insurance premiums given on page 3, make a frequency distribution that has 8 classes.

SOLUTION See page S-1.

Objective 6.1A Practice

Use the table below for Exercises 1 to 4.

Tuition per Term at 40 Colleges (in hundreds of dollars)

85	87	95	48	41	91	88	92
71	74	63	51	70	87	84	95
72	94	61	52	88	49	55	60
77	53	89	91	45	96	49	58
83	36	39	32	36	59	95	67

1. What is the range of the data in the tuition table?
2. Make a frequency distribution for the tuition table. Use 8 classes.
3. Which class has the greatest frequency?
4. What percent of the tuitions are between $9500 and $10,300 per term?

Solutions on pp. S-5–S-6.

Objective 6.1B Read histograms

A **histogram** is a bar graph that represents the data in a frequency distribution. The width of each bar represents a class, and the height of the bar corresponds to the frequency of the class.

A survey of 105 households is conducted, and the number of kilowatt-hours (kWh) of electricity used by each household in a 1-month period is recorded in the frequency distribution shown at the left below. The histogram for the frequency distribution is shown in Figure 1.

Take Note ➤
Note that the upper class limit of one class is the lower class limit of the next class. In this case, a class contains the range of data from the lower class limit up to, but not including, the upper class limit. Therefore, if a data value for the distribution at the right were 900, that data value would be included in the 900–950 class.

Classes (kWh)	*Frequency*
850–900	9
900–950	14
950–1000	17
1000–1050	25
1050–1100	16
1100–1150	14
1150–1200	10

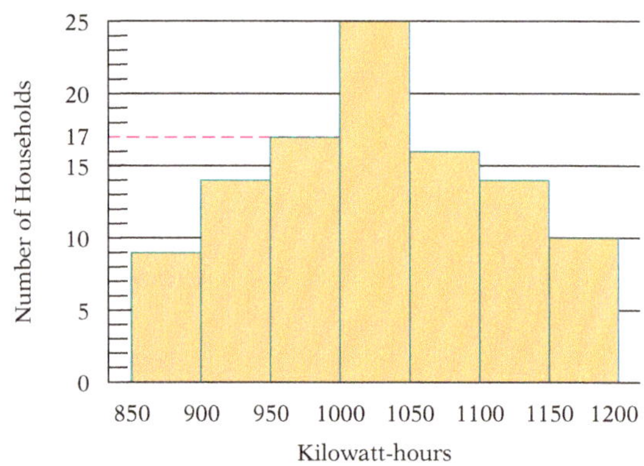

Figure 1 Electricity use of 105 households

From the frequency distribution or the histogram, we can see that 17 households used between 950 kWh and 1000 kWh during the 1-month period.

Focus on solving an application involving a histogram

Use the histogram in Figure 1 on page 4 to find the number of households that used 950 kWh of electricity or less during the month.

STRATEGY

To find the number of households:

- Read the histogram to find the number of households that used between 850 and 900 kWh and the number that used between 900 and 950 kWh.
- Add the two numbers.

SOLUTION

Number between 850 and 900 kWh: 9
Number between 900 and 950 kWh: 14

$9 + 14 = 23$

Twenty-three households used 950 kWh of electricity or less during the month.

Check your understanding 2

Use the histogram in Figure 1 on page 4 to find the number of households that used 1100 kWh of electricity or more during the month.

SOLUTION See pages S-1.

Objective 6.1B Practice

A total of 50 checking account balances were recorded. A histogram of these data is shown below. Use the histogram for Exercises 1 to 4.

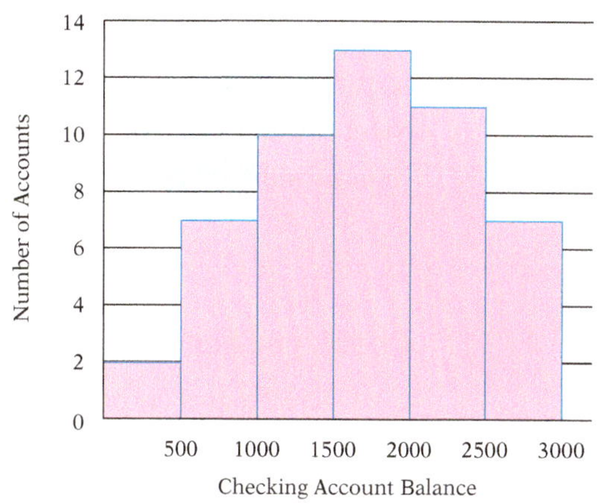

1. Find the number of account balances that were between \$2500 and \$3000.
2. How many account balances were less than \$2000?
3. What percent of the account balances were between \$2000 and \$2500?
4. What percent of the account balances were greater than \$1500?

Solutions on pp. S-6–S-7.

Objective 6.1C Read frequency polygons

A **frequency polygon** is a graph that displays information in a manner similar to a histogram. A dot is placed above the center of each class interval at a height corresponding to that class's frequency. The dots are then connected to form a broken-line graph. The center of a class interval is called the **class midpoint.**

The per capita incomes for the 50 states in a recent year are recorded in the frequency polygon in Figure 2. The number of states with a per capita income between \$40,000 and \$45,000 is 14.

The percent of states for which the per capita income is between \$40,000 and \$45,000 can be determined by solving the basic percent equation. The base is 50 and the amount is 14.

$$pB = A$$
$$p(50) = 14$$
$$\frac{50p}{50} = \frac{14}{50}$$
$$p = 0.28$$

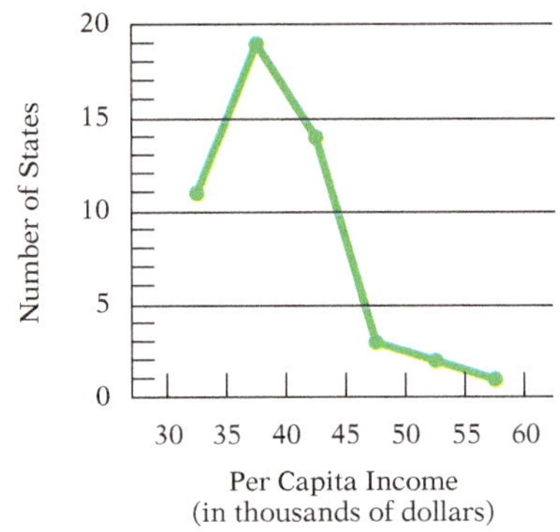

Figure 2 Per capita income for 50 states

Source: www.infoplease.com

28% of the states had a per capita income between \$40,000 and \$45,000.

Focus on solving an application involving a frequency polygon

Use Figure 2 to find the number of states for which the per capita income was \$50,000 or greater.

STRATEGY

To find the number of states:

- Read the frequency polygon to find the number of states with a per capita income between \$50,000 and \$55,000, and the number of states with a per capita income between \$55,000 and \$60,000.
- Add the numbers.

SOLUTION

Number with per capita income between \$50,000 and \$55,000: 2

Number with per capita income between \$55,000 and \$60,000: 1

$2 + 1 = 3$

The per capita income was \$50,000 or greater in 3 states.

Check your understanding 3

Use Figure 2 to find the ratio of the number of states with a per capita income between \$50,000 and \$55,000 to the number with a per capita income between \$40,000 and \$45,000.

SOLUTION See pages S-1–S-2.

Objective 6.1C Practice

The scores of 50 nurses taking a state board exam were recorded. A frequency polygon of these scores is shown below. Use the frequency polygon for Exercises 1 to 4.

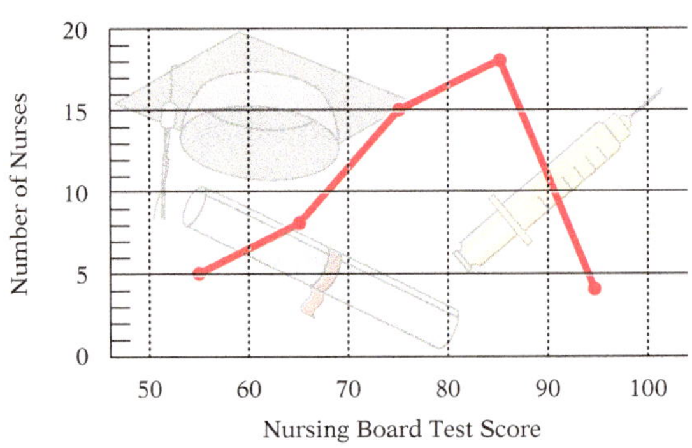

1. Find the number of test scores that were between 60 and 70.
2. How many nurses had scores that were greater than 80?
3. What percent of the nurses had scores between 70 and 90?
4. What percent of the nurses had scores greater than 70?

Solutions on pp. S-7–S-8.

SECTION 6.2 Statistical Measures

Objective 6.2A Find the mean, median, and mode of a distribution

The average annual rainfall in Mobile, Alabama, is 67 in. The average annual snowfall in Syracuse, New York, is 111 in. The average daily low temperature in January in Bismarck, North Dakota, is −4°F. Each of these statements uses one number to describe an entire collection of numbers. Such a number is called an *average.* In statistics, there are various ways to calculate an average. Three of the most common—*mean, median,* and *mode*—are discussed here.

An automotive engineer tests the miles-per-gallon ratings of 15 cars and records the results as follows:

Miles-per-Gallon Ratings of 15 Cars

25 22 21 27 25 35 29 31 25 26 21 39 34 32 28

The **mean** of the data is the sum of the measurements divided by the number of measurements. The symbol for the mean is $\bar{x}$.

Formula for the Mean

$$\text{Mean} = \bar{x} = \frac{\text{sum of all data values}}{\text{number of data values}}$$

EXAMPLE

The mean of the 5 data values 39, 43, 27, 51, and 40 is

$$\bar{x} = \frac{39 + 43 + 27 + 51 + 40}{5} = \frac{200}{5} = 40$$

To find the mean of the miles-per-gallon data, add the 15 numbers and then divide by 15.

$$\bar{x} = \frac{25 + 22 + 21 + 27 + 25 + 35 + 29 + 31 + 25 + 26 + 21 + 39 + 34 + 32 + 28}{15}$$

$$= \frac{420}{15} = 28$$

The mean number of miles per gallon for the 15 cars tested was 28 mi/gal.

The mean is one of the most frequently computed averages. It is the one that is commonly used to calculate a student's performance in a class.

The scores for a history student on 5 tests were 78, 82, 91, 87, and 93. What was the mean score for this student?

To find the mean, add the numbers. $\bar{x} = \dfrac{78 + 82 + 91 + 87 + 93}{5}$

Then divide by 5. $= \dfrac{431}{5} = 86.2$

The mean score for the history student was 86.2.

Focus on finding the mean of a data set

Twenty students were asked the number of units in which they were currently enrolled. The responses were

15 12 13 15 17 18 13 20 9 16

14 10 15 12 17 16 6 14 15 12

Find the mean number of units taken by these students.

STRATEGY

To find the mean number of units taken by the 20 students:
- Determine the sum of the numbers.
- Divide the sum by 20.

SOLUTION

$15 + 12 + 13 + 15 + 17 + 18 + 13 + 20 + 9 +$
$16 + 14 + 10 + 15 + 12 + 17 + 16 + 6 +$
$14 + 15 + 12 = 279$ • The sum of the numbers

$\bar{x} = \dfrac{279}{20} = 13.95$

The mean is 13.95 units.

Check your understanding 1

The amounts spent on lunch by the last 10 customers at a fast-food restaurant were

6.32 8.21 7.45 7.90 7.58 6.45 7.05 8.00 5.59 6.75

Find the mean amount spent on lunch by these customers.

SOLUTION See page S-2.

Focus on solving an application using the mean

A bowler has scores of 165, 172, 168, and 185 for four games. What score must the bowler achieve on the next game so that the mean for the five games is 174?

STRATEGY

To find the score, use the formula for the mean, letting n be the score on the fifth game.

SOLUTION

$$174 = \frac{165 + 172 + 168 + 185 + n}{5}$$

$$174 = \frac{690 + n}{5}$$

$870 = 690 + n$ • Multiply each side by 5.

$180 = n$ • Subtract 690 from each side.

The score on the fifth game must be 180.

Check your understanding 2

You have scores of 82, 91, 79, and 83 on four exams. What score must you receive on the fifth exam to have a mean of 84 for the five exams?

SOLUTION See page S-2.

The **median** of a data set is the number that separates the data into two equal parts when the numbers are arranged from smallest to largest (or largest to smallest). There are always an equal number of values above the median and below the median.

To find the median of a set of numbers, first arrange the numbers from smallest to largest. The median is the number in the middle. The data for the miles-per-gallon ratings given on page 7 are arranged below from smallest to largest.

Take Note ➤
Half the data values are less than 27 and half the data values are greater than 27. The median indicates the center, or middle, of the set of data.

21 21 22 25 25 25 26 27 28 29 31 32 34 35 39

7 values below ↑ 7 values above

middle number
Median

The median is 27.

If the data contain an *even* number of values, the median is the sum of the two middle numbers, divided by 2.

Take Note ➤
If the data contain an even number of values, the median is the mean of the two middle values.

The selling prices of the last six homes sold by a real estate agent were $175,000, $150,000, $250,000, $130,000, $245,000, and $190,000. Find the median selling price of these homes.

Arrange the numbers from smallest to largest. Because there are an even number of values, the median is the sum of the two middle numbers, divided by two.

130,000 150,000 175,000 190,000 245,000 250,000

middle 2 numbers: 175,000 190,000

$$\text{Median} = \frac{175{,}000 + 190{,}000}{2} = 182{,}500$$

The median selling price of a home was $182,500.

Focus on finding the median of a data set

Twenty students were asked the number of units in which they were currently enrolled. The responses were

15 12 13 15 17 18 13 20 9 16

14 10 15 12 17 16 6 14 15 12

Find the median number of units taken by these students. (Recall that we found the mean of these data in the first example.)

STRATEGY

To find the median number of units taken by the 20 students:
- Arrange the numbers from smallest to largest.
- Because there is an even number of values, the median is the sum of the two middle numbers, divided by 2.

SOLUTION

6 9 10 12 12 12 13 13 14 14

15 15 15 15 16 16 17 17 18 20

$$\text{Median} = \frac{14 + 15}{2} = 14.5$$

The median is 14.5 units.

Check your understanding 3

The amounts spent on lunch by the last 10 customers at a fast-food restaurant were

6.32 8.21 7.45 7.90 7.58 6.45 7.05 8.00 5.59 6.75

Find the median amount spent on lunch by these customers. (Recall that we found the mean of these data in Check your understanding 1.)

SOLUTION See page S-2.

The **mode** of a set of numbers is the value that occurs most frequently. If a set of numbers has no number that occurs more than once, then the data have no mode.

Here again are the data for the gasoline mileage ratings of cars.

Miles-per-Gallon Ratings of 15 Cars

25 22 21 27 25 35 29 31 25 26 21 39 34 32 28

↑ (25) ↑ (25) ↑ (25)

25 is the number that occurs most frequently. The mode is 25.

The gasoline mileage ratings data show that the mean, median, and mode of a set of numbers do not have to be the same value. For the data on the miles-per-gallon ratings of 15 cars,

Mean = 28 Median = 27 Mode = 25

Although any of the averages can be used when the data collected consist of numbers, the mean and median are not appropriate for *qualitative* data. Examples of qualitative data include recording a person's favorite color or recording a person's preference from among classical, hard rock, jazz, rap, and country western music. It does not make sense to say that the *average* favorite color is red or the *average* musical choice is jazz. The mode is used to indicate the most frequently chosen color or musical category. The **modal response** is the category that receives the greatest number of responses.

A survey asked people to state whether they strongly disagreed, disagreed, had no opinion, agreed, or strongly agreed with the state governor's position on increasing taxes to pay for health care. What was the modal response for these data?

Strongly disagreed	57
Disagreed	68
No opinion	12
Agreed	45
Strongly agreed	58

Because a response of "disagreed" was recorded most frequently, the modal response was "disagreed."

Objective 6.2A Practice

1. The numbers of big-screen televisions sold each month for one year were recorded by an electronics store. The results were 15, 12, 20, 20, 19, 17, 22, 24, 17, 20, 15, and 27. Calculate the mean and median number of televisions sold per month.
2. The numbers of seats occupied on a jet for 16 transatlantic flights were recorded. The numbers were 309, 422, 389, 412, 401, 352, 367, 319, 410, 391, 330, 408, 399, 387, 411, and 398. Calculate the mean and median number of occupied seats.
3. Your scores on six history tests were 78, 92, 95, 77, 94, and 88. If an "average" score of 90 receives an A for the course, which average, the mean or the median, would you prefer the instructor use?
4. A survey by an ice cream store asked people to name their favorite ice cream from five flavors. The responses were mint chocolate chip, 34; pralines and cream, 27; German chocolate cake, 44; chocolate raspberry swirl, 34; and rocky road, 42. What was the modal response?
5. A survey asked people to rate the performance of the city's mayor. The responses were very unsatisfactory, 230; unsatisfactory, 403; satisfactory, 1237; very satisfactory, 403. What was the modal response for this survey?

Solutions on pp. S-8–S-10.

Objective 6.2B Find the standard deviation of a distribution

Consider two students, each of whom has taken five exams.

Student A

84 86 83 85 87

$$\bar{x} = \frac{84 + 86 + 83 + 85 + 87}{5} = \frac{425}{5} = 85$$

The mean for Student A is 85.

Student B

90 75 94 68 98

$$\bar{x} = \frac{90 + 75 + 94 + 68 + 98}{5} = \frac{425}{5} = 85$$

The mean for Student B is 85.

For each of these students, the mean (average) for the five tests is 85. However, Student A has a more consistent record of scores than Student B. One measure of the consistency, or "clustering," of data near the mean is the **standard deviation.**

To calculate the standard deviation:

1. Sum the squares of the differences between each data value and the mean.
2. Divide the result in Step 1 by the number of data values.
3. Take the square root of the result in Step 2.

Here is the calculation for Student A. The symbol for standard deviation is the Greek letter *sigma*, denoted by σ.

Step 1

x	$(x - \bar{x})$	$(x - \bar{x})^2$
84	$(84 - 85)$	$(-1)^2 = 1$
86	$(86 - 85)$	$1^2 = 1$
83	$(83 - 85)$	$(-2)^2 = 4$
85	$(85 - 85)$	$0^2 = 0$
87	$(87 - 85)$	$2^2 = 4$
		Total $= 10$

Step 2 $\frac{10}{5} = 2$

Step 3 $\sigma = \sqrt{2} \approx 1.414$

Take Note ➤ The standard deviation of Student B's scores is greater than the standard deviation of Student A's scores, and the range of Student B's scores $(98 - 68 = 30)$ is greater than the range of Student A's scores $(87 - 83 = 4)$.

The standard deviation of Student A's scores is approximately 1.414.

Following a similar procedure for Student B, we find that the standard deviation of Student B's scores is approximately 11.524. Because the standard deviation of Student B's scores is greater than that of Student A's scores $(11.524 > 1.414)$, Student B's scores are not as consistent as those of Student A.

In this text, standard deviations are rounded to the nearest thousandth.

Focus on finding the standard deviation of a data set

The weights in pounds of the players on the five-man front line of a college football team are 210, 245, 220, 230, and 225. Find the standard deviation of the weights.

STRATEGY

To calculate the standard deviation:

- Find the mean of the weights.
- Use the procedure for calculating standard deviation.

SOLUTION

$$\bar{x} = \frac{210 + 245 + 220 + 230 + 225}{5} = 226$$

Step 1

x	$(x - \bar{x})^2$
210	$(210 - 226)^2 = 256$
245	$(245 - 226)^2 = 361$
220	$(220 - 226)^2 = 36$
230	$(230 - 226)^2 = 16$
225	$(225 - 226)^2 = 1$
	Total $= 670$

Step 2

$\frac{670}{5} = 134$

Step 3

$\sigma = \sqrt{134} \approx 11.576$

The standard deviation of the weights is approximately 11.576 lb.

Check your understanding 4

The numbers of miles logged by a runner for the last six days of running were 5, 7, 3, 6, 9, and 6. Find the standard deviation of the numbers of miles run. Round to the nearest thousandth.

SOLUTION See page S-3.

Objective 6.2B Practice

For Exercises 1 to 3, round to the nearest thousandth.

1. An airline recorded the times for a ground crew to unload the baggage from an airplane. The recorded times, in minutes, were 12, 18, 20, 14, and 16. Find the standard deviation of these times.
2. The numbers of rooms occupied in a hotel on six consecutive days were 234, 321, 222, 246, 312, and 396. Find the standard deviation of the numbers of rooms occupied.
3. Seven coins were tossed 100 times. The numbers of heads recorded were 56, 63, 49, 50, 48, 53, and 52. Find the standard deviation of the numbers of heads.

Solutions on pp. S-10–S-11.

SECTION 6.3 Measures of Relative Position

Objective 6.3A Calculate percentiles and quartiles

Most standardized examinations provide scores in terms of *percentiles*, which are defined as follows:

***p*th Percentile**
A value x is called the ***p*th percentile** of a data set provided $p\%$ of the data values are less than x.

Focus on using percentiles

According to the U.S. Department of Labor, as of May 2009 the median annual salary for a physical therapist was \$74,480. If the 90th percentile for the annual salary of a physical therapist was \$105,900, find the percent of physical therapists whose annual salary was

a. more than \$74,480.
b. less than \$105,900.
c. between \$74,480 and \$105,900.

SOLUTION

a. By definition, the median is the 50th percentile. Therefore, 50% of the physical therapists earned more than \$74,480 per year.
b. Because \$105,900 is the 90th percentile, 90% of all physical therapists made less than \$105,900.
c. From parts (a) and (b), $90\% - 50\% = 40\%$ of the physical therapists earned between \$74,480 and \$105,900.

Check your understanding 1

According to the U.S. Department of Labor, as of May 2009 the median annual salary for a police dispatcher was \$34,790. If the 25th percentile for the annual salary of a police dispatcher was \$27,550, find the percent of police dispatchers whose annual salary was

a. less than \$34,790.
b. more than \$27,550.
c. between \$27,550 and \$34,790.

SOLUTION See page S-3.

The following formula can be used to find the percentile that corresponds to a particular data value in a set of data.

Percentile for a Given Data Value

Given a set of data and a data value x,

$$\text{Percentile of score } x = \frac{\text{number of data values less than } x}{\text{total number of data values}} \cdot 100$$

Focus on finding a percentile

On a reading examination given to 900 students, Elaine's score of 602 was higher than the scores of 576 of the students who took the examination. What is the percentile for Elaine's score?

SOLUTION

$$\begin{aligned}\text{Percentile} &= \frac{\text{number of data values less than 602}}{\text{total number of data values}} \cdot 100\\ &= \frac{576}{900} \cdot 100\\ &= 64\end{aligned}$$

Elaine's score of 602 places her at the 64th percentile.

Check your understanding 2

On an examination given to 8600 students, Hal's score of 405 was higher than the scores of 3952 of the students who took the examination. What is the percentile for Hal's score?

SOLUTION See page S-3.

The three numbers Q_1, Q_2, and Q_3 that partition a data set into four (approximately) equal groups are called the **quartiles** of the data. For instance, for the data set below, the values $Q_1 = 11$, $Q_2 = 29$, and $Q_3 = 104$ are the quartiles of the data.

2, 5, 5, 8, 11, 12, 19, 22, 23, 29, 31, 45, 83, 91, 104, 159, 181, 312, 354

(11 ↑ Q_1; 29 ↑ Q_2; 104 ↑ Q_3)

The quartile Q_1 is called the *first quartile*. The quartile Q_2 is called the *second quartile*. It is the median of the data. The quartile Q_3 is called the *third quartile*. The following method of finding quartiles makes use of medians.

The Median Procedure for Finding Quartiles

1. Arrange the data from smallest to largest or largest to smallest. This is called *ranking* the data.
2. Find the median of the data. This is the second quartile, Q_2.
3. The first quartile, Q_1, is the median of the data values less than Q_2. The third quartile, Q_3, is the median of the data values greater than Q_2.

Focus on using medians to find the quartiles of a data set

The numbers of calories per 100 ml of 25 popular sodas are listed below. Find the quartiles for the data.

43	37	42	40	53	62	36	32	50	49
26	53	73	48	45	39	45	48	40	56
41	36	58	42	39					

SOLUTION

Step 1: Rank the data as shown below.

1) 26	**2)** 32	**3)** 36	**4)** 36	**5)** 37	**6)** 39	**7)** 39	**8)** 40	**9)** 40
10) 41	**11)** 42	**12)** 42	**13)** 43	**14)** 45	**15)** 45	**16)** 48	**17)** 48	**18)** 49
19) 50	**20)** 53	**21)** 53	**22)** 56	**23)** 58	**24)** 62	**25)** 73		

Step 2: The median of these 25 data values has a rank of 13. Thus the median is 43. The second quartile Q_2 is the median of the data, so $Q_2 = 43$.

Step 3: There are 12 data values less than the median and 12 data values greater than the median. The first quartile is the median of the data values less than the median. Thus Q_1 is the mean of the data values with ranks of 6 and 7.

$$Q_1 = \frac{39 + 39}{2} = 39$$

The third quartile is the median of the data values greater than the median. Thus Q_3 is the mean of the data values with ranks of 19 and 20.

$$Q_3 = \frac{50 + 53}{2} = 51.5$$

Check your understanding 3

The weights, in ounces, of 15 avocados in a random sample are listed below. Find the quartiles for the data.

12.4	10.8	14.2	7.5	10.2	11.4	12.6	12.8	13.1	15.6
9.8	11.4	12.2	16.4	14.5					

SOLUTION See page S-3.

Objective 6.3A Practice

1. On a reading test, Shaylen's score of 455 was higher than the scores of 4256 of the 7210 students who took the test. Find the percentile, rounded to the nearest percent, for Shaylen's score.
2. On a placement examination, Rick scored lower than 1210 of the 12,860 students who took the exam. Find the percentile, rounded to the nearest percent, for Rick's score.

3. Kevin scored at the 65th percentile on a test given to 9840 students. How many students scored lower than Kevin?
4. A survey was given to 18 students. One question asked about the one-way distance the student had to travel to attend college. The results, in miles, are shown in the following table. Use the median procedure for finding quartiles to find the first, second, and third quartiles for the data.

Miles Traveled to Attend College								
12	18	4	5	26	41	1	8	10
10	3	28	32	10	85	7	5	15

5. The following table shows the number of prescriptions a doctor wrote each day for a 36-day period. Use the median procedure for finding quartiles to find the first, second, and third quartiles for the data.

Number of Prescriptions Written per Day					
8	12	14	10	9	16
7	14	10	7	11	16
11	12	8	14	13	10
9	14	15	12	10	8
10	14	8	7	12	15
14	10	9	15	10	12

Solutions on pp. S-11–S-12.

Objective 6.3B Create a box-and-whiskers plot

A **box-and-whiskers plot** is a graph that gives a more comprehensive picture of a set of data. A box-and-whiskers plot shows five numbers: the smallest value, the first quartile, the median, the third quartile, and the largest value. The **interquartile range** is the difference between Q_3 and Q_1.

A box-and-whiskers plot shows the data in the interquartile range as a box. Suppose we are given the following data on the prices of 15 half-gallon cartons of ice cream. The data are arranged in order from lowest to highest price.

2.39 2.66 3.18 3.26 3.58 3.86 4.18 4.29

4.36 4.45 4.56 4.71 4.83 5.44 5.49

From these data, we can see that the smallest value is 2.39, the largest value is 5.49, and the median is 4.29. We can also calculate $Q_1 = 3.26$ and $Q_3 = 4.71$. Now think of a number line that includes the five values just listed. With this in mind, mark off the five values. Draw a box that spans the distance from Q_1 to Q_3. Draw a vertical line the height of the box at the median, Q_2.

The box-and-whiskers plot for the data on the prices of ice cream is shown below.

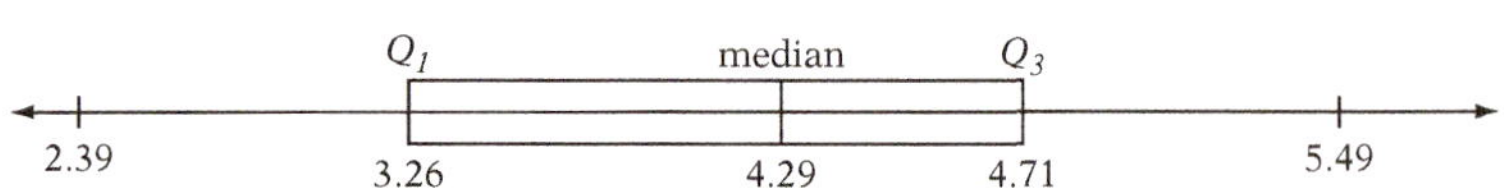

Note that the box-and-whiskers plot labels five values: the smallest, 2.39; the first quartile Q_1, 3.26; the median, 4.29; the third quartile Q_3, 4.71; and the largest value, 5.49.

Focus on drawing a box-and-whiskers plot

The numbers of people registered for 14 dance classes at a dance studio are listed below.

30 45 54 24 48 38 43

38 46 53 62 64 40 35

Draw a box-and-whiskers plot for the data.

STRATEGY

To draw the box-and-whiskers plot:

- Arrange the data from smallest to largest.
- Find the median.
- Find Q_1, the median of the lower half of the data, and Q_3, the median of the upper half of the data.
- Draw the box-and-whiskers plot using the following five values: the smallest value, Q_1, the median, Q_3, and the largest value.

SOLUTION

24 30 35 **38** 38 40 43

45 46 48 **53** 54 62 **64**

$$\text{Median} = \frac{43 + 45}{2} = 44$$

$Q_1 = 38$ • The median of the top row of data

$Q_3 = 53$ • The median of the bottom row of data

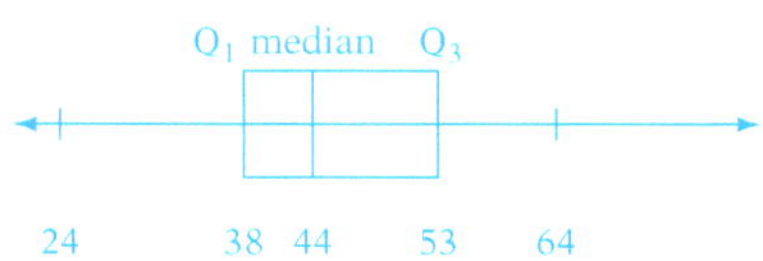

Check your understanding 4

The numbers of people attending a quilt exhibit during an 18-day period are listed below.

86 77 58 45 94 96 83 76 75

65 68 72 78 85 87 92 55 61

Draw a box-and-whiskers plot for the data.

SOLUTION See page S-4.

Objective 6.3B Practice

1. The hourly wages for entry-level positions at various firms were recorded by a labor research firm. The results follow. Find the first and third quartiles, and draw a box-and-whiskers plot of the data.

Starting Hourly Wages for 16 Companies

8.09	11.50	7.46	7.70	9.85	9.03	11.40	9.31
10.35	7.45	7.35	8.64	9.02	8.12	10.05	8.94

2. The cholesterol levels for 14 adults are recorded in the table below. Find the first and the third quartiles, and draw a box-and-whiskers plot of the data.

Cholesterol Levels for 14 Adults

375	185	254	221	183	251	258
292	214	172	233	208	198	211

3. The gasoline consumption of 19 cars was tested, and the results were recorded in the following table. Find the first and the third quartiles, and draw a box-and-whiskers plot of the data.

Miles-per-Gallon Ratings for 19 Cars

33	21	30	32	20	31	25	20	16	24
22	31	30	28	26	19	21	17	26	

Solutions on pp. S-12–S-13.

SECTION 6.4 Introduction to Probability

Objective 6.4A Calculate the probability of simple events

A weather forecaster estimates that there is a 75% chance of rain. A state lottery director claims that there is a $\frac{1}{9}$ chance of winning a prize offered by the lottery. Each of these statements involves some degree of uncertainty. The degree of uncertainty is called **probability.** For the statements above, the probability of rain is 75% and the probability of winning a prize in the lottery is $\frac{1}{9}$.

A probability is determined from an **experiment,** which is any activity that has an observable outcome. Examples of experiments are

Tossing a coin and observing whether it lands heads or tails

Interviewing voters to determine their preference for a political candidate

Recording the percent change in the price of a stock

All of the possible outcomes of an experiment are called the **sample space** of the experiment. The outcomes of an experiment are listed between braces and frequently designated by S.

For each experiment, list all of the possible outcomes.

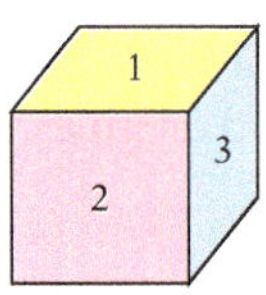

1. A number cube, which has the numbers from 1 to 6 written on its sides, is rolled once.

Any of the numbers from 1 to 6 could show on the top of the cube.

$S = \{1, 2, 3, 4, 5, 6\}$

2. A fair coin is tossed once.

A fair coin is one for which heads and tails have an equal chance of being tossed.

$S = \{H, T\}$, where H represents heads and T represents tails.

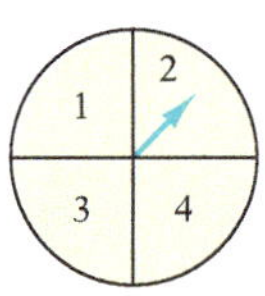

3. The spinner at the left is spun once.

Assuming the spinner does not come to rest on a line, the arrow could come to rest in any one of the four sectors. $S = \{1, 2, 3, 4\}$

An **event** is one or more outcomes of an experiment. Events are denoted by capital letters. Consider the experiment of rolling the number cube given on page 18. Some possible events are

The number is even. $E = \{2, 4, 6\}$

The number is a prime number. $P = \{2, 3, 5\}$

The number is less than 10. $T = \{1, 2, 3, 4, 5, 6\}$. Note that in this case, the event is the entire sample space.

The number is greater than 20. This event is impossible for the given sample space. The impossible event is symbolized by $\emptyset$.

When discussing experiments and events, it is convenient to refer to the *favorable outcomes* of an experiment. These are the outcomes of the experiment that satisfy the requirements of the particular event. For instance, consider the experiment of rolling a fair die once. The sample space is $\{1, 2, 3, 4, 5, 6\}$, and one possible event E would be rolling a number that is divisible by 3. The outcomes of the experiment that are favorable to E are 3 and 6, and $E = \{3, 6\}$.

Probability Formula

The probability of an event E, written $P(E)$, is the ratio of the number of favorable outcomes of an experiment to the total number of possible outcomes of the experiment.

$$P(E) = \frac{\text{number of favorable outcomes}}{\text{number of possible outcomes}}$$

The outcomes of the experiment of tossing a fair coin are *equally likely*. Either one of the outcomes is just as likely as the other. If a fair coin is tossed once, the probability of a head or a tail is $\frac{1}{2}$. Each event, heads or tails, is equally likely. The probability formula applies to experiments for which the outcomes are equally likely.

Not all experiments have equally likely outcomes. Consider an exhibition baseball game between a professional team and a college team. Although either team *could* win the game, the probability that the professional team will win is greater than the probability that the college team will win. The outcomes are not equally likely. For the experiments in this section, we will assume that the outcomes are equally likely.

There are five choices, *a* through *e*, for each question on a multiple-choice test. By just guessing, what is the probability of choosing the correct answer for a certain question?

It is possible to select any of the letters *a*, *b*, *c*, *d*, or *e*. There are 5 possible outcomes of the experiment.

The event E is guessing the correct answer. There is 1 favorable outcome, guessing the correct answer.

Use the probability formula. $P(E) = \dfrac{\text{number of favorable outcomes}}{\text{number of possible outcomes}} = \dfrac{1}{5}$

The probability of guessing the correct answer is $\frac{1}{5}$.

Each of the letters in the word *Tennessee* is written on a card, and the cards are placed in a hat. If one card is drawn at random from the hat, what is the probability that the card has the letter *e* on it?

The phrase "at random" means that each card has an equal chance of being drawn.

There are 9 letters in *Tennessee*. Therefore, there are 9 possible outcomes of the experiment.

There are 4 cards with an *e* on them.

There are 4 favorable outcomes of the experiment, the 4 *e*'s.

Use the probability formula.

$$P(E) = \frac{\text{number of favorable outcomes}}{\text{number of possible outcomes}} = \frac{4}{9}$$

The probability is $\frac{4}{9}$.

Focus on calculating a probability

A large box contains 25 red, 35 blue, and 40 white balls. If one ball is randomly selected from the box, what is the probability that it is blue? Write the answer as a percent.

STRATEGY

To find the probability:
- Count the number of outcomes of the experiment.
- Count the number of outcomes of the experiment favorable to the event E that the ball selected is blue.
- Use the probability formula.

SOLUTION

There are 100 $(25 + 35 + 40)$ balls in the box.

35 balls of the 100 are blue.

$$P(E) = \frac{35}{100} = 0.35 = 35\%$$

There is a 35% chance of selecting a blue ball.

Check your understanding 1

There are eight covered circles on a "scratch card" that is given to each customer at a fast-food restaurant. Under one of the circles on each card is a symbol for a free soft drink. If the customer scratches off one circle, what is the probability that the soft drink symbol will be uncovered?

SOLUTION See page S-4.

Calculating the probability of an event requires counting the number of possible outcomes of an experiment and the number of outcomes that are favorable to the event. One way to do this is to list the outcomes of the experiment in some systematic way. Using a table is often very helpful.

Q_1	Q_2	Q_3
T	T	T
T	T	F
T	F	T
T	F	F
F	T	T
F	T	F
F	F	T
F	F	F

A professor writes three true/false questions for a test. If the professor randomly chooses which questions will have an answer of true and which will have an answer of false, what is the probability that the test will have 2 true questions and 1 false question?

The experiment S consists of choosing T or F for each of the three questions. The possible outcomes of the experiment are shown in the table at the left.

$S = \{TTT, TTF, TFT, TFF, FTT, FTF, FFT, FFF\}$

There are 8 outcomes for S.

The event E consists of 2 true questions and 1 false question.

$E = \{TTF, TFT, FTT\}$

There are 3 outcomes for E.

Use the probability formula.

$$P(E) = \frac{3}{8}$$

The probability of 2 true questions and 1 false question is $\frac{3}{8}$.

Two dice are rolled, one after the other. The sample space is shown below. There are 36 possible outcomes.

Possible Outcomes of Rolling Two Dice

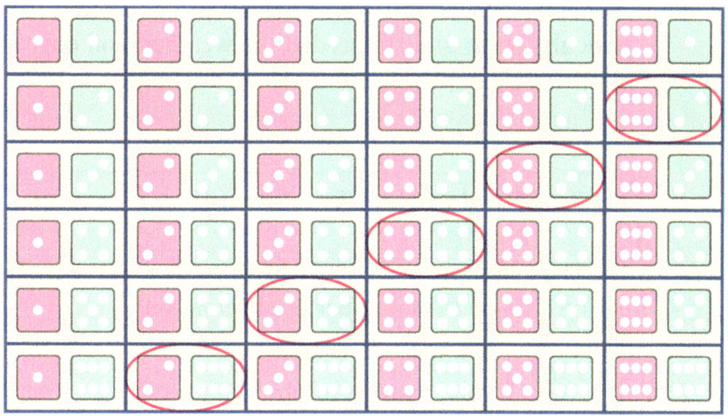

The outcomes in which the sum of the numbers rolled is 8 are circled above. There are 5 possible outcomes that have a sum of 8. The probability that the sum of the numbers rolled is 8 is $P(E) = \frac{5}{36}$.

Focus on calculating a probability involving dice

Two dice are rolled once. Calculate the probability that the sum of the numbers rolled is 7.

STRATEGY

To find the probability:

- Count the number of possible outcomes of the experiment.
- Count the outcomes of the experiment that are favorable to the event "the sum is 7."
- Use the probability formula.

SOLUTION

There are 36 possible outcomes.

There are 6 outcomes favorable to E: (1, 6), (2, 5), (3, 4), (4, 3), (5, 2), and (6, 1).

$$P(E) = \frac{6}{36} = \frac{1}{6}$$

The probability that the sum is 7 is $\frac{1}{6}$.

Check your understanding 2

Two dice are rolled once. Calculate the probability that the two numbers showing on the upward faces are equal.

SOLUTION See page S-4.

The probabilities that we have calculated so far are referred to as *mathematical* or *theoretical probabilities*. The calculations are based on theory—for example, that either side of a coin is equally likely to land face up or that each of the six sides of a fair die is equally likely to land face up. Not all probabilities arise from such assumptions.

Empirical probabilities are based on observations. For instance, a weather forecast of a 75% chance of rain is an empirical probability. From historical records kept by the weather bureau, when a similar weather pattern existed, rain occurred 75% of the time. It is theoretically impossible to predict the weather, and only observations of past weather patterns can be used to predict future weather conditions.

Empirical Probability Formula

The empirical probability of an event E is the ratio of the number of observations of E to the total number of observations.

$$P(E) = \frac{\text{number of observations of } E}{\text{total number of observations}}$$

Records from an insurance company show that of 2549 claims for theft filed by policyholders, 927 were claims for more than \$5000. What is the empirical probability that the next claim for theft this company receives will be a claim for more than \$5000?

The empirical probability of E is the ratio of the number of claims for over \$5000 to the total number of claims.

$$P(E) = \frac{927}{2549} \approx 0.36$$

The probability is approximately 0.36.

Objective 6.4A Practice

1. Three cards—one red, one green, and one blue—are to be arranged in a stack. Using R for red, G for green, and B for blue, make a list of all the different stacks that can be formed. (Some computer monitors are called RGB monitors for the colors red, green, and blue.)
2. If two dice are rolled, what is the probability that the sum of the dots showing on the upward faces is 5?

3. A dodecahedral die has 12 sides labeled with the numbers 1 to 12. If the die is rolled once, what is the probability that the upward face shows an 11?

4. A tetrahedral die has four sides labeled with the numbers 1 to 4. If two tetrahedral dice are rolled, what is the probability that the sum of the numbers showing on the upward faces is 4?

5. A survey of 725 people showed that 587 had a group health insurance plan provided by their employer. Based on this survey, what is the empirical probability that an employee has a group health insurance plan? Write the answer as a percent rounded to the nearest tenth.

Solutions on pp. S-14–S-15.

Objective 6.4B Calculate the probability of compound events

Suppose you draw a single card from a standard deck of playing cards. The sample space S consists of the 52 cards of the deck. We can express the number of elements in the sample space using the notation $n(S) = 52$. Now consider the events

$$E_1 = \text{a 4 is drawn} = \{\spadesuit 4, \heartsuit 4, \diamondsuit 4, \clubsuit 4\}$$

$$E_2 = \text{a spade is drawn}$$
$$= \{\spadesuit A, \spadesuit 2, \spadesuit 3, \spadesuit 4, \spadesuit 5, \spadesuit 6, \spadesuit 7, \spadesuit 8, \spadesuit 9, \spadesuit 10, \spadesuit J, \spadesuit Q, \spadesuit K\}$$

It is possible, on one draw, to satisfy the conditions of both events: the $\spadesuit 4$ could be drawn. This card is an element of both E_1 and E_2.

Now consider the events

$$E_3 = \text{a 5 is drawn} = \{\spadesuit 5, \heartsuit 5, \diamondsuit 5, \clubsuit 5\}$$

$$E_4 = \text{a king is drawn} = \{\spadesuit K, \heartsuit K, \diamondsuit K, \clubsuit K\}$$

In this case, it is not possible to draw one card that satisfies the conditions of both events. There are no elements common to both sets. Two events that cannot occur at the same time are called **mutually exclusive events.** The events E_3 and E_4 are mutually exclusive events, whereas E_1 and E_2 are not.

Mutually Exclusive Events

Two events A and B are mutually exclusive if they cannot occur at the same time. That is, A and B are mutually exclusive when $A \cap B = \varnothing$.

The probability that either of two mutually exclusive events will occur can be determined by adding the probabilities of the individual events.

Probability of Mutually Exclusive Events

If A and B are two mutually exclusive events, then the probability that either A or B will occur is

$$P(A \text{ or } B) = P(A) + P(B)$$

Focus on finding the probability of mutually exclusive events

Suppose a single card is drawn from a standard deck of playing cards. Find the probability of drawing a 5 or a king.

SOLUTION

Let $A = \{♠5, ♥5, ♦5, ♣5\}$ and $B = \{♠K, ♥K, ♦K, ♣K\}$. There are 52 cards in a standard deck of playing cards; thus $n(S) = 52$. Because the events are mutually exclusive, we can use the formula for the probability of mutually exclusive events.

$$P(A \text{ or } B) = P(A) + P(B)$$

• Formula for the probability of mutually exclusive events

$$= \frac{1}{13} + \frac{1}{13} = \frac{2}{13}$$

• $P(A) = \frac{4}{52} = \frac{1}{13}$, $P(B) = \frac{4}{52} = \frac{1}{13}$

The probability of drawing a 5 or a king is $\frac{2}{13}$.

Check your understanding 3

Two fair dice are tossed once. What is the probability of rolling a 7 or an 11? For the sample space for this experiment, see page 23.

SOLUTION See page S-5.

Consider the experiment of rolling two dice. Let A be the event of rolling a sum of 8 and let B be the event of rolling a double (the same number on both dice).

$A =$ 

$B =$ 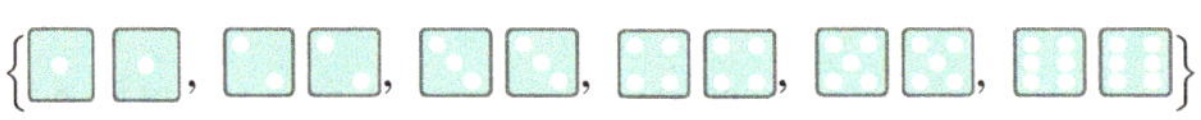

These events are *not* mutually exclusive because it is possible to satisfy the conditions of both events on one toss of the dice—a  could be rolled. Therefore, $P(A \text{ or } B)$, the probability of a sum of 8 or a double, cannot be calculated using the formula for the probability of mutually exclusive events. However, a modification of that formula can be used.

Addition Rule for Probabilities

If A and B are two events in a sample space S, then

$$P(A \text{ or } B) = P(A) + P(B) - P(A \text{ and } B)$$

In this formula, $P(A \text{ and } B) = \frac{n(A \cap B)}{n(S)}$, where $n(A \cap B)$ represents the number of elements favorable to *both* A and B, and $n(S)$ represents the number of possible outcomes (the number of elements in the sample space).

Using this formula with

$A =$ 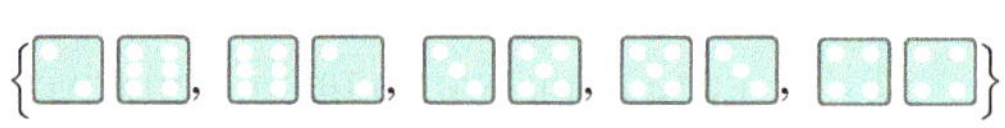

$B =$ 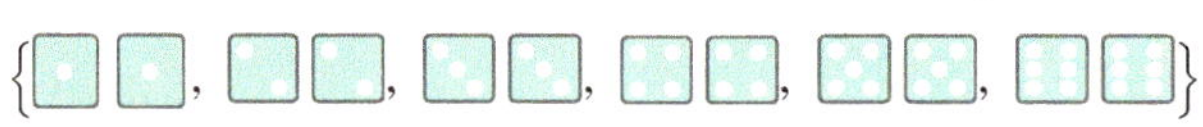

$A \cap B =$ 

the probability of A or B can be calculated.

$$P(A \text{ or } B) = P(A) + P(B) - P(A \text{ and } B)$$

$$= \frac{5}{36} + \frac{6}{36} - \frac{1}{36}$$

• $P(A) = \frac{5}{36}, P(B) = \frac{6}{36}, P(A \cap B) = \frac{1}{36}$

$$= \frac{10}{36} = \frac{5}{18}$$

On a single roll of two dice, the probability of rolling a sum of 8 or a double is $\frac{5}{18}$.

Focus on using the addition rule for probabilities

The table at the left shows data from an experiment conducted to test the effectiveness of a flu vaccine. If one person is selected from this population, what is the probability that the person was vaccinated or contracted the flu?

	F	No *F*	Total
V	21	198	219
No *V*	76	195	271
Total	97	393	490

V: Vaccinated
F: Contracted the flu

SOLUTION

Let $V = \{\text{people who were vaccinated}\}$ and $F = \{\text{people who contracted the flu}\}$. These events are not mutually exclusive because there are 21 people who were vaccinated and who contracted the flu. The sample space S consists of the 490 people who participated in the experiment. From the table, $n(V) = 219$, $n(F) = 97$, and $n(V \text{ and } F) = 21$.

$$P(V \text{ or } F) = P(V) + P(F) - P(V \text{ and } F)$$

$$= \frac{219}{490} + \frac{97}{490} - \frac{21}{490}$$

$$= \frac{295}{490} \approx 0.602$$

The probability of selecting a person who was vaccinated or who contracted the flu is approximately 60.2%.

Check your understanding 4

The data in the following table show the starting salaries of college graduates with selected degrees. If one person is chosen from this population, what is the probability that the person has a degree in business or has a starting salary between \$20,000 and \$24,999?

	Degree			
Salary (in \$)	**Engineering**	**Business**	**Chemistry**	**Psychology**
Less than 20,000	0	4	1	12
20,000–24,999	4	16	3	16
25,000–29,999	7	21	5	15
30,000–34,999	12	35	5	7
35,000 or more	12	22	4	5

SOLUTION See page S-5.

Consider the experiment of tossing a single die once. The sample space is

$S =$ {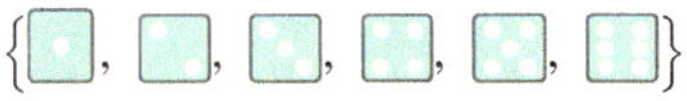}

Now consider the event $E =$ {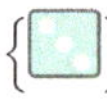}, that is, the event of tossing a 3. Recall that the probability of E is

$$P(E) = \frac{1}{6} \quad \begin{array}{l} \leftarrow \text{Number of elements in } E \\ \leftarrow \text{Number of elements in the sample space} \end{array}$$

The **complement** of an event E, symbolized by E^c, is the "opposite" of event E. The complement includes all those outcomes of S that are not in E and excludes the outcomes in E. For the event E above, E^c is the event of not tossing a 3. Thus

$E =$ {}

Note that because E and E^c are opposite events, they are mutually exclusive, and their union is the entire sample space S. Thus $P(E) + P(E^c) = P(S)$. But $P(S) = 1$, so $P(E^c) = 1 - P(E)$.

Probability of the Complement of an Event
If E is an event and E^c is the complement of the event, then $$P(E^c) = 1 - P(E)$$

Continuing our example, the probability of not tossing a 3 is given by

$$\begin{aligned} P(\text{not tossing a 3}) &= 1 - P(\text{tossing a 3}) \\ &= 1 - \frac{1}{6} = \frac{5}{6} \end{aligned}$$

You can also verify the probability of E^c directly:

$$P(\text{not tossing a 3}) = \frac{5}{6} \quad \begin{array}{l} \leftarrow \text{Number of elements in } E^c \\ \leftarrow \text{Number of elements in the sample space} \end{array}$$

Focus on finding the probability of the complement of an event

The probability of tossing a sum of 11 on the toss of two dice is $\frac{1}{18}$. What is the probability of not tossing a sum of 11 on the toss of two dice?

SOLUTION

Use the formula for the probability of the complement of an event. Let $E = \{\text{toss a sum of 11}\}$. Then $E^c = \{\text{toss a sum that is not 11}\}$.

$$\begin{aligned} P(E^c) &= 1 - P(E) \\ &= 1 - \frac{1}{18} = \frac{17}{18} \qquad \bullet\ P(E) = \frac{1}{18} \end{aligned}$$

The probability of not tossing a sum of 11 is $\frac{17}{18}$.

Check your understanding 5

The probability that a person has type A blood is 34%. What is the probability that a person does not have type A blood?

SOLUTION See page S-5.

Take Note ➤
The phrase "at least one" means one or more. Tossing a coin three times and asking the probability of getting at least one head means to calculate the probability of getting one, two, or three heads.

Suppose we toss a coin three times and want to calculate the probability of tossing heads *at least once.* We could list all the possibilities of tossing a coin three times, as shown below, and then find the ones that contain at least one head.

$$\{\underbrace{\text{HHH, HHT, HTH, HTT, THH, THT, TTH}}_{\text{at least one head}}, \text{TTT}\}$$

The probability of at least one head is $\frac{7}{8}$.

Another way to calculate this result is to use the formula for the probability of the complement of an event. Let $E = \{\text{at least one head}\}$. From the list above, note that E contains every outcome except TTT (no heads). Thus $E^c = \{\text{TTT}\}$ and we have

$$P(E) = 1 - P(E^c)$$

$$= 1 - \frac{1}{8} = \frac{7}{8} \qquad \bullet\ P(E^c) = \frac{n(E^c)}{n(S)} = \frac{1}{8}$$

This is the same result that we calculated above. Sometimes working with a complement is much easier than proceeding directly.

Objective 6.4B Practice

1. Two dice are rolled. Find the probability of rolling a 2 or a 10.
2. If $P(A) = 0.2$, $P(B) = 0.5$, and $P(A \text{ and } B) = 0.1$, find $P(A \text{ or } B)$.
3. Suppose you ask a friend to randomly choose an integer between 1 and 10, inclusive. What is the probability that the number your friend chooses will be more than 6 or odd?
4. Suppose you roll two fair dice. Find the probability of rolling a 6 or rolling "doubles" (both dice show the same number).
5. The table below shows the unemployment status of individuals in a particular town by age group.

Age	Full-time	Part-time	Unemployed
0–17	24	164	371
18–25	185	203	148
26–34	348	67	27
35–49	581	179	104
50+	443	162	173

   If the person is randomly chosen from the town's population, what is the probability that the person is under 18 or employed part time?

6. Suppose the probability that it will rain tomorrow is 0.38. What is the probability that it will not rain tomorrow?

Solutions on pp. S-15–S-16.

Objective 6.4C Calculate the odds of an event

Sometimes the chance of an event occurring is given in terms of *odds*. This concept is closely related to probability.

Odds in Favor of an Event

The **odds in favor** of an event is the ratio of the number of favorable outcomes of an experiment to the number of unfavorable outcomes.

$$\text{Odds in favor} = \frac{\text{number of favorable outcomes}}{\text{number of unfavorable outcomes}}$$

Odds Against an Event

The **odds against** an event is the ratio of the number of unfavorable outcomes of an experiment to the number of favorable outcomes.

$$\text{Odds against} = \frac{\text{number of unfavorable outcomes}}{\text{number of favorable outcomes}}$$

To find the odds in favor of rolling a 4 when a die is rolled once, list the favorable outcomes and the unfavorable outcomes.

Favorable outcomes: 4 Unfavorable outcomes: 1, 2, 3, 5, 6

$$\text{Odds in favor of rolling a 4} = \frac{\text{number of favorable outcomes}}{\text{number of unfavorable outcomes}} = \frac{1}{5}$$

Frequently, the odds of an event are expressed as a ratio using the word *to*. For the last problem, the odds in favor of rolling a 4 are 1 to 5.

It is possible to compute the probability of an event from the odds-in-favor fraction. The probability of an event is the ratio of the numerator to the sum of the numerator and denominator.

The odds in favor of winning a prize in a charity drawing are 1 to 19. What is the probability of winning a prize?

Write the ratio 1 to 19 as a fraction. $1 \text{ to } 19 = \frac{1}{19}$

The probability of winning a prize is the ratio of the numerator to the sum of the numerator and denominator. $\text{Probability} = \frac{1}{1+19} = \frac{1}{20}$

The probability of winning a prize is $\frac{1}{20}$.

Focus on calculating the probability of an event by using odds

In a horse race, the odds against a particular horse winning the race are posted as 9 to 2. What is the probability of the horse winning the race?

STRATEGY

To calculate the probability of the horse winning:

- Restate the odds against as odds in favor.
- Using the odds-in-favor fraction, the probability of winning is the ratio of the numerator to the sum of the numerator and denominator.

SOLUTION

The odds against winning are 9 to 2. Therefore, the odds in favor of winning are 2 to 9, or $\frac{2}{9}$.

$$\text{Probability of winning} = \frac{2}{2+9} = \frac{2}{11}$$

The probability of the horse winning the race is $\frac{2}{11}$.

Check your understanding 6

The odds in favor of contracting the flu during a flu epidemic are 2 to 13. Calculate the probability of getting the flu.

SOLUTION See page S-5.

Objective 6.4C Practice

1. A coin is tossed once. What are the odds in favor of its showing heads?
2. A coin is tossed twice. What are the odds in favor of its showing tails both times?
3. On a board game, the odds in favor of winning \$5000 are 3 to 7. What is the probability of winning \$5000?
4. Two dice are rolled. What are the odds in favor of rolling a 7?
5. A stock market analyst estimates that the odds in favor of a stock going up in value are 2 to 1. What is the probability of the stock *not* going up in value?

Solutions on pp. S-16–S-17.

SOLUTIONS TO MODULE 6

Solutions to Check Your Understanding

Section 6.1 Check your understanding 1

STRATEGY

To make the frequency distribution:

- Find the range.
- Divide the range by 8, the number of classes. The quotient is the class width.
- Tabulate the data for each class.

SOLUTION

Range = 998 − 118 = 880

Class width $= \dfrac{880}{8} = 110$

Dollar Amount of Insurance Premiums

Classes	Frequency
118–228	3
229–339	4
340–450	5
451–561	8
562–672	6
673–783	7
784–894	10
895–1005	7

Check your understanding 2

STRATEGY

To find the number of households:

- Read the histogram to find the number of households using between 1100 and 1150 kWh and the number using between 1150 and 1200 kWh.
- Add the two numbers.

SOLUTION

Number between 1100 and 1150 kWh: 14
Number between 1150 and 1200 kWh: 10

14 + 10 = 24

Twenty-four households used 1100 kWh of electricity or more during the month.

Check your understanding 3

STRATEGY

To find the ratio:

- Read the frequency polygon to find the number of states with a per capita income between \$50,000 and \$55,000 and between \$40,000 and \$45,000.
- Write the ratio of the number of states with a per capita income between \$50,000 and \$55,000 to the number of states with a per capita income between \$40,000 and \$45,000.

SOLUTION

Number of states with a per capita income between \$50,000 and \$55,000: 2

Number of states with a per capita income between \$40,000 and \$45,000: 14

$$\frac{\text{Income between \$50,000 and \$55,000}}{\text{Income between \$40,000 and \$45,000}}$$
$$= \frac{2}{14} = \frac{1}{7}$$

The ratio is $\frac{1}{7}$ or 1 to 7.

Section 6.2

Check your understanding 1

STRATEGY

To calculate the mean amount spent:

- Calculate the sum of the amounts spent by the customers.
- Divide the sum by the number of customers.

SOLUTION

The sum of the amounts spent is 71.30.

$$\bar{x} = \frac{71.30}{10} = 7.13$$

The mean amount spent by the customers was \$7.13.

Check your understanding 2

STRATEGY

To find the score, use the formula for the mean, letting n be the score on the fifth exam.

SOLUTION

$$84 = \frac{82 + 91 + 79 + 83 + n}{5}$$
$$84 = \frac{335 + n}{5}$$
$$5 \cdot 84 = 5\left(\frac{335 + n}{5}\right)$$
$$420 = 335 + n$$
$$85 = n$$

The score on the fifth exam must be 85.

Check your understanding 3

STRATEGY

To calculate the median amount spent by the customers:

- Arrange the numbers from smallest to largest.
- Because there is an even number of values, the median is the sum of the two middle numbers, divided by 2.

SOLUTION

Arrange the numbers from smallest to largest.

5.59 6.32 6.45 6.75 7.05
7.45 7.58 7.90 8.00 8.21

$$\text{Median} = \frac{7.05 + 7.45}{2} = 7.25$$

The median amount spent on lunch was \$7.25.

Check your understanding 4

STRATEGY

To calculate the standard deviation:

- Find the mean number of miles run.
- Use the procedure for calculating standard deviation.

SOLUTION

$$\bar{x} = \frac{5 + 7 + 3 + 6 + 9 + 6}{6} = 6$$

Step 1

x	$(x - \bar{x})^2$
5	$(5 - 6)^2 = 1$
7	$(7 - 6)^2 = 1$
3	$(3 - 6)^2 = 9$
6	$(6 - 6)^2 = 0$
9	$(9 - 6)^2 = 9$
6	$(6 - 6)^2 = 0$
	Total $= 20$

Step 2 $\dfrac{20}{6} = \dfrac{10}{3}$

Step 3 $\sigma = \sqrt{\dfrac{10}{3}} \approx 1.826$

The standard deviation is about 1.826 mi.

Section 6.3 Check your understanding 1

a. By definition, the median is the 50th percentile. Therefore, 50% of the police dispatchers earned less than $34,790 per year.

b. Because $27,550 is in the 25th percentile, 100% − 25% = 75% of all police dispatchers made more than $27,550.

c. From parts (a) and (b), 50% − 25% = 25% of the police dispatchers earned between $27,550 and $34,790.

Check your understanding 2

$$\begin{aligned}\text{Percentile} &= \frac{\text{number of data values less than 405}}{\text{total number of data values}} \cdot 100\\ &= \frac{3952}{8600} \cdot 100\\ &\approx 46\end{aligned}$$

Hal's score of 405 places him at the 46th percentile.

Check your understanding 3

Rank the data.

7.5 9.8 10.2 10.8 11.4 11.4 12.2 12.4 12.6 12.8 13.1 14.2 14.5
15.6 16.4

The median of these 15 data values has a rank of 8. Thus the median is 12.4. The second quartile, Q_2, is the median of the data, so $Q_2 = 12.4$.

The first quartile is the median of the seven values less than Q_2. Thus Q_1 has a rank of 4, so $Q_1 = 10.8$.

The third quartile is the median of the values greater than Q_2. Thus Q_3 has a rank of 12, so $Q_3 = 14.2$.

Check your understanding 4

STRATEGY

To draw the box-and-whiskers plot:

- Arrange the data from smallest to largest.
- Find the median.
- Find Q_1, the median of the lower half of the data, and Q_3, the median of the upper half of the data.
- Draw the box-and-whiskers plot using the following five values: the smallest value, Q_1, the median, Q_3, and the largest value.

SOLUTION

45	55	58	61	65	68	72	75	76
77	78	83	85	86	87	92	94	96

$$\text{Median} = \frac{76 + 77}{2} = 76.5$$

$Q_1 = 65$ • The median of the lower half of the data

$Q_3 = 86$ • The median of the upper half of the data

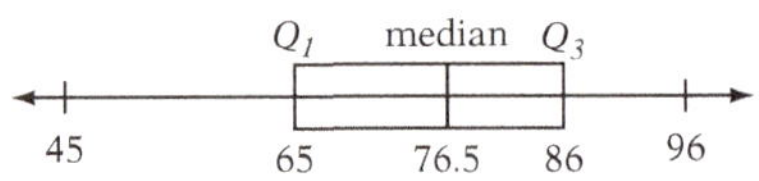

Section 6.4 Check your understanding 1

STRATEGY

To calculate the probability:

- Count the number of possible outcomes of the experiment.
- Count the number of outcomes of the experiment that are favorable to the event E that the soft drink symbol is uncovered.
- Use the probability formula.

SOLUTION

There are 8 possible outcomes of the experiment.
There is 1 outcome favorable to E, uncovering the soft drink symbol.

$$P(E) = \frac{1}{8}$$

The probability is $\frac{1}{8}$ that the soft drink symbol will be uncovered.

Check your understanding 2

STRATEGY

To calculate the probability:

- Count the number of possible outcomes of the experiment.
- Count the number of outcomes of the experiment that are favorable to the event “the two numbers rolled are the same.”
- Use the probability formula.

SOLUTION

There are 36 possible outcomes.
There are 6 favorable outcomes: (1, 1), (2, 2), (3, 3), (4, 4), (5, 5), and (6, 6).

$$P(E) = \frac{6}{36} = \frac{1}{6}$$

The probability that the two numbers are equal is $\frac{1}{6}$.

Check your understanding 3

Let A be the event of rolling a 7, and let B be the event of rolling an 11. From the sample space on page 21, $P(A) = \frac{6}{36} = \frac{1}{6}$ and $P(B) = \frac{2}{36} = \frac{1}{18}$. Because A and B are mutually exclusive events,

$$P(A \text{ or } B) = P(A) + P(B) = \frac{1}{6} + \frac{1}{18} = \frac{4}{18} = \frac{2}{9}$$

The probability of rolling a 7 or an 11 is $\frac{2}{9}$.

Check your understanding 4

Let $A = \{\text{people with a degree in business}\}$ and

$B = \{\text{people with a starting salary between \$20,000 and \$24,999}\}$.

Then, from the table, $n(A) = 4 + 16 + 21 + 35 + 22 = 98$,

$n(B) = 4 + 16 + 3 + 16 = 39$, and $n(A \text{ and } B) = 16$. The total number of people represented in the table is 206.

$$\begin{aligned} P(A \text{ or } B) &= P(A) + P(B) - P(A \text{ and } B) \\ &= \frac{98}{206} + \frac{39}{206} - \frac{16}{206} = \frac{121}{206} \approx 0.587 \end{aligned}$$

The probability of choosing a person who has a degree in business or a starting salary between \$20,000 and \$24,999 is about 58.7%.

Check your understanding 5

If E is the event that a person has type A blood, then E^C is the event that the person does not have type A blood, and

$$P(E^c) = 1 - P(E) = 1 - 0.34 = 0.66$$

The probability that a person does not have type A blood is 66%.

Check your understanding 6

STRATEGY

To calculate the probability:

- Write the odds in favor of contracting the flu as a fraction.
- The probability of contracting the flu is the numerator of the odds-in-favor fraction over the sum of the numerator and the denominator.

SOLUTION

Odds in favor of contracting the flu $= \dfrac{2}{13}$

Probability of contracting the flu $= \dfrac{2}{2 + 13} = \dfrac{2}{15}$

The probability of contracting the flu is $\frac{2}{15}$.

Solutions to Objective Practice Exercises

Objective 6.1A

1. Range $= 96 - 32 = 64$

2. STRATEGY

To prepare the frequency distribution table, first find the range. Then divide the range by 8, the number of classes. If necessary, round the quotient to a whole number. This is the class width.

SOLUTION

Range = 96 − 32 = 64

Class width $= \frac{64}{8} = 8$

Per-Term Tuition at 40 Colleges

Classes	Tally	Frequency
32–40	////	4
41–49	/////	5
50–58	/////	5
59–67	/////	5
68–76	////	4
77–85	////	4
86–94	/////////	9
95–103	////	4

3. STRATEGY

To find the class with the greatest frequency, refer to the frequency table in Exercise 2.

SOLUTION

The class with the greatest frequency is 86–94.

4. STRATEGY

To find the percent:

- Use Exercise 2 to find the number of colleges that charge a tuition between \$9500 and \$10,300 per term.
- Use the basic percent equation.

SOLUTION

Number of colleges = 4

$$\text{Percent} \cdot \text{base} = \text{amount}$$
$$p(40) = 4$$
$$p = \frac{4}{40} = 0.10$$

10% of the tuitions are between \$9500 and \$10,300 per term.

Objective 6.1B

1. STRATEGY

To find the number of account balances between \$2500 and \$3000, read the histogram.

SOLUTION

There are 7 account balances between \$2500 and \$3000.

2. STRATEGY

To find the number of account balances less than \$2000:

- Read the histogram to find the number of account balances between \$0–\$500, \$500–\$1000, \$1000– \$1500, and \$1500–\$2000.
- Add the numbers.

SOLUTION

Number of account balances between:

\$0–\$500: 2

\$500–\$1000: 7

\$1000–\$1500: 10

\$1500–\$2000: 13

$2 + 7 + 10 + 13 = 32$

There were 32 account balances less than \$2000.

3. STRATEGY

To find the percent:

- Read the histogram to find the number of account balances between \$2000 and \$2500.
- Use the basic percent equation.

SOLUTION

Number of account balances between \$2000 and \$2500: 11

$$\begin{aligned} \text{Percent} \cdot \text{base} &= \text{amount} \\ p(50) &= 11 \\ p &= \frac{11}{50} = 0.22 \end{aligned}$$

22% of the account balances were between \$2000 and \$2500.

4. STRATEGY

To find the percent:

- Read the histogram to find the number of account balances between \$1500–\$2000, \$2000–\$2500, and \$2500–\$3000.
- Add the numbers.
- Use the basic percent equation.

SOLUTION

Number of account balances between:

\$1500–\$2000: 13

\$2000–\$2500: 11

\$2500–\$3000: 7

$13 + 11 + 7 = 31$

$$\begin{aligned} \text{Percent} \cdot \text{base} &= \text{amount} \\ p(50) &= 31 \\ p &= \frac{31}{50} = 0.62 \end{aligned}$$

62% of the account balances were greater than \$1500.

Objective 6.1C

1. STRATEGY

To find the number of test scores between 60 and 70, read the frequency polygon.

SOLUTION

There are 8 test scores between 60 and 70.

2. STRATEGY

To find the number of nurses:

- Read the frequency polygon to find the number of nurses whose score was between 80 and 90 and the number whose score was between 90 and 100.
- Add the numbers.

SOLUTION

Number of nurses whose score was between:

80 and 90: 18

90 and 100: 4

$18 + 4 = 22$

22 of the nurses had scores greater than 80.

3. STRATEGY

To find the percent:

- Read the frequency polygon to find the number of nurses who scored between 70 and 80 and the number who scored between 80 and 90.
- Add the numbers.
- Use the basic percent equation.

SOLUTION

Number of nurses who scored between:

70 and 80: 15

80 and 90: 18

$15 + 18 = 33$

$$\begin{aligned} \text{Percent} \cdot \text{base} &= \text{amount} \\ p(50) &= 33 \\ p &= \frac{33}{50} = 0.66 \end{aligned}$$

66% of the nurses had scores between 70 and 90.

4. STRATEGY

To find the percent:

- Read the frequency polygon to find the number of nurses who scored between 70 and 80, between 80 and 90, and between 90 and 100.
- Add the numbers.
- Use the basic percent equation.

SOLUTION

Number of nurses whose score was between:

70 and 80: 15

80 and 90: 18

90 and 100: 4

$15 + 18 + 4 = 37$

$$\begin{aligned} \text{Percent} \cdot \text{base} &= \text{amount} \\ p(50) &= 37 \\ p &= \frac{37}{50} = 0.74 \end{aligned}$$

74% of the nurses had scores greater than 70.

Objective 6.2A

1. STRATEGY

To find the mean number of televisions sold per month:

- Determine the sum of the numbers sold.
- Divide the sum by 12.

To find the median number of televisions sold per month:

- Arrange the numbers from smallest to largest.
- Because there is an even number of values, the median is the sum of the two middle numbers, divided by 2.

SOLUTION

The sum of the numbers is 228.

$$\bar{x} = \frac{228}{12} = 19$$

The mean number of televisions sold is 19.

12 15 15 17 17 19

20 20 20 22 24 27

$$\text{Median} = \frac{19 + 20}{2} = 19.5$$

The median number of televisions sold is 19.5.

2. STRATEGY

To find the mean number of occupied seats:

- Determine the sum of the numbers of occupied seats.
- Divide the sum by 16.

To find the median number of occupied seats:

- Arrange the numbers from smallest to largest.
- Because there is an even number of values, the median is the sum of the two middle numbers, divided by 2.

SOLUTION

The sum of the numbers is 6105.

$$\bar{x} = \frac{6105}{16} = 381.5625$$

The mean number of occupied seats is 381.5625.

309 319 330 352 367 387 389 391

398 399 401 408 410 411 412 422

$$\text{Median} = \frac{391 + 398}{2} = 394.5$$

The median number of occupied seats is 394.5.

3. STRATEGY

To find the mean score:

- Determine the sum of the numbers.
- Divide the sum by 6.

To find the median score:

- Arrange the numbers from smallest to largest.
- The median is the sum of the two middle numbers, divided by 2.

Compare the mean and median scores.

SOLUTION

The sum of the numbers is 524.

$$\bar{x} = \frac{524}{6} \approx 87.33$$

The mean is approximately 87.33.

77 78 88 92 94 95

$$\text{Median} = \frac{88 + 92}{2} = 90$$

The median score is 90.

The mean score is 87.33, and the median score is 90. Using the mean score, the instructor would give the student a B. Using the median score, the instructor would give the student an A. The median is the preferable average.

4. STRATEGY

To find the modal response, write down the category that received the most responses.

SOLUTION

Because a response of German chocolate cake was recorded most frequently, the modal response was German chocolate cake.

5. STRATEGY

To find the modal response, write down the category that received the most responses.

SOLUTION

Because a response of satisfactory was recorded most frequently, the modal response was satisfactory.

Objective 6.2B

1. STRATEGY

To calculate the standard deviation:

- Find the mean of the times.
- Use the procedure for calculating standard deviation.

SOLUTION

$$\bar{x} = \frac{12 + 18 + 20 + 14 + 16}{5} = 16$$

x	$(x - \bar{x})^2$
12	$(12 - 16)^2 = 16$
18	$(18 - 16)^2 = 4$
20	$(20 - 16)^2 = 16$
14	$(14 - 16)^2 = 4$
16	$(16 - 16)^2 = 0$
	Total $= 40$

$$\frac{40}{5} = 8$$

$$\sigma = \sqrt{8} \approx 2.828$$

The standard deviation of the times is approximately 2.828 min.

2. STRATEGY

To calculate the standard deviation:

- Find the mean number of rooms occupied.
- Use the procedure for calculating standard deviation.

SOLUTION

$$\bar{x} = \frac{234 + 321 + 222 + 246 + 312 + 396}{6} = \frac{1731}{6} = 288.5$$

x	$(x - \bar{x})^2$
234	$(234 - 288.5)^2 = 2{,}970.25$
321	$(321 - 2.885)^2 = 1{,}056.25$
222	$(222 - 288.5)^2 = 4{,}422.25$
246	$(246 - 288.5)^2 = 1{,}806.25$
312	$(312 - 288.5)^2 = 552.25$
396	$(396 - 288.5)^2 = 11{,}556.25$
	Total $= 22{,}363.50$

$$\frac{22{,}363.5}{6} \approx 3727.25$$

$$\sigma = \sqrt{3727.25} \approx 61.051$$

The standard deviation of the numbers of rooms occupied is approximately 61.051 rooms.

3. STRATEGY

To calculate the standard deviation:

- Find the mean number of heads.
- Use the procedure for calculating standard deviation.

SOLUTION

$$\bar{x} = \frac{56 + 63 + 49 + 50 + 48 + 53 + 52}{7} = \frac{371}{7} = 53$$

x	$(x - \bar{x})^2$
56	$(56 - 53)^2 = 9$
63	$(63 - 53)^2 = 100$
49	$(49 - 53)^2 = 16$
50	$(50 - 53)^2 = 9$
48	$(48 - 53)^2 = 25$
53	$(53 - 53)^2 = 0$
52	$(52 - 53)^2 = 1$
	Total $= 160$

$$\frac{160}{7} \approx 22.857$$

$$\sigma = \sqrt{22.857} \approx 4.781$$

The standard deviation of the numbers of heads is approximately 4.781 heads.

Objective 6.3A

1.
$$\text{Percentile} = \frac{\text{number of scores less than 455}}{\text{total number of scores}} \cdot 100 = \frac{4256}{7210} \cdot 100 \approx 59$$

Shaylen's score was at the 59th percentile.

2. Number of scores less than Rick's score $= 12{,}860 - 1210 = 11{,}650$

$$\text{Percentile} = \frac{\text{number of scores less than Rick's score}}{\text{total number of scores}} \cdot 100$$
$$= \frac{11{,}650}{12{,}860} \cdot 100 \approx 91$$

Rick's score was at the 91st percentile.

3. $\text{Kevin's percentile} = \dfrac{x}{9840} \cdot 100 = 65$

$$x = \frac{65 \cdot 9840}{100} = 6396$$

6396 students scored lower than Kevin.

4. Arrange the data values from smallest to largest.

1, 3, 4, 5, 5, 7, 8, 10, 10,
10, 12, 15, 18, 26, 28, 32, 41, 85

The median (or second quartile) is the average of the 9th and 10th data values: $\dfrac{10 + 10}{2} = 10$.

The first quartile Q_1 is the median of the 1st through 9th data values, 5.

The third quartile Q_3 is the median of the 10th through 18th data values, 26.

$Q_1 = 5, Q_2 = 10, Q_3 = 26$

5. Arrange the data values from smallest to largest.

7, 7, 7, 8, 8, 8, 8, 9, 9, 9, 10, 10,
10, 10, 10, 10, 10, 11, 11, 12, 12, 12, 12, 12,
13, 14, 14, 14, 14, 14, 14, 15, 15, 15, 16, 16

The median (or second quartile) is the average of the 18th and 19th data values: $\dfrac{11 + 11}{2} = 11$.

The first quartile Q_1 is the median of the 1st through 18th data values, or the average of the 9th and 10th data values: $\dfrac{9 + 9}{2} = 9$.

The third quartile Q_3 is the median of the 19th through 36th data values, or the average of the 27th and 28th data values: $\dfrac{14 + 14}{2} = 14$.

$Q_1 = 9, Q_2 = 11, Q_3 = 14$

Objective 6.3B

1. STRATEGY

To draw the box-and-whiskers plot:

- Arrange the data from smallest to largest, and then find the median.
- Find Q_1, the median of the lower half of the data.
- Find Q_3, the median of the upper half of the data.
- Determine the smallest and largest data values.
- Draw the box-and-whiskers plot.

SOLUTION

7.35 7.45 7.46 7.70 8.09 8.12 8.64 8.94

9.02 9.03 9.31 9.85 10.05 10.35 11.40 11.50

$$\text{Median} = \frac{8.94 + 9.02}{2} = 8.98$$

$$Q_1 = \frac{7.70 + 8.09}{2} = 7.895, \ Q_3 = \frac{9.85 + 10.05}{2} = 9.95$$

Smallest value: 7.35

Largest value: 11.50

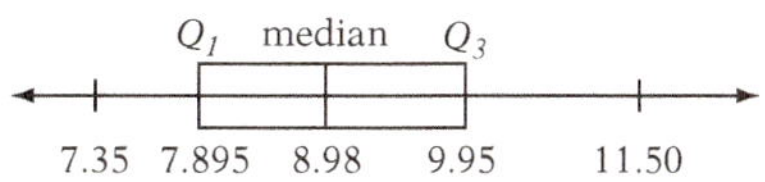

2. STRATEGY

To draw the box-and-whiskers plot:

- Arrange the data from smallest to largest, and then find the median.
- Find Q_1, the median of the lower half of the data.
- Find Q_3, the median of the upper half of the data.
- Determine the smallest and largest data values.
- Draw the box-and-whiskers plot.

SOLUTION

172 183 185 198 208 211 214 221 233 251 254 258 292 375

$$\text{Median} = \frac{214 + 221}{2} = 217.5$$

$Q_1 = 198, Q_3 = 254$

Smallest value: 172

Largest value: 375

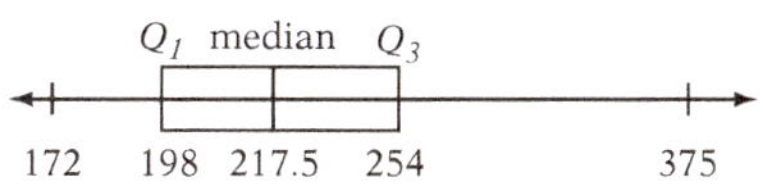

3. STRATEGY

To draw the box-and-whiskers plot:

- Arrange the data from smallest to largest, and then find the median.
- Find Q_1, the median of the lower half of the data.
- Find Q_3, the median of the upper half of the data.
- Determine the smallest and largest data values.
- Draw the box-and-whiskers plot.

SOLUTION

16 17 19 20 20 21 21 22 24 25

26 26 28 30 30 31 31 32 33

Median = 25

$Q_1 = 20$

$Q_3 = 30$

Smallest value: 16

Largest value: 33

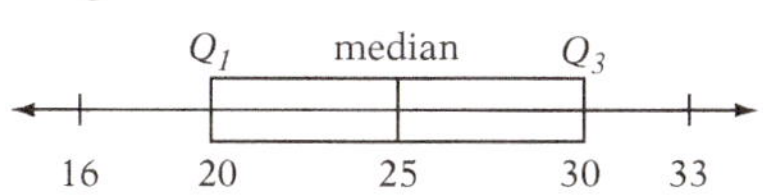

Objective 6.4A

1. Possible outcomes of stacks:

C_1	C_2	C_3
R	G	B
R	B	G
G	R	B
G	B	R
B	R	G
B	G	R

2. STRATEGY

To find the probability:

- Refer to the table on page 21 to count the number of possible outcomes of the experiment.
- Count the number of outcomes of the experiment that are favorable to the event "the sum is 5."
- Use the probability formula.

SOLUTION

There are 36 possible outcomes.

There are 4 outcomes favorable to E: (1, 4), (2, 3), (3, 2), (4, 1)

$$P(E) = \frac{\text{number of favorable outcomes}}{\text{number of possible outcomes}}$$

$$P(E) = \frac{4}{36} = \frac{1}{9}$$

The probability that the sum is 5 is $\frac{1}{9}$.

3. STRATEGY

To find the probability:

- Count the number of possible outcomes of the experiment.
- Count the number of outcomes of the experiment favorable to a roll of 11.
- Use the probability formula.

SOLUTION

A dodecahedral die has 12 sides.

There is 1 outcome favorable to E: 11

$$P(E) = \frac{\text{number of favorable outcomes}}{\text{number of possible outcomes}}$$

$$P(E) = \frac{1}{12}$$

The probability that the number rolled is 11 is $\frac{1}{12}$.

4. STRATEGY

To find the probability:

- Refer to the table on page 21 to count the number of possible outcomes of the experiment.
- Count the number of outcomes of the experiment that are favorable to the event "the sum is 4."
- Use the probability formula.

SOLUTION

There are 16 possible outcomes.

There are 3 outcomes favorable to E: (1, 3), (2, 2), (3,1)

$$P(E) = \frac{\text{number of favorable outcomes}}{\text{number of possible outcomes}}$$

$$P(E) = \frac{3}{16}$$

The probability that the sum is 4 is $\frac{3}{16}$.

5. STRATEGY

To find the empirical probability, use the probability formula and divide the number of observations of E (587) by the total number of observations (725).

SOLUTION

$$P(E) = \frac{\text{number of observations of } E}{\text{total number of observations}}$$

$$P(E) = \frac{587}{725} \approx 0.81$$

The probability is 81.0% that an employee has a group health insurance plan.

Objective 6.4B

1. Let A be the event of rolling a 2, and let B be the event of rolling a 10. The events in A and B cannot occur at the same time, so they are mutually exclusive. From the sample space on page 21,

$$P(A) = \frac{1}{36} \text{ and } P(B) = \frac{3}{36}$$

Because A and B are mutually exclusive events,

$$P(A \text{ or } B) = P(A) + P(B) = \frac{1}{36} + \frac{3}{36} = \frac{4}{36} = \frac{1}{9}$$

The probability of rolling a 2 or a 10 is $\frac{1}{9}$.

2. $$\begin{aligned} P(A \text{ or } B) &= P(A) + P(B) - P(A \text{ and } B) \\ &= 0.2 + 0.5 - 0.1 \\ &= 0.6 \end{aligned}$$

3. Let A be the event of the number being more than 6, and let B be the event of the number being odd. The sample space consists of 10 numbers. $n(A) = 4$, $n(B) = 5$, and $n(A \text{ and } B) = 2$.

$$\begin{aligned} P(A \text{ or } B) &= P(A) + P(B) - P(A \text{ and } B) \\ &= \frac{4}{10} + \frac{5}{10} - \frac{2}{10} = \frac{7}{10} \end{aligned}$$

The probability that the number will be more than 6 or odd is $\frac{7}{10}$.

4. Let A be the event of rolling a 6, and let B be the event of rolling doubles. From page 21, there are 36 elements in the sample space. $n(A) = 5$, $n(B) = 6$, and $n(A \text{ and } B) = 1$.

$$\begin{aligned} P(A \text{ or } B) &= P(A) + P(B) - P(A \text{ and } B) \\ &= \frac{5}{36} + \frac{6}{36} - \frac{1}{36} = \frac{10}{36} = \frac{5}{18} \end{aligned}$$

The probability of rolling a 6 or rolling doubles is $\frac{5}{18}$.

5. Let $A = \{\text{people under 18}\}$, and let $B = \{\text{people employed part-time}\}$. Then, from the table, $n(A) = 24 + 164 + 371 = 559$, $n(B) = 164 + 203 + 67 + 179 + 162 = 775$, and $n(A \text{ and } B) = 164$. The total number of people represented in the table is 3179.

$$P(A \text{ or } B) = P(A) + P(B) - P(A \text{ and } B)$$
$$= \frac{559}{3179} + \frac{775}{3179} - \frac{164}{3179} = \frac{1170}{3179}$$

The probability that the person is under 18 or employed part-time is $\frac{1170}{3179}$.

6. If E is the event of rain tomorrow, then E^c is the event of no rain tomorrow.

$$P(E^c) = 1 - P(E) = 1 - 0.38 = 0.62$$

The probability that it will not rain tomorrow is 0.62, or 62%.

Objective 6.4C

1. STRATEGY
To find the odds in favor:
- Count the number of favorable outcomes.
- Count the number of unfavorable outcomes.
- Use the formula for the odds in favor of an event.

SOLUTION
Number of favorable outcomes: 1

Number of unfavorable outcomes: 1

$$\text{Odds in favor} = \frac{\text{number of favorable outcomes}}{\text{number of unfavorable outcomes}}$$

$$\text{Odds in favor} = \frac{1}{1}$$

The odds in favor of the coin showing heads are 1 to 1.

2. STRATEGY
To find the odds in favor:
- Count the number of favorable outcomes.
- Count the number of unfavorable outcomes.
- Use the formula for the odds in favor of an event.

SOLUTION
There is 1 favorable outcome: TT

There are 3 unfavorable outcomes: TH, HT, HH

$$\text{Odds in favor} = \frac{\text{number of favorable outcomes}}{\text{number of unfavorable outcomes}}$$

$$\text{Odds in favor} = \frac{1}{3}$$

The odds in favor of the coin showing tails twice are 1 to 3.

3. STRATEGY
To calculate the probability of winning, use the odds-in-favor fraction. The probability of winning is the ratio of the numerator to the sum of the numerator and denominator.

SOLUTION

$$\text{Probability of winning} = \frac{3}{3 + 7} = \frac{3}{10}$$

The probability of winning the board game is $\frac{3}{10}$.

4. STRATEGY

To find the odds in favor:

- Count the number of favorable outcomes.
- Count the number of unfavorable outcomes.
- Use the formula for the odds in favor of an event.

SOLUTION

Use the table on page 21:

Number of favorable outcomes: 6

Number of unfavorable outcomes: 30

Odds in favor $= \dfrac{6}{30} = \dfrac{1}{5}$

The odds in favor of rolling a 7 are $\frac{1}{5}$, or 1 to 5.

5. STRATEGY

To calculate the probability of the stock going down:

- Restate the odds of the stock going up in terms of the odds of the stock going down.
- Using the odds in favor of the stock going down, the probability of the stock going down is the ratio of the numerator to the sum of the numerator and denominator.

SOLUTION

The odds in favor of the stock going up: 2 to 1

The odds in favor of the stock going down: 1 to 2

Probability of the stock going down: $= \dfrac{1}{1 + 2} = \dfrac{1}{3}$

The probability of the stock going down is $\frac{1}{3}$.

MODULE

7 The Mathematics of Finance

SECTION 7.1

Simple Interest

Objective 7.1A Calculate simple interest

When you deposit money in a bank—for example, in a savings account—you are permitting the bank to use your money. The bank pays you for this privilege in the form of **interest.** If you borrow money from a bank, the amount you pay for the privilege of using that money is also called interest.

The amount deposited in a bank or borrowed from a bank is called the **principal.** The amount of interest paid is usually given as a percent of the principal. The percent used to determine the amount of interest is called the **interest rate.**

Interest paid on the original principal is called **simple interest.** The formula used to calculate simple interest is given below.

Simple Interest Formula

The simple interest formula is

$$I = Prt$$

where I is the interest, P is the principal, r is the interest rate, and t is the time period.

Interest rates are generally given as percents. Before performing calculations involving an interest rate, write the interest rate as a decimal.

Focus on calculating simple interest

Calculate the simple interest due on a 3-month loan of \$2000 if the annual simple interest rate is 6.5%.

SOLUTION

Use the simple interest formula. Substitute the values $P = 2000$ and $r = 6.5\% = 0.065$ into the formula. Because the interest rate is an annual rate, the time must be measured in years: $t = \frac{3 \text{ months}}{1 \text{ year}} = \frac{3 \text{ months}}{12 \text{ months}} = \frac{3}{12}$.

$I = Prt$

$I = 2000(0.065)\left(\frac{3}{12}\right)$

$I = 32.5$

The simple interest due is \$32.50.

Check your understanding 1

Calculate the simple interest due on a 4-month loan of \$1500 if the annual simple interest rate is 5.25%.

SOLUTION See page S-1.

Focus on calculating simple interest

Calculate the simple interest due on a 6-month loan of \$500 if the annual simple interest rate is 1.5%.

SOLUTION

Use the simple interest formula. Substitute the values $P = 500$ and $r = 1.5\% = 0.015$ into the formula. Because the interest rate is an annual rate, the time must be measured in years: $t = \frac{6 \text{ months}}{1 \text{ year}} = \frac{6 \text{ months}}{12 \text{ months}}$.

$$I = Prt$$
$$I = 500(0.015)\left(\frac{6}{12}\right)$$
$$I = 3.75$$

The simple interest due is \$3.75.

Check your understanding 2

Calculate the simple interest due on a 9-month loan of \$700 if the annual simple interest rate is 1.25%. Round to the nearest cent.

SOLUTION See page S-1.

Remember that in the simple interest formula, time t is measured in the same period as the interest rate. Therefore, if the time period of a loan with an annual interest rate is given in days, it is necessary to convert the time period of the loan to a fractional part of a year. There are two methods for converting time from days to years: the exact method and the ordinary method. Using the exact method, the number of days of the loan is divided by 365, the number of days in a year.

Exact method: $t = \dfrac{\text{number of days}}{365}$

The ordinary method is based on there being an average of 30 days in a month and 12 months in a year $(30 \cdot 12 = 360)$. Using this method, the number of days of the loan is divided by 360.

Ordinary method: $t = \dfrac{\text{number of days}}{360}$

The ordinary method is used by most businesses. Therefore, unless otherwise stated, the ordinary method will be used in this text.

Focus on calculating simple interest

Calculate the simple interest due on a 45-day loan of \$3500 if the annual simple interest rate is 8%.

SOLUTION

Use the simple interest formula. Substitute the following values into the formula: $P = 3500$, $r = 8\% = 0.08$, and $t = \frac{\text{number of days}}{360} = \frac{45}{360}$.

$$I = Prt$$
$$I = 3500(0.08)\left(\frac{45}{360}\right)$$
$$I = 35$$

The simple interest due is \$35.

Check your understanding 3

Calculate the simple interest due on a 120-day loan of \$7000 if the annual simple interest rate is 5.25%.

SOLUTION See page S-1.

The simple interest formula can be used to find the interest rate on a loan when the interest, principal, and time period of the loan are known. An example is given below.

Focus on calculating an annual simple interest rate

The simple interest charged on a 6-month loan of \$3000 is \$150. Find the annual simple interest rate.

SOLUTION

Use the simple interest formula. Solve the equation for r.

$$
\begin{aligned}
I &= Prt \\
150 &= 3000(r)\left(\frac{6}{12}\right) \\
150 &= 1500r && \bullet\ 3000\left(\frac{6}{12}\right) = 1500 \\
0.10 &= r && \bullet\ \text{Divide each side of the equation by 1500.} \\
r &= 10\% && \bullet\ \text{Write the decimal as a percent.}
\end{aligned}
$$

The annual simple interest rate on the loan is 10%.

Check your understanding 4

The simple interest charged on a 6-month loan of \$12,000 is \$462. Find the annual simple interest rate.

SOLUTION See page S-2.

Objective 7.1A Practice

For Exercises 1 to 3, calculate the simple interest earned on the investment.

1. $P = \$7000$, $r = 6.5\%$, $t = 6$ months.
2. $P = \$9000$, $r = 6.75\%$, $t = 4$ months.
3. $P = \$2000$, $r = 1.25\%$, $t = 9$ months.
4. Calculate the annual simple interest rate if $P = \$2000$, $I = \$80$, $t = 6$ months.
5. You deposit \$1500 in an account earning 5.2% annual simple interest. Calculate the simple interest in 6 months.
6. Calculate the simple interest due on a 45-day loan of \$1600 if the annual simple interest rate is 9%.

Solutions on pp. S-8–S-9.

Objective 7.1B Calculate future value and maturity value

When you borrow money, the total amount to be repaid to the lender is the sum of the principal and interest. This sum is calculated using the following future value or maturity value formula for simple interest.

Future Value or Maturity Value Formula for Simple Interest

The future value or maturity value formula for simple interest is

$$A = P + I$$

where A is the amount after the interest I has been added to the principal P.

This formula can be used for loans or investments. When used for a loan, A is the total amount to be repaid to the lender; this sum is called the **maturity value** of the loan. In the last "Focus on" example in the preceding objective, the simple interest charged on the loan of \$3000 was \$150. The maturity value of the loan is therefore \$3000 + \$150 = \$3150.

For an investment, such as a deposit in a bank savings account, A is the total amount on deposit after the interest earned has been added to the principal. This sum is called the **future value** of the investment.

Focus on calculating a maturity value

Calculate the maturity value of a simple interest, 8-month loan of \$8000 if the annual simple interest rate is 9.75%.

SOLUTION

Step 1: Find the interest. Use the simple interest formula. Substitute the values $P = 8000$, $r = 9.75\% = 0.0975$, and $t = \frac{8}{12}$ into the formula.

$$I = Prt$$

$$I = 8000(0.0975)\left(\frac{8}{12}\right)$$

$$I = 520$$

Step 2: Find the maturity value. Use the maturity value formula for simple interest. Substitute the values $P = 8000$ and $I = 520$ into the formula.

$$A = P + I$$

$$A = 8000 + 520$$

$$A = 8520$$

The maturity value of the loan is \$8520.

Check your understanding 5

Calculate the maturity value of a simple interest, 9-month loan of \$4000 if the annual simple interest rate is 8.75%.

SOLUTION See page S-2.

Recall that the simple interest formula states that $I = Prt$. We can substitute Prt for I in the future value or maturity value formula as follows.

$$A = P + I$$

$$A = P + Prt$$

$$\mathbf{A = P(1 + rt)} \quad \text{• Factor } P \text{ from each term.}$$

In the final equation, A is the future value of an investment or the maturity value of a loan, P is the principal, r is the interest rate, and t is the time period.

We used the formula $A = P + I$ in the preceding "Focus on" example. The formula $A = P(1 + rt)$ is used in the two "Focus on" examples that follow. Note that two steps were required to find the solution in the preceding example, but only one step is required in the next two examples.

Focus on calculating a maturity value

Calculate the maturity value of a simple interest, 3-month loan of \$3800. The annual simple interest rate is 6%.

SOLUTION

Substitute the following values into the formula $A = P(1 + rt)$: $P = 3800$, $r = 6\% = 0.06$, and $t = \frac{3}{12}$.

$$A = P(1 + rt)$$
$$A = 3800\left[1 + 0.06\left(\frac{3}{12}\right)\right]$$
$$A = 3800(1 + 0.015)$$
$$A = 3800(1.015)$$
$$A = 3857$$

The maturity value of the loan is \$3857.

Check your understanding 6

Calculate the maturity value of a simple interest, 1-year loan of \$6700. The annual simple interest rate is 8.9%.

SOLUTION See page S-2.

Focus on calculating a future value

Find the future value after 1 year of \$850 in an account earning 8.2% annual simple interest.

SOLUTION

Because $t = 1$, $rt = r(1) = r$. Therefore, $1 + rt = 1 + r = 1 + 0.082 = 1.082$.

$$A = P(1 + rt)$$
$$A = 850(1.082)$$
$$A = 919.7$$

The future value of the account after 1 year is \$919.70.

Check your understanding 7

Find the future value after 1 year of \$680 in an account earning 6.4% annual simple interest.

SOLUTION See page S-2.

Recall that the formula $A = P + I$ states that A is the amount after the interest has been added to the principal. Subtracting P from each side of this equation yields the following formula.

$$I = A - P$$

This formula states that the amount of interest paid is equal to the total amount minus the principal.

Focus on calculating a simple interest rate

The maturity value of a 3-month loan of $4000 is $4085. What is the annual simple interest rate?

SOLUTION

First find the amount of interest paid. Subtract the principal from the maturity value.

$$I = A - P$$
$$I = 4085 - 4000$$
$$I = 85$$

Find the annual simple interest rate by solving the simple interest formula for r.

$$I = Prt$$
$$85 = 4000(r)\left(\frac{3}{12}\right)$$
$$85 = 1000r \qquad \bullet\ 4000\left(\frac{3}{12}\right) = 1000$$
$$0.085 = r \qquad \bullet\ \text{Divide each side of the equation by 1000.}$$
$$r = 8.5\% \qquad \bullet\ \text{Write the decimal as a percent.}$$

The annual simple interest rate on the loan is 8.5%.

Check your understanding 8

The maturity value of a 4-month loan of $9000 is $9240. What is the annual simple interest rate?

SOLUTION See page S-3.

Objective 7.1B Practice

For Exercises 1 and 2, use the formula $A = P(1 + rt)$ to calculate the maturity value of the simple interest loan.

1. $P = \$15{,}000$, $r = 8.9\%$, $t = 6$ months
2. $P = \$2800$, $r = 9.2\%$, $t = 3$ months
3. Calculate the maturity value of a simple-interest, 8-month loan of $7000 if the annual simple interest rate is 8.7%.
4. You deposit $750 in an account paying 7.3% annual simple interest. Find the future value of the investment after 1 year.
5. Find the annual simple interest rate on a 3-month loan of $5000 if the maturity value of the loan is $5125.

Solutions on pp. S-9–S-10.

SECTION 7.2 Compound Interest

Objective 7.2A Calculate the future value of an investment using compound interest

Simple interest is generally used for loans of 1 year or less. For loans of more than 1 year, the interest paid on the money borrowed is called *compound interest.* **Compound interest** is interest calculated not only on the original principal but also on any interest that has already been earned.

To illustrate compound interest, suppose you deposit \$1000 in a savings account earning 5% interest, compounded annually (once a year).

During the first year, the interest earned is calculated as follows.

$$I = Prt$$
$$I = \$1000(0.05)(1) = \$50$$

At the end of the first year, the total amount in the account is

$$A = P + I$$
$$A = \$1000 + \$50 = \$1050$$

During the second year, the interest earned is calculated using the amount in the account at the end of the first year.

$$I = Prt$$
$$I = \$1050(0.05)(1) = \$52.50$$

Note that the interest earned during the second year (\$52.50) is greater than the interest earned during the first year (\$50). This is because the interest earned during the first year was added to the original principal, and the interest for the second year was calculated using this sum. If the account earned simple interest rather than compound interest, the interest earned each year would be the same (\$50).

At the end of the second year, the total amount in the account is the sum of the amount in the account at the end of the first year and the interest earned during the second year.

$$A = P + I$$
$$A = \$1050 + 52.50 = \$1102.50$$

The interest earned during the third year is calculated using the amount in the account at the end of the second year (\$1102.50).

$$I = Prt$$
$$I = \$1102.50(0.05)(1) = \$55.125 \approx \$55.13$$

The interest earned each year keeps increasing. This is the effect of compound interest.

In this example, the interest is compounded annually (once a year). However, interest can be compounded semiannually (twice a year), quarterly (four times a year), monthly, or daily. The frequency with which the interest is compounded is called the **compounding period.**

Recall that the future value of an investment is the value of the investment after the original principal has been invested for a period of time. In other words, it is the principal plus the interest earned. The future value formula enables us to calculate the future value of an investment for any number of compounding periods per year.

Future Value Formula

The **future value** A of an investment is given by

$$A = P\left(1 + \frac{r}{n}\right)^{nt}$$

where P is the amount of money invested or deposited (also called the *present value*), r is the annual interest rate as a decimal, n is the number of compounding periods per year, and t is the number of years.

To illustrate how to determine the values of the variables in this formula, consider depositing \$5000 in an account earning 6% annual interest, compounded quarterly, for a period of 3 years.

The amount deposited is P:	$P = 5000$
The annual interest rate is 6%:	$r = 6\% = 0.06$
When interest is compounded quarterly, there are four compounding periods per year:	$n = 4$
The time is 3 years:	$t = 3$

We stated previously that interest can be compounded annually, semiannually, quarterly, monthly, or daily. The possible values of n (the number of compounding periods per year) are recorded in the table at the left.

Values of n (Number of Compounding Periods per Year)

If Interest Is Compounded	Then n =
annually	1
semiannually	2
quarterly	4
monthly	12
daily	360

Focus on calculating future value

Calculate the future value of \$10,000 deposited in an account earning 8% annual interest, compounded semiannually, for 4 years.

SOLUTION

Use the future value formula.
$P = 10{,}000$, $r = 8\% = 0.08$, $n = 2$, $t = 4$

$$A = P\left(1 + \frac{r}{n}\right)^{nt}$$
$$A = 10{,}000\left(1 + \frac{0.08}{2}\right)^{2\cdot 4}$$
$$A = 10{,}000(1 + 0.04)^{8}$$
$$A = 10{,}000(1.04)^{8}$$
$$A \approx 10{,}000(1.368569)$$
$$A \approx 13{,}685.69$$

The future value after 4 years is approximately \$13,685.69.

Calculator Note

When using a scientific calculator to solve the future value formula for A, use the keystroking sequence

1 [+] r [÷] n [=] [y^x] [(] n [×] t [)] [×] P [=]

When using a graphing calculator, use the sequence

P [(] 1 [+] r [÷] n [)] [^] [(] n [×] t [)] [ENTER]

Check your understanding 1

Calculate the future value of \$4000 deposited in an account earning 6% annual interest, compounded monthly, for 2 years.

SOLUTION See page S-3.

Focus on calculating future value

Calculate the future value of \$5000 earning 9% annual interest, compounded daily, for 3 years.

SOLUTION

Use the future value formula.

$P = 5000, r = 9\% = 0.09, n = 360, t = 3$

$$A = P\left(1 + \frac{r}{n}\right)^{nt}$$
$$A = 5000\left(1 + \frac{0.09}{360}\right)^{360\cdot 3}$$
$$A = 5000(1 + 0.00025)^{1080}$$
$$A = 5000(1.00025)^{1080}$$
$$A \approx 5000(1.3099202)$$
$$A \approx 6549.60$$

The future value after 3 years is approximately \$6549.60.

Check your understanding 2

Calculate the future value of \$2500 earning 9% annual interest, compounded daily, for 4 years.

SOLUTION See page S-3.

The formula $I = A - P$ was used in Section 7.1 to find the interest earned on an investment or the interest paid on a loan. This same formula is used to calculate compound interest.

Focus on calculating compound interest

How much interest is earned in 2 years on \$4000 deposited in an account paying 6% annual interest, compounded quarterly?

SOLUTION

Calculate the future value. Use the future value formula.

$P = 4000, r = 6\% = 0.06, n = 4, t = 2$

$$A = P\left(1 + \frac{r}{n}\right)^{nt}$$
$$A = 4000\left(1 + \frac{0.06}{4}\right)^{4\cdot 2}$$
$$A = 4000(1 + 0.015)^{8}$$
$$A = 4000(1.015)^{8}$$
$$A \approx 4000(1.264926)$$
$$A \approx 4505.97$$

Calculate the interest earned. Use the formula $I = A - P$.

$I = A - P$
$I = 4505.97 - 4000$
$I = 505.97$

The amount of interest earned is approximately \$505.97.

Check your understanding 3

How much interest is earned in 6 years on \$8000 deposited in an account paying 9% annual interest, compounded monthly?

SOLUTION See page S-4.

Objective 7.2A Practice

For Exercises 1 to 3, use the future value formula to calculate the future value of the investment.

1. $P = \$500$, $r = 9\%$ compounded quarterly, $t = 6$ years
2. $P = \$8500$, $r = 9\%$ compounded monthly, $t = 10$ years
3. $P = \$9000$, $r = 5.5\%$ compounded quarterly, $t = 3$ years
4. How much interest is earned in 5 years on \$8500 deposited in an account paying 9% annual interest, compounded semiannually?
5. \$15,000 is deposited for 4 years into an account earning 8% annual interest.
   a. Calculate the future value of the investment if the interest is compounded semiannually.
   b. Calculate the future value if the interest is compounded quarterly.
   c. How much greater is the future value of the investment when the interest is compounded quarterly?

Solutions on pp. S-10–S-11.

Objective 7.2B Calculate the present value of an investment using compound interest

The **present value** of an investment is the original principal invested, or the value of the investment before it earns any interest. Therefore, it is the principal P in the future value formula. Present value is used to determine how much money must be invested today in order for an investment to have a specific value at a future date.

The formula for the present value of an investment is found by solving the future value formula for P.

Present Value Formula

The present value formula is

$$P = \frac{A}{\left(1 + \frac{r}{n}\right)^{nt}}$$

where P is the present value of an investment or deposit, A is the future value of the investment or deposit, r is the annual interest rate as a decimal, n is the number of compounding periods per year, and t is the number of years.

Focus on calculating present value

How much money should be invested in an account that earns 8% annual interest, compounded quarterly, in order to have $30,000 in 5 years?

SOLUTION

Use the present value formula.

$A = 30{,}000$, $r = 8\% = 0.08$, $n = 4$, $t = 5$

Calculator Note

To calculate P in the present value formula using a calculator, you can first calculate $\left(1 + \frac{r}{n}\right)^{nt}$. Next use the [1/x] key or the [x^{-1}] key. This will place the value of $\left(1 + \frac{r}{n}\right)^{nt}$ in the denominator. Then multiply by the value of A.

$$P = \frac{A}{\left(1 + \frac{r}{n}\right)^{nt}}$$

$$P = \frac{30{,}000}{\left(1 + \frac{0.08}{4}\right)^{4\cdot 5}} = \frac{30{,}000}{1.02^{20}} \approx \frac{30{,}000}{1.485947396}$$

$$P \approx 20{,}189.14$$

$20,189.14 should be invested in the account in order to have $30,000 in 5 years.

Check your understanding 4

How much money should be invested in an account that earns 9% annual interest, compounded semiannually, in order to have $20,000 in 5 years?

SOLUTION See page S-4.

Focus on calculating present value

Suppose you wish to have $4500 in 4 years to take a trip to help build clean water facilities in an impoverished neighborhood. How much money would you have to deposit into an account that earns 6.5% annual interest, compounded daily, to reach your goal? Round to the nearest cent.

SOLUTION

Use the present value formula.

$A = 4500$, $r = 6.5\% = 0.065$, $n = 365$ (interest is compounded daily), $t = 4$

$$P = \frac{A}{\left(1 + \frac{r}{n}\right)^{nt}}$$

$$= \frac{4500}{\left(1 + \frac{0.065}{365}\right)^{365(4)}} \approx \frac{4500}{(1 + 0.000178082918)^{1460}}$$

$$\approx \frac{4500}{1.296900066} \approx 3469.81$$

It is necessary to deposit $3469.81 into the account.

Check your understanding 5

A physician's assistant wishes to purchase a car in 3 years and plans on paying $18,000 for the car. How much money would the physician's assistant have to deposit into an account that earns 4% annual interest, compounded monthly, to reach the goal? Round to the nearest cent.

SOLUTION See page S-5.

Objective 7.2B Practice

1. How much money should be invested in an account that earns 6% annual interest, compounded monthly, in order to have $15,000 in 5 years?
2. You borrow $6000 to help pay your college expenses. You agree to repay the loan at the end of 5 years at 10% annual interest, compounded quarterly.
   a. What is the maturity value of the loan?
   b. How much interest are you paying on the loan?
3. A couple plans to save for their child's education. What principal must be deposited by the parents when the child is born if they wish to have $40,000 by the time the child reaches the age of 18? Assume the money earns 8% annual interest, compounded quarterly.
4. How much money would you have to invest today at 9% annual interest, compounded daily, in order to have $1,000,000 in 30 years?
5. What principal must be deposited by the parents of a newborn if they wish to have $51,000 in 18 years? Assume the money earns 2.5% annual interest, compounded daily.

Solutions on pp. S-12–S-13.

Objective 7.2C Calculate the effects of inflation

We have discussed compound interest and its effect on the growth of an investment. After your money has been invested for a period of time in an account that pays interest, you will have more money than you originally deposited. But does that mean you will be able to buy more with the final amount than you were able to buy with the original investment at the time you deposited the money? The answer is *not necessarily,* and the reason is inflation.

Inflation is an economic condition in which there are increases in the costs of goods and services. Inflation is expressed as a percent; for example, we speak of an annual inflation rate of 7%.

To calculate the effects of inflation, we use the same procedure we used to calculate future value. Although inflation rates vary dramatically, in this objective we will assume constant annual inflation rates, and we will use annual compounding in solving inflation problems. In other words, $n = 1$ for these exercises.

Focus on calculating the effect of inflation on salary

Suppose your annual salary today is $35,000. You want to know what an equivalent salary will be in 20 years—that is, a salary that will have the same purchasing power. Assume a 6% inflation rate.

SOLUTION

Use the future value formula with $P = 35{,}000$, $r = 6\% = 0.06$, and $t = 20$. The inflation rate is an annual rate, so $n = 1$.

$$A = P\left(1 + \frac{r}{n}\right)^{nt}$$

$$A = 35{,}000\left(1 + \frac{0.06}{1}\right)^{1\cdot 20}$$

$$A = 35{,}000(1.06)^{20}$$

$$A \approx 35{,}000(3.20713547)$$

$$A \approx 112{,}249.74$$

Twenty years from now, you need to earn an annual salary of approximately $112,249.74 in order to have the same purchasing power.

Check your understanding 6

Assume the average new car sticker price in 2013 was \$28,000. Use an annual inflation rate of 5% to estimate the average new car sticker price in 2030.

SOLUTION See page S-5.

The present value formula can be used to determine the effect of inflation on the future purchasing power of a given amount of money. Substitute the inflation rate for the interest rate in the present value formula. The compounding period is 1 year. Again we will assume a constant rate of inflation.

Focus on calculating the effect of inflation on future purchasing power

Suppose you purchase an insurance policy in 2015 that will provide you with \$250,000 when you retire in 2050. Assuming an annual inflation rate of 8%, what will be the purchasing power of the \$250,000 in 2050?

SOLUTION

Use the present value formula. $A = 250{,}000$, $r = 8\% = 0.08$, $t = 35$

The inflation rate is an annual rate, so $n = 1$.

$$P = \frac{A}{\left(1 + \frac{r}{n}\right)^{nt}}$$

$$P = \frac{250{,}000}{\left(1 + \frac{0.08}{1}\right)^{1 \cdot 35}}$$

$$P = \frac{250{,}000}{(1 + 0.08)^{35}}$$

$$P \approx \frac{250{,}000}{14.785344}$$

$$P \approx 16{,}908.64$$

Assuming an annual inflation rate of 8%, the purchasing power of \$250,000 will be about \$16,908.64 in 2050.

Check your understanding 7

Suppose you purchase an insurance policy in 2015 that will provide you with \$500,000 when you retire in 40 years. Assuming an annual inflation rate of 7%, what will be the purchasing power of half a million dollars in 2055?

SOLUTION See page S-5.

Objective 7.2C Practice

1. Suppose your current salary is \$40,000. Assuming an annual inflation rate of 7%, what salary will you need to earn 3 years from now in order to have the same purchasing power? Round to the nearest dollar.
2. The median price of a house in a certain city is \$230,000. If the annual inflation rate for the price of a house is 4%, find the median price of a house in the same city 5 years from now. Round to the nearest dollar.
3. The annual health care premium for an employee at a corporation is \$11,000. If the annual inflation rate for health care premiums at this corporation is 3.5%, what will be the annual health care premium for the same employee in 6 years? Round to the nearest dollar.
4. A retired couple has a fixed income of \$46,000 per year. Assuming an annual inflation rate of 6%, what will be the purchasing power of the couple's annual income in 10 years? Round to the nearest dollar.

Solutions on pp. S-13–S-14.

SECTION 7.3 Credit Cards and Consumer Loans

Objective 7.3A Calculate interest on a credit card bill

When a customer uses a credit card to make a purchase, the customer is actually receiving a loan. Therefore, there is frequently an added cost to the consumer who purchases on credit. This added cost may be in the form of an annual fee or interest charges on purchases. A **finance charge** is an amount paid in excess of the cash price; it is the cost to the customer for the use of credit.

Most credit card companies issue monthly bills. The due date on the bill is usually 1 month after the billing date (the date the bill is prepared and sent to the customer). If the bill is paid in full by the due date, the customer pays no finance charge. If the bill is not paid in full by the due date, a finance charge is added to the next bill.

Suppose a credit card billing date is the 10th day of each month. If a credit card purchase is made on April 15, then May 10 is the billing date (the 10th day of the month following April). The due date is June 10 (one month from the billing date). If the bill is paid in full before June 10, no finance charge is added. However, if the bill is not paid in full, interest charges on the outstanding balance will start to accrue (be added) on June 10, and any purchase made after June 10 will immediately start accruing interest.

The most common method of determining finance charges is the **average daily balance method.**

Average Daily Balance

$$\text{Average daily balance} = \frac{\text{sum of the total amounts owed each day of the month}}{\text{number of days in the billing period}}$$

Suppose an unpaid bill for \$315 had a due date of April 10. A purchase of \$28 was made on April 12, and \$123 was charged on April 24. A payment of \$50 was made on April 15. The next billing date is May 10. The interest on the average daily balance is 1.5% per month. Find the finance charge on the May 10 bill.

SOLUTION

To find the finance charge, first prepare a table showing the unpaid balance at the beginning of the billing cycle and at the time of each purchase, the number of days the balance is owed, and the product of these numbers. A negative sign in the "payments or purchases" column of the table indicates that a payment was made on that date.

Date	Payments or Purchases	Balance Each Day	Number of Days until Balance Changes	Unpaid Balance Times Number of Days
April 10–11		\$315	2	\$630
April 12–14	\$28	\$343	3	\$1029
April 15–23	− \$50	\$293	9	\$2637
April 24–May 9	\$123	\$416	16	\$6656
Total				\$10,952

The sum of the total amounts owed each day of the month is \$10,952.

Find the average daily balance.

$$\text{Average daily balance} = \frac{\text{sum of the total amounts owed each day of the month}}{\text{number of days in the billing period}}$$

$$= \frac{10{,}952}{30}$$ • There are 30 days from April 10 to May 10.

$$\approx 365.07$$

Calculate the finance charge by using the simple interest formula.

$I = Prt$

$I = 365.07(0.015)(1)$

$I \approx 5.48$

The finance charge on the May 10 bill is \$5.48.

Focus on calculating interest on a credit card bill

An unpaid bill of \$620 had a due date of March 10. A purchase of \$214 was made on March 15, and \$67 was charged on March 30. A payment of \$200 was made on March 22. The interest on the average daily balance is 1.5% per month. Find the finance charge on the April 10 bill.

SOLUTION

First calculate the sum of the total amounts owed each day of the month.

Date	Payments or Purchases	Balance Each Day	Number of Days until Balance Changes	Unpaid Balance Times Number of Days
March 10–14		\$620	5	$\$620 \cdot 5 = \$3100$
March 15–21	\$214	$\$620 + \$214 = \$834$	7	$\$834 \cdot 7 = \$5838$
March 22–29	− \$200	$\$834 - \$200 = \$634$	8	$\$634 \cdot 8 = \$5072$
March 30–April 9	\$67	$\$634 + \$67 = \$701$	11	$\$701 \cdot 11 = \$7711$
Total				\$21,721

The sum of the total amounts owed each day of the month is \$21,721.

Find the average daily balance.

$$\text{Average daily balance} = \frac{\text{sum of the total amounts owed each day of the month}}{\text{number of days in the billing period}}$$

$$= \frac{21{,}721}{31}$$

• There are 31 days from March 10 to April 10.

$$\approx \$700.68$$

Calculate the finance charge by using the simple interest formula.

$I = Prt$

$I = 700.68(0.015)(1)$

$I \approx 10.51$

The finance charge on the April 10 bill is \$10.51.

Check your understanding 1

A bill for \$1024 was due on July 1. Purchases of \$315 were made on July 7, and \$410 was charged on July 22. A payment of \$400 was made on July 15. The interest on the average daily balance is 1.2% per month. Find the finance charge on the August 1 bill.

SOLUTION See page S-6.

Objective 7.3A Practice

1. Calculate the finance charge for a credit card with an average daily balance of \$118.72 and a monthly interest rate of 1.25%.
2. A credit card account had a \$244 balance on March 5. A purchase of \$152 was made on March 12, and a payment of \$100 was made on March 28. Find the average daily balance if the billing date is April 5.
3. On August 10, a credit card account had a balance of \$345. A purchase of \$56 was made on August 15, and \$157 was charged on August 27. A payment of \$75 was made on August 15. The interest on the average daily balance is 1.25% per month. Find the finance charge on the September 10 bill.

4. The activity date, company, and amount for a credit card bill are shown below. The due date of the bill is July 10. On June 10, there was an unpaid balance of \$987.81. Find the finance charge if the interest rate is 1.8% per month.

Activity Date	Company	Amount
June 10	Unpaid balance	987.81
June 11	Jan's Surf Shop	156.33
June 12	Albertson's	45.61
June 15	The Down Shoppe	59.84
June 16	News Mart	18.54
June 20	Cardiff Delicatessen	23.09
June 22	The Olde Golf Mart	126.92
June 28	Lee's Hawaiian Restaurant	41.78
June 30	City Food Drive	100.00
July 2	Credit card payment	−1000.00
July 8	Safeway Stores	161.38

Solutions on pp. S-15–S-16.

Objective 7.3B Calculate the monthly payment on an APR loan

The stated interest rate for most consumer loans is normally the annual percentage rate, APR, as required by the Truth in Lending Act. This rate is based on compound interest rather than simple interest. The payment amount for an APR loan is given by the following formula.

Payment Formula for an APR Loan

The payment formula for an APR loan is given by

$$PMT = Ai\left(\frac{(1+i)^{nt}}{(1+i)^{nt}-1}\right)$$

where PMT is the payment; A is the loan amount; t is the number of years; and $i = \frac{r}{n}$, where r is the annual interest rate as a decimal and n is the number of payments per year.

The payment formula given above is used to calculate monthly payments on most consumer loans, such as car loans or home mortgage loans.

Focus on calculating a monthly payment

An electronics store is offering a 55-inch television for \$5995 and financing at an annual interest rate of 4.25% for 4 years. Find the monthly payment.

SOLUTION

Use the payment formula for an APR loan. We will calculate the results in steps.

Step 1: Calculate i. Because payments are made monthly, $n = 12$, the number of months in 1 year.

$$\begin{aligned} i &= \frac{r}{n} \\ &= \frac{0.0425}{12} && \bullet\ r = 0.0425;\ n = 12 \\ &\approx 0.0035416667 \end{aligned}$$

Step 2: Calculate $(1 + i)^{nt}$. Use i from Step 1; $n = 12$; $t = 4$, the number of years

$$\begin{aligned} (1 + i)^{nt} &= (1 + 0.0035416667)^{12(4)} \\ &= (1.0035416667)^{48} \\ &\approx 1.184948921 \end{aligned}$$

Step 3: Substitute the calculated values into the formula.

$$\begin{aligned} PMT &= Ai\left(\frac{(1 + i)^{nt}}{(1 + i)^{nt} - 1}\right) \\ &= 5995(0.0035416667)\left(\frac{1.184948921}{1.184948921 - 1}\right) \\ &= 21.23229187\left(\frac{1.184948921}{0.184948921}\right) \\ &\approx 136.033 \end{aligned}$$

The monthly payment is \$136.03.

Check your understanding 2

A pharmacist purchases a car and obtains a loan of \$25,000 at an annual interest rate of 2.5% for 5 years. Find the monthly car payment.

SOLUTION See pages S-6–S-7.

Focus on calculating a monthly car payment

A web designer purchases a car for \$18,395.

a. If the sales tax is 6.5% of the purchase price, find the amount of the sales tax.
b. If the car license fee is 1.2% of the purchase price, find the amount of the license fee.
c. If the designer makes a \$2500 down payment, find the amount of the loan the designer needs.
d. Assuming the designer gets the loan in part (c) at an annual interest rate of 7.5% for 4 years, determine the monthly car payment.

SOLUTION

a. Sales tax $= 0.065(18{,}395) = 1195.675$

The sales tax is \$1195.68.

b. License fee $= 0.012(18{,}395) = 220.74$

The license fee is \$220.74.

c. $$\begin{aligned} \text{Loan amount} &= \text{purchase price} + \text{sales tax} + \text{license fee} - \text{down payment} \\ &= 18{,}395 + 1195.68 + 220.74 - 2500 \\ &= 17{,}311.42 \end{aligned}$$

The loan amount is \$17,311.42.

d. Use the payment formula for an APR loan.

Step 1: Calculate i. Because payments are made monthly, $n = 12$, the number of months in 1 year.

$$\begin{aligned} i &= \frac{r}{n} \\ &= \frac{0.075}{12} \qquad \bullet\ r = 0.075;\ n = 12 \\ &= 0.00625 \end{aligned}$$

Step 2: Calculate $(1 + i)^{nt}$. Use i from Step 1; $n = 12$; $t = 4$, the number of years.

$$\begin{aligned} (1 + i)^{nt} &= (1 + 0.00625)^{12(4)} \\ &= (1.00625)^{48} \\ &\approx 1.348599151 \end{aligned}$$

Step 3: Substitute the calculated values into the formula.

$$\begin{aligned} PMT &= Ai\left(\frac{(1 + i)^{nt}}{(1 + i)^{nt} - 1}\right) \\ &= 17{,}311.42(0.00625)\left(\frac{1.348599151}{1.348599151 - 1}\right) \\ &= 108.196375\left(\frac{1.348599151}{0.348599151}\right) \\ &\approx 108.196375(3.868624313) \\ &\approx 418.571 \end{aligned}$$

The monthly car payment is \$418.57.

Check your understanding 3

A school superintendent purchases a new sedan for \$26,788.

a. If the sales tax is 5.25% of the purchase price, find the amount of the sales tax.

b. The superintendent makes a \$2500 down payment and the license fee is \$145. Find the amount the superintendent must finance.

c. Assuming the superintendent gets the loan in part (b) at an annual interest rate of 8.1% for 5 years, determine the superintendent's monthly car payment.

SOLUTION See page S-7.

Objective 7.3B Practice

1. Arrowood's Camera Store advertises a Canon Power Shot S95 camera for \$400, including taxes. If you finance the purchase of this camera for 1 year at an annual percentage rate of 6.9%, find the monthly payment.
2. After becoming a commercial pilot, Lorna Kao decides to purchase a 1978 Cessna 182 for \$64,995. Assuming the sales tax is 5.5% of the purchase price, find each of the following.
   a. What is the total cost, including sales tax?
   b. If Lorna makes a down payment of 20% of the total cost, find the down payment.
   c. Assuming Lorna finances the remaining cost at an annual interest rate of 7.15% for 10 years, find the monthly payment.

3. Luis Mahla purchases a used Porsche Boxster for \$42,600 and finances the entire amount at an annual interest rate of 5.7% for 5 years. Find the monthly payment. Assume the sales tax is 6% of the purchase price and the license fee is 1% of the purchase price.
4. Suppose you negotiate a selling price of \$26,995 for a Ford Explorer. You make a down payment of 10% of the selling price and finance the remaining balance for 3 years at an annual interest rate of 7.5%. The sales tax is 7.5% of the selling price, and the license fee is 0.9% of the selling price. Find the monthly payment.

Solutions on pp. S-16–S-18.

SOLUTIONS TO MODULE 7

Solutions to Check Your Understanding

Section 7.1 Check your understanding 1

Use the simple interest formula.

$P = 1500,\ r = 5.25\% = 0.0525$

$$t = \frac{4 \text{ months}}{1 \text{ year}} = \frac{4 \text{ months}}{12 \text{ months}} = \frac{4}{12}$$

$$I = Prt$$
$$I = 1500(0.0525)\left(\frac{4}{12}\right)$$
$$I = 26.25$$

The simple interest due is \$26.25.

Check your understanding 2

Use the simple interest formula.

$P = 700,\ r = 1.25\% = 0.0125$

$$t = \frac{9 \text{ months}}{1 \text{ year}} = \frac{9 \text{ months}}{12 \text{ months}} = \frac{9}{12}$$

$$I = Prt$$
$$I = 700(0.0125)\left(\frac{9}{12}\right)$$
$$I = 6.5625$$

The simple interest due is \$6.56.

Check your understanding 3

Use the simple interest formula.

$P = 7000,\ r = 5.25\% = 0.0525$

$$t = \frac{\text{number of days}}{360} = \frac{120}{360}$$

$$I = Prt$$
$$I = 7000(0.0525)\left(\frac{120}{360}\right)$$
$$I = 122.5$$

The simple interest due is \$122.50.

Check your understanding 4

Use the simple interest formula. Solve for r.

$$I = Prt$$
$$462 = 12{,}000(r)\left(\frac{6}{12}\right)$$
$$462 = 6000r$$
$$0.077 = r$$
$$r = 7.7\%$$

The annual simple interest rate on the loan is 7.7%.

Check your understanding 5

Find the interest. Use the simple interest formula.

$$P = 4000,\ r = 8.75\% = 0.0875,\ t = \frac{9}{12}$$

$$I = Prt$$
$$I = 4000(0.0875)\left(\frac{9}{12}\right)$$
$$I = 262.50$$

Find the maturity value. Use the maturity value formula for simple interest.

$$A = P + I$$
$$A = 4000 + 262.50$$
$$A = 4262.50$$

The maturity value of the loan is \$4262.50.

Check your understanding 6

$$P = 6700,\ r = 8.9\% = 0.089,\ t = 1$$

$$A = P(1 + rt)$$
$$A = 6700[1 + 0.089(1)]$$
$$A = 6700(1 + 0.089)$$
$$A = 6700(1.089)$$
$$A = 7296.30$$

The maturity value of the loan is \$7296.30.

Check your understanding 7

$$P = 680,\ r = 6.4\% = 0.064,\ t = 1$$

$$A = P(1 + rt)$$
$$A = 680[1 + 0.064(1)]$$
$$A = 680(1 + 0.064)$$
$$A = 680(1.064)$$
$$A = 723.52$$

After 1 year, \$723.52 is in the account.

Check your understanding 8

Find the amount of interest paid.

$I = A - P$

$I = 9240 - 9000$

$I = 240$

Solve the simple interest formula for r.

$$I = Prt$$

$$240 = 9000(r)\left(\frac{4}{12}\right)$$

$$240 = 3000r$$

$$0.08 = r$$

$$r = 8\%$$

The annual simple interest rate on the loan is 8%.

Section 7.2

Check your understanding 1

Use the future value formula.

$P = 4000,\ r = 6\% = 0.06,\ n = 12,\ t = 2$

$$A = P\left(1 + \frac{r}{n}\right)^{nt}$$

$$A = 4000\left(1 + \frac{0.06}{12}\right)^{12\cdot 2}$$

$$A = 4000(1 + 0.005)^{24}$$

$$A = 4000(1.005)^{24}$$

$$A \approx 4000(1.127160)$$

$$A \approx 4508.64$$

The future value after 2 years is approximately \$4508.64.

Check your understanding 2

Use the future value formula.

$P = 2500,\ r = 9\% = 0.09,\ n = 360,\ t = 4$

$$A = 2500\left(1 + \frac{0.09}{360}\right)^{360\cdot 4}$$

$$A = 2500(1 + 0.00025)^{1440}$$

$$A = 2500(1.00025)^{1440}$$

$$A \approx 2500(1.4332649)$$

$$A \approx 3583.16$$

The future value after 4 years is approximately \$3583.16.

Check your understanding 3

Calculate the future value. Use the future value formula.

$P = 8000,\ r = 9\% = 0.09,\ n = 12,\ t = 6$

$$A = P\left(1 + \frac{r}{n}\right)^{nt}$$

$$A = 8000\left(1 + \frac{0.09}{12}\right)^{12\cdot 6}$$

$$A = 8000(1 + 0.0075)^{72}$$

$$A = 8000(1.0075)^{72}$$

$$A \approx 8000(1.7125527)$$

$$A \approx 13{,}700.42$$

Calculate the interest earned.

$$I = A - P$$

$$I = 13{,}700.42 - 8000$$

$$I = 5700.42$$

The amount of interest earned is approximately \$5700.42.

Check your understanding 4

Use the present value formula.

$A = 20{,}000,\ r = 9\% = 0.09,\ n = 2,\ t = 5$

$$P = \frac{A}{\left(1 + \frac{r}{n}\right)^{nt}}$$

$$P = \frac{20{,}000}{\left(1 + \frac{0.09}{2}\right)^{2\cdot 5}}$$

$$P = \frac{20{,}000}{(1 + 0.045)^{10}}$$

$$P \approx \frac{20{,}000}{1.552969}$$

$$P \approx 12{,}878.55$$

\$12,878.55 should be invested in the account.

Check your understanding 5

Use the present value formula.

$A = 18{,}000,\ r = 4\% = 0.04,\ n = 12$ (compounded monthly), $t = 3$

$$P = \frac{A}{\left(1 + \frac{r}{n}\right)^{nt}}$$

$$= \frac{18{,}000}{\left(1 + \frac{0.04}{12}\right)^{12(3)}} \approx \frac{18{,}000}{(1 + 0.0033333333)^{36}}$$

$$= \frac{18{,}000}{1.127271875} \approx 15{,}967.75$$

The physician's assistant must deposit \$15,967.75.

Check your understanding 6

Use the future value formula.

$P = 28{,}000,\ r = 5\% = 0.05,\ t = 17$

The inflation rate is an annual rate, so $n = 1$.

$$A = P\left(1 + \frac{r}{n}\right)^{nt}$$

$$A = 28{,}000\left(1 + \frac{0.05}{1}\right)^{1 \cdot 17}$$

$$A = 28{,}000(1 + 0.05)^{17}$$

$$A = 28{,}000(1.05)^{17}$$

$$A \approx 28{,}000(2.2920183)$$

$$A \approx 64{,}176.51$$

The average new car sticker price in 2030 will be approximately \$64,176.51.

Check your understanding 7

Use the present value formula.

$A = 500{,}000,\ r = 7\% = 0.07,\ t = 40$

The inflation rate is an annual rate, so $n = 1$.

$$P = \frac{A}{\left(1 + \frac{r}{n}\right)^{nt}}$$

$$P = \frac{500{,}000}{\left(1 + \frac{0.07}{1}\right)^{1 \cdot 40}}$$

$$P = \frac{500{,}000}{(1 + 0.07)^{40}}$$

$$P \approx \frac{500{,}000}{14.9744578}$$

$$P \approx 33{,}390.19$$

In 2055, the purchasing power of \$500,000 will be approximately \$33,390.19.

Section 7.3 Check your understanding 1

Calculate the sum of the total amounts owed each day of the month.

Date	Payments or Purchases	Balance Each Day	Number of Days until Balance Changes	Unpaid Balance Times Number of Days
July 1–6		\$1024	6	$\$1024 \cdot 6 = \$6144$
July 7–14	\$315	$\$1024 + \$315 = \$1339$	8	$\$1339 \cdot 8 = \$10,712$
July 15–21	− \$400	$\$1339 - \$400 = \$939$	7	$\$939 \cdot 7 = \$6573$
July 22–31	\$410	$\$939 + \$410 = \$1349$	10	$\$1349 \cdot 10 = \$13,490$
Total				\$36,919

$$\text{Average daily balance} = \frac{\text{sum of the total amounts owed each day of the month}}{\text{number of days in the billing period}}$$

$$= \frac{36,919}{31} \approx \$1190.94$$

Use the simple interest formula to calculate the finance charge.

$I = Prt$

$I = 1190.94(0.012)(1)$

$I \approx 14.29$

The finance charge on the August 1 bill is \$14.29.

Check your understanding 2

Use the payment formula for an APR loan.

Step 1: Calculate i. Because payments are made monthly, $n = 12$, the number of months in 1 year.

$$i = \frac{r}{n}$$

$$= \frac{0.025}{12} \qquad \bullet\ r = 0.025;\ n = 12$$

$$\approx 0.0020833333$$

Step 2: Calculate $(1 + i)^{nt}$. Use i from Step 1; $n = 12$; $t = 5$, the number of years.

$$(1 + i)^{nt} = (1 + 0.0020833333)^{12(5)}$$

$$= (1.0020833333)^{60}$$

$$\approx 1.1133001099$$

Step 3: Substitute the calculated values into the formula.

$$PMT = Ai\left(\frac{(1+i)^{nt}}{(1+i)^{nt}-1}\right)$$

$$= 25{,}000(0.0020833333)\left(\frac{1.1133001099}{1.1133001099 - 1}\right)$$

$$= 52.0833325\left(\frac{1.1133001099}{0.1133001099}\right)$$

$$\approx 52.0833325(9.82611677)$$

$$\approx 511.777$$

The monthly payment is \$511.78.

Check your understanding 3

a. Sales tax $= 26{,}788(0.0525) = 1406.37$

The sales tax is \$1406.37.

b. Loan amount = purchase price + sales tax + license fee − down payment

$= 26{,}788 + 1406.37 + 145 - 2500 = 25{,}839.37$

The loan amount is \$25,839.37.

c. Use the payment formula for an APR loan.

Step 1: Calculate i. Because payments are made monthly, $n = 12$, the number of months in 1 year.

$$i = \frac{r}{n}$$

$$= \frac{0.081}{12} \qquad \bullet\ r = 0.081;\ n = 12$$

$$= 0.00675$$

Step 2: Calculate $(1 + i)^{nt}$. Use i from Step 1; $n = 12$; $t = 5$, the number of years.

$$(1+i)^{nt} = (1 + 0.00675)^{12(5)}$$

$$= (1.00675)^{60}$$

$$\approx 1.497263704$$

Step 3: Substitute the calculated values into the formula.

$$PMT = Ai\left(\frac{(1+i)^{nt}}{(1+i)^{nt}-1}\right)$$

$$= 25{,}839.37(0.00675)\left(\frac{1.497263704}{1.497263704 - 1}\right)$$

$$\approx 174.4157475\left(\frac{1.497263704}{0.497263704}\right)$$

$$\approx 174.4157475(3.011005412)$$

$$\approx 525.167$$

The monthly car payment is \$525.17.

Solutions to Objective Practice Exercises

Objective 7.1A

1. Use the simple interest formula.

$P = 7000, r = 6.5\% = 0.065$

Because the interest rate is an annual rate, the time must be measured in years:

$$t = \frac{6 \text{ months}}{1 \text{ year}} = \frac{6 \text{ months}}{12 \text{ months}} = \frac{6}{12}$$

$$\begin{aligned} I &= Prt \\ &= 7000(0.065)\left(\frac{6}{12}\right) = 227.50 \end{aligned}$$

The interest earned is \$227.50.

2. Use the simple interest formula.

$P = 9000, r = 6.75\% = 0.0675$

Because the interest rate is an annual rate, the time must be measured in years:

$$t = \frac{4 \text{ months}}{1 \text{ year}} = \frac{4 \text{ months}}{12 \text{ months}} = \frac{4}{12}$$

$$\begin{aligned} I &= Prt \\ &= 9000(0.0675)\left(\frac{4}{12}\right) = 202.50 \end{aligned}$$

The interest earned is \$202.50.

3. Use the simple interest formula.

$P = 2000, r = 1.25\% = 0.0125$

Because the interest rate is an annual rate, the time must be measured in years:

$$t = \frac{9 \text{ months}}{1 \text{ year}} = \frac{9 \text{ months}}{12 \text{ months}} = \frac{9}{12}$$

$$\begin{aligned} I &= Prt \\ &= 2000(0.0125)\left(\frac{9}{12}\right) = 18.75 \end{aligned}$$

The interest earned is \$18.75.

4. Solve the simple interest formula for r.

$$\begin{aligned} I &= Prt \\ 80 &= 2000(r)\left(\frac{6}{12}\right) \\ 80 &= 1000r \\ \frac{80}{1000} &= r \\ 0.08 &= r \end{aligned}$$

The simple interest rate is 8%.

5. Use the simple interest formula.

$P = 1500,\ r = 5.2\% = 0.052,\ t = \dfrac{6\text{ months}}{1\text{ year}} = \dfrac{6\text{ months}}{12\text{ months}} = \dfrac{6}{12}$

$$\begin{aligned} I &= Prt \\ &= 1500(0.052)\left(\frac{6}{12}\right) = 39 \end{aligned}$$

The simple interest earned is \$39.

6. Use the simple interest formula.

$P = 1600,\ r = 9\% = 0.09,\ t = \dfrac{45\text{ days}}{1\text{ year}} = \dfrac{45\text{ days}}{360\text{ days}} = \dfrac{45}{360}$

$$\begin{aligned} I &= Prt \\ &= 1600(0.09)\left(\frac{45}{360}\right) = 18 \end{aligned}$$

The simple interest earned is \$18.

Objective 7.1B

1. Since the interest rate is per year, the time period of the loan is $t = \dfrac{6\text{ months}}{1\text{ year}} = \dfrac{6\text{ months}}{12\text{ months}} = \dfrac{6}{12}$.

$$\begin{aligned} A &= P(1 + rt) \\ &= 15{,}000\left[1 + 0.089\left(\frac{6}{12}\right)\right] \\ &= 15{,}000(1.0445) = 15{,}667.50 \end{aligned}$$

The maturity value of the loan is \$15,667.50.

2. Since the interest rate is per year, the time period of the loan is $t = \dfrac{3\text{ months}}{1\text{ year}} = \dfrac{3\text{ months}}{12\text{ months}} = \dfrac{3}{12}$.

$$\begin{aligned} A &= P(1 + rt) \\ &= 2800\left[1 + 0.092\left(\frac{3}{12}\right)\right] \\ &= 2800(1.023) = 2864.40 \end{aligned}$$

The maturity value of the loan is \$2864.40.

3. $$\begin{aligned} A &= P(1 + rt) \\ &= 7000\left[1 + 0.087\left(\frac{8}{12}\right)\right] \\ &= 7000(1.058) = 7406 \end{aligned}$$

The maturity value of the loan is \$7406.

4. $$\begin{aligned} A &= P(1 + rt) \\ &= 750[1 + 0.073(1)] \\ &= 750(1.073) = 804.75 \end{aligned}$$

The future value of the investment is \$804.75.

5. First find the amount of interest paid. Subtract the principal from the maturity value.

$I = A - P = 5125 - 5000 = 125$

Find the simple interest rate by solving the simple interest formula for r.

$$\begin{aligned} I &= Prt \\ 125 &= 5000(r)\left(\frac{3}{12}\right) \\ 125 &= 1250r \\ 0.10 &= r \\ r &= 10\% \end{aligned}$$

The simple interest rate on the loan is 10%.

Objective 7.2A **1.** $P = 500$, $r = 9\% = 0.09$, $n = 4$, $t = 6$

$$\begin{aligned} A &= P\left(1 + \frac{r}{n}\right)^{nt} \\ &= 500\left(1 + \frac{0.09}{4}\right)^{(4\cdot 6)} \\ &= 500(1.0225)^{24} \approx 852.88 \end{aligned}$$

The future value after 6 years is approximately $852.88.

2. $P = 8500$, $r = 9\% = 0.09$, $n = 12$, $t = 10$

$$\begin{aligned} A &= P\left(1 + \frac{r}{n}\right)^{nt} \\ &= 8500\left(1 + \frac{0.09}{12}\right)^{(12\cdot 10)} \\ &= 8500(1.0075)^{120} \approx 20{,}836.54 \end{aligned}$$

The future value after 10 years is approximately $20,836.54.

3. $P = 9000$, $r = 5.5\% = 0.055$, $n = 4$, $t = 3$

$$\begin{aligned} A &= P\left(1 + \frac{r}{n}\right)^{nt} \\ &= 9000\left(1 + \frac{0.055}{4}\right)^{(4\cdot 3)} \\ &= 9000(1.01375)^{12} \approx 10{,}602.61 \end{aligned}$$

The future value is approximately $10,602.61.

4. Calculate the future value.

$P = 8500$, $r = 9\% = 0.09$, $n = 2$, $t = 5$

$$\begin{aligned} A &= P\left(1 + \frac{r}{n}\right)^{nt} \\ &= 8500\left(1 + \frac{0.09}{2}\right)^{(2\cdot 5)} \\ &= 8500(1.045)^{10} \approx 13{,}200.24 \end{aligned}$$

Calculate the interest earned.

$I = A - P = 13{,}200.24 - 8500 = 4700.24$

The amount of interest earned is approximately \$4700.24.

5. **a.** $P = 15{,}000$, $r = 8\% = 0.08$, $n = 2$, $t = 4$

$$\begin{aligned} A &= P\left(1 + \frac{r}{n}\right)^{nt} \\ &= 15{,}000\left(1 + \frac{0.08}{2}\right)^{(2\cdot 4)} \\ &= 15{,}000(1.04)^{8} \approx 20{,}528.54 \end{aligned}$$

The future value of the investment is approximately \$20,528.54.

b. $P = 15{,}000$, $r = 8\% = 0.08$, $n = 4$, $t = 4$

$$\begin{aligned} A &= P\left(1 + \frac{r}{n}\right)^{nt} \\ &= 15{,}000\left(1 + \frac{0.08}{4}\right)^{(4\cdot 4)} \\ &= 15{,}000(1.02)^{16} \approx 20{,}591.79 \end{aligned}$$

The future value of the investment is approximately \$20,591.79.

c. $20{,}591.79 - 20{,}528.54 = 63.25$

The future value of the investment is \$63.25 greater when interest is compounded quarterly.

Objective 7.2B

1. Use the present value formula.

$A = 15{,}000,\ r = 6\% = 0.06,\ n = 12,\ t = 5$

$$P = \frac{A}{\left(1 + \frac{r}{n}\right)^{nt}}$$
$$= \frac{15{,}000}{\left(1 + \frac{0.06}{12}\right)^{(12 \cdot 5)}}$$
$$= \frac{15{,}000}{(1.005)^{60}}$$
$$\approx 11{,}120.58$$

\$11,120.58 should be invested in the account in order to have \$15,000 in 5 years.

2. a. Use the future value formula.

$P = 6000,\ r = 10\% = 0.10,\ n = 4,\ t = 5$

$$A = P\left(1 + \frac{r}{n}\right)^{nt}$$
$$= 6000\left(1 + \frac{0.10}{4}\right)^{(4 \cdot 5)}$$
$$= 6000(1.025)^{20} \approx 9831.70$$

The maturity value of the loan is approximately \$9831.70.

b. $9831.70 - 6000 = 3831.70$

You are paying \$3831.70 in interest.

3. Use the present value formula.

$A = 40{,}000,\ r = 8\% = 0.08,\ n = 4,\ t = 18$

$$P = \frac{A}{\left(1 + \frac{r}{n}\right)^{nt}}$$
$$= \frac{40{,}000}{\left(1 + \frac{0.08}{4}\right)^{(4 \cdot 18)}}$$
$$= \frac{40{,}000}{(1.02)^{72}}$$
$$\approx 9612.75$$

\$9612.75 should be invested in the account in order to have \$40,000 in 18 years.

4. Use the present value formula.

$A = 1{,}000{,}000,\ r = 9\% = 0.09,\ n = 360,\ t = 30$

$$P = \frac{A}{\left(1 + \frac{r}{n}\right)^{nt}}$$
$$= \frac{1{,}000{,}000}{\left(1 + \frac{0.09}{360}\right)^{(360\cdot 30)}}$$
$$= \frac{1{,}000{,}000}{(1.00025)^{10{,}800}}$$
$$\approx 67{,}228.19$$

\$67,228.19 should be invested in the account in order to have \$1,000,000 in 30 years.

5. Use the present value formula.

$A = 51{,}000,\ r = 2.5\% = 0.025,\ n = 360,\ t = 18$

$$P = \frac{A}{\left(1 + \frac{r}{n}\right)^{nt}}$$
$$= \frac{51{,}000}{\left(1 + \frac{0.025}{360}\right)^{(360\cdot 18)}}$$
$$= \frac{51{,}000}{(1.000069444)^{6480}}$$
$$\approx 32{,}519.64$$

\$32,519.64 must be deposited to have \$51,000 in 18 years.

Objective 7.2C

1. Use the future value formula.

$P = 40{,}000,\ r = 7\% = 0.07,\ n = 1$ (because the inflation rate is an annual rate), and $t = 3$

$$A = P\left(1 + \frac{r}{n}\right)^{nt}$$
$$A = 40{,}000\left(1 + \frac{0.07}{1}\right)^{1\cdot 3}$$
$$= 40{,}000(1.07)^3$$
$$= 40{,}000(1.225043)$$
$$= 49{,}001.72$$

You will need an annual salary of \$49,002.

2. Use the future value formula.

$P = 230{,}000$, $r = 4\% = 0.04$, $n = 1$ (because the inflation rate is an annual rate), and $t = 5$

$$A = P\left(1 + \frac{r}{n}\right)^{nt}$$

$$\begin{aligned} A &= 230{,}000\left(1 + \frac{0.04}{1}\right)^{1\cdot 5} \\ &= 230{,}000(1.04)^5 \\ &\approx 230{,}000(1.216652902) \\ &\approx 279{,}830.1676 \end{aligned}$$

The median price of a house will be about \$279,830.

3. Use the future value formula.

$P = 11{,}000$, $r = 3.5\% = 0.035$, $n = 1$ (because the inflation rate is an annual rate), and $t = 6$

$$A = P\left(1 + \frac{r}{n}\right)^{nt}$$

$$\begin{aligned} A &= 11{,}000\left(1 + \frac{0.035}{1}\right)^{1\cdot 6} \\ &= 11{,}000(1.035)^6 \\ &\approx 11{,}000(1.229255326) \\ &\approx 13{,}521.81 \end{aligned}$$

The annual premium will be approximately \$13, 522.

4. Use the future value formula.

$A = 46{,}000$, $r = 6\% = 0.06$, $n = 1$ (because the inflation rate is an annual rate), and $t = 10$

$$A = P\left(1 + \frac{r}{n}\right)^{nt}$$

$$\begin{aligned} 46{,}000 &= P\left(1 + \frac{0.06}{1}\right)^{1\cdot 10} \\ 46{,}000 &= P(1.06)^{10} \\ 46{,}000 &\approx P(1.790847697) \\ \frac{46{,}000}{1.790847697} &= P \\ 25{,}686.16 &\approx P \end{aligned}$$

The purchasing power of the couple's income in 10 years will be about \$25,686.

Objective 7.3A

1. Use the simple interest formula.

   $P = \$118.72$, $r = 1.25\% = 0.0125$, $n = 1$ (because the interest rate is given as a monthly rate)

   $$I = Prt$$
   $$= 118.72(0.0125)(1)$$
   $$= 1.484$$

   The finance charge is $1.48.

2. Calculate the sum of the total amounts owed each day of the month.

Date	Payments or Purchases	Balance Each Day	Days until Balance Changes	Unpaid Balance Times Number of Days
Mar. 5–11		$244	7	$1708
Mar. 12–27	$152	$396	16	$6336
Mar. 28–Apr. 4	− $100	$296	8	$2368
Total				$10,412

   $$\text{Average daily balance} = \frac{10{,}412}{31} \approx 335.87$$

   The average daily balance is $335.87.

3. Calculate the sum of the total amounts owed each day of the month.

Date	Payments or Purchases	Balance Each Day	Days until Balance Changes	Unpaid Balance Times Number of Days
Aug. 10–14		$345	5	$1725
Aug. 15	$56	$401	0	$0
Aug. 15–26	− $75	$326	12	$3912
Aug. 27–Sep. 9	$157	$483	14	$6762
Total				$12,399

   $$\text{Average daily balance} = \frac{12{,}399}{31} \approx 399.97$$

   Use the simple interest formula to calculate the finance charge.

   $$I = Prt$$
   $$= (399.97)(0.0125)(1) \approx 5.00$$

   The finance charge is $5.00.

4. Calculate the sum of the total amounts owed each day of the month.

Date	Payments or Purchases	Balance Each Day	Days until Balance Changes	Unpaid Balance Times Number of Days
Jun. 10		\$987.81	1	\$987.81
Jun. 11	\$156.33	\$1144.14	1	\$1144.14
Jun. 12–14	\$45.61	\$1189.75	3	\$3569.25
Jun. 15	\$59.84	\$1249.59	1	\$1249.59
Jun. 16–19	\$18.54	\$1268.13	4	\$5072.52
Jun. 20–21	\$23.09	\$1291.22	2	\$2582.44
Jun. 22–27	\$126.92	\$1418.14	6	\$8508.84
Jun. 28–29	\$41.78	\$1459.92	2	\$2919.84
Jun. 30–Jul. 1	\$100	\$1559.92	2	\$3119.84
Jul. 2–7	− \$1000	\$559.92	6	\$3359.52
Jul. 8–9	\$161.38	\$721.30	2	\$1442.60
Total				\$33,956.39

$$\text{Average daily balance} = \frac{33{,}956.39}{30} \approx 1131.88.$$

Use the simple interest formula to calculate the finance charge.

$$\begin{aligned} I &= Prt \\ &= (1131.88)(0.018)(1) \approx 20.37 \end{aligned}$$

The finance charge is \$20.37.

Objective 7.3B

1. Use the payment formula for an APR loan. Because payments are made monthly, $n = 12$, the number of months in 1 year. Calculate i.

$$i = \frac{r}{n} = \frac{0.069}{12} = 0.00575$$

Calculate $(1 + i)^{nt}$.

$$\begin{aligned} (1 + i)^{nt} &= (1 + 0.00575)^{12(1)} \\ &= 1.00575^{12} \\ &\approx 1.071224495 \end{aligned}$$

Substitute the calculated values into the formula.

$$\begin{aligned} PMT &= Ai\left(\frac{(1 + i)^{nt}}{(1 + i)^{nt} - 1}\right) \\ &\approx 400(0.00575)\left(\frac{1.071224495}{1.071224495 - 1}\right) \\ &= 2.3\left(\frac{1.071224495}{0.071224495}\right) \\ &\approx 2.3(15.04011359) \\ &\approx 34.592 \end{aligned}$$

The monthly payment is \$34.59.

2. **a.** Sales tax $= 0.055(64{,}995) = 3574.725$

The sales tax is \$3574.73.

Total cost $= 64{,}995 + 3574.73 = 68{,}569.73$

The total cost is \$68,569.73.

b. Down payment $= 0.20(68{,}569.73)$

$= 13{,}713.946$

The down payment is \$13,713.95.

c. Loan amount $=$ total cost $-$ down payment

Loan amount $= 68{,}569.73 - 13{,}713.95$

$= 54{,}855.78$

Use the payment formula for an APR loan. Because payments are made monthly, $n = 12$, the number of months in 1 year. Calculate i.

$$i = \frac{r}{n} = \frac{0.0715}{12} = 0.0059583333$$

Calculate $(1 + i)^{nt}$.

$$\begin{aligned}(1 + i)^{nt} &= (1 + 0.0059583333)^{12(10)}\\ &= 1.0059583333^{120}\\ &\approx 2.039854161\end{aligned}$$

Substitute the calculated values into the formula.

$$\begin{aligned}PMT &= Ai\left(\frac{(1 + i)^{nt}}{(1 + i)^{nt} - 1}\right)\\ &\approx 54{,}855.78(0.0059583333)\left(\frac{2.039854161}{2.039854161 - 1}\right)\\ &= 326.8490207\left(\frac{2.039854161}{1.039854161}\right)\\ &\approx 326.8490207(1.96167332)\\ &\approx 641.171\end{aligned}$$

The monthly payment is \$641.17.

3. Sales tax $= 0.06(42{,}600) = 2556$

License fee $= 0.01(42{,}600) = 426$

Loan amount $= 42{,}600 + 2556 + 426 = 45{,}582$

Use the payment formula for an APR loan. Because payments are made monthly, $n = 12$, the number of months in 1 year. Calculate i.

$$i = \frac{r}{n} = \frac{0.057}{12} = 0.00475$$

Calculate $(1 + i)^{nt}$.

$$\begin{aligned}(1 + i)^{nt} &= (1 + 0.00475)^{12(5)}\\ &= 1.00475^{60}\\ &\approx 1.328865088\end{aligned}$$

Substitute the calculated values into the formula.

$$\begin{aligned} PMT &= Ai\left(\frac{(1+i)^{nt}}{(1+i)^{nt}-1}\right) \\ &\approx 45{,}582(0.00475)\left(\frac{1.328865088}{1.328865088-1}\right) \\ &= 216.5145\left(\frac{1.328865088}{0.328865088}\right) \\ &\approx 216.5145(4.040760593) \\ &\approx 874.883 \end{aligned}$$

The monthly payment is \$874.88.

4. Sales tax $= 0.075(26{,}995) = 2024.63$

License fee $= 0.009(26{,}995) = 242.96$

Down payment $= 0.10(26{,}995) = 2699.50$

Loan amount $=$ purchase price $-$ down payment $+$ sales tax $+$ license fee

$= 26{,}995 - 2699.50 + 2024.63 + 242.96$

$= 26{,}563.09$

Use the payment formula for an APR loan. Because payments are made monthly, $n = 12$, the number of months in 1 year. Calculate i.

$$i = \frac{r}{n} = \frac{0.075}{12} = 0.00625$$

Calculate $(1+i)^{nt}$.

$$\begin{aligned} (1+i)^{nt} &= (1+0.00625)^{12(3)} \\ &= 1.00625^{36} \\ &\approx 1.251446136 \end{aligned}$$

Substitute the calculated values into the formula.

$$\begin{aligned} PMT &= Ai\left(\frac{(1+i)^{nt}}{(1+i)^{nt}-1}\right) \\ &\approx 26{,}563.09(0.00625)\left(\frac{1.251446136}{1.251446136-1}\right) \\ &= 166.0193125\left(\frac{1.251446136}{0.251446136}\right) \\ &\approx 166.0193125(4.976994898) \\ &\approx 826.277 \end{aligned}$$

The monthly payment is \$826.28.

TI-84 Plus

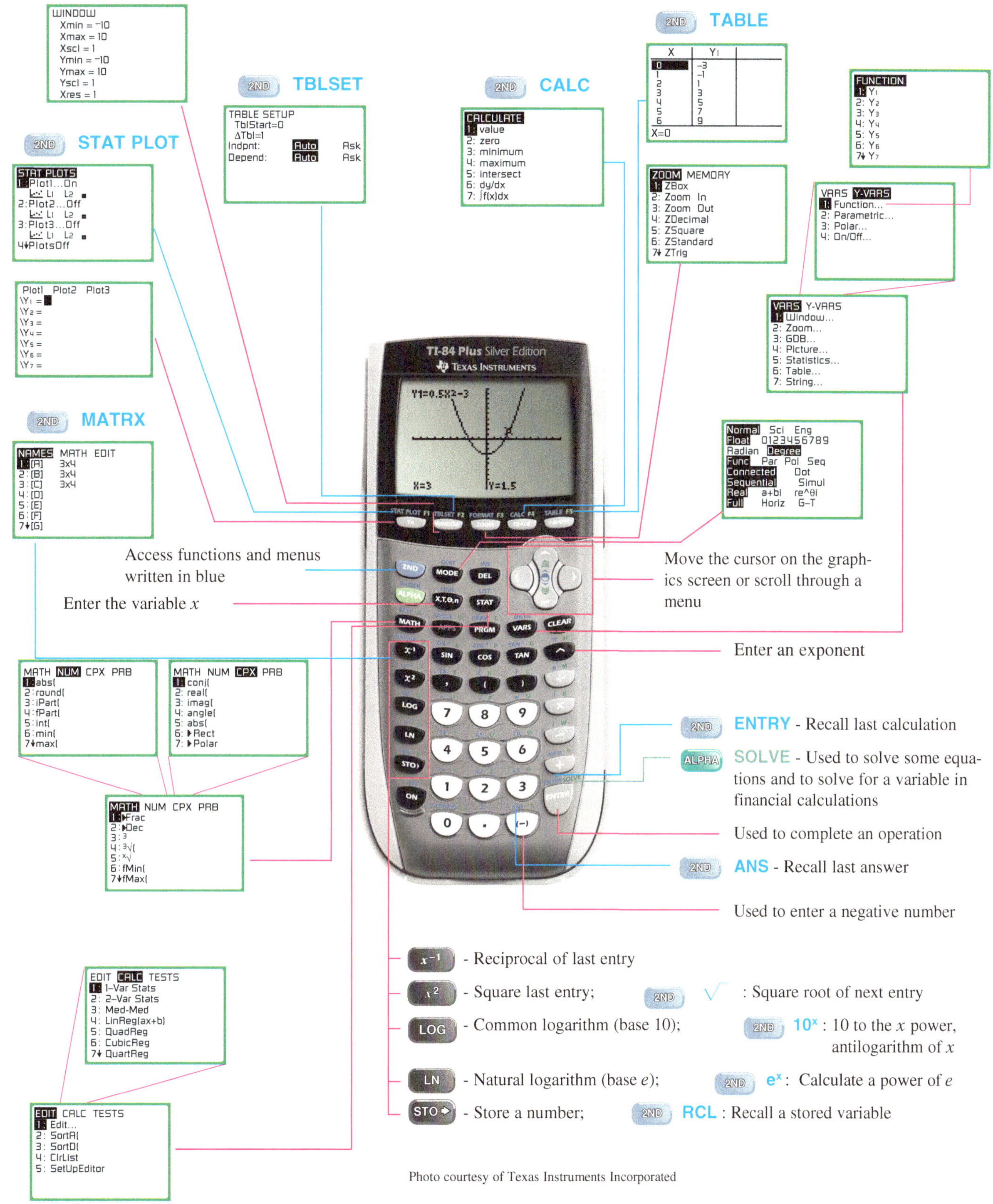

Photo courtesy of Texas Instruments Incorporated

Keystroke Guide for the TI-84 Plus

Basic Operations

Numerical calculations are performed on the **home screen.** You can always return to the home screen by pressing [2ND] QUIT. Pressing [CLEAR] erases the home screen.

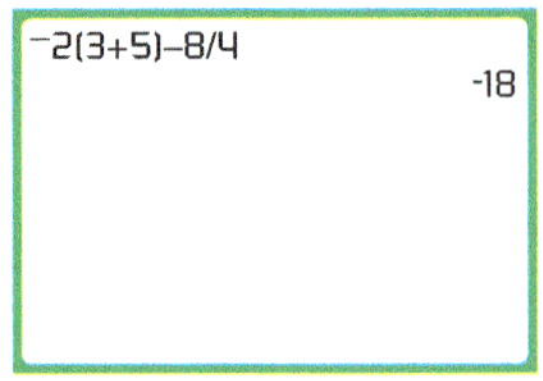

To evaluate the expression $-2(3 + 5) - 8 \div 4$, use the following keystrokes.

[(−)] 2 [(] 3 [+] 5 [)] [−] 8 [÷] 4 [ENTER]

Note: There is a difference between the key to enter a negative number, [(−)], and the key for subtraction, [−]. You cannot use these keys interchangeably.

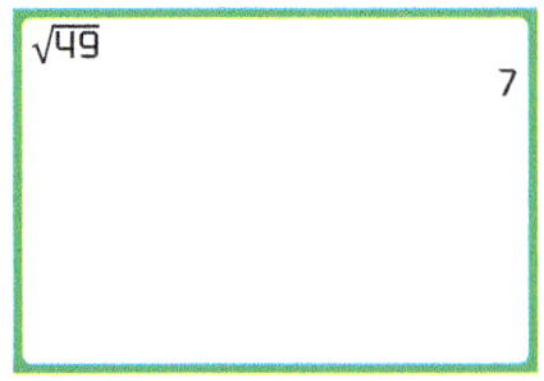

The [2ND] key is used to access the commands in gold writing above a key. For instance, to evaluate $\sqrt{49}$, press [2ND] √ 49 [ENTER].

The [ALPHA] key is used to place a letter on the screen. One reason to do this is to store a value of a variable. The following keystrokes give A the value of 5.

5 [STO▸] [ALPHA] A [ENTER]

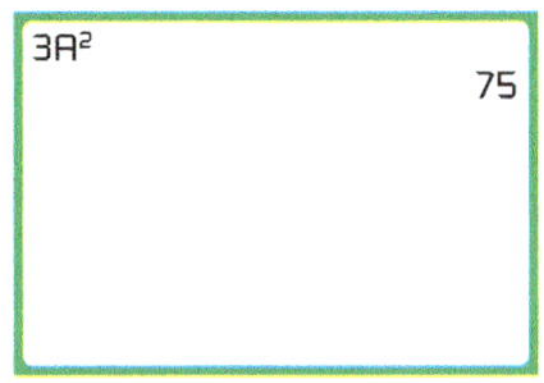

This value is now available in calculations. For instance, we can find the value of $3a^2$ by using the following keystrokes: 3 [ALPHA] A [x^2]. To display the value of the variable on the screen, press [2ND] RCL [ALPHA] A.

Note: When you use the [ALPHA] key, only capital letters are available on the TI-83 calculator.

Complex Numbers

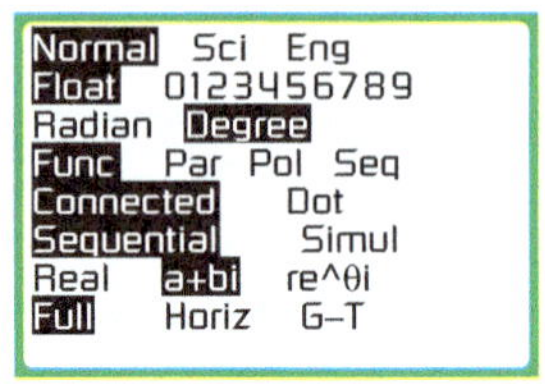

To perform operations on complex numbers, first press [MODE] and then use the arrow keys to select $a+bi$. Then press [ENTER] [2ND] QUIT.

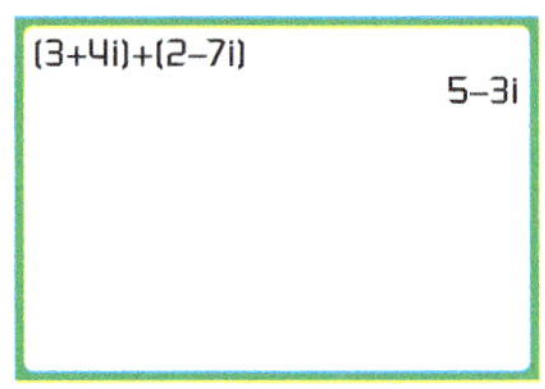

Addition of complex numbers To add $(3 + 4i) + (2 - 7i)$, use the keystrokes

[(] 3 [+] 4 [2ND] i [)] [+]
[(] 2 [−] 7 [2ND] i [)] [ENTER].

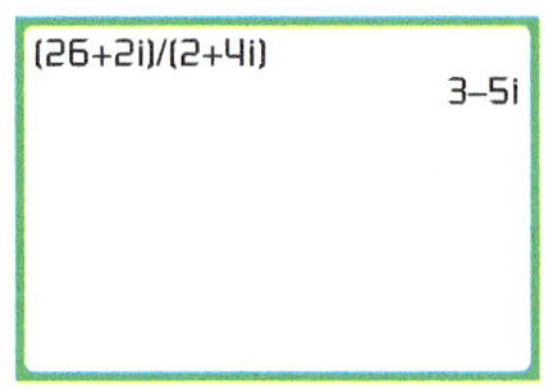

Division of complex numbers To divide $\dfrac{26 + 2i}{2 + 4i}$, use the keystrokes [(] 26 [+] 2 [2ND] i [)] [÷] [(] 2 [+] 4 [2ND] i [)] [ENTER].

Note: Operations for subtraction and multiplication are similar.

Additional operations on complex numbers can be found by selecting CPX under the MATH key.

To find the absolute value of $2 - 5i$, press MATH (scroll to CPX) (scroll to abs) ENTER 2 − 5 2ND i ENTER.

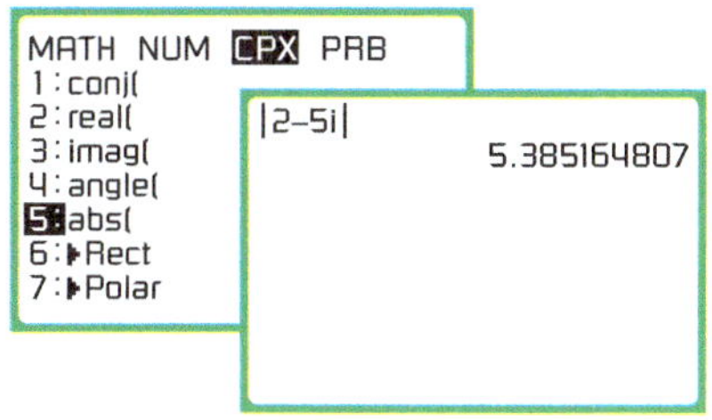

Evaluating Functions

There are various methods of evaluating a function but all methods require that the expression be entered as one of the ten functions Y_1 to Y_0. To evaluate $f(x) = \frac{x^2}{x-1}$ when $x = -3$, enter the expression into, for instance, Y_1, and then press VARS ▸ 1 1 ((−) 3) ENTER.

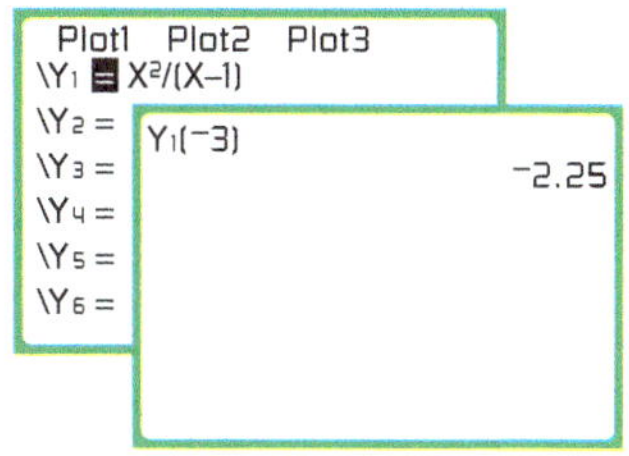

Take Note ➤
Use the down arrow key to scroll past Y_7 to see Y_8, Y_9, and Y_0.

Note: If you try to evaluate a function at a number that is not in the domain of the function, you will get an error message. For instance, 1 is not in the domain of $f(x) = \frac{x^2}{x-1}$. If we try to evaluate the function at 1, the error screen at the right appears.

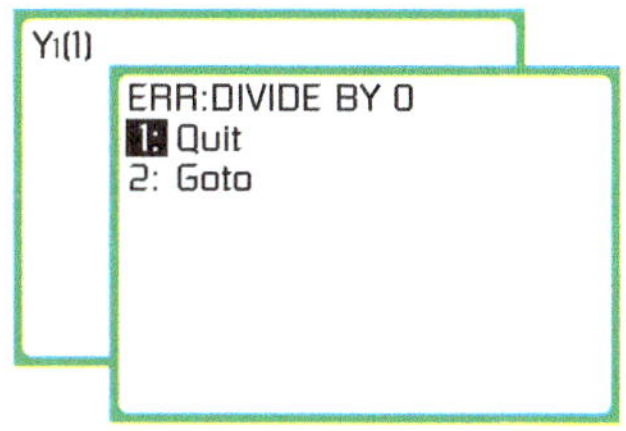

Evaluating Variable Expressions

To evaluate a variable expression, first store the values of each variable. Then enter the variable expression. For instance, to evaluate $s^2 + 2sl$ when $s = 4$ and $l = 5$, use the following keystrokes.

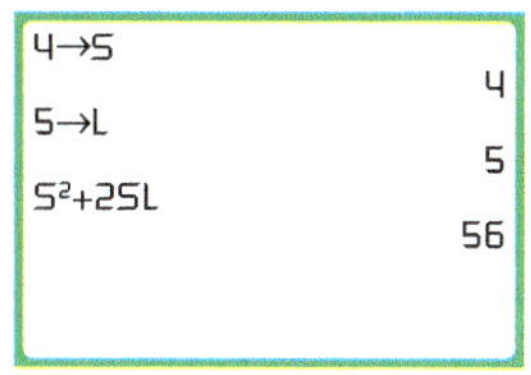

4 STO▸ ALPHA S ENTER 5 STO▸ ALPHA L ENTER ALPHA S x^2 + 2 ALPHA S ALPHA L ENTER

Graph

To graph a function, use the Y= key to enter the expression for the function, select a suitable viewing window, and then press GRAPH. For instance, to graph $f(x) = 0.1x^3 - 2x - 1$ in the standard viewing window, use the following keystrokes.

Y= 0.1 X,T,θ,n ^ 3 ▸ − 2 X,T,θ,n − 1 ZOOM (scroll to 6) ENTER

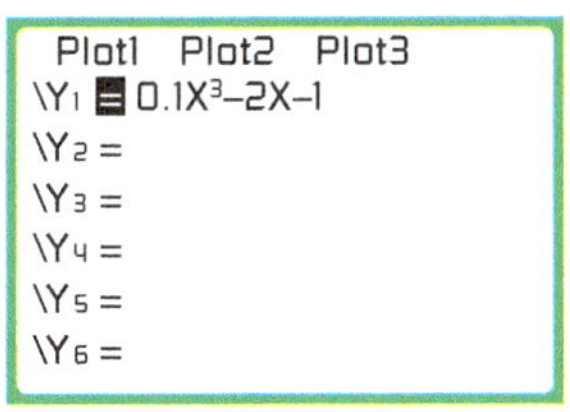

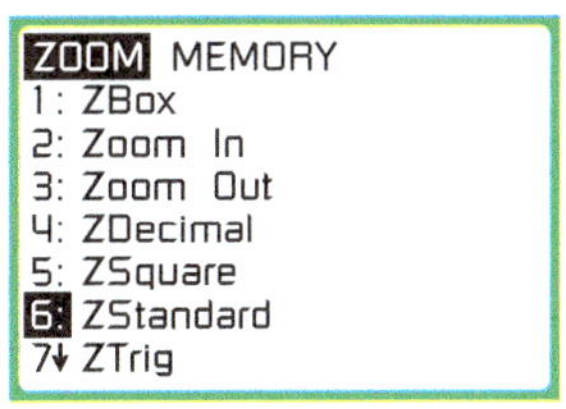

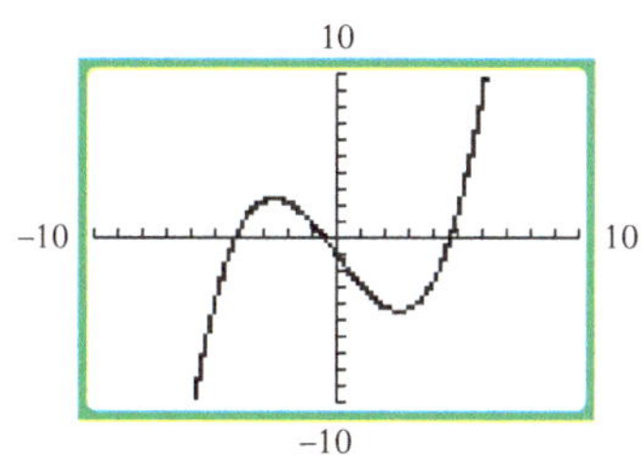

Note: For the keystrokes above, you do not have to scroll to 6. Alternatively, use ZOOM 6. This will select the standard viewing window and automatically start the graph. Use the WINDOW key to create a custom window for a graph.

Graphing Inequalities

To illustrate this feature, we will graph $y \le 2x - 1$. Enter $2x - 1$ into Y_1. Because $y \le 2x - 1$, we want to shade below the graph. Move the cursor to the left of Y_1 and press ENTER three times. Press GRAPH.

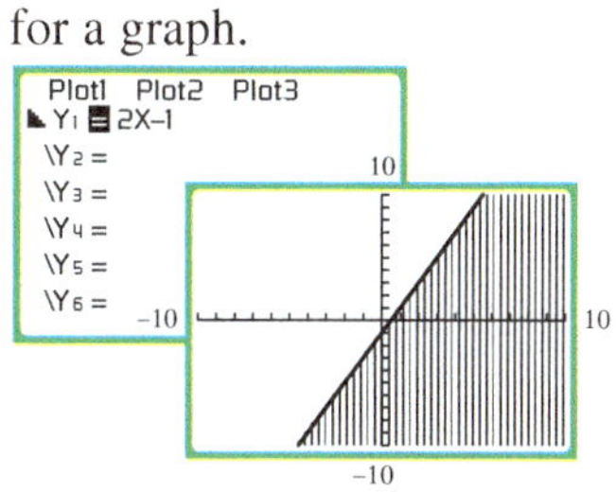

Note: To shade above the graph, move the cursor to the left of Y1 and press ENTER two times. An inequality with the symbol $\leq$ or $\geq$ should be graphed with a solid line, and an inequality with the symbol $<$ or $>$ should be graphed with a dashed line. However, a graphing calculator does not distinguish between a solid line and a dashed line.

To graph the solution set of a system of inequalities, solve each inequality for y and graph each inequality. The solution set is the intersection of the two inequalities. The solution set of

$$\begin{aligned} 3x + 2y &> 10 \\ 4x - 3y &\leq 5 \end{aligned}$$

is shown at the right.

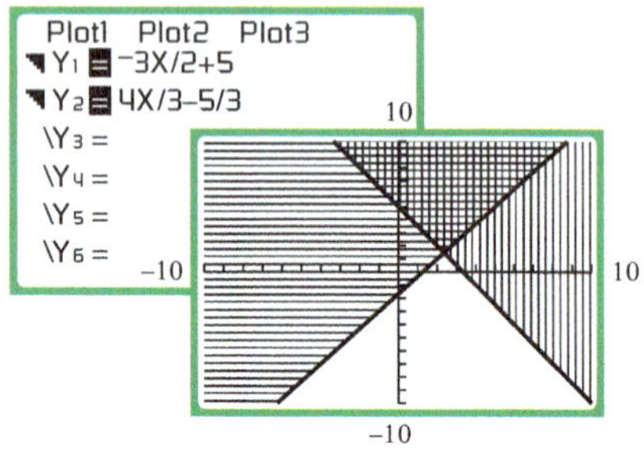

Intersect The INTERSECT feature is used to solve a system of equations. To illustrate this feature, we will use the system of equations $\begin{aligned} 2x - 3y &= 13 \\ 3x + 4y &= -6 \end{aligned}$.

Note: Some equations can be solved by this method. See the section "Solve an equation" on the next page. Also, this method is used to find a number in the domain of a function for a given number in the range. See the section "Find a domain element" on the next page.

Solve each of the equations in the system of equations for y. In this case, we have $y = \frac{2}{3}x - \frac{13}{3}$ and $y = -\frac{3}{4}x - \frac{3}{2}$.

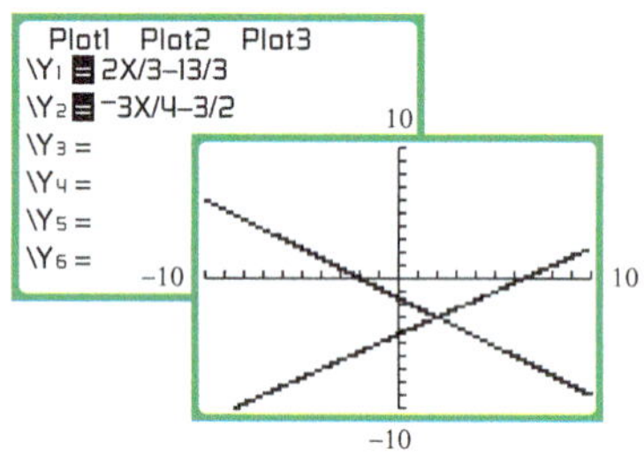

Use the Y-editor to enter $\frac{2}{3}x - \frac{13}{3}$ into Y1 and $-\frac{3}{4}x - \frac{3}{2}$ into Y2. Graph the two functions in the standard viewing window. (If the window does not show the point of intersection of the two graphs, adjust the window until you can see the point of intersection.)

Press 2ND CALC (scroll to 5, **intersect**) ENTER.

Alternatively, you can just press 2ND CALC 5.

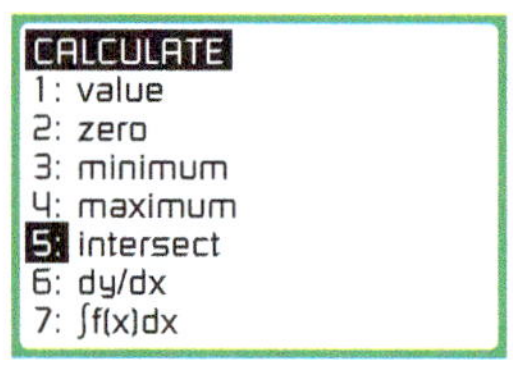

First curve? is shown at the bottom of the screen and identifies one of the two graphs on the screen. Press ENTER.

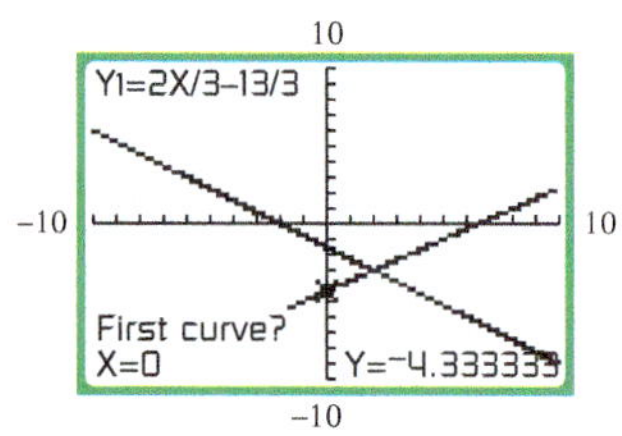

Second curve? is shown at the bottom of the screen and identifies the second of the two graphs on the screen. Press ENTER.

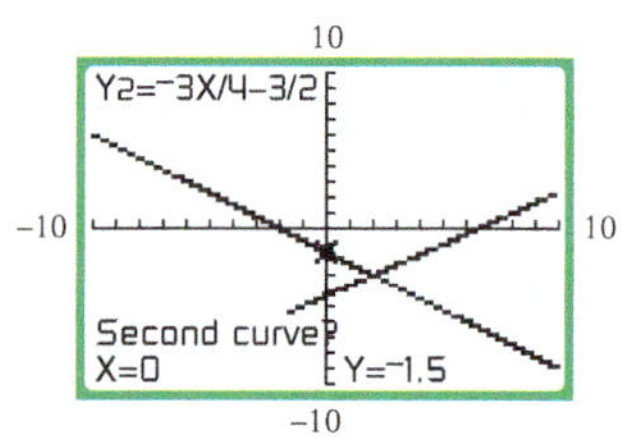

Guess? shown at the bottom of the screen asks you to use the left or right arrow key to move the cursor to the *approximate* location of the point of intersection. (If there are two or more points of intersection, it does not matter which one you choose first.) Press ENTER.

The solution of the system of equations is (2, −3).

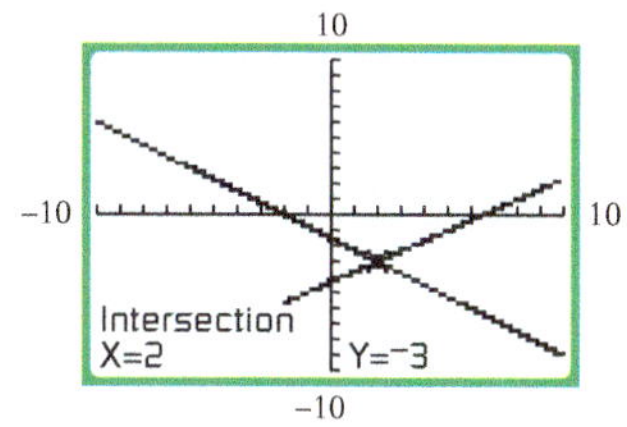

Solve an equation To illustrate the steps involved, we will solve the equation $2x + 4 = -3x - 1$. The idea is to write the equation as the system of equations $y = 2x + 4$, $y = -3x - 1$ and then use the steps for solving a system of equations.

Use the Y-editor to enter the left and right sides of the equation into Y_1 and Y_2. Graph the two functions and then follow the steps for Intersect.

The solution is −1, the x-coordinate of the point of intersection.

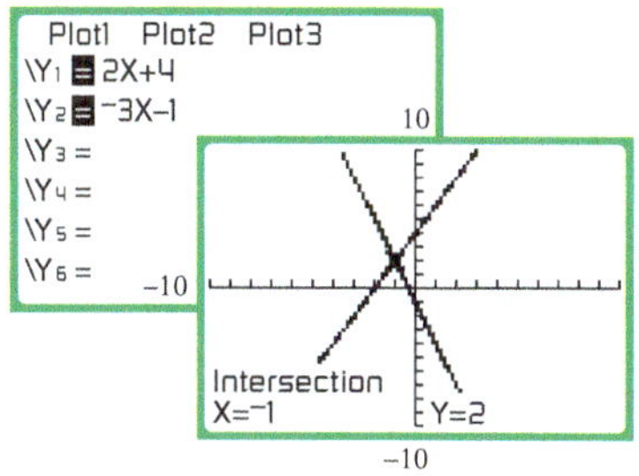

Find a domain element For this example, we will find a number in the domain of $f(x) = -\frac{2}{3}x + 2$ that corresponds to 4 in the range of the function. This is like solving the system of equations $y = -\frac{2}{3}x + 2$ and $y = 4$.

Use the Y= editor to enter the expression for the function in Y_1 and the desired output, 4, in Y_2. Graph the two functions and then follow the steps for Intersect.

The point of intersection is (−3, 4). The number −3 in the domain of f produces an output of 4 in the range of f.

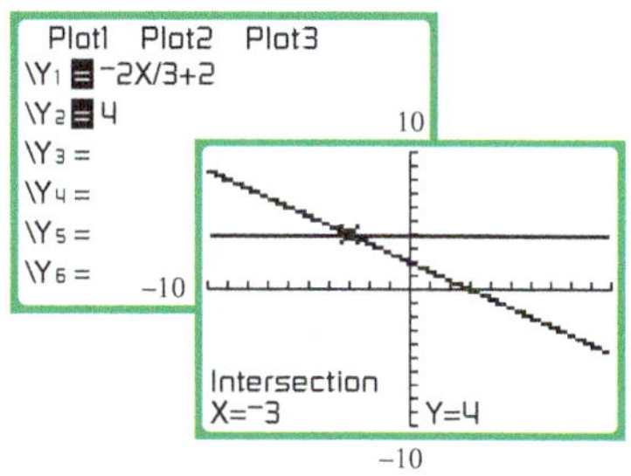

Math

Pressing MATH gives you access to many built-in functions. The following keystrokes will convert 0.125 to a fraction: .125 MATH 1 ENTER.

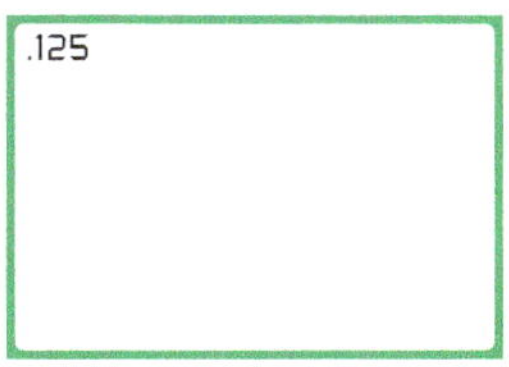

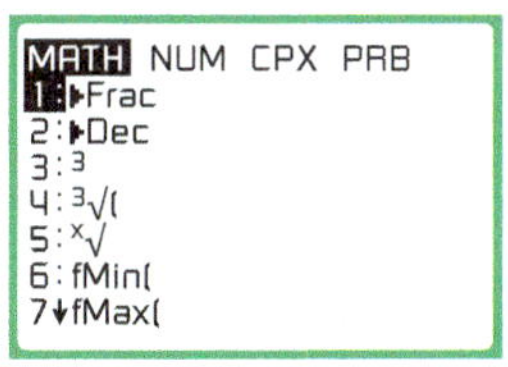

Additional built-in functions under MATH can be found by pressing MATH ▶. For instance, to evaluate $-|-25|$, press (−) MATH ▶ 1 (−) 25 ENTER.

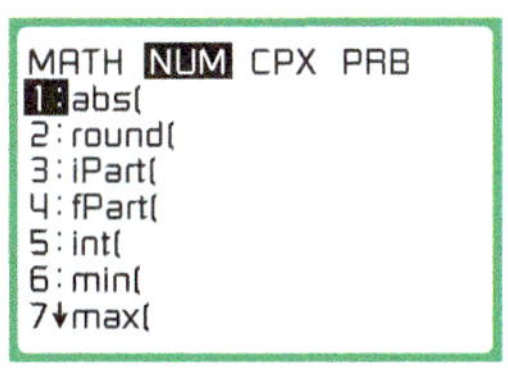

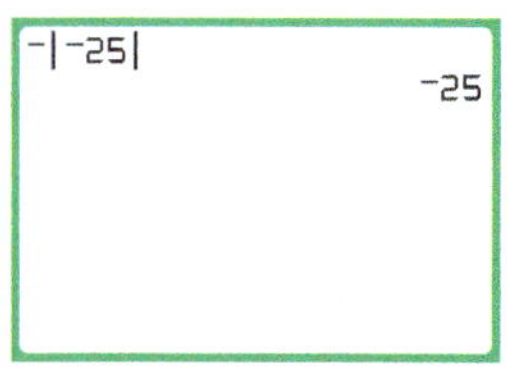

See your owner's manual for assistance with other functions under the MATH key.

Min and Max

The local minimum and the local maximum values of a function are calculated by accessing the CALC menu. For this demonstration, we will find the minimum value and the maximum value of $f(x) = 0.2x^3 + 0.3x^2 - 3.6x + 2$.

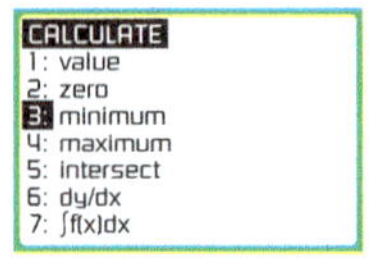

Enter the function into Y1. Press 2ND CALC (scroll to 3 for minimum of the function) ENTER.

Alternatively, you can just press 2ND CALC 3.

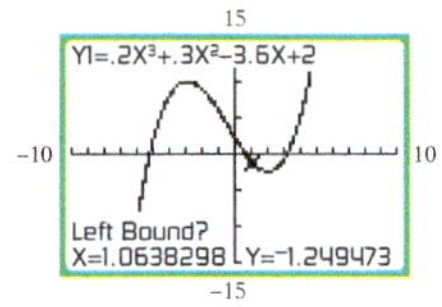

Left Bound? shown at the bottom of the screen asks you to use the left or right arrow key to move the cursor to the *left* of the minimum. Press ENTER.

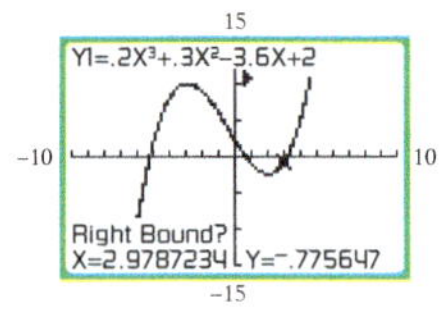

Right Bound? shown at the bottom of the screen asks you to use the left or right arrow key to move the cursor to the *right* of the minimum. Press ENTER.

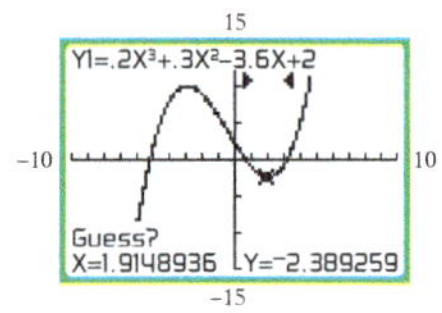

Guess? shown at the bottom of the screen asks you to use the left or right arrow key to move the cursor to the *approximate* location of the minimum. Press ENTER.

The minimum value of the function is the y-coordinate. For this example, the minimum value of the function is -2.4.

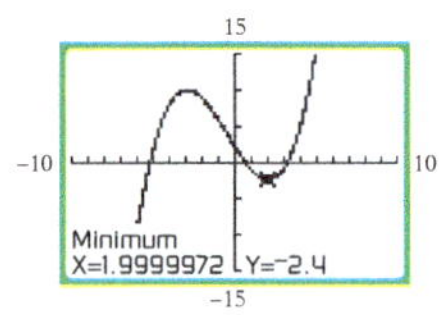

The x-coordinate for the minimum is 2. However, because of rounding errors in the calculation, it is shown as a number close to 2.

To find the maximum value of the function, follow the same steps as above except select maximum under the CALC menu. The screens for this calculation are shown below.

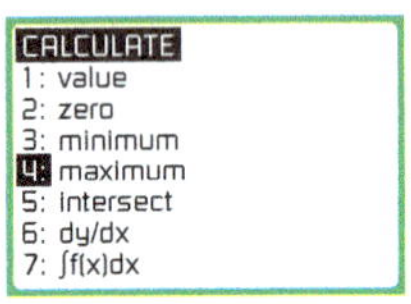

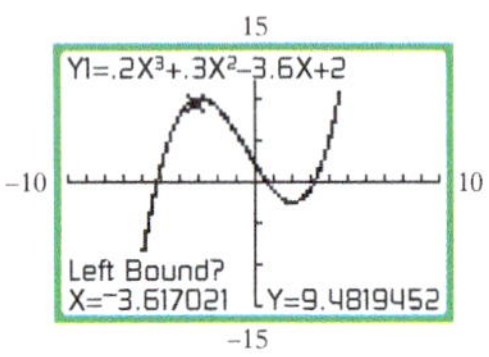

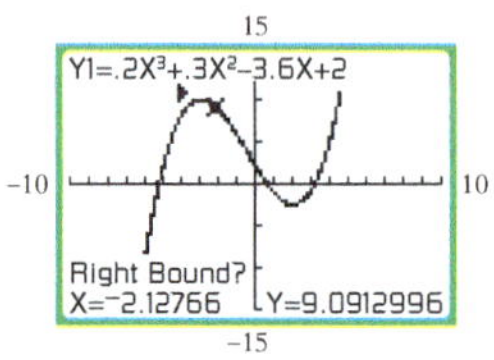

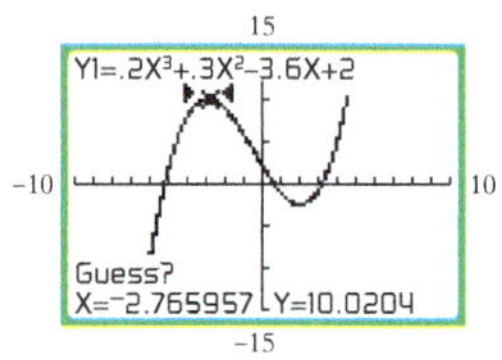

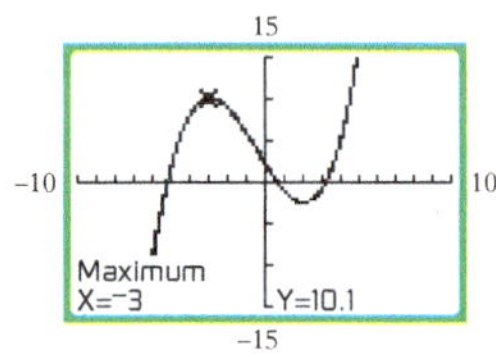

The maximum value of the function is 10.1.

Radical Expressions

To evaluate a square-root expression, press 2ND $\sqrt{\ }$.

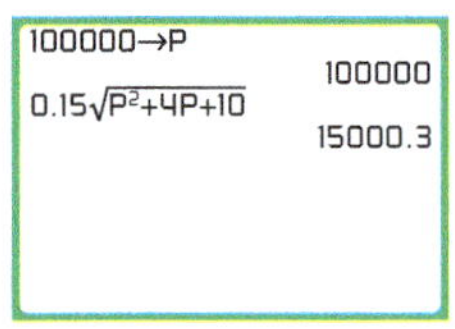

For instance, to evaluate $0.15\sqrt{p^2 + 4p + 10}$ when $p = 100{,}000$, first store 100,000 in P. Then press 0.15 2ND $\sqrt{\ }$ ALPHA P x^2 + 4 ALPHA P + 10 ENTER.

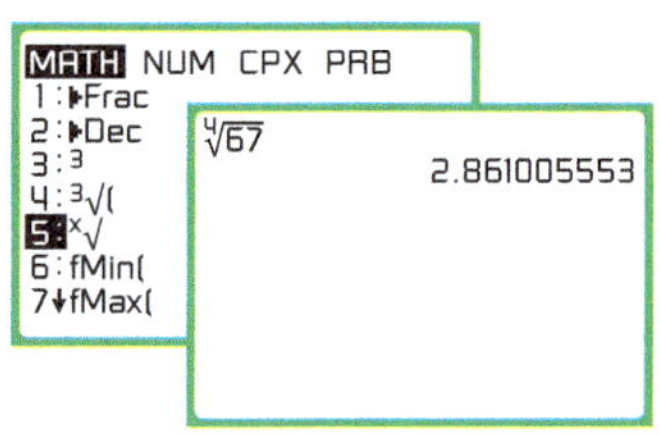

To evaluate a radical expression other than a square root, access $\sqrt[x]{\ }$ by pressing MATH. For instance, to evaluate $\sqrt[4]{67}$, press 4 (the index of the radical) MATH (scroll to 5) ENTER 67 ENTER.

Scientific Notation

To enter a number in scientific notation, use 2ND EE. For instance, to find $\frac{3.45 \times 10^{-12}}{1.5 \times 10^{25}}$, press 3.45 2ND EE (-) 12 ÷ 1.5 2ND EE 25 ENTER. The answer is 2.3×10^{-37}.

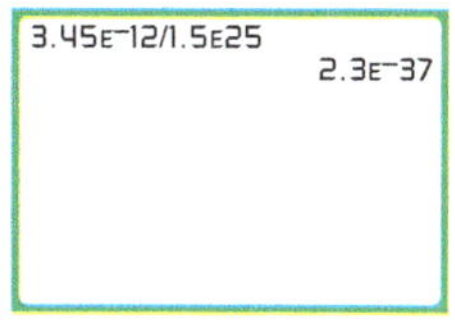

Sequences and Series

The terms of a sequence and the sum of a series can be calculated by using the 2ND LIST feature.

Store a sequence A sequence is stored in one of the lists L1 through L6. For instance, to store the sequence 1, 3, 5, 7, 9 in L1, use the following keystrokes.

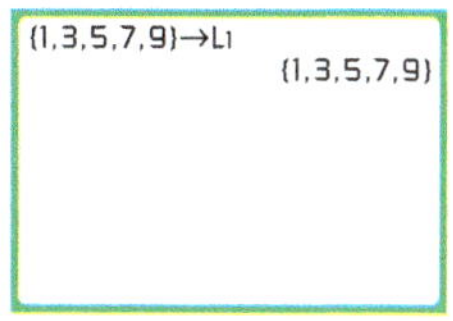

2ND { 1 , 3 , 5 , 7 ,
9 2ND } STO▸ 2ND L1 ENTER

Display the terms of a sequence The terms of a sequence are displayed by using the function seq(expression, variable, begin, end, increment). For instance, to display the 3rd through 8th terms of the sequence given by $a_n = n^2 + 6$, enter the following keystrokes.

2ND LIST ▸ (scroll to 5)
ENTER X,T,θ,n x^2 + 6
, X,T,θ,n , 3 , 8
, 1 ENTER STO▸ 2ND L1 ENTER

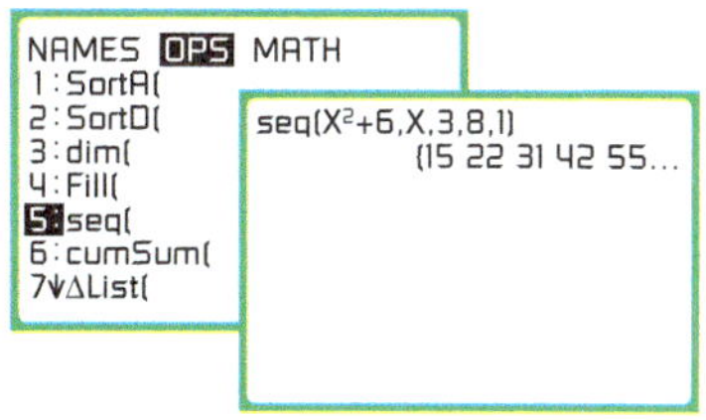

The keystrokes STO▸ 2ND L1 ENTER store the terms of the sequence in L1. This is not necessary but is sometimes helpful if additional work will be done with that sequence.

Find a sequence of partial sums To find a sequence of partial sums, use the cumSum(function. For instance, to find the sequence of partial sums for 2, 4, 6, 8, 10, use the following keystrokes.

2ND LIST ▸ (scroll to 6)
ENTER 2ND { 2 , 4 , 6
, 8 , 10 2ND }) ENTER

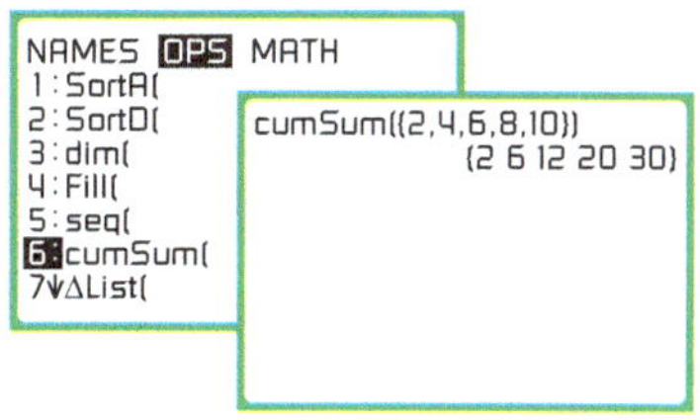

If a sequence is stored as a list in L1, then the sequence of partial sums can be calculated by pressing 2ND LIST ▸ (scroll to 6 [or press 6]) ENTER 2ND L1) ENTER.

Find the sum of a series The sum of a series is calculated using sum<list, start, end>. For instance, to find $\sum_{n=3}^{6} (n^2 + 2)$, enter the following keystrokes.

2ND LIST ▸ ▸ (scroll to 5)
ENTER 2ND LIST ▸ (scroll to 5 [or press 5])
ENTER X,T,θ,n x^2 + 2 ,
X,T,θ,n , 3 , 6 , 1) ENTER

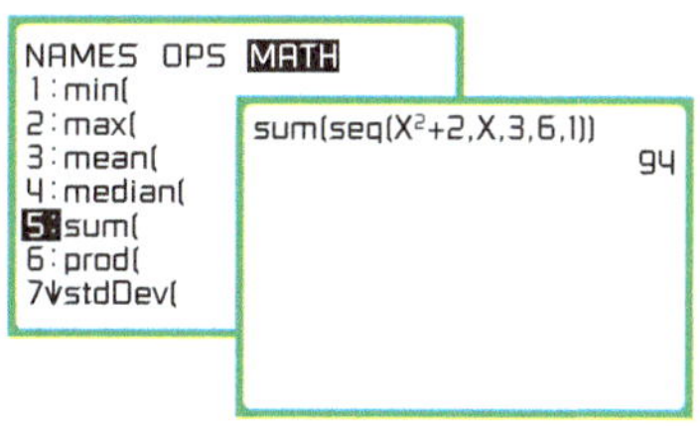

Table There are three steps in creating an input/output table for a function. First use the Y= editor to input the function. The second step is setting up the table, and the third step is displaying the table.

To set up the table, press 2ND TBLSET. TblStart is the first value of the independent variable in the input/output table. ΔTbl is the difference between successive values. Setting this to 1 means that, for this table, the input values are $-2, -1, 0, 1, 2, \ldots$ If ΔTbl $= 0.5$, then the input values are $-2, -1.5, -1, -0.5, 0, 0.5, \ldots$

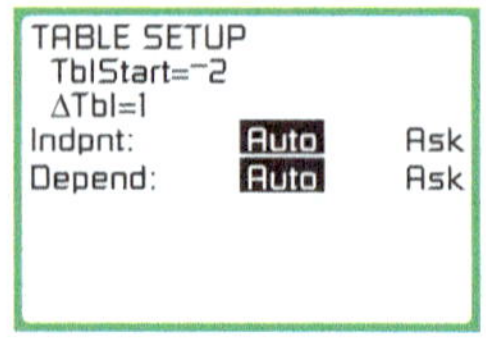

Indpnt is the independent variable. When this is set to Auto, values of the independent variable are automatically entered into the table. Depend is the dependent variable. When this is set to Auto, values of the dependent variable are automatically entered into the table.

To display the table, press 2ND TABLE. An input/output table for $f(x) = x^2 - 1$ is shown at the right.

Once the table is on the screen, the up and down arrow keys can be used to display more values in the table. For the table at the right, we used the up arrow key to move to $x = -7$.

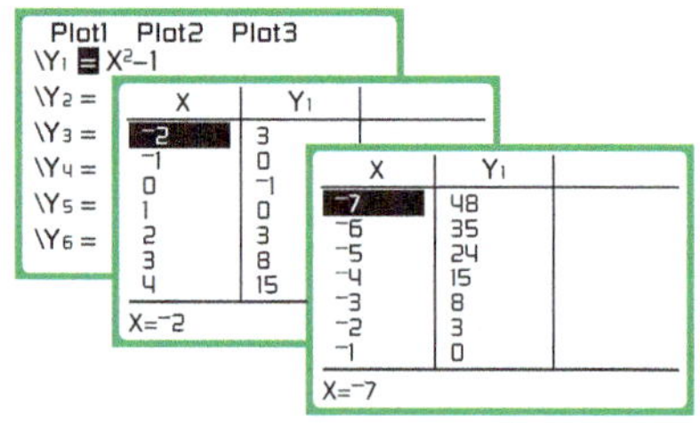

An input/output table for any given input can be created by selecting Ask for the independent variable. The table at the right shows an input/output table for $f(x) = \frac{4x}{x-2}$ for selected values of x. Note the word ERROR when 2 was entered. This occurred because f is not defined when $x = 2$.

Note: Using the table feature in Ask mode is the same as evaluating a function for given values of the independent variable. For instance, from the table at the right, we have $f(4) = 8$.

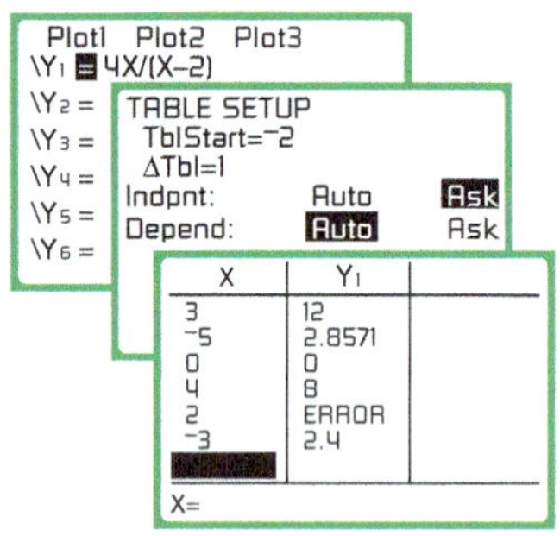

Test The TEST feature has many uses, one of which is to graph the solution set of a linear inequality in one variable. To illustrate this feature, we will graph the solution set of $x - 1 < 4$. Press Y= X,T,θ,n − 1 2ND TEST (scroll to 5) ENTER 4 GRAPH.

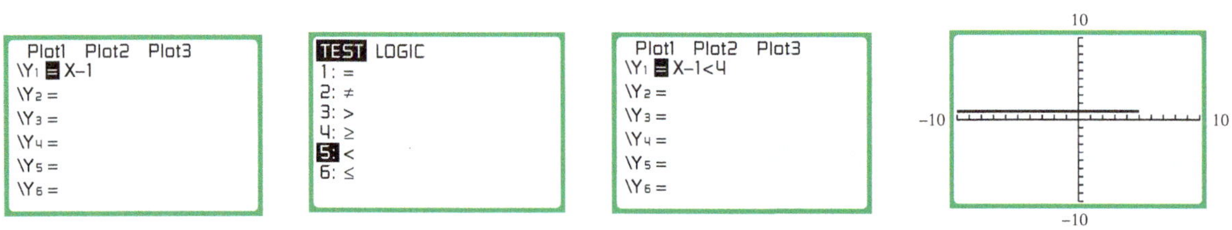

Trace Once a graph is drawn, pressing TRACE will place a cursor on the screen, and the coordinates of the point below the cursor are shown at the bottom of the screen. Use the left and right arrow keys to move the cursor along the graph. For the graph at the right, we have $f(4.8) = 3.4592$, where $f(x) = 0.1x^3 - 2x + 2$ is shown at the top left of the screen.

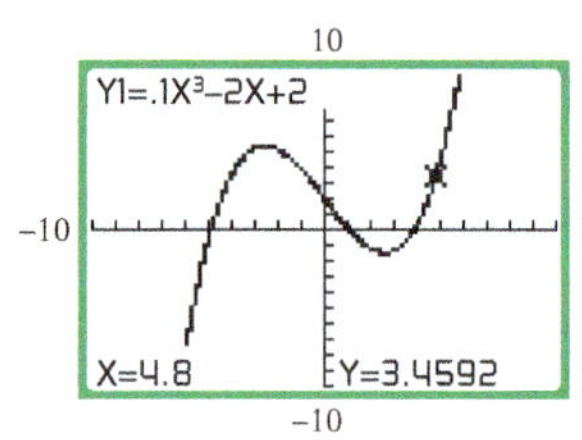

In TRACE mode, you can evaluate a function at any value of the independent variable that is within Xmin and Xmax. To do this, first graph the function. Now press TRACE (the value of x) ENTER. For the graph at the left below, we used $x = -3.5$. If a value of x is chosen outside the window, an error message is displayed.

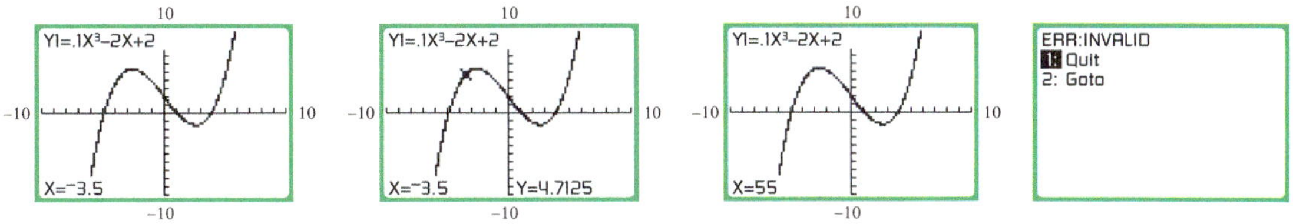

In the example above in which we entered -3.5 for x, the value of the function was calculated as 4.7125. This means that $f(-3.5) = 4.7125$. The keystrokes 2ND QUIT VARS ▷ 1 1 MATH 1 ENTER will convert the decimal value to a fraction.

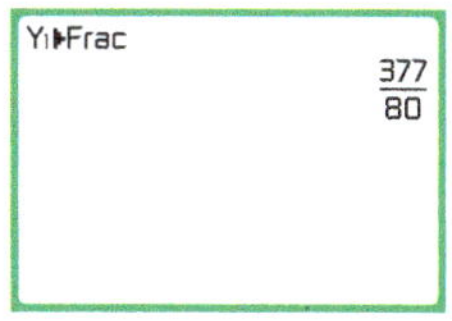

When the TRACE feature is used with two or more graphs, the up and down arrow keys are used to move between the graphs. The graphs below are for the functions $f(x) = 0.1x^3 - 2x + 2$ and $g(x) = 2x - 3$. By using the up and down arrows, we can place the cursor on either graph. The right and left arrows are used to move along the graph.

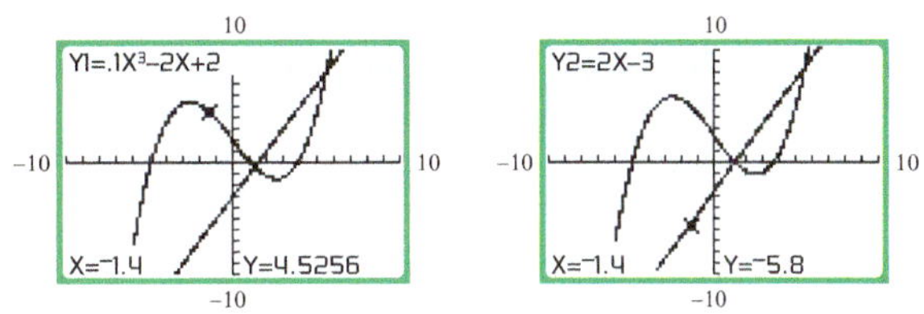

Window The viewing window for a graph is controlled by pressing WINDOW. Xmin and Xmax are the minimum value and maximum value, respectively, of the independent variable shown on the graph. Xscl is the distance between tic marks on the x-axis. Ymin and Ymax are the minimum value and maximum value, respectively, of the dependent variable shown on the graph. Yscl is the distance between tic marks on the y-axis. Leave Xres as 1.

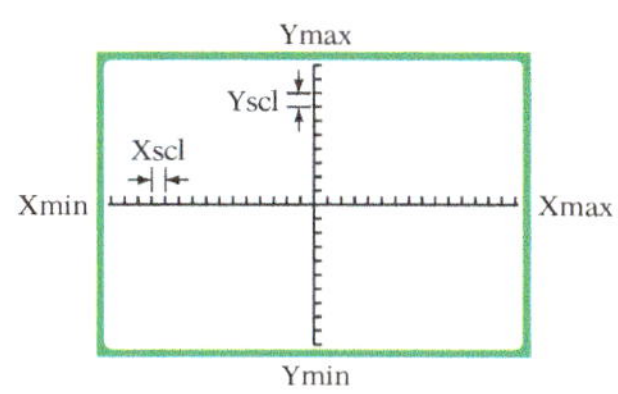

Note: In the standard viewing window, the distance between tic marks on the x-axis is different from the distance between tic marks on the y-axis. This will distort a graph. A more accurate picture of a graph can be created by using a square viewing window. See ZOOM.

Y= The Y= editor is used to enter the expression for a function. There are ten possible functions, labeled Y1 to Y0, that can be active at any one time. For instance, to enter $f(x) = x^2 + 3x - 2$ as Y1, use the following keystrokes.

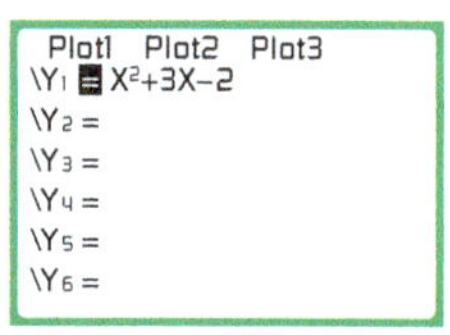

Y= X,T,θ,n x^2 + 3 X,T,θ,n – 2

Note: If an expression is already entered for Y1, place the cursor anywhere on that expression and press CLEAR.

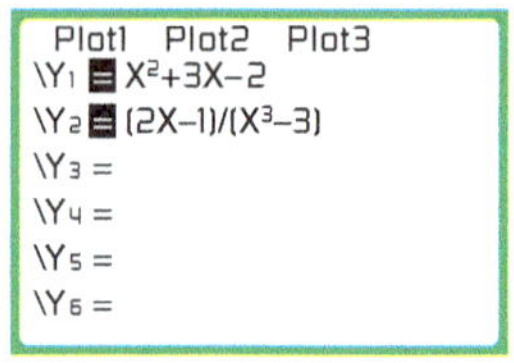

To enter $s = \frac{2v - 1}{v^3 - 3}$ into Y2, place the cursor to the right of the equals sign for Y2. Then press (2 X,T,θ,n − 1) ÷ (X,T,θ,n ^ 3 ▷ − 3) ENTER.

Note: When we enter an equation, the independent variable, v in the expression above, is entered using X,T,θ,n. The dependent variable, s in the expression above, is one of Y1 to Y0. Also note the use of parentheses to ensure the correct order of operations.

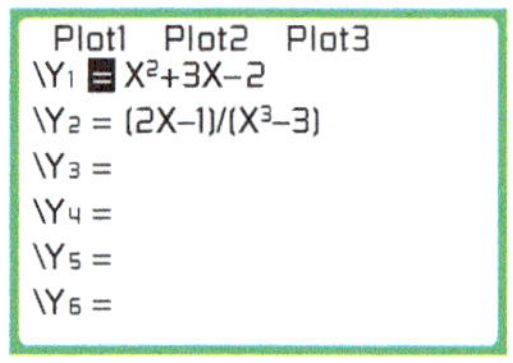

Observe the black rectangle that covers the equals sign for the two examples we have shown. This rectangle means that the function is "active." If we were to press GRAPH, then the graph of both functions would appear. You can make a function inactive by using the arrow keys to move the cursor over the equals sign of that function and then pressing ENTER. This will remove the black rectangle. We have done that for Y2, as shown at the right. Now if GRAPH is pressed, only Y1 will be graphed.

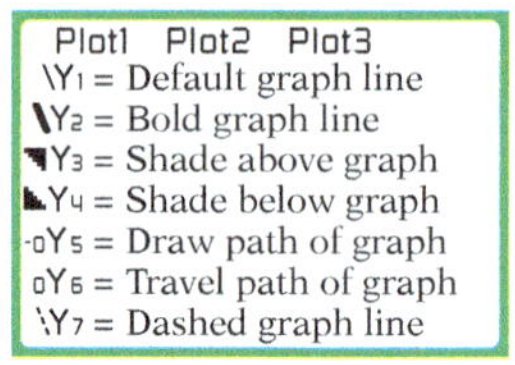

It is also possible to control the appearance of the graph by moving the cursor on the Y= screen to the left of any Y. With the cursor in this position, pressing ENTER will change the appearance of the graph. The options are shown at the right.

Zero The ZERO feature of a graphing calculator is used for various calculations: to find the x-intercepts of a function, to solve some equations, and to find the zero of a function.

***x*-intercepts** To illustrate the procedure for finding x-intercepts, we will use $f(x) = x^2 + x - 2$.

First, use the Y-editor to enter the expression for the function and then graph the function in the standard viewing window. (It may be necessary to adjust this window so that the intercepts are visible.) Once the graph is displayed, use the keystrokes below to find the x-intercepts of the graph of the function.

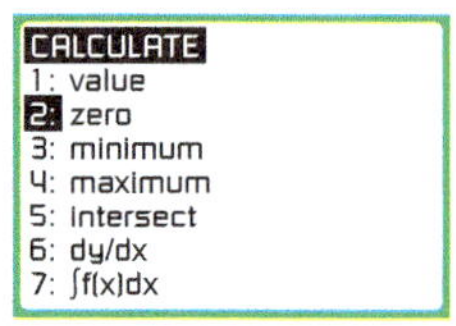

Press 2ND CALC (scroll to 2 for **zero** of the function) ENTER.

Alternatively, you can just press 2ND CALC 2.

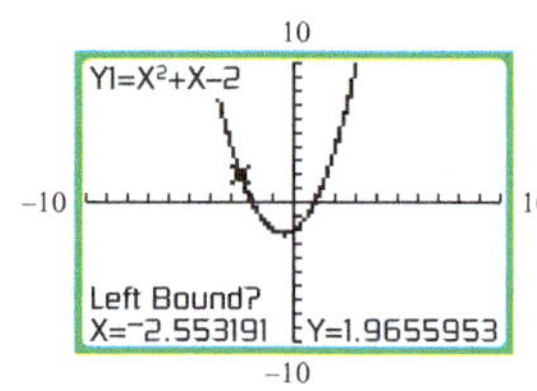

Left Bound? shown at the bottom of the screen asks you to use the left or right arrow key to move the cursor to the *left* of the desired x-intercept. Press ENTER.

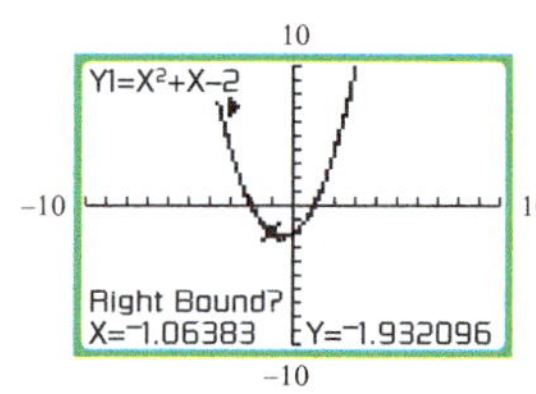

Right Bound? shown at the bottom of the screen asks you to use the left or right arrow key to move the cursor to the *right* of the desired x-intercept. Press ENTER.

Guess? shown at the bottom of the screen asks you to use the left or right arrow key to move the cursor to the *approximate* location of the desired x-intercept. Press ENTER.

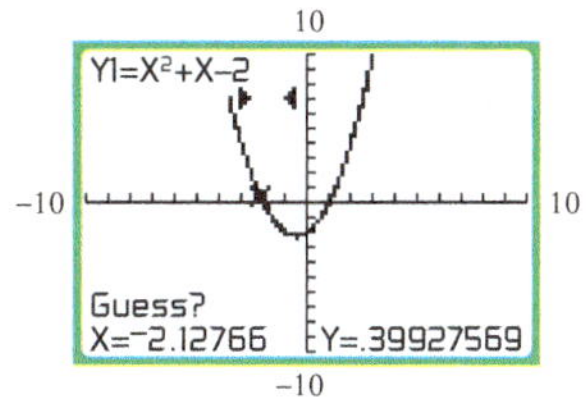

The x-coordinate of an x-intercept is -2. Therefore, an x-intercept is $(-2, 0)$.

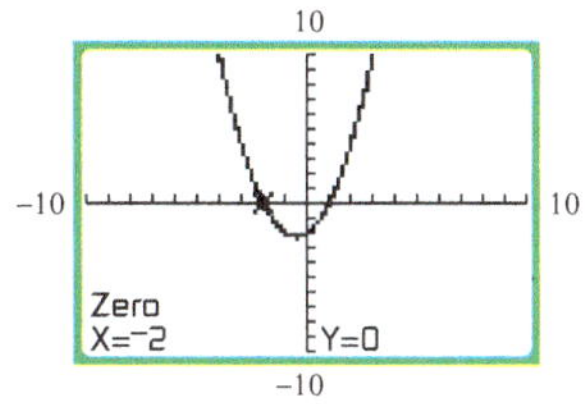

To find the other x-intercept, follow the same steps as above. The screens for this calculation are shown below.

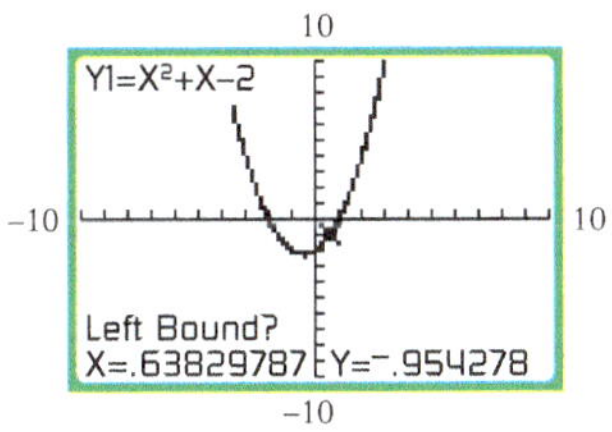

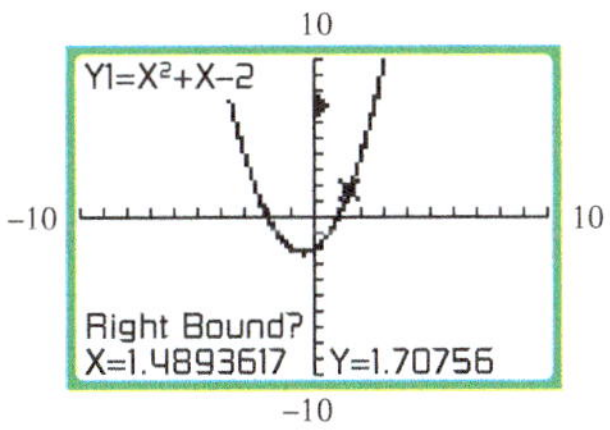

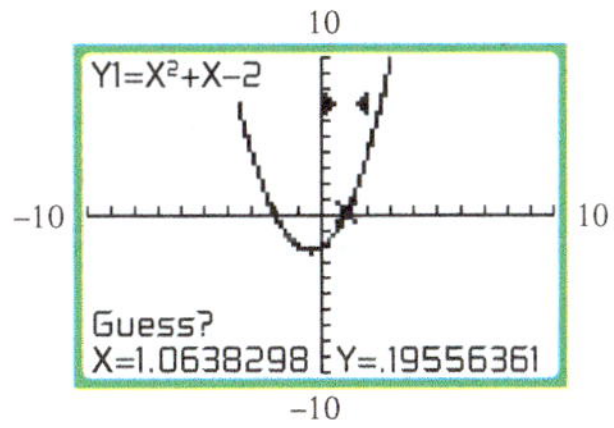

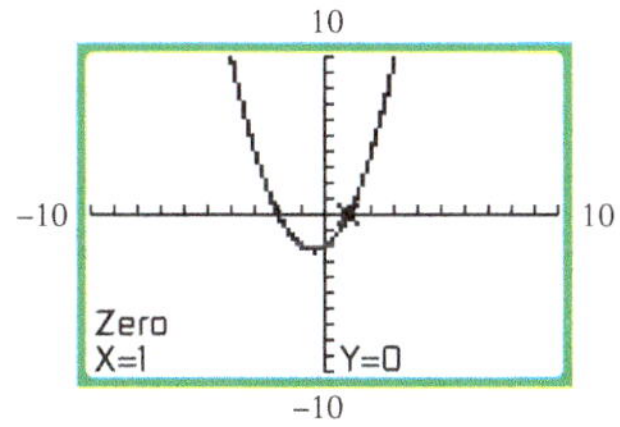

A second x-intercept is $(1, 0)$.

Solve an equation To use the ZERO feature to solve an equation, first rewrite the equation with all terms on one side. For instance, one way to solve the equation $x^3 - x + 1 = -2x + 3$ is first to rewrite it as $x^3 + x - 2 = 0$. Enter $x^3 + x - 2$ into Y1 and then follow the steps for finding x-intercepts.

Find the real zeros of a function To find the real zeros of a function, follow the steps for finding x-intercepts.

Zoom

Pressing ZOOM allows you to select some preset viewing windows. This key also gives you access to ZBox, Zoom In, and Zoom Out. These functions enable you to redraw a selected portion of a graph in a new window. Some windows used frequently in this text are shown below.

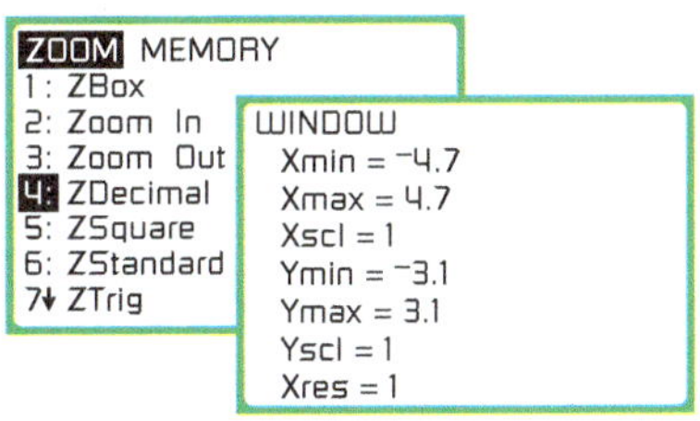

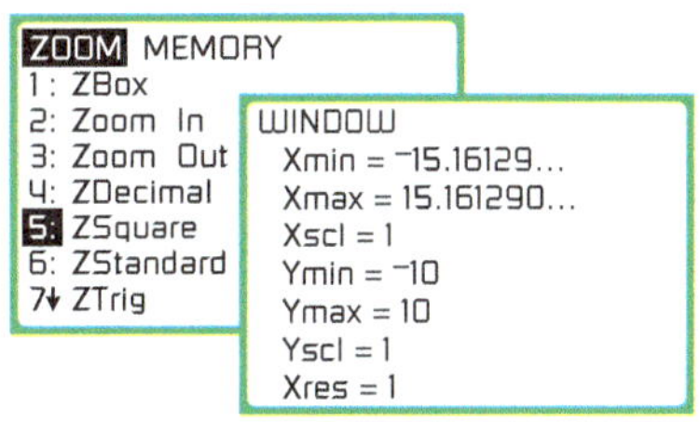

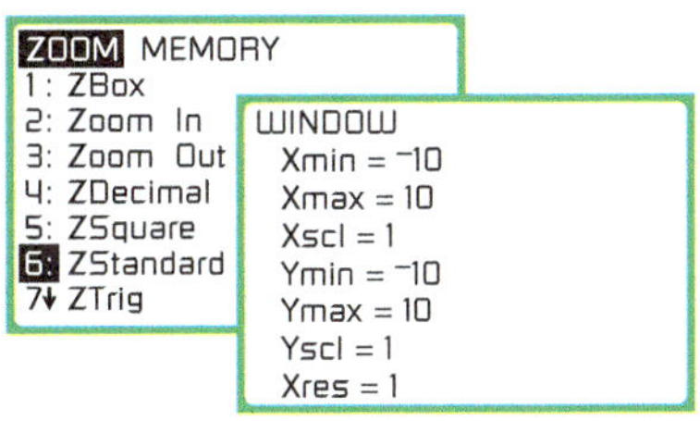

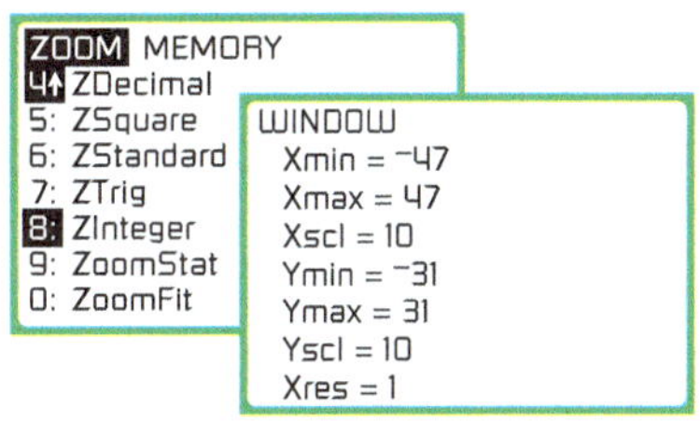

Table of Measurement Abbreviations

U.S. Customary System

Length		Capacity		Weight		Area	
in.	inches	oz	fluid ounces	oz	ounces	in^2	square inches
ft	feet	c	cups	lb	pounds	ft^2	square feet
yd	yards	qt	quarts				
mi	miles	gal	gallons				

Metric System

Length		Capacity		Weight/Mass		Area	
mm	millimeter (0.001 m)	ml	milliliter (0.001 L)	mg	milligram (0.001 g)	cm^2	square centimeters
cm	centimeter (0.01 m)	cl	centiliter (0.01 L)	cg	centigram (0.01 g)	m^2	square meters
dm	decimeter (0.1 m)	dl	deciliter (0.1 L)	dg	decigram (0.1 g)		
m	meter	L	liter	g	gram		
dam	decameter (10 m)	dal	decaliter (10 L)	dag	decagram (10 g)		
hm	hectometer (100 m)	hl	hectoliter (100 L)	hg	hectogram (100 g)		
km	kilometer (1000 m)	kl	kiloliter (1000 L)	kg	kilogram (1000 g)		

Time

h	hours	min	minutes	s	seconds

Table of Symbols

Symbol	Meaning
$+$	add
$-$	subtract
$\cdot$, $\times$, $(a)(b)$	multiply
$\frac{a}{b}$, $\div$	divide
( )	parentheses, a grouping symbol
[ ]	brackets, a grouping symbol
π	pi, a number approximately equal to $\frac{22}{7}$ or 3.14
$-a$	the opposite, or additive inverse, of a
$\frac{1}{a}$	the reciprocal, or multiplicative inverse, of a
$=$	is equal to
$\approx$	is approximately equal to
$\neq$	is not equal to
$<$	is less than
$\leq$	is less than or equal to
$>$	is greater than
$\geq$	is greater than or equal to
(a, b)	an ordered pair whose first component is a and whose second component is b
$^\circ$	degree (for angles)
$\sqrt{a}$	the principal square root of a
$\varnothing$, { }	the empty set
$\lvert a \rvert$	the absolute value of a
$\cup$	union of two sets
$\cap$	intersection of two sets
$\in$	is an element of (for sets)
$\notin$	is not an element of (for sets)

Table of Properties

Properties of Real Numbers

The Associative Property of Addition
If a, b, and c are real numbers, then
$(a + b) + c = a + (b + c)$.

The Associative Property of Multiplication
If a, b, and c are real numbers, then
$(a \cdot b) \cdot c = a \cdot (b \cdot c)$.

The Commutative Property of Addition
If a and b are real numbers, then
$a + b = b + a$.

The Commutative Property of Multiplication
If a and b are real numbers, then
$a \cdot b = b \cdot a$.

The Addition Property of Zero
If a is a real number, then
$a + 0 = 0 + a = a$.

The Multiplication Property of One
If a is a real number, then
$a \cdot 1 = 1 \cdot a = a$.

The Multiplication Property of Zero
If a is a real number, then
$a \cdot 0 = 0 \cdot a = 0$.

The Inverse Property of Multiplication
If a is a real number and $a \neq 0$, then
$a \cdot \frac{1}{a} = \frac{1}{a} \cdot a = 1$.

The Inverse Property of Addition
If a is a real number, then
$a + (-a) = (-a) + a = 0$.

Distributive Property
If a, b, and c are real numbers, then
$a(b + c) = ab + ac$.

Properties of Equations

Addition Property of Equations
If $a = b$, then $a + c = b + c$.

Multiplication Property of Equations
If $a = b$ and $c \neq 0$, then $a \cdot c = b \cdot c$.

Properties of Inequalities

Addition Property of Inequalities
If $a > b$, then $a + c > b + c$.
If $a < b$, then $a + c < b + c$.

Multiplication Property of Inequalities
If $a > b$ and $c > 0$, then $ac > bc$.
If $a < b$ and $c > 0$, then $ac < bc$.
If $a > b$ and $c < 0$, then $ac < bc$.
If $a < b$ and $c < 0$, then $ac > bc$.

Properties of Exponents

If m and n are integers, then $x^m \cdot x^n = x^{m+n}$.
If m and n are integers, then $(x^m)^n = x^{mn}$.

If $x \neq 0$, then $x^0 = 1$.

If m and n are integers and $x \neq 0$, then $\frac{x^m}{x^n} = x^{m-n}$.

If m, n, and p are integers, then $(x^m \cdot y^n)^p = x^{mp}y^{np}$.
If n is a positive integer and $x \neq 0$, then
$x^{-n} = \frac{1}{x^n}$ and $\frac{1}{x^{-n}} = x^n$.

If m, n, and p are integers and $y \neq 0$, then $\left(\frac{x^m}{y^n}\right)^p = \frac{x^{mp}}{y^{np}}$.

Principle of Zero Products

If $a \cdot b = 0$, then $a = 0$ or $b = 0$.

Properties of Radical Expressions

If a and b are positive real numbers, then $\sqrt{ab} = \sqrt{a}\sqrt{b}$.

If a and b are positive real numbers, then $\sqrt{\frac{a}{b}} = \frac{\sqrt{a}}{\sqrt{b}}$.

Property of Squaring Both Sides of an Equation

If a and b are real numbers and $a = b$, then $a^2 = b^2$.

Properties of Logarithms

If x, y, and b are positive real numbers and $b \neq 1$, then
$\log_b(xy) = \log_b x + \log_b y$.
If x, y, and b are positive real numbers and $b \neq 1$, then
$\log_b \frac{x}{y} = \log_b x - \log_b y$.

If x and b are positive real numbers, $b \neq 1$, and r is any real number, then $\log_b x^r = r \log_b x$.
If x and b are positive real numbers and $b \neq 1$, then
$\log_b b^x = x$.

Table of Algebraic and Geometric Formulas

Slope of a Line

$$m = \frac{y_2 - y_1}{x_2 - x_1},\ x_1 \neq x_2$$

Point-slope Formula for a Line

$$y - y_1 = m(x - x_1)$$

Quadratic Formula

$$x = \frac{-b \pm \sqrt{b^2 - 4ac}}{2a}$$

$$\text{discriminant} = b^2 - 4ac$$

Perimeter and Area of a Triangle, and Sum of the Measures of the Angles

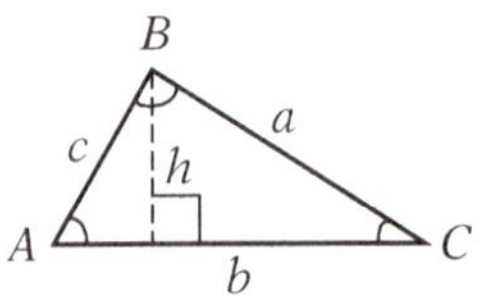

$P = a + b + c$

$A = \frac{1}{2}bh$

The sum of the measures of the angles in a triangle is 180°.

Pythagorean Theorem

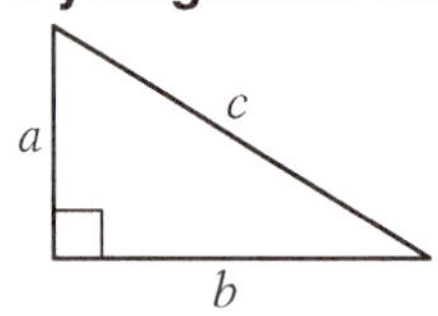

$$a^2 + b^2 = c^2$$

Perimeter and Area of a Rectangle

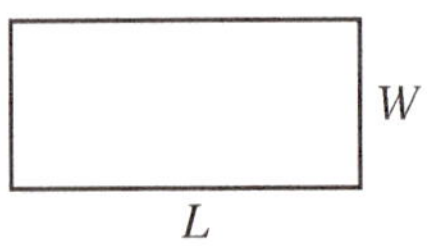

$P = 2L + 2W$

$A = LW$

Perimeter and Area of a Square

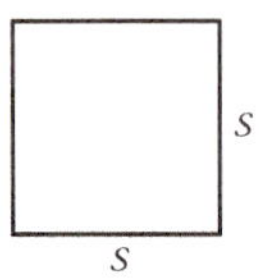

$P = 4s$

$A = s^2$

Area of a Trapezoid

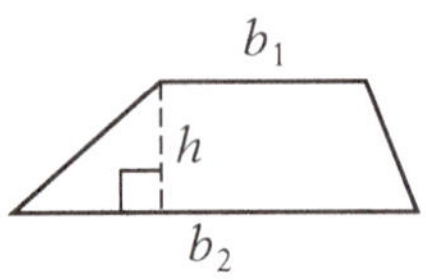

$$A = \frac{1}{2}h(b_1 + b_2)$$

Circumference and Area of a Circle

$C = 2\pi r$

$A = \pi r^2$

Volume and Surface Area of a Rectangular Solid

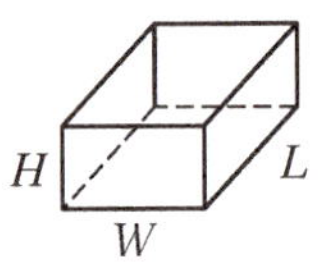

$V = LWH$

$SA = 2LW + 2LH + 2WH$

Volume and Surface Area of a Sphere

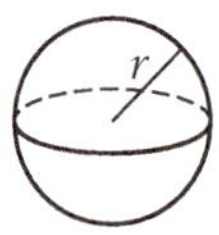

$V = \frac{4}{3}\pi r^3$

$SA = 4\pi r^2$

Volume and Surface Area of a Right Circular Cylinder

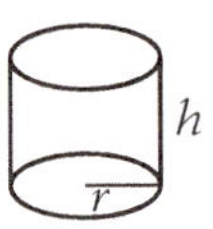

$V = \pi r^2 h$

$SA = 2\pi r^2 + 2\pi rh$

Volume and Surface Area of a Right Circular Cone

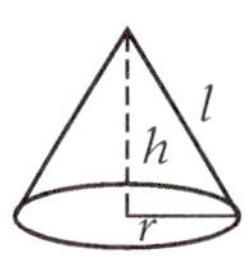

$V = \frac{1}{3}\pi r^2 h$

$SA = \pi r^2 + \pi rl$

Index of Applications

Index